S. Gardner

THIRD EDITION

ECONOMICS OF THE ENVIRONMENT
Selected Readings

Edited by

ROBERT DORFMAN

NANCY S. DORFMAN

W · W · NORTON & COMPANY

New York   London

D0082189

The text of this book is composed in Times Roman
Composition by ComCom
Manufacturing by Haddon Craftsmen
Book design by Natasha Sylvester

Library of Congress Cataloging-in-Publication Data

Economics of the environment : selected readings / edited by Robert
    Dorfman and Nancy S. Dorfman. — 3rd ed.
        p.   cm.
        Includes index.
        1. Pollution—Economic aspects.   2. Environmental
policy—Costs.
        I. Dorfman, Robert.   II. Dorfman, Nancy S.
        HC79.P55D65   1993
        363.7—dc20   92–17707

ISBN 0-393-96310-1

W. W. Norton & Company, Inc., 500 Fifth Avenue
New York, N.Y. 10110

W. W. Norton & Company Ltd., 10 Coptic Street
London WC1A 1PU

1 2 3 4 5 6 7 8 9 0

To Joni and Loren

CONTENTS

III
—·—

DESIGNING AND IMPLEMENTING
ENVIRONMENTAL POLICIES   **199**

IV
—·—

BENEFIT-COST ANALYSIS AND MEASUREMENT   **293**

CONTENTS   **vii**

PREFACE to the Third Edition

Concern over the profligate way in which people throughout the globe have been exploiting the natural environment has not abated since the First Edition of this collection appeared in 1972. If anything, it has grown more insistent as alarming evidence of potential and actual damages to the atmosphere and climate of the entire planet has captured the world's attention. Economists have a twofold interest in the matter: to preserve the viability and healthfulness of the environment, and to ensure that the measures taken to protect the environment be effective and at the same time not reduce the flow of ordinary useful goods and services any more than is necessary. This second concern is the special province of economists. The more severe the environmental damages that need to be contained, the greater the importance of targeting the ones that pose the greatest threat to the well-being of the planet and containing them in the least costly way possible. All of the papers in this collection are concerned in one way or another with meeting this challenge.

With few exceptions, the thirty papers presented here have appeared previously in other publications. Robert Dorfman's introduction to benefit-cost analysis (Chapter 18) is new, and "Some Concepts of Welfare Economics" (Chapter 5) has been revised and moved from the Introduction in the earlier editions of this collection. Most of the papers are presented in their entirety as they appeared originally. Exceptions to this rule are indicated.

Almost two thirds of the papers are new to this Third Edition, selected to take account of important recent contributions to environmental economics, suggestions from users of earlier editions, and shifts in the general focus of interest within the field. An entire section is now devoted to problems of the global environment, and more attention is paid than previously to the concept of tradable emissions permits. The presentation of benefit-cost analysis has been strengthened considerably, by popular demand. The ten papers that are retained from earlier editions have withstood the test of time to become what many readers describe as "classics."

As in previous editions, we begin in Section I with an overview of the sources of the problem of environmental degradation, the economist's approach to solutions, and a brief history of policy. Section II again develops the concepts and methods of economic analysis that are applied to environmental problems; and Section III evaluates alternative policy solutions. Section IV is now devoted to the difficult task of analyzing and measuring the benefits and costs of improving the environment, and Section V focuses on the global dimensions of the problem. Throughout, we have tried to strike a balance between articles of mainly analytical interest and those that provide insights into the nature of specific environmental problems.

In the twenty years since the First Edition of this collection appeared, a number of excellent textbooks on the economics of the environment have become available. Some may wonder whether there remains a need for such a collection. There are, indeed, drawbacks as well as advantages in trying to gain insight into a difficult social problem by studying a collection of source materials such as this. The main drawback is that the collection cannot be as neat, tidy, and coherent as a textbook or treatise. But collections can reflect the liveliness, the heat of controversy, the earnestness of firsthand testimony that placid texts and treatises cannot convey. Collections like this one permit the reader to explore in considerable depth some of the real live problems of the environment and to observe in detail the way that economists struggle to come up with solutions. Some might say that the purpose of such a collection is to "put the meat on the bones" of analysis that often can be arid and lacking in verisimilitude. So, in this volume we are allowing some of the main contenders to speak with their own voices, and by the same token we are allowing readers to draw their own conclusions from the assorted evidence and arguments.

We are grateful to the many users of the Second Edition who took the trouble to write us about what they liked and didn't like and what they wanted to see more of, especially to the thirty-five instructors who responded to our questionnaire. Their suggestions were invaluable to us in preparing this Third Edition.

March 1992

NSD
RD

I

AN OVERVIEW

In the very first paper in this collection, Garrett Hardin sets forth the essence of the problem of environmental degradation. "The Tragedy of the Commons," now a classic, was one of the earliest contributions to the environmental movement when it was written in 1968. Hardin, a biologist, observed that overgrazing unrestricted commonlands, prior to their enclosure, was a metaphor for the overexploitation of all of the earth's land, air, and water resources that are common property. The root cause of overgrazing was the absence of a mechanism for obliging herdsmen to take into account the harmful effects of their own herds' grazing on all of the other herdsmen who shared the common. The solution lay in assigning property rights so that owners could limit the use of the commons.

But Hardin recognizes that air, water, and many other environmental resources, unlike the traditional commons, cannot readily be fenced and parceled out to private owners who would be motivated to preserve them. This observation foreshadows the central question raised in succeeding chapters in this volume and, indeed, in the entire field of environmental economics: How can we oblige users to internalize the damages they inflict on environmental resources that, by their very nature, cannot be owned by anyone?

Larry E. Ruff focuses on solutions to the problem of environmental pollution. Hardin pointed out that the degradation of the commons arose from the fact that individual users bore only a small fraction of the cost of their own damages. Ruff's solution, a theme that will be repeated and elaborated in this volume, is to charge a price for using the environment that will cause its users to internalize all of the costs they impose on the rest of society.

1

Under normal circumstances, people have to pay for the resources and commodities they use. They are led thereby to use only the amounts that will yield them marginal benefits that are worth the price. When we are not required to pay for venting fumes into the atmosphere, pouring pollutants into lakes and rivers, or depositing garbage in landfills, we disregard the costs of these activities to society. We clutter the environment with our wastes when the benefits we enjoy are far less than the external costs we impose on other people. The price system fails when there are external effects like these because it doesn't apply to such costs.

Hardin saw the emergence of the "tragedy" as a consequence of increasing resource scarcity due to unrestrained population growth. But we now know that population growth is not the only culprit. Assaults on the environment, some of them of global proportions, can be traced as well to increases in consumption and changes in the nature of consumption and production in societies whose populations are relatively stable.

Along with most other economists, Ruff argues that "putting a price on pollution" would incorporate the environment into the normal operation of the price system. Charging for using the environment as a depository for wastes would lead firms and individuals to weigh their marginal benefits against the cost, resulting in the same relatively efficient use of the environment that charging for privately owned resources and commodities induces. As a practical matter, it is virtually impossible to achieve the same level of efficiency through direct governmental regulation as through charging, especially when the pollution comes from multiple sources.

Ruff emphasizes the importance of finding the right price to charge. The objective is not to prohibit the discharge of all wastes, but to limit total emissions to the amount that equates the marginal damages of emissions to the marignal cost of abatement. In the twenty years or so since Ruff wrote this article, economists have discovered how hard it is to identify the right price, and attention has turned more recently to schemes that use marketable emissions permits, which set the total level of emissions by all dischargers in some region and allow the price of permits to be determined in the market. The policy alternatives are discussed at greater length in Section III.

The third paper, by Allen V. Kneese, reminds us that the generation of waste is not an anomaly that can simply be outlawed. It is a natural and unavoidable by-product of all human activity. Unless they are continuously recycled or permanently stored in inventories or structures, all goods that are produced must sooner or later be returned to the natural environment in one form or another. Thus the total volume of residuals that is eventually discharged into the waste stream will be approximately equal to the weight of basic fuels, food, and other raw materials that enter the processing and production system, plus gases taken from the atmosphere. The rate at which virgin materials are turned into wastes can, however, be slowed by improvements in the efficiency of energy conversion and of materials utilization and reutilization (recycling).

The recipients of these wastes are the earth's land, air, and water resources. Kneese's materials balance model shows that, given the volume of residuals, any effort to restrict their flow into one of these receptors will necessarily increase the flow into others. Therefore the protection of land, air, and water resources cannot be viewed as separate problems. They are closely intertwined.

In the final paper in this section, Paul R. Portney takes us into the practical world of environmental policy-making as it has evolved in the United States from the early 1970s to the present. Portney leads us through the series of policy decisions that must be wrestled with once a government has decided to control a class of emissions, and then surveys the manner in which Congress has dealt with the decisions. The decisions of most interest to economists concern the basis to be used for setting policy goals, and the method for allocating responsibility for attaining these goals among different dischargers.

Portney describes the environmental policy that has actually evolved in the United States as a hybrid. There is not much consistency among the various standard-setting frameworks that have been embodied in laws governing the discharge of different wastes. Some reflect a zero-risk philosophy, others require the "best possible" abatement technology, and occasionally a benefit-cost balancing is called for. But until passage of the 1990 Clean Air Act, Congress was fairly consistent in one respect: it almost uniformly eschewed the recommendations of economists to employ price incentives for allocating responsibility for abatement among dischargers. Centralized command and control systems that either set standards of performance or mandate the use of specific technologies have generally been the rule. The Clean Air Act of 1990 introduced the first incentive system on a scale of any significance when it specified a program of marketable discharge permits for the control of acid rain deposition caused by sulfur dioxide (SO_2) emissions from the nation's major electric utilities.

1

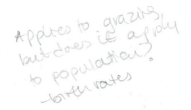

Applies to grazing, but does it apply to population/birth rates

The Tragedy of the Commons

GARRETT HARDIN

Are kids an economic boost or drain?

Garrett Hardin is Professor of Human Ecology Emeritus at
the University of California at Santa Barbara.

At the end of a thoughtful article on the future of nuclear war, J. B.
Wiesner and H. F. York concluded that "Both sides in the arms race
are . . . confronted by the dilemma of steadily increasing military power
and steadily decreasing national security. *It is our considered professional
judgment that this dilemma has no technical solution.* If the great powers
continue to look for solutions in the area of science and technology only,
the result will be to worsen the situation."[1]

I would like to focus your attention not on the subject of the article
(national security in a nuclear world) but on the kind of conclusion they
reached, namely, that there is no technical solution to the problem. An
implicit and almost universal assumption of discussions published in professional and semipopular scientific journals is that the problem under
discussion has a technical solution. A technical solution may be defined as
one that requires a change only in the techniques of the natural sciences,
demanding little or nothing in the way of change in human values or ideas
of morality.

In our day (though not in earlier times) technical solutions are always
welcome. Because of previous failures in prophecy, it takes courage to
assert that a desired technical solution is not possible. Wiesner and York
exhibited this courage; publishing in a science journal, they insisted that
the solution to the problem was not to be found in the natural sciences.

"The Tragedy of the Commons," by Garret Hardin, from *Science,* v.
162 (1968), pp. 1243–48. Copyright 1968 by the AAAS. Reprinted by
permission.

[1]J. B. Wiesner and H. F. York, *Scientific American* 211 (No. 4), 27 (1964).

They cautiously qualified their statement with the phrase, "It is our considered professional judgment. . . ." Whether they were right or not is not the concern of the present article. Rather, the concern here is with the important concept of a class of human problems which can be called "no technical solution problems," and more specifically, with the identification and discussion of one of these.

It is easy to show that the class is not a null class. Recall the game of tick-tack-toe. Consider the problem, "How can I win the game of tick-tack-toe?" It is well known that I cannot, if I assume (in keeping with the conventions of game theory) that my opponent understands the game perfectly. Put another way, there is no "technical solution" to the problem. I can win only by giving a radical meaning to the word "win." I can hit my opponent over the head; or I can falsify the records. Every way in which I "win" involves, in some sense, an abandonment of the game, as we intuitively understand it. (I can also, of course, openly abandon the game—refuse to play it. This is what most adults do.)

The class of "no technical solution problems" has members. My thesis is that the "population problem," as conventionally conceived, is a member of this class. How it is conventionally conceived needs some comment. It is fair to say that most people who anguish over the population problem are trying to find a way to avoid the evils of overpopulation without relinquishing any of the privileges they now enjoy. They think that farming the seas or developing new strains of wheat will solve the problem—technologically. I try to show here that the solution they seek cannot be found. The population problem cannot be solved in a technical way, any more than can the problem of winning the game of tick-tack-toe.

WHAT SHALL WE MAXIMIZE?

Population, as Malthus said, naturally tends to grow "geometrically," or, as we would now say, exponentially. In a finite world this means that the per-capita share of the world's goods must decrease. Is ours a finite world?

A fair defense can be put forward for the view that the world is infinite; or that we do not know that it is not. But, in terms of the practical problems that we must face in the next few generations with the foreseeable technology, it is clear that we will greatly increase human misery if we do not, during the immediate future, assume that the world available to the terrestrial human population is finite. "Space" is no escape.[2]

A finite world can support only a finite population; therefore, population growth must eventually equal zero. (The case of perpetual wide fluctuations above and below zero is a trivial variant that need not be

[2]G. Hardin, *Journal of Heredity* 50, 68 (1959), S. von Hoernor, *Science* 137, 18 (1962).

discussed.) When this condition is met, what will be the situation of mankind? Specifically, can Bentham's goal of "the greatest good for the greatest number" be realized?

No—for two reasons, each sufficient by itself. The first is a theoretical one. It is not mathematically possible to maximize for two (or more) variables at the same time. This was clearly stated by von Neumann and Morgenstern,[3] but the principle is implicit in the theory of partial differential equations, dating back at least to D'Alembert (1717–1783).

The second reason springs directly from biological facts. To live, any organism must have a source of energy (for example, food). This energy is utilized for two purposes: mere maintenance and work. For man, maintenance of life requires about 1600 kilocalories a day ("maintenance calories"). Anything that he does over and above merely staying alive will be defined as work, and is supported by "work calories" which he takes in. Work calories are used not only for what we call work in common speech; they are also required for all forms of enjoyment, from swimming and automobile racing to playing music and writing poetry. If our goal is to maximize population it is obvious what we must do: We must make the work calories per person approach as close to zero as possible. No gourmet meals, no vacations, no sports, no music, no literature, no art. . . . I think that everyone will grant, without argument or proof, that maximizing population does not maximize goods. Bentham's goal is impossible.

In reaching this conclusion I have made the usual assumption that it is the acquisition of energy that is the problem. The appearance of atomic energy has led some to question this assumption. However, given an infinite source of energy, population growth still produces an inescapable problem. The problem of the acquisition of energy is replaced by the problem of its dissipation, as J. H. Fremlin has so wittily shown.[4] The arithmetic signs in the analysis are, as it were, reversed; but Bentham's goal is unobtainable.

The optimum population is, then, less than the maximum. The difficulty of defining the optimum is enormous; so far as I know, no one has seriously tackled this problem. Reaching an acceptable and stable solution will surely require more than one generation of hard analytical work—and much persuasion.

We want the maximum good per person; but what is good? To one person it is wilderness, to another it is ski lodges for thousands. To one it is estuaries to nourish ducks for hunters to shoot; to another it is factory land. Comparing one good with another is, we usually say, impossible because goods are incommensurable. Incommensurables cannot be compared.

[3] J. von Neumann and O. Morgenstern, *Theory of Games and Economic Behavior* (Princeton University Press, Princeton, N.J., 1947), p. 11.
[4] J. H. Fremlin, *New Scientist,* No. 415 (1964), p. 285.

Theoretically this may be true; but in real life incommensurables *are* commensurable. Only a criterion of judgment and a system of weighting are needed. In nature the criterion is survival. Is it better for a species to be small and hideable, or large and powerful? Natural selection commensurates the incommensurables. The compromise achieved depends on a natural weighting of the values of the variables.

Man must imitate this process. There is no doubt that in fact he already does, but unconsciously. It is when the hidden decisions are made explicit that the arguments begin. The problem for the years ahead is to work out an acceptable theory of weighting. Synergistic effects, nonlinear variation, and difficulties in discounting the future make the intellectual problem difficult, but not (in principle) insoluble.

Has any cultural group solved this practical problem at the present time, even on an intuitive level? One simple fact proves that none has: there is no prosperous population in the world today that has, and has had for some time, a growth rate of zero. Any people that has intuitively identified its optimum point will soon reach it, after which its growth rate becomes and remains zero.

Of course, a positive growth rate might be taken as evidence that a population is below its optimum. However, by any reasonable standards, the most rapidly growing populations on earth today are (in general) the most miserable. This association (which need not be invariable) casts doubt on the optimistic assumption that the positive growth rate of a population is evidence that it has yet to reach its optimum.

We can make little progress in working toward optimum population size until we explicitly exorcise the spirit of Adam Smith in the field of practical demography. In economic affairs, *The Wealth of Nations* (1776) popularized the "invisible hand," the idea that an individual who "intends only his own gain," is, as it were, "led by an invisible hand to promote . . . the public interest."[5] Adam Smith did not assert that this was invariably true, and perhaps neither did any of his followers. But he contributed to a dominant tendency of thought that has ever since interfered with positive action based on rational analysis, namely, the tendency to assume that decisions reached individually will, in fact, be the best decisions for an entire society. If this assumption is correct it justifies the continuance of our present policy of *laissez faire* in reproduction. If it is correct we can assume that men will control their individual fecundity so as to produce the optimum population. If the assumption is not correct, we need to reexamine our individual freedoms to see which ones are defensible.

[5]A. Smith, *The Wealth of Nations* (Modern Library, New York, 1937), p. 423.

TRAGEDY OF FREEDOM IN A COMMONS

The rebuttal to the invisible hand in population control is to be found in a scenario first sketched in a little-known pamphlet in 1833 by a mathematical amateur named William Forster Lloyd (1794–1852).[6] We may well call it "the tragedy of the commons," using the word "tragedy" as the philosopher Whitehead used it: "The essence of dramatic tragedy is not unhappiness. It resides in the solemnity of the remorseless working of things." He then goes on to say, "This inevitableness of destiny can only be illustrated in terms of human life by incidents which in fact involve unhappiness. For it is only by them that the futility of escape can be made evident in the drama."[7]

The tragedy of the commons develops in this way. Picture a pasture open to all. It is to be expected that each herdsman will try to keep as many cattle as possible on the commons. Such an arrangement may work reasonably satisfactorily for centuries because tribal wars, poaching, and disease keep the numbers of both man and beast well below the carrying capacity of the land. Finally, however, comes the day of reckoning, that is, the day when the long-desired goal of social stability becomes a reality. At this point, the inherent logic of the commons remorselessly generates tragedy.

As a rational being, each herdsman seeks to maximize his gain. Explicitly or implicitly, more or less consciously, he asks, "What is the utility *to me* of adding one more animal to my herd?" This utility has one negative and one positive component.

1. The positive component is a function of the increment of one animal. Since the herdsman receives all the proceeds from the sale of the additional animal, the positive utility is nearly $+1$.

2. The negative component is a function of the additional overgrazing created by one more animal. Since, however, the effects of overgrazing are shared by all the herdsmen, the negative utility for any particular decision-making herdsman is only a fraction of -1.

Adding together the component partial utilities, the rational herdsman concludes that the only sensible course for him to pursue is to add another animal to his herd. And another. . . . But this is the conclusion reached by each and every rational herdsman sharing a commons. Therein is the tragedy. Each man is locked into a system that compels him to increase his herd without limit—in a world that is limited. Ruin is the destination toward which all men rush, each pursuing his own best interest in a society that believes in the freedom of the commons. Freedom in a commons brings ruin to all.

[6]W. F. Lloyd, *Two Lectures on the Checks to Population* (Oxford University Press, Oxford, England, 1833).
[7]A. N. Whitehead, *Science and the Modern World* (Mentor, New York, 1948), p. 17.

Some would say that this is a platitude. Would that it were! In a sense, it was learned thousands of years ago, but natural selection favors the forces of psychological denial.[8] The individual benefits as an individual from his ability to deny the truth even though society as a whole, of which he is a part, suffers. Education can counteract the natural tendency to do the wrong thing, but the inexorable succession of generations requires that the basis for this knowledge be constantly refreshed.

A simple incident that occurred a few years ago in Leominster, Massachusetts, shows how perishable the knowledge is. During the Christmas shopping season the parking meters downtown were covered with plastic bags that bore tags reading: "Do not open until after Christmas. Free parking courtesy of the mayor and city council." In other words, facing the prospect of an increased demand for already scarce space, the city fathers reinstituted the system of the commons. (Cynically, we suspect that they gained more votes than they lost by this retrogressive act.)

In an approximate way, the logic of the commons has been understood for a long time, perhaps since the discovery of agriculture or the invention of private property in real estate. But it is understood mostly only in special cases which are not sufficiently generalized. Even at this late date, cattlemen leasing national land on the Western ranges demonstrate no more than an ambivalent understanding, in constantly pressuring federal authorities to increase the head count to the point where overgrazing produces erosion and weed-dominance. Likewise, the oceans of the world continue to suffer from the survival of the philosophy of the commons. Maritime nations still respond automatically to the shibboleth of the "freedom of the seas." Professing to believe in the "inexhaustible resources of the oceans," they bring species after species of fish and whales closer to extinction.[9]

The National Parks present another instance of the working out of the tragedy of the commons. At present, they are open to all, without limit. The parks themselves are limited in extent—there is only one Yosemite Valley—whereas population seems to grow without limit. The values that visitors seek in the parks are steadily eroded. Plainly, we must soon cease to treat the parks as commons or they will be of no value to anyone.

What shall we do? We have several options. We might sell them off as private property. We might keep them as public property, but allocate the right to enter them. The allocation might be on the basis of wealth, by the use of an auction system. It might be on the basis of merit, as defined by some agreed-upon standards. It might be by lottery. Or it might be on a first-come, first-served basis, administered to long queues. These, I think, are all objectionable. But we must choose—or acquiesce in the destruction of the commons that we call our National Parks.

[8]G. Hardin, Ed., *Population, Evolution, and Birth Control* (Freeman, San Francisco, 1964), p. 56.
[9]S. McVay, *Scientific American* 216 (No. 8), 13 (1966).

POLLUTION

In a reverse way, the tragedy of the commons reappears in problems of pollution. Here it is not a question of taking something out of the commons, but of putting something in—sewage, or chemical, radioactive, and heat wastes into water; noxious and dangerous fumes into the air; and distracting and unpleasant advertising signs into the line of sight. The calculations of utility are much the same as before. The rational man finds that his share of the cost of the wastes he discharges into the commons is less than the cost of purifying his wastes before releasing them. Since this is true for everyone, we are locked into a system of "fouling our own nest," so long as we behave only as independent, rational, free-enterprisers.

The tragedy of the commons as a food basket is averted by private property, or something formally like it. But the air and waters surrounding us cannot readily be fenced, and so the tragedy of the commons as a cesspool must be prevented by different means, by coercive laws or taxing devices that make it cheaper for the polluter to treat his pollutants than to discharge them untreated. We have not progressed as far with the solution of this problem as we have with the first. Indeed, our particular concept of private property, which deters us from exhausting the positive resources of the earth, favors pollution. The owner of a factory on the bank of a stream—whose property extends to the middle of the stream— often has difficulty seeing why it is not his natural right to muddy the waters flowing past his door. The law, always behind the times, requires elaborate stitching and fitting to adapt it to this newly perceived aspect of the commons.

The pollution problem is a consequence of population. It did not much matter how a lonely American frontiersman disposed of his waste. "Flowing water purifies itself every ten miles," my grandfather used to say, and the myth was near enough to the truth when he was a boy, for there were not too many people. But as population became denser, the natural chemical and biological recycling processes became overloaded, calling for a redefinition of property rights.

HOW TO LEGISLATE TEMPERANCE?

Analysis of the pollution problem as a function of population density uncovers a not generally recognized principle of morality, namely: *the morality of an act is a function of the state of the system at the time it is performed.*[10] Using the commons as a cesspool does not harm the general public under frontier conditions, because there is no public; the same

[10]J. Fletcher, *Situation Ethics* (Westminster, Philadelphia, 1966).

behavior in a metropolis is unbearable. A hundred and fifty years ago a plainsman could kill an American bison, cut out only the tongue for his dinner, and discard the rest of the animal. He was not in any important sense being wasteful. Today, with only a few thousand bison left, we would be appalled at such behavior.

In passing, it is worth nothing that the morality of an act cannot be determined from a photograph. One does not know whether a man killing an elephant or setting fire to the grassland is harming others until one knows the total system in which his act appears. "One picture is worth a thousand words," said an ancient Chinese; but it may take ten thousand words to validate it. It is as tempting to ecologists as it is to reformers in general to try to persuade others by way of the photographic shortcut. But the essence of an argument cannot be photographed: it must be presented rationally—in words.

That morality is system-sensitive escaped the attention of most codifiers of ethics in the past. "Thou shalt not . . ." is the form of traditional ethical directives which make no allowance for particular circumstances. The laws of our society follow the pattern of ancient ethics, and therefore are poorly suited to governing a complex, crowded, changeable world. Our epicyclic solution is to augment statutory law with administrative law. Since it is practically impossible to spell out all the conditions under which it is safe to burn trash in the backyard or to run an automobile without smog control, by law we delegate the details to bureaus. The result is administrative law, which is rightly feared for an ancient reason—*Quis custodiet ipsos custodes?*—Who shall watch the watchers themselves? John Adams said that we must have a "government of laws and not men." Bureau administrators, trying to evaluate the morality of acts in the total system, are singularly liable to corruption, producing a government by men, not laws.

Prohibition is easy to legislate (though not necessarily to enforce); but how do we legislate temperance? Experience indicates that it can be accomplished best through the mediation of administrative law. We limit possibilities unnecessarily if we suppose that the sentiment of *Quis custodiet* denies us the use of administrative law. We should rather retain the phrase as a perpetual reminder of fearful dangers we cannot avoid. The great challenge facing us now is to invent the corrective feedbacks that are needed to keep custodians honest. We must find ways to legitimate the needed authority of both the custodians and the corrective feedbacks.

FREEDOM TO BREED IS INTOLERABLE

The tragedy of the commons is involved in population problems in another way. In a world governed solely by the principle of "dog eat dog"—
if indeed there ever was such a world—how many children a family had
would not be a matter of public concern. Parents who bred too exuber-
antly would leave fewer descendants, not more, because they would be
unable to care adequately for their children. David Lack and others have
found that such a negative feedback demonstrably controls the fecundity
of birds.[11] But men are not birds, and have not acted like them for
millenniums, at least.

If each human family were dependent only on its own resources; *if* the
children of improvident parents starved to death; *if,* thus, overbreeding
brought its own "punishment" to the germ line—*then* there would be no
public interest in controlling the breeding of families. But our society is
deeply committed to the welfare state,[12] and hence is confronted with
another aspect of the tragedy of the commons.

In a welfare state, how shall we deal with the family, the religion, the
race, or the class (or indeed any distinguishable and cohesive group) that
adopts overbreeding as a policy to secure its own aggrandizement?[13] To
couple the concept of freedom to breed with the belief that everyone born
has an equal right to the commons is to lock the world into a tragic course
of action.

Unfortunately this is just the course of action that is being pursued by
the United Nations. In late 1967, some thirty nations agreed to the following: "The Universal Declaration of Human Rights describes the family as
the natural and fundamental unit of society. It follows that any choice and
decision with regard to the size of the family must irrevocably rest with the
family itself, and cannot be made by anyone else."[14]

It is painful to have to deny categorically the validity of this right;
denying it, one feels as uncomfortable as a resident of Salem, Massachusetts, who denied the reality of witches in the seventeenth century. At the
present time, in liberal quarters, something like a taboo acts to inhibit
criticism of the United Nations. There is a feeling that the United Nations
is "our last and best hope," that we shouldn't find fault with it; we
shouldn't play into the hands of the archconservatives. However, let us not
forget what Robert Louis Stevenson said: "The truth that is suppressed by
friends is the readiest weapon of the enemy." If we love the truth we must
openly deny the validity of the Universal Declaration of Human Rights,

[11]D. Lack, *The Natural Regulation of Animal Numbers* (Clarendon Press, Oxford, England,
1954).
[12]H. Girvetz, *From Wealth to Welfare* (Stanford University Press, Stanford, Calif., 1950).
[13]G. Hardin, *Perspectives in Biology and Medicine* 6, 366 (1963).
[14]U Thant, *International Planned Parenthood News,* No. 168 (February 1968), p. 3.

Does contraception allow families to choose size? If contraception avail. to all who want it, much lower b.r.

even though it is promoted by the United Nations. We should also join with Kingsley Davis[15] in attempting to get Planned Parenthood–World Population to see the error of its ways in embracing the same tragic ideal.

decisions about family size rest w/ family — Hardin disagrees. Wants to legislate family size?

CONSCIENCE IS SELF-ELIMINATING

It is a mistake to think that we can control the breeding of mankind in the long run by an appeal to conscience. Charles Galton Darwin made this point when he spoke on the centennial of the publication of his grandfather's great book. The argument is straightforward and Darwinian.

People vary. Confronted with appeals to limit breeding, some people will undoubtedly respond to the plea more than others. Those who have more children will produce a larger fraction of the next generation than those with more susceptible consciences. The differences will be accentuated, generation by generation.

In C. G. Darwin's words: "It may well be that it would take hundreds of genrations for the progenitive instinct to develop in this way, but if it should do so, nature would have taken her revenge, and the variety *Homo contracipiens* would become extinct and would be replaced by the variety *Homo progenitivus.*"[16]

The argument assumes that conscience or the desire for children (no matter which) is hereditary—but hereditary only in the most general formal sense. The result will be the same whether the attitude is transmitted through germ cells, or exosomatically, to use A. J. Lotka's term. (If one denies the latter possibility as well as the former, then what's the point of education?) The argument has here been stated in the context of the population problem, but it applies equally well to any instance in which society appeals to an individual exploiting a commons to restrain himself for the general good—by means of his conscience. To make such an appeal is to set up a selective system that works toward the elimination of conscience from the race.

PATHOGENIC EFFECTS OF CONSCIENCE

The long-term disadvantage of an appeal to conscience should be enough to condemn it; but it has serious short-term disadvantages as well. If we ask a man who is exploiting a commons to desist "in the name of con-

[15]K. Davis, *Science* 158, 730 (1967).
[16]S. Tax, Ed., *Evolution After Darwin* (University of Chicago Press, Chicago, 1960), vol. 2, p. 469.

science," what are we saying to him? What does he hear?—not only at the moment but also in the wee small hours of the night when, half asleep, he remembers not merely the words we used but also the nonverbal communication cues we gave him unawares? Sooner or later, consciously or subconsciously, he senses that he has received two communications, and that they are contradictory: 1. (intended communication) "If you don't do as we ask, we will openly condemn you for not acting like a responsible citizen"; 2. (the unintended communication) "If you *do* behave as we ask, we will secretly condemn you for a simpleton who can be shamed into standing aside while the rest of us exploit the commons."

Everyman then is caught in what Bateson has called a "double bind." Bateson and his co-workers have made a plausible case for viewing the double bind as an important causative factor in the genesis of schizophrenia. [17] The double bind may not always be so damaging, but it always endangers the mental health of anyone to whom it is applied. "A bad conscience," said Nietzsche, "is a kind of illness."

To conjure up a conscience in others is tempting to anyone who wishes to extend his control beyond the legal limits. Leaders at the highest level succumb to this temptation. Has any president during the past generation failed to call on labor unions to moderate voluntarily their demands for higher wages, or to steel companies to honor voluntary guidelines on prices? I can recall none. The rhetoric used on such occasions is designed to produce feelings of guilt in noncooperators.

For centuries it was assumed without proof that guilt was a valuable, perhaps even an indispensable, ingredient of the civilized life. Now, in this post-Freudian world, we doubt it.

Paul Goodman speaks from the modern point of view when he says: "No good has ever come from feeling guilty, neither intelligence, policy, nor compassion. The guilty do not pay attention to the object but only to themselves, and not even to their own interests, which might make sense, but to their anxieties." [18]

One does not have to be a professional psychiatrist to see the consequences of anxiety. We in the Western world are just emerging from a dreadful two centuries-long Dark Ages of Eros that was sustained partly by prohibition laws, but perhaps more effectively by the anxiety-generating mechanisms of education. Alex Comfort has told the story well in *The Anxiety Makers;* [19] it is not a pretty one.

Since proof is difficult, we may even concede that the results of anxiety may sometimes, from certain points of view, be desirable. The larger question we should ask is whether, as a matter of policy, we should ever encourage the use of a technique the tendency (if not the intention) of

[17] G. Bateson, D. D. Jackson, J. Haley, J. Weakland, *Behavioral Science* 1, 251 (1956).
[18] P. Goodman, *New York Review of Books* 10 (8), 22 (23 May 1968).
[19] A. Comfort, *The Anxiety Makers* (Nelson, London, 1967).

which is psychologically pathogenic. We hear much talk these days of responsible parenthood; the coupled words are incorporated into the titles of some organizations devoted to birth control. Some people have proposed massive propaganda campaigns to instill responsibility into the nation's (or the world's) breeders. But what is the meaning of the word conscience? When we use the word responsibility in the absence of substantial sanctions, are we not trying to browbeat a free man in a commons into acting against his own interest? Responsibility is a verbal counterfeit for a substantial quid pro quo. It is an attempt to get something for nothing.

If the word responsibility is to be used at all, I suggest that it be in the sense Charles Frankel uses it.[20] "Responsibility," says this philosopher, "is the product of definite social arrangements." Notice that Frankel calls for social arrangements—not propaganda.

MUTUAL COERCION MUTUALLY AGREED UPON

The social arrangements that produce responsibility are arrangements that create coercion, of some sort. Consider bank robbing. The man who takes money from a bank acts as if the bank were a commons. How do we prevent such action? Certainly not by trying to control his behavior solely by a verbal appeal to his sense of responsibility. Rather than rely on propaganda we follow Frankel's lead and insist that a bank is not a commons; we seek the definite social arrangements that will keep it from becoming a commons. That we thereby infringe on the freedom of would-be robbers we neither deny nor regret.

The morality of bank robbing is particularly easy to understand because we accept complete prohibition of this activity. We are willing to say, "Thou shalt not rob banks," without providing for exceptions. But temperance also can be created by coercion. Taxing is a good coercive device. To keep downtown shoppers temperate in their use of parking space we introduce parking meters for short periods, and traffic fines for longer ones. We need not actually forbid a citizen to park as long as he wants to; we need merely make it increasingly expensive for him to do so. Not prohibition, but carefully biased options are what we offer him. A Madison Avenue man might call this persuasion; I prefer the greater candor of the word coercion.

Coercion is a dirty word to most liberals now, but it need not forever be so. As with the four-letter words, its dirtiness can be cleansed away by exposure to the light, by saying it over and over without apology or

[20]C. Frankel, *The Case for Modern Man* (Harper & Row, New York, 1955), p. 203.

embarrassment. To many, the word coercion implies arbitrary decisions of distant and irresponsible bureaucrats; but this is not a necessary part of its meaning. The only kind of coercion I recommend is mutual coercion, mutually agreed upon by the majority of the people affected.

To say that we mutually agree to coercion is not to say that we are required to enjoy it, or even to pretend we enjoy it. Who enjoys taxes? We all grumble about them. But we accept compulsory taxes because we recognize that voluntary taxes would favor the conscienceless. We institute and (grumblingly) support taxes and other coercive devices to escape the horror of the commons.

An alternative to the commons need not be perfectly just to be preferable. With real estate and other material goods, the alternative we have chosen is the institution of private property coupled with legal inheritance. Is this system perfectly just? As a genetically trained biologist I deny that it is. It seems to me that, if there are to be differences in individual inheritance, legal possession should be perfectly correlated with biological inheritance—that those who are biologically more fit to be the custodians of property and power should legally inherit more. But genetic recombination continually makes a mockery of the doctrine of "like father, like son" implicit in our laws of legal inheritance. An idiot can inherit millions, and a trust fund can keep his estate intact. We must admit that our legal system of private property plus inheritance is unjust—but we put up with it because we are not convinced, at the moment, that anyone has invented a better system. The alternative of the commons is too horrifying to contemplate. Injustice is preferable to total ruin.

It is one of the peculiarities of the warfare between reform and the status quo that it is thoughtlessly governed by a double standard. Whenever a reform measure is proposed it is often defeated when its opponents triumphantly discover a flaw in it. As Kingsley Davis has pointed out,[21] worshipers of the status quo sometimes imply that no reform is possible without unanimous agreement, an implication contrary to historical fact. As nearly as I can make out, automatic rejection of proposed reforms is based on one of two unconscious assumptions: (1) that the status quo is perfect; or (2) that the choice we face is between reform and no action; if the proposed reform is imperfect, we presumably should take no action at all, while we wait for a perfect proposal.

But we can never do nothing. That which we have done for thousands of years is also action. It also produces evils. Once we are aware that the status quo is action, we can then compare its discoverable advantages and disadvantages with the predicted advantages and disadvantages of the proposed reform, discounting as best we can for our lack of experience.

[21]See J. D. Roslansky, *Genetics and the Future of Man* (Appleton-Century-Crofts, New York, 1966), p. 177.

On the basis of such a comparison, we can make a rational decision which will not involve the unworkable assumption that only perfect systems are tolerable.

RECOGNITION OF NECESSITY

Perhaps the simplest summary of this analysis of man's population problems is this: the commons, if justifiable at all, is justifiable only under conditions of low-population density. As the human population has increased, the commons has had to be abandoned in one aspect after another.

First we abandoned the commons in food gathering, enclosing farmland and restricting pastures and hunting and fishing areas. These restrictions are still not complete throughout the world.

Somewhat later we saw that the commons as a place for waste disposal would also have to be abandoned. Restrictions on the disposal of domestic sewage are widely accepted in the Western world; we are still struggling to close the commons to pollution by automobiles, factories, insecticide sprayers, fertilizing operations, and atomic energy installations.

In a still more embryonic state is our recognition of the evils of the commons in matters of pleasure. There is almost no restriction on the propagation of sound waves in the public medium. The shopping public is assaulted with mindless music, without its consent. Our government has paid out billions of dollars to create a supersonic transport which would disturb 50,000 people for every one person whisked from coast to coast 3 hours faster. Advertisers muddy the airwaves of radio and television and pollute the view of travelers. We are a long way from outlawing the commons in matters of pleasure. Is this because our Puritan inheritance makes us view pleasure as something of a sin, and pain (that is, the pollution of advertising) as the sign of virtue?

Every new enclosure of the commons involves the infringement of somebody's personal liberty. Infringements made in the distant past are accepted because no contemporary complains of a loss. It is the newly proposed infringements that we vigorously oppose; cries of "rights" and "freedom" fill the air. But what does "freedom" mean? When men mutually agreed to pass laws against robbing, mankind became more free, not less so. Individuals locked into the logic of the commons are free only to bring on universal ruin; once they see the necessity of mutual coercion, they become free to pursue other goals. I believe it was Hegel who said, "Freedom is the recognition of necessity."

The most important aspect of necessity that we must now recognize is the necessity of abandoning the commons in breeding. No technical solution can rescue us from the misery of overpopulation. Freedom to breed

why is breeding a tragedy of commons?   *-educate women*
what does he advocate?   *-promote contraception*

who bears the burden of children
parents, society, env.?

will bring ruin to all. At the moment, to avoid hard decisions many of us *agree?* are tempted to propagandize for conscience and responsible parenthood. The temptation must be resisted, because an appeal to independently acting consciences selects for the disappearance of all conscience in the long run, and an increase in anxiety in the short.

The only way we can preserve and nurture other and more precious freedoms is by relinquishing the freedom to breed, and that very soon. "Freedom is the recognition of necessity"—and it is the role of education to reveal to all the necessity of abandoning the freedom to breed. Only so can we put an end to this aspect of the tragedy of the commons.

? How do my children decrease the utility of commons?

read

People abuse the commons.

- Can't expect people to respect common property out of conscience or responsibility to common good - guilt doesn't change behavior. Need policy.

- everyone acts rationally to maximize their benefit

therefore must

- close commons by regulating their use, (e.g. kids per family).

- each closure limits personal liberty

- laws are mutual coercion - nec. to protect man from tragedy of commons - ruin

- must abandon commons in breeding

- cannot depend on conscience + responsibility

- must legislate "breeding"

2

The Economic Common Sense of Pollution

LARRY E. RUFF

Larry E. Ruff is a director of Putnam, Hayes and Bartlett, Inc., in Washington, D.C.

We are going to make very little real progress in solving the problem of pollution until we recognize it for what, primarily, it is: an economic problem, which must be understood in economic terms. Of course, there are *noneconomic* aspects of pollution, as there are with all economic problems, but all too often, such secondary matters dominate discussion. Engineers, for example, are certain that pollution will vanish once they find the magic gadget or power source. Politicians keep trying to find the right kind of bureaucracy; and bureaucrats maintain an unending search for the correct set of rules and regulations. Those who are above such vulgar pursuits pin their hopes on a moral regeneration or social revolution, apparently in the belief that saints and socialists have no garbage to dispose of. But as important as technology, politics, law, and ethics are to the pollution question, all such approaches are bound to have disappointing results, for they ignore the primary fact that pollution is an economic problem.

Before developing an economic analysis of pollution, however, it is necessary to dispose of some popular myths.

First, pollution is not new. Spanish explorers landing in the sixteenth century noted that smoke from Indian campfires hung in the air of the Los Angeles basin, trapped by what is now called the inversion layer. Before

"The Economic Common Sense of Pollution," by Larry E. Ruff. Reprinted with permission of the author from: *The Public Interest,* no. 19 (Spring 1970), pp. 69–85. ©1970 by National Affairs, Inc.

the first century B.C., the drinking waters of Rome were becoming polluted.

Second, most pollution is not due to affluence, despite the current popularity of this notion. In India, the pollution runs in the streets, and advice against drinking the water in exotic lands is often well taken. Nor can pollution be blamed on the self-seeking activities of greedy capitalists. Once-beautiful rivers and lakes which are now open sewers and cesspools can be found in the Soviet Union as well as in the United States, and some of the world's dirtiest air hangs over cities in Eastern Europe, which are neither capitalist nor affluent. In many ways, indeed, it is much more difficult to do anything about pollution in noncapitalist societies. In the Soviet Union, there is no way for the public to become outraged or to exert any pressure, and the polluters and the courts there work for the same people, who often decide that clean air and water, like good clothing, are low on their list of social priorities.

In fact, it seems probable that affluence, technology, and slow-moving, inefficient democracy will turn out to be the cure more than the cause of pollution. After all, only an affluent, technological society can afford such luxuries as moon trips, three-day weekends, and clean water, although even our society may not be able to afford them all; and only in a democracy can the people hope to have any real influence on the choice among such alternatives.

What *is* new about pollution is what might be called the *problem* of pollution. Many unpleasant phenomena—poverty, genetic defects, hurricanes—have existed forever without being considered problems; they are, or were, considered to be facts of life, like gravity and death, and a mature person simply adjusted to them. Such phenomena become problems only when it begins to appear that something can and should be done about them. It is evident that pollution had advanced to the problem stage. Now the question is what can and should be done?

Most discussions of the pollution problem begin with some startling facts: Did you know that 15,000 tons of filth are dumped into the air of Los Angeles County every day? But by themselves, such facts are meaningless, if only because there is no way to know whether 15,000 tons is a lot or a little. It is much more important for clear thinking about the pollution problem to understand a few economic concepts than to learn a lot of sensational-sounding numbers.

MARGINALISM

One of the most fundamental economic ideas is that of *marginalism,* which entered economic theory when economists became aware of the differential calculus in the 19th century and used it to formulate economic problems as problems of "maximization." The standard economic problem came to be viewed as that of finding a level of operation of some activity which would maximize the net gain from that activity, where the net gain is the difference between the benefits and the costs of the activity. As the level of activity increases, both benefits and costs will increase; but because of diminishing returns, costs will increase faster than benefits. When a certain level of the activity is reached, any further expansion increases costs more than benefits. At this "optimal" level, "marginal cost"—or the cost of expanding the activity—equals "marginal benefit," or the benefit from expanding the activity. Further expansion would cost more than it is worth, and reduction in the activity would reduce benefits more than it would save costs. The net gain from the activity is said to be maximized at this point.

This principle is so simple that it is almost embarrassing to admit it is the cornerstone of economics. Yet intelligent men often ignore it in discussion of public issues. Educators, for example, often suggest that, if it is better to be literate than illiterate, there is no logical stopping point in supporting education. Or scientists have pointed out that the benefits derived from "science" obviously exceed the costs and then have proceeded to infer that their particular project should be supported. The correct comparison, of course, is between *additional* benefits created by the proposed activity and the *additional* costs incurred.

The application of marginalism to questions of pollution is simple enough conceptually. The difficult part lies in estimating the cost and benefits functions, a question to which I shall return. But several important qualitative points can be made immediately. The first is that the choice facing a rational society is *not* between clean air and dirty air, or between clear water and polluted water, but rather between various *levels* of dirt and pollution. The aim must be to find that level of pollution abatement where the costs of further abatement begin to exceed the benefits.

The second point is that the optimal combination of pollution control methods is going to be a very complex affair. Such steps as demanding a 10 per cent reduction in pollution from all sources, without considering the relative difficulties and costs of the reduction, will certainly be an inefficient approach. Where it is less costly to reduce pollution, we want a greater reduction, to a point where an additional dollar spent on control anywhere yields the same reduction in pollution levels.

MARKETS, EFFICIENCY, AND EQUITY

A second basic economic concept is the idea—or the ideal—of the self-regulating economic system. Adam Smith illustrated this ideal with the example of bread in London: the uncoordinated, selfish actions of many people—farmer, miller, shipper, baker, grocer—provide bread for the city dweller, without any central control and at the lowest possible cost. Pure self-interest, guided only by the famous "invisible hand" of competition, organizes the economy efficiently.

The logical basis of this rather startling result is that, under certain conditions, competitive prices convey all the information necessary for making the optimal decision. A builder trying to decide whether to use brick or concrete will weigh his requirements and tastes against the prices of the materials. Other users will do the same, with the result that those whose needs and preferences for brick are relatively the strongest will get brick. Further, profit-maximizing producers will weigh relative production costs, reflecting society's productive capabilities, against relative prices, reflecting society's tastes and desires, when deciding how much of each good to produce. The end result is that users get brick and cement in quantities and proportions that reflect their individual tastes and society's production opportunities. No other solution would be better from the standpoint of all the individuals concerned.

This suggests what it is that makes pollution different. The efficiency of competitive markets depends on the identity of *private* costs and *social* costs. As long as the brick-cement producer must compensate somebody for every cost imposed by his production, his profit-maximizing decisions about how much to produce, and how, will also be socially efficient decisions. Thus, if a producer dumps wastes into the air, river, or ocean; if he pays nothing for such dumping; and if the disposed wastes have no noticeable effect on anyone else, living or still unborn; then the private and social costs of disposal are identical and nil, and the producer's private decisions are socially efficient. *But if these wastes do affect others, then the social costs of waste disposal are not zero. Private and social costs diverge, and private profit-maximizing decisions are not socially efficient.* Suppose, for example, that cement production dumps large quantities of dust into the air, which damages neighbors, and that the brick-cement producer pays these neighbors nothing. In the social sense, cement will be over-produced relative to brick and other products because users of the products will make decisions based on market prices which do not reflect true social costs. They will use cement when they should use brick, or when they should not build at all.

This divergence between private and social costs is the fundamental cause of pollution of all types, and it arises in any society where decisions are at all decentralized—which is to say, in any economy of any size which hopes to function at all. Even the socialist manager of the brick-cement plant, told to maximize output given the resources at his disposal, will use the People's Air to dispose of the People's Wastes; to do otherwise would be to violate his instructions. And if instructed to avoid pollution "when possible," he does not know what to do: how can he decide whether more brick or cleaner air is more important for building socialism? The capitalist manager is in exactly the same situation. Without prices to convey the needed information, he does not know what action is in the public interest, and certainly would have no incentive to act correctly even if he did know.

Although markets fail to perform efficiently when private and social costs diverge, this does not imply that there is some inherent flaw in the idea of acting on self-interest in response to market prices. Decisions based on private cost calculations are typically correct from a social point of view; and even when they are not quite correct, it often is better to accept this inefficiency than to turn to some alternative decision mechanism, which may be worse. Even the modern economic theory of socialism is based on the high correlation between managerial self-interest and public good. There is no point in trying to find something—some omniscient and omnipotent *deus ex machina*—to replace markets and self-interest. Usually it is preferable to modify existing institutions, where necessary, to make private and social interest coincide.

And there is a third relevant economic concept: the fundamental distinction between questions of efficiency and questions of equity or fairness. A situation is said to be efficient if it is not possible to rearrange things so as to benefit one person without harming any others. That is the *economic* equation for efficiency. *Politically,* this equation can be solved in various ways; though most reasonable men will agree that efficiency is a good thing, they will rarely agree about which of the many possible efficient states, each with a different distribution of "welfare" among individuals, is the best one. Economics itself has nothing to say about which efficient state is the best. That decision is a matter of personal and philosophical values, and ultimately must be decided by some political process. Economics can suggest ways of achieving efficient states, and can try to describe the equity considerations involved in any suggested social policy; but the final decisions about matters of "fairness" or "justice" cannot be decided on economic grounds.

ESTIMATING THE COSTS OF POLLUTION

Both in theory and practice, the most difficult part of an economic approach to pollution is the measurement of the cost and benefits of its abatement. Only a small fraction of the costs of pollution can be estimated straightforwardly. If, for example, smog reduces the life of automobile tires by 10 per cent, one component of the cost of smog is 10 per cent of tire expenditures. It has been estimated that, in a moderately polluted area of New York City, filthy air imposes extra costs for painting, washing, laundry, etc., of $200 per person per year. Such costs must be included in any calculation of the benefits of pollution abatement, and yet they are only a part of the relevant costs—and often a small part. Accordingly it rarely is possible to justify a measure like river pollution control solely on the basis of costs to individuals or firms of treating water because it usually is cheaper to process only the water that is actually used for industrial or municipal purposes, and to ignore the river itself.

The costs of pollution that cannot be measured so easily are often called "intangible" or "noneconomic," although neither term is particularly appropriate. Many of these costs are as tangible as burning eyes or a dead fish, and all such costs are relevant to a valid economic analysis. Let us therefore call these costs "nonpecuniary."

The only real difference between nonpecuniary costs and the other kind lies in the difficulty of estimating them. If pollution in Los Angeles Harbor is reducing marine life, this imposes costs on society. The cost of reducing commercial fishing could be estimated directly: it would be the fixed cost of converting men and equipment from fishing to an alternative occupation, plus the difference between what they earned in fishing and what they earn in the new occupation, plus the loss to consumers who must eat chicken instead of fish. But there are other, less straightforward costs: the loss of recreation opportunities for children and sportsfishermen and of research facilities for marine biologists, etc. Such costs are obviously difficult to measure and may be very large indeed; but just as surely as they are not zero, so too are they not infinite. Those who call for immediate action and damn the cost, merely because the spiney starfish and furry crab populations are shrinking, are putting an infinite marginal value on these creatures. This strikes a disinterested observer as an overestimate.

The above comments may seem crass and insensitive to those who, like one angry letter-writer to the Los Angeles *Times,* want to ask: "If conservation is not for its own sake, then what in the world *is* it for?" Well, what *is* the purpose of pollution control? Is it for its own sake? Of course not. If we answer that it is to make the air and water clean and quiet, then the question arises: what is the purpose of clean air and water? If the answer is, to please the nature gods, then it must be conceded that all pollution must cease immediately because the cost of angering the gods is presumably infinite. But if the answer is that the purpose of clean air and water is to further human enjoyment of life on this planet, then we are faced with the economists' basic question: given the limited alternatives that a niggardly nature allows, how can we best further human enjoyment of life? And the answer is, by making intelligent marginal decisions on the basis of costs and benefits. Pollution control is for lots of things: breathing comfortably, enjoying mountains, swimming in water, for health, beauty, and the general delectation. But so are many other things, like good food and wine, comfortable housing and fast transportation. The question is not which of these desirable things we should have, but rather what combination is most desirable. To determine such a combination, we must know the rate at which individuals are willing to substitute more of one desirable thing for less of another desirable thing. Prices are one way of determining those rates.

But if we cannot directly observe market prices for many of the costs of pollution, we must find another way to proceed. One possibility is to infer the costs from other prices, just as we infer the value of an ocean view from real estate prices. In principle, one could estimate the value people put on clean air and beaches by observing how much more they are willing to pay for property in nonpolluted areas. Such information could be obtained; but there is little of it available at present.

Another possible way of estimating the costs of pollution is to ask people how much they would be willing to pay to have pollution reduced. A resident of Pasadena might be willing to pay $100 a year to have smog reduced 10 or 20 per cent. In Barstow, where the marginal cost of smog is much less, a resident might not pay $10 a year to have smog reduced 10 per cent. If we knew how much it was worth to everybody, we could add up these amounts and obtain an estimate of the cost of a marginal amount of pollution. The difficulty, of course, is that there is no way of guaranteeing truthful responses. Your response to the question, how much is pollution costing *you,* obviously will depend on what you think will be done with this information. If you think you will be compensated for these costs, you will make a generous estimate; if you think that you will be charged for the control in proportion to these costs, you will make a small estimate.

In such cases it becomes very important how the questions are asked. For example, the voters could be asked a question of the form: Would you like to see pollution reduced x per cent if the result is a y per cent increase in the cost of living? Presumably a set of questions of this form could be used to estimate the costs of pollution, including the so-called "unmeasurable" costs. But great care must be taken in formulating the questions. For one thing, if the voters will benefit differentially from the activity, the questions should be asked in a way which reflects this fact. If, for example, the issue is cleaning up a river, residents near the river will be willing to pay more for the cleanup and should have a means of expressing this. Ultimately, some such political procedure probably will be necessary, at least until our more direct measurement techniques are greatly improved.

Let us assume that, somehow, we have made an estimate of the social cost function for pollution, including the marginal cost associated with various pollution levels. We now need an estimate of the benefits of pollution—or, if you prefer, of the costs of pollution abatement. So we set the Pollution Control Board (PCB) to work on this task.

The PCB has a staff of engineers and technicians, and they begin working on the obvious question: for each pollution source, how much would it cost to reduce pollution by 10 per cent, 20 per cent, and so on. If the PCB has some economists, they will know that the cost of reducing total pollution by 10 per cent is *not* the total cost of reducing each pollution source by 10 per cent. Rather, they will use the equimarginal principle and find the pattern of control such that an additional dollar spent on control of any pollution source yields the same reduction. This will minimize the cost of achieving any given level of abatement. In this way the PCB can generate a "cost of abatement" function, and the corresponding marginal cost function.

While this procedure seems straightforward enough, the practical difficulties are tremendous. The amount of information needed by the PCB is staggering; to do this job right, the PCB would have to know as much about each plant as the operators of the plant themselves. The cost of gathering these data is obviously prohibitive, and, since marginal principles apply to data collection too, the PCB would have to stop short of complete information, trading off the resulting loss in efficient control against the cost of better information. Of course, just as fast as the PCB obtained the data, a technological change would make it obsolete.

The PCB would have to face a further complication. It would not be correct simply to determine how to control existing pollution sources given their existing locations and production methods. Although this is almost certainly what the PCB would do, the resulting cost functions will overstate the true social cost of control. Muzzling existing plants is only one method of control. Plants can move, or switch to a new process, or even to a new product. Consumers can switch to a less polluting substitute. There are any number of alternatives, and the poor PCB engineers can never know them all. This could lead to some costly mistakes. For example, the PCB may correctly conclude that the cost of installing effective dust control at the cement plant is very high and hence may allow the pollution to continue, when the best solution is for the cement plant to switch to brick production while a plant in the desert switches from brick to cement. The PCB can never have all this information and therefore is doomed to inefficiency, sometimes an inefficiency of large proportions.

Once cost and benefit functions are known, the PCB should choose a level of abatement that maximizes net gain. This occurs where the marginal cost of further abatement just equals the marginal benefit. If, for example, we could reduce pollution damages by $2 million at a cost of $1 million, we should obviously impose that $1 million cost. But if the damage reduction is only $½ million, we should not and in fact should reduce control efforts.

This principle is obvious enough but is often overlooked. One author, for example, has written that the national cost of air pollution is $11 billion a year but that we are spending less than $50 million a year on control; he infers from this that "we could justify a tremendous strengthening of control efforts on purely economic grounds." That *sounds* reasonable, if all you care about are sounds. But what is the logical content of the statement? Does it imply we should spend $11 billion on control just to make things even? Suppose we were spending $11 billion on control and thereby succeeded in reducing pollution costs to $50 million. Would this imply we were spending too *much* on control? Of course not. We must compare the *marginal* decrease in pollution costs to the *marginal* increase in abatement costs.

DIFFICULT DECISIONS

Once the optimal pollution level is determined, all that is necessary is for the PCB to enforce the pattern of controls which it has determined to be optimal. (Of course, this pattern will not really be the best one, because the PCB will not have all the information it should have.) But now a new problem arises: how should the controls be enforced?

The most direct and widely used method is in many ways the least efficient: direct regulation. The PCB can decide what each polluter must do to reduce pollution and then simply require that action under penalty of law. But this approach has many shortcomings. The polluters have little incentive to install the required devices or to keep them operating properly. Constant inspection is therefore necessary. Once the polluter has complied with the letter of the law, he has no incentive to find better methods of pollution reduction. Direct control of this sort has a long history of inadequacy; the necessary bureaucracies rarely manifest much vigor, imagination, or devotion to the public interest. Still, in some situations there may be no alternative.

A slightly better method of control is for the PCB to set an acceptable level of pollution for each source and let the polluters find the cheapest means of achieving this level. This reduces the amount of information the PCB needs, but not by much. The setting of the acceptable levels becomes a matter for negotiation, political pull, or even graft. As new plants are built and new control methods invented, the limits should be changed; but if they are, the incentive to find new designs and new techniques is reduced.

A third possibility is to subsidize the reduction of pollution, either by subsidizing control equipment or by paying for the reduction of pollution below standard levels. This alternative has all the problems of the above methods, plus the classic shortcoming which plagues agricultural subsidies: the old joke about getting into the not-growing-cotton business is not always so funny.

The PCB will also have to face the related problem of deciding *who* is going to pay the costs of abatement. Ultimately, this is a question of equity or fairness which economics cannot answer; but economics can suggest ways of achieving equity without causing inefficiency. In general, the economist will say: if you think polluter A is deserving of more income at polluter B's expense, then by all means give A some of B's income; but do *not* try to help A by allowing him to pollute freely. For example, suppose A and B each operate plants which produce identical amounts of pollution. Because of different technologies, however, A can reduce his pollution 10 per cent for $100, while B can reduce his pollution 10 per cent for $1,000. Suppose your goal is to reduce total pollution 5 per cent. Surely it is obvious that the best (most efficient) way to do this is for A to reduce his pollution 10 per cent while B does nothing. But suppose B is rich and A is poor. Then many would demand that B reduce his pollution 10 per cent while A does nothing because B has a greater "ability to pay." Well, perhaps B does have greater ability to pay, and perhaps it is "fairer" that he pay the costs of pollution control; but if so, B should pay the $100 necessary to reduce A's pollution. To force B to reduce his own pollution 10 per cent is equivalent to taxing B $1,000 and then blowing the $1,000 on an extremely inefficient pollution control method. Put this way, it is obviously a stupid thing to do; but put in terms of B's greater ability to pay, it will get considerable support though it is no less stupid. The more efficient alternative is not always available, in which case it may be acceptable to use the inefficient method. Still, it should not be the responsibility of the pollution authorities to change the distribution of welfare in society; this is the responsibility of higher authorities. The PCB should concentrate on achieving economic efficiency without being grossly unfair in its allocation of costs.

Clearly, the PCB has a big job which it will never be able to handle with any degree of efficiency. Some sort of self-regulating system, like a market, is needed, which will automatically adapt to changes conditions, provide incentives for development and adoption of improved control methods, reduce the amount of information the PCB must gather and the amount of detailed control it must exercise, and so on. This, by any standard, is a tall order.

PUTTING A PRICE ON POLLUTION

And yet there is a very simple way to accomplish all this. *Put a price on pollution.* A price-based control mechanism would differ from an ordinary market transaction system only in that the PCB would set the prices, instead of their being set by demand-supply forces, and that the state would force payment. Under such a system, anyone could emit any amount of pollution so long as he pays the price which the PCB sets to approximate the marginal social cost of pollution. Under this circumstance, private decisions based on self-interest are efficient. If pollution consists of many components, each with its own social cost, there should be different prices for each component. Thus, extremely dangerous materials must have an extremely high price, perhaps stated in terms of "years in jail" rather than "dollars," although a sufficiently high dollar price is essentially the same thing. In principle, the prices should vary with geographical location, season of the year, direction of the wind, and even day of the week, although the cost of too many variations may preclude such fine distinctions.

Once the prices are set, polluters can adjust to them any way they choose. Because they act on self-interest they will reduce their pollution by every means possible up to the point where further reduction would cost more than the price. Because all face the same price for the same type of pollution, the marginal cost of abatement is the same everywhere. If there are economies of scale in pollution control, as in some types of liquid waste treatment, plants can cooperate in establishing joint treatment facilities. In fact, some enterprising individual could buy these wastes from various plants (at negative prices—i.e., they would get paid for carting them off), treat them, and then sell them at a higher price, making a profit in the process. (After all, this is what rubbish removal firms do now.) If economies of scale are so substantial that the provider of such a service becomes a monopolist, then the PCB can operate the facilities itself.

Obviously, such a scheme does not eliminate the need for the PCB. The board must measure the output of pollution from all sources, collect the fees, and so on. But it does not need to know anything about any plant except its total emission of pollution. It does not control, negotiate, threaten, or grant favors. It does not destroy incentive because development of new control methods will reduce pollution payments.

As a test of this price system of control, let us consider how well it would work when applied to automobile pollution, a problem for which direct control is usually considered the only feasible approach. If the price system can work here, it can work anywhere.

Suppose, then, that a price is put on the emissions of automobiles. Obviously, continuous metering of such emissions is impossible. But it should be easy to determine the average output of pollution for cars of different makes, models, and years, having different types of control devices and using different types of fuel. Through graduated registration fees and fuel taxes, each car owner would be assessed roughly the social cost of his car's pollution, adjusted for whatever control devices he has chosen to install and for his driving habits. If the cost of installing a device, driving a different car, or finding alternative means of transportation is less than the price he must pay to continue his pollution, he will presumably take the necessary steps. But each individual remains free to find the best adjustment to his particular situation. It would be remarkable if everyone decided to install the same devices which some states currently require; and yet that is the effective assumption of such requirements.

Even in the difficult case of auto pollution, the price system has a number of advantages. Why should a person living in the Mojave Desert, where pollution has little social cost, take the same pains to reduce air pollution as a person living in Pasadena? Present California law, for example, makes no distinction between such areas; the price system would. And what incentive is there for auto manufacturers to design a less polluting engine? The law says only that they must install a certain device in every car. If GM develops a more efficient engine, the law will eventually be changed to require this engine on all cars, raising costs and reducing sales. But will such development take place? No collusion is needed for manufacturers to decide unanimously that it would be foolish to devote funds to such development. But with a pollution fee paid by the consumer, there is a real advantage for any firm to be first with a better engine, and even a collusive agreement wouldn't last long in the face of such an incentive. The same is true of fuel manufacturers, who now have no real incentive to look for better fuels. Perhaps most important of all, the present situation provides no real way of determining whether it is cheaper to reduce pollution by muzzling cars or industrial plants. The experts say that most smog comes from cars; but *even if true, this does not imply that it is more efficient to control autos rather than other pollution sources.* How can we decide which is more efficient without mountains of information? The answer is, by making drivers and plants pay the same price for the same pollution, and letting self-interest do the job.

In situations where pollution outputs can be measured more or less directly (unlike the automobile pollution case), the price system is clearly superior to direct control. A study of possible control methods in the Delaware estuary, for example, estimated that, compared to a direct control scheme requiring each polluter to reduce his pollution by a fixed percentage, an effluent charge which would achieve the same level of pollution abatement would be only half as costly—a saving of about $150 million. Such a price system would also provide incentive for further improvements, a simple method of handling new plants, and revenue for the control authority.

In general, the price system allocates costs in a manner which is at least superficially fair: those who produce and consume goods which cause pollution, pay the costs. But the superior efficiency in control and apparent fairness are not the only advantages of the price mechanism. Equally important is the case with which it can be put into operation. It is not necessary to have detailed information about all the techniques of pollution reduction, or estimates of all costs and benefits. Nor is it necessary to determine whom to blame or who should pay. All that is needed is a mechanism for estimating, if only roughly at first, the pollution output of all polluters, together with a means of collecting fees. Then we can simply pick a price—any price—for each category of pollution, and we are in business. The initial price should be chosen on the basis of some estimate of its effects but need not be the optimal one. If the resulting reduction in pollution is not "enough," the price can be raised until there is sufficient reduction. A change in technology, number of plants, or whatever, can be accommodated by a change in the price, even without detailed knowledge of all the technological and economic data. Further, once the idea is explained, the price system is much more likely to be politically acceptable than some method of direct control. Paying for a service, such as garbage disposal, is a well-established tradition, and is much less objectionable than having a bureaucrat nosing around and giving arbitrary orders. When businessmen, consumers, and politicians understand the alternatives, the price system will seem very attractive indeed.

WHO SETS THE PRICES?

An important part of this method of control obviously is the mechanism that sets and changes the pollution price. Ideally, the PCB could choose this price on the basis of an estimate of the benefits and costs involved, in effect imitating the impersonal workings of ordinary market forces. But because many of the costs and benefits cannot be measured, a less "objective," more political procedure is needed. This political procedure could take the form of a referendum, in which the PCB would present to the voters alternative schedules of pollution prices, together with the estimated effects of each. There would be a massive propaganda campaign waged by the interested parties, of course. Slogans such as "Vote NO on 12 and Save Your Job," or "Proposition 12 Means Higher Prices," might be overstatements but would contain some truth, as the individual voter would realize when he considered the suggested increase in gasoline taxes and auto registration fees. But the other side, in true American fashion, would respond by overstating *their* case: "Smog Kills, YES on 12," or "Stop *Them* from Ruining *Your* Water." It would be up to the PCB to inform the public about the true effects of the alternatives; but ultimately, the voters would make the decision.

It is fashionable in intellectual circles to object to such democratic procedures on the ground that the uncultured masses will not make correct decisions. If this view is based on the fact that the technical and economic arguments are likely to be too complex to be decided by direct referendum, it is certainly a reasonable position; one obvious solution is to set up an elective or appointive board to make the detailed decisions, with the expert board members being ultimately responsible to the voters. But often there is another aspect to the antidemocratic position—a feeling that it is impossible to convince the people of the desirability of some social policy, not because the issues are too complex but purely because their values are "different" and inferior. To put it bluntly: many ardent foes of pollution are not so certain that popular opinion is really behind them, and they therefore prefer a more bureaucratic and less political solution.

The question of who should make decisions for whom, or whose desires should count in a society, is essentially a noneconomic question that an economist cannot answer with authority, whatever his personal views on the matter. The political structures outlined here, when combined with the economic suggestions, can lead to a reasonably efficient solution of the pollution problem in a society where the tastes and values of all men are given some consideration. In such a society, when any nonrepresentative group is in a position to impose its particular evaluation of the costs and benefits, an inefficient situation will result. The swimmer or tidepool enthusiast who wants Los Angeles Harbor converted into a crystal-clear swimming pool, at the expense of all the workers, consumers, and businessmen who use the harbor for commerce and industry, is indistinguishable from the stockholder in Union Oil who wants maximum output from offshore wells, at the expense of everyone in the Santa Barbara area. Both are urging an inefficient use of society's resources; both are trying to get others to subsidize their particular thing—a perfectly normal, if not especially noble, endeavor.

If the democratic principle upon which the above political suggestions are based is rejected, the economist cannot object. He will still suggest the price system as a tool for controlling pollution. With any method of decision—whether popular vote, representative democracy, consultation with the nature gods, or a dictate of the intellectual elite—the price system can simplify control and reduce the amount of information needed for decisions. It provides an efficient, comprehensive, easily understood, adaptable, and reasonably fair way of handling the problem. It is ultimately the only way the problem will be solved. Arbitrary, piecemeal, stop-and-go programs of direct control have not and will not accomplish the job.

SOME OBJECTIONS AREN'T AN ANSWER

There are some objections that can be raised against the price system as a tool of pollution policy. Most are either illogical or apply with much greater force to any other method of control.

For example, one could object that what has been suggested here ignores the difficulties caused by fragmented political jurisdictions; but this is true for any method of control. The relevant question is: what method of control makes interjurisdictional cooperation easier and more likely? And the answer is: a price system, for several reasons. First, it is probably easier to get agreement on a simple schedule of pollution prices than on a complex set of detailed regulations. Second, a uniform price

schedule would make it more difficult for any member of the "cooperative" group to attract industry from the other areas by promising a more
lenient attitude toward pollution. Third, and most important, a price
system generates revenues for the control board, which can be distributed
to the various political entities. While the allocation of these revenues
would involve some vigorous discussion, any alternative methods of control would require the various governments to raise taxes to pay the costs,
a much less appealing prospect; in fact, there would be a danger that the
pollution prices might be considered a device to generate revenue rather
than to reduce pollution, which could lead to an overly clean, inefficient
situation.

Another objection is that the Pollution Control Board might be captured by those it is supposed to control. This danger can be countered by
having the board members subject to election or by having the pollution
prices set by referendum. With any other control method, the danger of
the captive regulator is much greater. A uniform price is easy for the public
to understand, unlike obscure technical arguments about boiler temperatures and the costs of electrostatic collectors versus low-sulfur oil from
Indonesia; if pollution is too high, the public can demand higher prices,
pure and simple. And the price is the same for all plants, with no excuses.
With direct control, acceptable pollution levels are negotiated with each
plant separately and in private, with approved delays and special permits
and other nonsense. The opportunities for using political influence and
simple graft are clearly much larger with direct control.

A different type of objection occasionally has been raised against the
price system, based essentially on the fear that it will solve the problem.
Pollution, after all, is a hot issue with which to assault The Establishment,
Capitalism, Human Nature, and Them; any attempt to remove the issue
by some minor change in institutions, well within The System, must be
resisted by The Movement. From some points of view, of course, this is
a perfectly valid objection. But one is hopeful that there still exists a
majority more concerned with finding solutions than with creating issues.

There are other objections which could be raised and answered in a
similar way. But the strongest argument for the price system is not found
in idle speculation but in the real world, and in particular, in Germany.
The Rhine River in Germany is a dirty stream, made notorious when an
insecticide spilled into the river and killed millions of fish. One tributary
of the Rhine, a river called the Ruhr, is the sewer for one of the world's
most concentrated industrial areas. The Ruhr River valley contains 40 per
cent of German industry, including 80 per cent of coal, iron, steel and
heavy chemical capacity. The Ruhr is a small river, with a low flow of less
than half the flow on the Potomac near Washington. The volume of wastes
is extremely large—actually exceeding the flow of the river itself in the dry
season! *Yet people and fish swim in the Ruhr River.*

This amazing situation is the result of over 40 years of control of the

Ruhr and its tributaries by a hierarchy of regional authorities. These authorities have as their goal the maintenance of the quality of the water in the area at minimum cost, and they have explicitly applied the equimarginal principle to accomplish this. Water quality is formally defined in a technological rather than an economic way; the objective is to "not kill the fish." Laboratory tests are conducted to determine what levels of various types of pollution are lethal to fish, and from these figures an index is constructed which measures the "amount of pollution" from each source in terms of its fish-killing capacity. This index is different for each source, because of differences in amount and composition of the waste, and geographical locale. Although this physical index is not really a very precise measure of the real economic *cost* of the waste, it has the advantage of being easily measured and widely understood. Attempts are made on an *ad hoc* basis to correct the index if necessary—if, for example, a nonlethal pollutant gives fish an unpleasant taste.

Once the index of pollution is constructed, a price is put on the pollution, and each source is free to adjust its operation any way it chooses. Geographical variation in prices, together with some direct advice from the authorities, encourage new plants to locate where pollution is less damaging. For example, one tributary of the Ruhr has been converted to an open sewer; it has been lined with concrete and landscaped, but otherwise no attempt is made to reduce pollution in the river itself. A treatment plant at the mouth of the river processes all these wastes at low cost. Therefore, the price of pollution on this river is set low. This arrangement, by the way, is a rational, if perhaps unconscious, recognition of marginal principles. The loss caused by destruction of *one* tributary is rather small, if the nearby rivers are maintained, while the benefit from having this inexpensive means of waste disposal is very large. However, if *another* river were lost, the cost would be higher and the benefits lower; one open sewer may be the optimal number.

The revenues from the pollution charges are used by the authorities to measure pollution, conduct tests and research, operate dams to regulate stream flow, and operate waste treatment facilities where economies of scale make this desirable. These facilities are located at the mouths of some tributaries, and at several dams in the Ruhr. If the authorities find pollution levels are getting too high, they simply raise the price, which causes polluters to try to reduce their wastes, and provides increased revenues to use on further treatment. Local governments influence the authorities, which helps to maintain recreation values, at least in certain stretches of the river.

This classic example of water management is obviously not exactly the price system method discussed earlier. There is considerable direct control, and the pollution authorities take a very active role. Price regulation is not used as much as it could be; for example, no attempt is made to vary the price over the season, even though high flow on the Ruhr is more than ten times larger than low flow. If the price of pollution were reduced during high flow periods, plants would have an incentive to regulate their production and/or store their wastes for release during periods when the river can more easily handle them. The difficulty of continuously monitoring wastes means this is not done; as automatic, continuous measurement techniques improve and are made less expensive, the use of variable prices will increase. Though this system is not entirely regulated by the price mechanism, prices are used more here than anywhere else, and the system is much more successful than any other.[1] So, both in theory and in practice, the price system is attractive, and ultimately must be the solution to pollution problems.

"IF WE CAN GO TO THE MOON, WHY . . . ETC?"

"If we can go to the moon, why can't we eliminate pollution?" This new, and already trite, rhetorical question invites a rhetorical response: "If physical scientists and engineers approached their tasks with the same kind of wishful thinking and fuzzy moralizing which characterizes much of the pollution discussion, we would never have gotten off the ground." Solving the pollution problem is no easier than going to the moon, and therefore requires a comparable effort in terms of men and resources and the same sort of logical hard-headedness that made Apollo a success. Social scientists, politicians, and journalists who spend their time trying to find someone to blame, searching for a magic device or regulation, or complaining about human nature, will be as helpful in solving the pollution problem as they were in getting us to the moon. The price system outlined here is no magic formula, but it attacks the problem at its roots, and has a real chance of providing a long-term solution.

[1]For a more complete discussion of the Ruhr Valley system, see Allen V. Kneese, *The Economics of Regional Water Quality Management* (Baltimore, Md.: Johns Hopkins Press, 1964).

3

Analysis of Environmental Pollution*

ALLEN V. KNEESE

Allen V. Kneese is a Senior Fellow at Resources for the Future.

INTRODUCTION

Economic theory has provided a conceptual structure indispensable for understanding contemporary environmental problems and for formulating effective and efficient policy approaches toward them. Concepts like external diseconomies and public goods provide enormously useful insights. But economic theorizing and research that take place without being well informed about the substantive character of the problems under study is in danger of being somewhat arid because of extreme abstraction or of expending scarce energy and talent in the pursuit of relatively unimportant matters. The objective of the present essay is to provide an introduction to the substantive aspects of one of the major environmental problems facing both developed and some developing economies—environmental pollution.

Environmental pollution has existed for many years in one form or another. It is an old phenomenon,[1] and yet in its contemporary forms it

"Analysis of Environmental Pollution," by Allen V. Kneese, from *The Swedish Journal of Economics* (March 1971), pp. 59–81. Reprinted by permission.

*Some of the numbers in the 1971 article have been updated by the editors with the approval of the author, and tables presenting earlier data have been omitted.
[1]Many accounts attest that severe environmental degradation has existed for a long time in the western countries. In fact, the immediate surroundings of most of mankind in this part

seems to have crept up on governments and even on pertinent professional disciplines such as biology, chemistry, most of engineering,[2] and, of course, economics. A few economists, such as Pigou, wrote intelligently and usefully on the matter a long time ago, but generally even that subset of economists especially interested in externalities seems to have regarded them as rather freakish anomalies in an otherwise smoothly functioning exchange system. Even the examples commonly used in the literature have a whimsical air about them. We have heard much of bees and apple orchards and a current favorite example is sparks from a steam locomotive—this being some eighty years after the introduction of the spark arrester and twenty years after the abandonment of the steam locomotive.

Moreover, air and water continued until very recently to serve the economist as examples of free goods. A whole new set of scarce environmental resources presenting unusually difficult allocation problems seems to have appeared on the scene with the profession having hardly noticed. Fortunately, this situation is changing fast and much good work is appearing in the current economics literature.

Substantial and thoughtful attention from economists is especially needed because the economic and institutional sources of the problem are either neglected or thoroughly misunderstood by most of those currently engaged in the rather frantic discussion of it.

I. "GLOBAL" PROBLEMS

I will begin this discussion of the substantive aspects of pollution problems by concentrating first on those problems, or potential problems, which affect the entire planet.* Thereafter I will focus on "regional" problems. By regional I mean all those other than global. One must use a word like

---

of the world were much worse a century ago than they are now. The following account of statements from an address of Charles Dickens may be interesting in this connection, especially to those who know contemporary London: "He knew of many places in it [London] unsurpassed in the accumulated horrors of their long neglect by the dirtiest old spots in the dirtiest old towns, under the worst old governments of Europe." He also said that the surroundings and conditions of life were such that "infancy was made stunted, ugly and full of pain—maturity made old—and old age imbecile." These statements are from *The Public Health a Public Question: First Report of the Metropolitan Sanitary Association,* address of Charles Dickens, Esq., London, 1850. Great achievements in the elementary sanitation of the close-in environment have been made, as well as impressive gains in public health. The distinguishing feature of contemporary environmental pollution seems to be the large-scale and subtle degradation of common property resources. This point is developed in the text below.

[2]There has been a relatively small group of sanitary engineers that has given close attention to environmental problems for a long time. I am here referring to the mainstream of work in these professions.

*Global problems are discussed more fully in the papers in Section V of this volume—*Eds.*

regional rather than terms pertaining to political jurisdictions such as nations, states, or cities because the scale of pollution resulting from the emissions of materials and energy follows the patterns, pulses, and rhythms of meteorological and hydrological systems rather than the boundaries of political systems—and therein lies one of the main problems.

The global problems to be discussed here pertain largely to the atmosphere because the marks of man have already been seen on that entire thin film of life-sustaining substance. It seems to have come as something of a shock to the natural science community that man not only can, *but has,* changed the chemical composition of the whole atmosphere. Other large-scale problems, or potential problems, particularly those related to the "biosphere," will be discussed more briefly.

Before proceeding to what may be real global problems, it will be desirable to dispose of one red herring. One of the spectres raised by the more alarmist school of ecologists is that man will deplete the world's oxygen supply by converting it into carbon dioxide in the process of burning fossil fuels for energy. This idea has now been thoroughly discredited by two separate pieces of evidence. The first is measurement of changes in the oxygen supply over a period of years. There is currently *one* monitoring station in the world whose objective it is to identify long-term changes in the atmosphere. The station is operated at a high elevation in Hawaii by the U.S. Weather Bureau. Observations there have shown the oxygen content of the atmosphere to be remarkably stable. The other piece of evidence—perhaps more persuasive—is in the form of a "gedankenexperiment." If one burns, on paper, the entire known world supply of fossil fuels and all the present plant biomass, the impact on the oxygen supply is to reduce it by about 3%. This is much too small to be noticed in most areas of the earth.

Potentially real effects on the atmosphere and climate are thought to be connected with changes in carbon dioxide and particulate matter (including aerosols) in the atmosphere, petroleum in the oceans, waste energy rejection to the atmosphere, and the widespread presence of toxic agents in the coastal waters and oceans. I will discuss each of these briefly in turn.

The production of carbon dioxide is an inevitable result of the combustion of fossil fuels. In contrast to O_2, the relative quantity of CO_2 in the atmosphere has increased measurably. The CO_2 concentration is now about 25% greater than it was at the beginning of the industrial revolution, an unprecedented rate of increase. The significance of this is that CO_2 absorbs infrared, or heat, radiations, and therefore an increasing concentration of it in the atmosphere would tend to raise the temperature of the surface of the earth. How much? Estimates vary. The most careful estimates indicate that when the concentration reaches double the pre-industrial level, which many scientists expect to occur in less than a century, the average surface temperature will be somewhere between 1.5 and 4.5 degrees Celsius higher than at present. The lower limit of this range is not regarded as very ominous, but the upper limit is comparable to the increase since the last Ice Age and could produce comparable climatic and ecological changes.

The uncertainty about the greenhouse gas–world temperature connection is as good an occasion for worry as for complacency. There is no doubt that the CO_2 concentration has been increasing rapidly for the past century or more, and has reached an unprecedented level. No one doubts the simplified theoretical models which show that this increasing concentration would warm the globe significantly. The doubts arise when one confronts the real globe, with its varying cloud cover, ocean currents, volcanic activity, etc., etc. Those complications might well cancel or more than cancel the effects of the unmodified physical principles. But then, again, they might not. Meteorologists assure us that the observations and analyses needed to clear up these doubts will not be available for another generation or so.

The increasing concentration of CO_2 in the atmosphere explains only about half of the greenhouse warming effect. Methane, a gaseous by-product of raising livestock and cultivating rice, also traps heat in the atmosphere. Though methane is not nearly as abundant as CO_2, there is enough to contribute significantly to the greenhouse effect. Modern chemistry has contributed a third important group of "greenhouse gases": the chlorofluorocarbons. The CFCs are synthetic chemicals widely used in industry, in fire extinguishing, and other applications because of their inertness and other valuable properties. There are, however, two exceptions to their general inertness. One is that in the atmosphere they are powerful heat-trappers; one molecule of a CFC does the work of several of CO_2. The other is that they combine enthusiastically with ozone. For this reason, the CFCs have been the principal agent in thinning the stratospheric ozone shield that reduces the amount of destructive ultraviolet radiation reaching the earth's surface. Primarily in order to protect the ozone layer, more than a hundred nations (including the U.S.) have subscribed to the Montreal Protocol, which obligates them to limit and, eventually, to phase out the production and use of CFCs.

The greenhouse gases, led by CO_2, and the CFCs are the principal man-made contaminants that affect the atmosphere on a worldwide scale. But sulfur oxides and nitrogen oxides, emitted largely by power plants and metallurgical industries, travel long distances, ignoring national boundaries, and return to earth as acid precipitation hundreds of miles from their sources. When they land in lakes, they acidify them and render them unsuitable for many forms of aquatic life, including most species of game and commercial fish. When they land in forests, they acidify the soil and appear to be the main causes of widespread "die-back" in the forests of North America and Europe.

Data from the atmospheric chemical network stations in Europe document the long-range travel of nitrogen and sulfur oxides. These data record the acidity and sulfur content of precipitation, and their effects on soils, surface waters, and biological systems in various European regions. In 1958, pH values[3] below 5 were found only in limited areas over the Netherlands; eight years later, values below 5 were widespread in Central Europe, and the values in the Netherlands were less than 4.

The world's oceans also are exposed to injurious side effects of human activity. For one thing, we annually spill on the order of 1.5 million tons of oil directly into the oceans, with perhaps another 4 million tons being delivered by terrestrial streams. A few million tons of oil scattered over the vast ocean may not seem like much, but unfortunately they are not scattered widely but discharged into the shallow, ecologically fertile and sensitive, continental shelves, where they are likely to kill phytoplankton and to enter food chains. There is cause for concern also about the wanton slaughter of marine mammals. Some species of whale have been driven close to extinction despite a succession of international agreements to protect them.

A final category of substances of possibly global significance are persistent organic toxins. DDT is a good example of these and has been found in living creatures all over the world. How it got to remote places like the Antarctic is still somewhat mysterious, but apparently substantial amounts are transmitted through the atmosphere as well as through the oceans. Aside from possible large-scale effects on ecological systems, these persistent toxins could affect the O_2–CO_2 balance by poisoning the phytoplankton which are involved in one of the important CO_2–O_2 conversion processes. We do not know whether this is happening or not.

Clearly, we are operating in a context of great uncertainty. Equally clearly, man's activities now and in the relatively near-term future may affect the world's climactic and biological regimes in a substantial way. It seems beyond question that a serious effort to understand man's effects on the planet and to monitor those effects is indicated. Should we need to control such things as the production of energy and CO_2 in the world, we will face an economic and political resource allocation problem of unprecedented difficulty and complexity.

The discussion of global effects of pollution was necessarily somewhat speculative, but now we turn to problems on a less grand scale. These regional problems are clear and present. A discussion is first presented under the traditional categories of waterborne, airborne, and solid residuals.[4] In the final section, I point explicitly to the interdependencies

---

[3]pH is a measure of acidity. The lower the pH, the higher the acidity.

[4]Due to limitations of space, I will concentrate on material residuals as sources of environmental pollution. There is some discussion of energy residuals—especially where they interact in important ways with material residuals. Noise, an important energy residual, is not treated at all. A good introductory discussion of noise can be found in chapter 1 of the *Handbook of Noise Control* (ed. Cyril M. Harris). McGraw-Hill, New York, 1957.

among these residuals streams and the implications of this for economic analysis. Unfortunately, most of the numbers given are from the United States. This is because I am simply not familiar with the data from other countries. The relationships in the United States may, however, be reasonably representative of those found in other industrialized countries.

II. WATERBORNE RESIDUALS

Degradable Residuals

A somewhat oversimplified but useful distinction for understanding what happens when residuals are discharged to watercourses is between *degradable* and *non-degradable* materials. The most widespread and best known degradable residual is domestic sewage, but, in the aggregate, industry produces greater amounts of degradable organic residuals almost all of which are generated by the food processing, meat packing, pulp and paper, petroleum refining, and chemicals industries. Some industrial plants are fantastic producers of degradable organic residuals: a single uncontrolled pulp mill, for example, can produce wastes equivalent to the sewage flow of a large city.

When an effluent bearing a substantial load of degradable organic residuals is expelled into an otherwise "clean" stream, a process known as "aerobic degradation" begins immediately. Stream biota, primarily bacteria, feed on the wastes and break them down into their inorganic forms of nitrogen, phosphorus, and carbon, which are basic plant nutrients. In the breaking down of degradable organic material, some of the oxygen which is dissolved in any "clean" water is utilized by the bacteria. But this depletion tends to be offset by reoxygenation which occurs through the air-water interface and also as a consequence of photosynthesis by the plants in the water. If the waste load is not too heavy, dissolved oxygen in the stream first will drop to a limited extent (say, to 4 or 5 parts per million from a saturation level of perhaps 8–10 ppm, depending upon temperature) and then rise again. This process can be described by a characteristically shaped curve or function known as the "oxygen sag." The differential equations which characterize this process were first introduced by Streeter and Phelps in 1925 and are often called the Streeter-Phelps equations.

If the degradable organic residual discharged to a stream becomes great enough, the process of degradation may exhaust the dissolved oxygen. In such cases, degradation is still carried forward but it takes place anaerobically, that is, through the action of bacteria which do not use free oxygen but organically or inorganically bound oxygen, common sources of which are nitrates and sulphates. Gaseous by-products result, among them carbon dioxide, methane, and hydrogen sulfide.

Water in which wastes are being degraded anaerobically emits foul odors, looks black and bubbly, and aesthetically is altogether offensive. Indeed, the unbelievably foul odors from the River Thames in mid-nineteenth century London caused the halls of Parliament to be hung with sheets soaked in quicklime and even induced recess upon occasion when the reek became too suffocating. So extreme a condition is rarely encountered nowadays, although it is by no means unknown. For example, a large lake near São Paulo, Brazil, is largely anaerobic, and most of the streams in the Japanese papermaking city Fuji are likewise lacking in oxygen. Other instances could be mentioned. But levels of dissolved oxygen low enough to kill fish and cause other ecological changes are a much more frequent and widespread problem.

High temperatures accelerate degradation. They also decrease the saturation level of dissolved oxygen in a body of water. So a waste load which would not induce low levels of dissolved oxygen at one temperature may do so if the temperature of the water rises. In such circumstances, heat may be considered a pollutant. Moreover, excess heat itself can be destructive to aquatic life. Huge amounts of heat are put into streams by the cooling water effluents of electric power plants and industry.

There is, in fact, increased concern about the impacts of heat residuals, particularly from power generation, in the face of the incessantly increasing demand for electric power and the development of nuclear power, the present "generation" of which requires more heat disposal per kwh generated than fossil fuel plants. Increasing use of cooling towers has been one response to this situation. But the use of cooling towers represents basically a transfer of the medium into which to reject the residual heat energy, that is, to the air instead of temporarily to the water. One author in the United States has discussed some aspects of what would happen over the central region of the United States under the alternative procedure, i.e., use of once-through cooling with discharge of waterborne heat to the main streams of the area, the Missouri and the Mississippi. About 540 million kilowatts of fossil fuel burning capacity are assumed installed and operating in this region by the year 2000. He writes:

> Imposing the requirement of at least 10 miles separation between stations and noting that such a generating capacity will raise the water temperature by about 20 deg F, we find approximately 3,000 miles of river spreading over the central region of the United States with a temperature 20 deg F higher than normal. [5]

Of course the ecological effects which would accompany such a large-scale heat discharge to our streams can only be speculated about at this time. If there were a substantial discharge of degradable organic residuals to those streams at the same time, they would almost certainly become

[5] S. M. Greenfield: *Science and the Natural Environment of Man,* p. 3, RAND Corporation, Santa Monica, Calif., February 17, 1969.

anaerobic in the summer time. The freshwater life forms we are accustomed to would be lost.

A conventional sewage treatment plant processing degradable organic residuals uses the same biochemical processes which occur naturally in a stream, but by careful control they are greatly speeded up. Under most circumstances, standard biological sewage treatment plants are capable of reducing the BOD (biochemical oxygen demand) in waste effluent by perhaps 90%. As with degradation occurring in a watercourse, plant nutrients are the end-product of the process.

Stretches of streams which persistently carry less than 4 or 5 ppm of oxygen will not support the higher forms of fish life. Even where they are not lethal, reduced levels of oxygen increase the sensitivity of fish to toxins. Water in which the degradable organic residuals have not been completely stabilized in more costly to treat for public or industrial supplies. Finally, the plant nutrients produced by bacterial degradation of degradable organic residuals, either in the stream or in treatment plants, may cause algae blooms. Up to a certain level, algae groth in a stream is not harmful and may even increase fish food, but larger amounts can be toxic to fish, produce odors, reduce the river's aesthetic appeals, and increase water supply treatment problems. Difficulties with algae are likely to become serious only when waste loads have become large enough to require high levels of treatment. Then residual plant nutrient products become abundant relative to streamflow and induce excessive plant growth.

Problems of this kind are particularly important in comparatively quiet waters such as lakes and tidal estuaries. In recent years certain Swiss and American lakes have changed their character radically because of the buildup of plant nutrients. The most widely known example is Lake Erie, although the normal "eutrophication" or aging process has been accelerated in many other lakes. The possibility of excessive algae growth is one of the difficult problems in planning for pollution control—especially in lakes, bays, and estuaries—for effective treatment processes today carry a high price tag.

In the United States, currently, BOD discharges by industry are apparently about twice as large as by municipalities. How fast BOD discharges grow depends on how effectively industrial wastes are controlled and municipal wastes treated. If current rates continue, BOD may grow about 3½% per year with plant nutrient discharges growing even faster.

Bacteria might also be included among what we have called the degradable pollutants since the enteric, infectious types tend to die off in watercourses, and treatment with chlorine or ozone is highly effective against them. Because of water supply treatment, the traditional scourges of polluted water—typhoid, paratyphoid, dysentery, gastroenteritis—have become almost unknown in advanced countries. One might say that public concern with environmental pollution peaked early in this century with the rapid spread of these diseases. But public health engineers were so successful in devising effective water supply treatment that attention to water pollution lapsed until its recent upsurge.

* * *

Non-degradable Pollution

BOD serves as a good indicator of pollution where one aspect is concerned—the degradable residuals. But many residuals are non-degradable. These are not attacked by stream biota and undergo no great change once they get into a stream. In other words, the stream does not "purify itself" of them. This category includes inorganic substances—such materials as inorganic colloidal matter, ordinary salt, and the salts of numerous heavy metals. When these substances are present in fairly large quantities, they result in toxicity, unpleasant taste, hardness, and, especially when chlorides are present, in corrosion. These residuals can be a public health problem—usually when they enter into food chains. Two particularly vicious instances of poisoning by heavy metals have stirred the population of Japan. These are mercury poisoning through eating contaminated fish (Minimata disease) and cadmium poisoning through eating contaminated rice (itai Itai disease). Several hundred people have been affected and more than a hundred have died. At the present time the Canadian government has forbidden the consumption of fish from both Lake Erie and Lake St. Clair because of feared mercury poisoning, and mercury has been discovered in many rivers in the United States.

Persistent Pollutants

There is a third group of pollutants, mostly of relatively recent origin, which does not fit comfortably into either the degradable or non-degradable categories. These "persistent" or "exotic" pollutants are best exemplified by the synthetic organic chemicals produced in profusion by modern chemical industry. They enter watercourses as effluents from industrial operations and also as waste residuals from many household and agricultural uses. These substances are termed "persistent" because stream biota cannot effectively attack their complex molecular chains. Some degradation does take place, but usually so slowly that the persistents travel long stream distances, and in groundwater, in virtually unchanged form. Detergents (e.g., ABS), pesticides (e.g., DDT), and phenols (resulting from the distillation of petroleum and coal products) are among the most common of these pollutants. Fortunately the recent development and successful manufacture of "soft" or degradable detergents has opened the way toward reduction or elimination of the problems associated with them, especially that of foaming. However, another problem associated with dry detergents has not been dealt with. These detergents contain phosphate "fillers" which may aggravate the nutrients problem.

Some of the persistent synthetic organics, like phenols and hard detergents, present primarily aesthetic problems. The phenols, for example, can cause an unpleasant taste in waters, especially when they are treated with chlorine to kill bacteria. Others are under suspicion as possible public health problems and are associated with periodic fish kills in streams. Some of the chemical insecticides are unbelievably toxic. The material endrin, which until recently was commonly used as an insecticide and rodenticide, is toxic to fish in minute concentrations. It has been calculated, for example, that 0.005 of a pound of endrin in three acres of water one foot deep is acutely toxic to fish.

Concentrations of the persistent organic substances have seldom if ever risen to levels in public water supplies high enough to present an *acute* danger to public health. The public health problem centers around the possible *chronic* effects of prolonged exposure to very low concentrations. Similarly, even in concentrations too low to be acutely poisonous to fish, these pollutants may have profound effects on stream ecology, especially through biological magnification in the food chain; higher creatures of other kinds—especially birds of prey—are now being seriously affected because persistent pesticides have entered their food chains. No solid evidence implicates present concentrations of organic chemicals in water supplies as a cause of health problems, but many experts are suspicious of them.

The long-lived radio-nuclides might also be included in the category of persistent pollutants. They are subject to degradation but at very low rates. Atomic power plants may be an increasingly important source of such pollutants. Generation of power by nuclear fission produces fission waste products which are contained in the fuel rods of reactors. In the course of time these fuels are separated by chemical processes to recover plutonium or to prevent waste products from "poisoning" the reactor and reducing its efficiency. Such atomic waste can impose huge external costs unless disposed of safely. A large volume of low-level waste resulting from the day-to-day operation of reactors can for the time being be diluted and discharged into streams, although the permissible standards for such discharge have recently been severely questioned in the United States, both outside and inside the Atomic Energy Commission.

"Hot" waste, containing long-lived substances such as radioactive strontium, cesium, and carbon, is in a different category from any other pollutant. So far, the only practical disposal method for high-level wastes is permanent storage. The "ultimate" solution to this contamination problem may be fusion energy which leaves no residuals except energy. But while some promising developments have occurred recently—especially in the Soviet Union—its development (even if possible) is, at least, decades away.

The Range of Alternatives

One of the most important features of the waterborne residuals problem, from the point of view of economic analysis, is the wide range of technical options which exist both for reducing the generation and discharge of wastes and for improving the assimilative capacity of watercourses. In industry, in addition to treatment, changes in the quality and type of inputs and outputs, the processes used, and by-product recovery are important ways of reducing residuals discharge. The capability of watercourses to assimilate residuals can often be increased by using releases from reservoirs to regulate low river flows and by the direct introduction of air or oxygen into them by mechanical means.[6]

III. AIRBORNE RESIDUALS[7]

Types, Sources, and Management Alternatives

There is virtually an infinity of airborne residuals that may be discharged to the atmosphere, but the ones most frequently responsible for local or regional air pollution, and most commonly measured, are carbon monoxide, hydrocarbons, sulfur dioxide, oxides of nitrogen, particulates and lead.

In the United States, by far the greatest tonnage of airborne residuals comes from the transportation sector, and virtually all of this is from internal combustion engines. They are especially important sources of carbon monoxide, hydrocarbons, and oxides of nitrogen. Emissions from automobiles and trucks can be reduced in a number of ways. In the United States, the EPA has relied mainly on requiring catalytic converters to be inserted in the exhaust systems of automobiles and light trucks. These converters are expensive, costing several hundred dollars each, and degrade the engines' efficiency somewhat. If properly maintained (which is not invariably the case in practice), they are very effective in eliminating carbon monoxide and hydrocarbons from automobile exhausts, less so in eliminating the nitrogen oxides. Many people outside the automobile industry believe that emissions can be reduced effectively and economically by introducing steam- or electrically powered cars, and by encouraging heavier use of mass transit.

[6]A fairly extensive discussion of technical options can be found in A. V. Kneese and B. T. Bower, *Managing Water Quality: Economics, Technology, Institutions.* The Johns Hopkins Press, Baltimore, 1968.
[7]In preparing the section on air pollution, I have benefitted from an unpublished memorandum by Blair Bower and Derrick Sewell, 1969.

Stationary sources (utility power, industry and households) are the main sources of sulfur oxides, particulates, and oxides of nitrogen. Control of emissions from these sources is a large and complex subject, but the main possibilities can be grouped into four categories: (1) fuel preparation (such as removing sulfur-bearing pyrites from coal before combustion), (2) fuel substitutions (such as substituting natural gas and low-sulfur oil and coal for high-sulfur coal), (3) redesigning burners (for example, in oil-burning furnaces two-stage combustion can reduce oxides of nitrogen), and (4) the treatment of stack gases (for example, stack gases can be scrubbed with water or dry removal processes can extract sulfur and particulates).[8]

Of course, all of these control technologies are likely to involve net costs even when they result in usable recovered materials. Furthermore, none of these processes inherently results in a reduction of CO_2. The possible significance of this was discussed in the opening section.

Assimilative Capacity of the Atmosphere

The capacity of the atmosphere to assimilate discharges of residuals varies with time, space, and the nature of the materials being discharged. From a resources management point of view it is necessary to be able to translate a specified time and location pattern of discharges of gaseous residuals into the resulting time and spatial pattern of ambient (environmental) concentrations, because in most cases there are multiple sources of discharge. With variations in type, quantity, and time pattern of discharge, the problem is compounded in complexity. However, imaginative applications of atmospheric diffusion models analogous to the water diffusion models described earlier have been used to help define "air sheds" for analysis of air quality management strategies.[9]

Impacts of Gaseous Residuals on Receptors

Perhaps of most immediate concern are direct effects on people, ranging in severity from the lethal to the merely annoying. Except for extreme air pollution episodes, fatalities are not, as a rule, traceable individually to the impact of air pollution, primarily because most of the effects are synergis-

[8]A good source on the technology of pollution control from stationary sources is Arthur B. Stern (ed.), *Air Pollution,* vols. I, II, III. Academic Press, New York, 1968.
[9]A. A. Teller: The Use of Linear Programming to Estimate the Cost of Some Alternative Air Pollution Abatement Policies," *Proceedings* of the IBM Scientific Computing Symposium on Water and Air Resource Management, held on Oct. 23–25, 1967, at the Thomas J. Watson Research Center, Yorktown Heights, N.Y.

tic. Thus, air pollution is an environmental stress which, in conjunction with a number of other environmental stresses, tends to increase the incidence and seriousness of a variety of pulmonary diseases, including lung cancer, emphysema, tuberculosis, pneumonia, bronchitis, asthma, and even the common cold. Clearly, however, acute air pollution episodes have raised death rates. Such occurrences have been observed in Belgium, Britain, Mexico, and the United States, among others. But the more important health effects appear to be associated with persistent exposure to the degraded air which exists in most cities.

The preponderance of evidence suggests that the relationship between such pollutants as SO_2, CO, particulates, and heavy metals and disease is real and large. [10] But one should not underrate the difficulties of establishing such relationships in an absolutely firm manner.

Direct effects on humans have parallels in the animal and plant worlds. Animals of commercial importance (livestock) are not located to any appreciable extent within cities, so effects on them are usually minor. Effects on pets (dogs, cats, and birds) almost certainly exist, although they have not been much documented.

As far as plants are concerned, much the same situation holds. Crops are mostly some distance away from cities, and hazards are likely to be rather special in nature (e.g., fluorides from superphosphate plants, or sulfur oxides from copper smelters). However, there are some districts where truck crops—mostly fruits and vegetables—are grown in close juxtaposition to major cities and are substantially affected by air pollution. In suburban gardens and city parks, there are deleterious effects on shrubs, flowers, shade trees, and even on forests in the air sheds of cities.

Damage to Property

A third category of effects comprises damage to property. Here again, sulfur oxides and oxidants are perhaps equally potent. Sulfur oxides combine with water to form sulfurous acid (H_2SO_3) and the much more corrosive sulfuric acid (H_2SO_4). These acids will damage virtually any exposed metal surface and will react especially strongly with limestone or marble (calcium carbonate). Thus many historic buildings and objects (like "Cleopatra's Needle" in New York) have suffered extremely rapid deterioration in modern times.

Sulfur oxides will also cause discoloration, hardening and embrittlement of rubber, plastic, paper, and other materials. Oxidants such as ozone will also produce the latter type of effect. Of course, the most widespread and noticeable of all forms of property damage is simple dirt

[10]See, for example, Mike Chappie and Lester B. Lave, "The health effects of air pollution: A Reanalysis," *J. of Urban Economics*, vol. 12 (1982), pp. 346–376.

(soot). Airborne dirt affects clothing, furniture, carpets, drapes, exterior paintwork, and automobiles. It leads to extra washing, vacuum cleaning, dry cleaning, and painting; and, of course, all of these activities do not entirely eliminate the dirt, so that people also must live in darker and dirtier surroundings.

A Few Comments Comparing Air and Water Pollution Problems

There are important parallels and contrasts between the effects and possible modes of management of water and air pollution.

1. In the United States and abroad, air pollution is heavily implicated as a factor affecting public health. Water pollution may be more costly in terms of non-human resources, but the current link of water pollution to public health problems on any large scale in advanced countries is a matter of suspicion concerning chronic effects rather than of firm evidence. Much stronger evidence links air pollution to public health problems.
2. As in the case of water pollution, a great many of the external costs imposed by air pollutants would appear to be measurable, but very little systematic measurement has yet been undertaken. The more straightforward effects are, for example, soiling, corrosion, reduction in property values, and agricultural losses.
3. Current technology apparently provides fewer classes of means of dealing with air pollution than with water pollution. In part this is because it is easier for man to control hydrological events than meteorological events. The assimilative capacity of the air mantle cannot be effectively augmented. In part it is because air is not delivered to users in pipes as water frequently is, so that it is only to a limited extent that polluted air is treatable before it is consumed. Therefore, we are in somewhat the same position in regard to polluted air as the fish are with polluted water. We live in it. Furthermore, it is also more difficult and costly to collect gaseous residuals for central treatment. Accordingly, control of air pollution is largely a matter of preventing pollutants from escaping from their source, eliminating the source, or of shifting location of the source or the recipient. Water pollution, on the other hand, is in general subject to a larger array of control measures. Nevertheless, both present intricate problems of devising optimal control systems.
4. To the extent that air sheds are definable, air shed authorities or compacts of districts are conceivable and may be useful administrative devices. In the United States the current federal policy approach points strongly in this direction and in this respect (but not others) is more advanced than the water pollution control programs.

IV. SOLID RESIDUALS

Just about every type of object made and used by man can and does eventually become a solid residual. Some of the main categories of importance are organic material, which includes garbage, and industrial solid wastes, such as from the canning industry, for example. Newspapers, wrappings, containers, and a great variety of other objects are found in household, commercial, office, industrial solid wastes. A very important source of solids in the United States is automobiles, which will be discussed separately, further on. In the United States, about 19 lb. per day per capita of solid wastes are collected of which about four are household and commercial. Industrial, demolition, and agricultural wastes constitute most of the other. Altogether the United States generates (exclusive of agricultural wastes) each year approximately 4–6 billion tons of solid residuals from household, commercial, animal, industrial, and mining activities and spends some $6 billion to handle and dispose of them. [11] In addition, there is a large amount of uncollected solids which litter the countryside.

The disposal of solid wastes can have a number of deleterious effects on society. Littering, dumps, and landfills produce visual disamenities. The disposal of solid wastes can cause adverse effects on air and water quality. Incineration of solid wastes is an obvious source of air pollution in many areas, as are burning dumps. Dumps tend to catch fire by spontaneous combustion unless they are relatively carefully controlled. Furthermore, drainage from disposal sites can reduce water quality in watercourses, and the sites may also provide a habitat for rodents and insect vectors. The disposal of municipal solid wastes (which excludes much industrial waste, automobiles, and all of agricultural waste) in the United States is roughly in the following proportions: about 73% goes into landfill operations of one kind or another, about 14% is incinerated, and 13% is recycled or used as fuel for generating energy. [12]

Clearly the major reliance for disposing of solid waste is placed on landfills, and that creates a problem. Fewer than a third of the 6,000-odd landfills now in operation meet the EPA standards for sanitary landfills. According to current rules, the non-complying landfills will not be allowed to receive waste after 1993, and will have to be replaced. In each case, two obstacles will have to be surmounted. First, the capital cost of a new landfill that satisfies the EPA requirements for protection against contaminating surface or subsurface waters and against emitting gaseous pollutants (especially methane) is substantial, upwards of $10 million excluding

[11]Based on World Resources Institute, *World Resources 1990–91,* Oxford University Press, New York, 1990, Table 21.6, p. 21.
[12]See Franklyn Associates, Ltd., *Characterization of Municipal Solid Waste in the United States: 1990 Update.* Prepared for the U.S. Environmental Protection Agency, Washington, D.C., June 1990.

land acquisition costs. Many, if not all, communities will have trouble raising such capital. Furthermore, these expensive new landfills will have to be large, too large for many small communities and counties to support individually. This gives rise to the second difficulty: There will have to be elaborate negotiations to arrange for regional solid waste facilities which, as the current phrase goes, nobody wants in his own backyard. As in other instances of technology forcing, EPA has posed a problem that nobody knows how to solve right now.

As previously mentioned, automobiles are a special problem. Of the 10–20 million junk cars in existence at any one time in the United States, about 73% are in the hands of wreckers, in other words, in junk yards; about 6% in the hands of scrap processors; and about 21% abandoned and littering the countryside. Recovery could be made much more economical by slight design changes, but presently there are no incentives to do so. Furthermore, unless it is managed to avoid them, the recycle or "secondary materials" industry itself can cause substantial external costs. For example, automobiles are usually burned prior to being prepared for scrap metal, and this can be an important source of local air pollution. Some of the processes involved in recycling automobiles have very high noise levels. While it seems clear that recycle of materials is underused under present circumstances, suffering as it does from tax, labelling, and other disadvantages with respect to new materials, it is also true that it is not a total panacea as one might gather from some of its more ecstatic adherents. For example, the paint on automobiles could not be recycled except at the expense of immense quantities of energy and other resources. Moreover, some materials such as paints, thinners, solvents, cleaners, fuels, etc., cannot perform their functions without being dissipated to the environment.

V. THE FLOW OF MATERIALS[13]

To tie together some of the points made in previous sections, it is useful to view environmental pollution and its control as a materials balance problem for the entire economy. Energy residuals could be treated in an entirely parallel fashion, but I will not discuss this here.[14]

The inputs to the system are fuels, foods, and raw materials which are

[13]This section is based heavily on R. U. Ayres and A. V. Kneese: "Production, consumption, and externalities." *American Economic Review 59,* no. 3 (June 1969).
[14]While very little direct exchange between material and energy occurs, it is important to note that there are significant tradeoffs between these residuals streams. For example, an effort to achieve complete recycle with present levels of materials flow would require monstrous amounts of energy to overcome entropy.

partly converted into final goods and partly become residuals. Except for increases in inventory, final goods also ultimately enter the residuals stream. Thus goods which are "consumed" really only render services temporarily. Their material substance remains in existence and must either be reused or discharged to the environment.

In an economy which is closed (no imports or exports) and where there is no net accumulation of stocks (plant, equipment, inventories, consumer durables, or residential buildings), the amount of residuals inserted into the natural environment must be approximately equal to the weight of basic fuels, food, and raw materials entering the processing and production system, plus gases taken from the atmosphere. This result, while obvious upon reflection, leads to the at first rather surprising corollary that residuals disposal involves a greater tonnage of material than basic materials processing, although many of the residuals, being gaseous, require no physical "handling."

Fig. 1 shows a materials flow of the type I have in mind in a little greater detail and relates it to a broad classification of economic sectors. In an open (regional or national) economy, it would be necessary to add flows representing imports and exports. In an economy undergoing stock or capital accumulation, the production of residuals in any given year would be less by that amount than the basic inputs. In the entire U.S. economy, accumulation accounts for about 10–15% of basic annual inputs, mostly in the form of construction materials, and there is some net importation of raw and partially processed materials amounting to 4 or 5% of domestic production.

Of the active inputs, perhaps three-quarters of the overall weight is eventually discharged to the atmosphere as carbon (combined with atmosphereic oxygen in the form of CO or CO_2) and hydrogen (combined with atmospheric oxygen as H_2O) under current conditions. This results from combustion of fossil fuels and from animal respiration. Discharge of carbon dioxide can be considered harmless in the short run, as we have seen, but may produce adverse climatic effects in the long run.

The remaining residuals are either gases (like carbon monoxide, nitrogen dioxide, and sulfur dioxide—all potentially harmful even in the short run), dry solids (like rubbish and scrap), or wet solids (like garbage, sewage, and industrial wastes suspended or dissolved in water). In a sense, the dry solids are an irreducible, limiting form of waste. By the application of appropriate equipment and energy, most undesirable substances can, in principle, be removed from water and air streams—but what is left must be disposed of in solid form, transformed, or reused. Looking at the matter in this way clearly reveals a primary interdependence among the various residuals streams which casts into doubt the traditional classification, which I have used earlier in this article, of air, water, and land pollution as individual categories for purposes of planning and control policy.

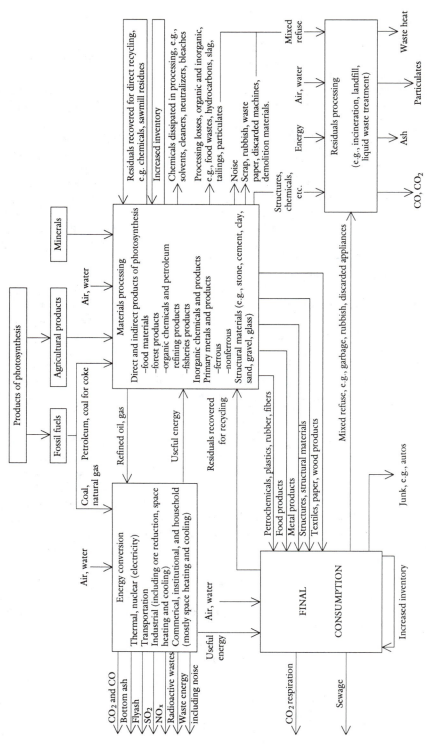

FIGURE 1. Materials Flow.

Residuals do not necessarily have to be discharged to the environment. In many instances, it is possible to recycle them back into the productive system. The materials balance view underlines the fact that the throughput of new materials necessary to maintain a given level of production and consumption decreases as the technical efficiency of energy conversion and materials utilization and reutilization increases. Similarly, other things being equal, the longer cars, buildings, machinery, and other durables remain in service, the fewer new materials are required to compensate for loss, wear, and obsolescence—although the use of old or worn machinery (e.g., automobiles) tends to increase other residuals problems. Technically efficient combustion of (desulfurized) fossil fuels would leave only water, ash, and carbon dioxide as residuals, while nuclear energy conversion need leave only negligible quantities of material residuals (although thermal pollution and radiation hazards cannot be dismissed by any means).

Given the population, industrial production, and transport service in an economy (a regional rather than a national economy would normally be the relevant unit), it is possible to visualize combinations of social policy which could lead to quite different relative burdens placed on the various residuals-receiving environmental media; or, given the possibilities for recycle and less residual-generating production processes, the overall burden to be placed upon the environment as a whole. To take one extreme, a region which went in heavily for electric space heating, electric transportation systems, and wet-scrubbing of stack gases (from steam plants and industries), which ground up its garbage and delivered it to the sewers and then discharged the raw sewage to watercourses, would protect its air resources to an exceptional degree. But this would come at the sacrifice of placing a heavy residuals load upon water resources. On the other hand, a region which treated municipal and industrial waste water streams to a high level and relied heavily on the incineration of sludges and solid wastes would protect its water and land resources at the expense of discharging waste residuals predominantly to the air. Finally, a region which practiced high-level recovery and recycle of waste materials and fostered low residual production processes to a far-reaching extent in each of the economic sectors might discharge very little residual waste to any of the environmental media.

Further complexities are added by the fact that sometimes it is, as we have seen, possible to modify an environmental medium through investment in control facilities so as to improve its assimilative capacity. The easiest to see but far from only example is with respect to watercourses, where reservoir storage can be used to augment low river flows that ordinarily are associated with critical pollution (high external cost situations). Thus, internalization of external costs associated with particular discharges, by means of taxes or other restrictions, even if done perfectly, cannot guarantee Pareto optimality. Collective investments involving public good aspects must enter into an optimal solution.

To recapitulate the main points these considerations raise for economic analysis briefly: (1) Technological external diseconomies are not freakish anomalies in the processes of production and consumption but an inherent and normal part of them. Residuals generation is inherent in virtually all production and consumption activities, and there are only two ways of handling them—recycle, or discharge into environmental media without or with modifications. (2) These external diseconomies are apt to be quantitatively negligible in a low-population or economically undeveloped setting, but they become progressively (nonlinearly) more important as the population rises and the level of output increases (i.e., as the natural reservoirs providing dilution and other assimilative properties become exhausted). (3) They cannot be properly dealt with by considering environmental media, such as air and water, in isolation. (4) Isolated and ad hoc taxes and other restrictions are not sufficient for their optimum control, although taxes and restrictions are essential elements in a more systematic and coherent program of environmental quality management. (5) Public investment programs, particularly including transportation systems, sewage disposal systems, and river flow regulation, are intimately related to the amounts and effects of residuals and must be planned in light of them. (6) There is a wide range of (technological) alternatives for coping with the environmental pollution problems stemming from liquid, gaseous, and solid residuals. Economic tools need to be selected and developed which can be used to approximate optimal combinations of these alternatives.

4

EPA and the Evolution of Federal Regulation

PAUL R. PORTNEY

Paul R. Portney is Vice President and Senior Fellow at
Resources for the Future.

By comparison with many other federal regulatory agencies, the EPA and
its siblings of the 1970s are relative newcomers. Indeed, it was more than
a century ago, in 1887, that Congress created the Interstate Commerce
Commission to regulate surface transportation industries, and it has been
more than seventy years since the Federal Reserve Board and Federal
Trade Commission were created to regulate commercial banks and decep-
tive trade practices, respectively. The first great burst of federal regulatory
activity took place in the 1930s, during which Congress created the Federal Power Commission, the Food and Drug Administration, the Federal
Home Loan Bank Board, the Federal Deposit Insurance Corporation, the
Securities and Exchange Commission, the Federal Communications Commission, the Federal Maritime Commission, and the Civil Aeronautics
Board.

Between 1938 and 1970 little took place in the way of new federal
regulatory activity.[1] Following this lull, however, came the second major
burst of federal regulation. In quick order were created the EPA, the
National Highway Traffic Safety Administration, the Consumer Product

"EPA and the Evolution of Federal Regulation," by Paul R. Portney,
from *Public Policies for Environmental Protection,* Paul R. Portney,
ed., pp. 7–25. Reprinted by permission.

[1]Exceptions were the creation of the Federal Aviation Administration, the Federal Highway
Administration, and the Federal Railroad Administration to oversee safety in the respective
industries.

Safety Commission, the Occupational Safety and Health Administration, the Mining Safety and Health Administration, the Nuclear Regulatory Commission, the Commodity Futures Trading Commission, and the Office of Surface Mining Reclamation and Enforcement. Along with the Food and Drug Administration (FDA) (and omitting the Commodity Futures Trading Commission), these have come to be known as the "social" regulatory agencies. Generally speaking, they are those having to do with environmental protection and the safety and health of consumers and workers.

Aside from age, there are at least three important distinctions between the "old-line" agencies and the newer, social regulators like the EPA.[2] The first has to do with their reasons for being. In principle at least, all regulatory agencies have been created to remedy a perceived failure of the free market to allocate resources efficiently. Except for the FDA, the older agencies were meant either to control "natural monopolies" or to protect individuals from fraudulent advertising or unsound financial practices on the part of financial intermediaries or depository institutions.[3] (The latter justification was clearly a reaction to the calamitous Great Depression, during which many of these agencies were created.)

Federal intervention in the areas of environmental protection and the safety and health of workers and consumers has a very different rationale. Here, it is argued, the government must intervene because of externalities or imperfect information. The former arise when the production of a good or service results in some costs (like pollution damage) which, in the absence of regulation, are unlikely to be borne by the producer. In such cases, the prices of products will not reflect what society must give up to have them, so that Adam Smith's "invisible hand" will steer us awry. In the case of imperfect information, workers or consumers may be only dimly aware of the health hazards associated with various occupations or consumer products or foodstuffs. If so, they will be unable to trade off higher risks for either higher wages or lower prices in an informed way, so that the unaided market would not necessarily result in either the right amount or the correct distribution of risk.

The newer regulatory agencies differ from their elders in another important way, having to do with the specificity of their focus. The older agencies can be thought of as dealing with a single industry—the Interstate

---

[2]See George C. Eads and Michael Fix, *Relief or Reform? Reagan's Regulatory Dilemma* (Washington, D.C., Urban Institute Press, 1984) pp. 12–15.

[3]A natural monopoly is said to exist when the per-unit cost of producing a good or service continues to fall with increases in output. In such a case, it is argued, consumers will enjoy the lowest prices if one firm serves the whole market rather than sharing it with two or more competitors, as long as the single provider is regulated so as not to abuse its monopoly position. Natural monopolies are thought to arise most often when the fixed costs of doing business are a large proportion of total costs. The traditional examples are local telephone service, electricity distribution, and natural gas pipelines.

Commerce Commission (ICC) with surface transportation, the Federal Communications Commission (FCC) with communications, the Civil Aeronautics Board (CAB) with airlines, and so on. This is not true of the newer social regulatory agencies. Thus, for instance, the Occupational Safety and Health Administration (OSHA) regulates workplace conditions in a wide variety of industries ranging from chemicals to agriculture. Similarly, the EPA regulates emissions of air and water pollution from the electric utility, steel, food processing, petroleum refining, and many other industries. The broader mandate of the social regulatory agencies may be a more difficult one to satisfy, since it requires each agency to become knowledgeable about and sensitive to the special problems and production technologies in many different industries.

A final distinction between the old-line agencies and their newer counterparts concerns recent developments in the scope of their activities. While it is always difficult to generalize over members of such a diverse group, the thrust of recent legislation and administrative actions at the older regulatory agencies has been in the direction of a sharp curtailment in the extent of their intervention in the markets they have regulated. This has been most pronounced in the case of the CAB, which was legislated out of existence in 1985 after it was recognized that the airline industry was ripe for competition. Similarly, the FCC has proposed significant reductions in the scope of its own activity, and both the trucking and railroad industries have been substantially deregulated through recent legislation and through administrative rulemakings at the ICC. The financial regulatory agencies and the Federal Energy Regulatory Commission (FERC, the successor to the Federal Power Commission) have also seen their powers eroded over time.

In general, this has not been the case at the newer, social regulatory agencies. While many have questioned the *way* these agencies have pursued their goals, relatively few have suggested that there is no need for them at all. It would be difficult to argue that unfettered competition among firms would lead to the right amount of pollution, product safety, or workplace risk so long as the problems of external effects or imperfect information characterize the conditions of production or consumption. Thus, while there is much agitation for regulatory reform—about which more is said below—there are very few calls for the abolition of the EPA, OSHA, FDA, or other social regulatory agencies. Indeed, in the case of the EPA at least, new responsibilities and regulatory programs have been added to the old almost continually since its creation. Trying to initiate new programs while struggling to master the existing ones is one of the problems with which the EPA has had to contend constantly.

THE CREATION AND GROWTH OF THE EPA

1970 EPA

At the vanguard of the new social regulatory agencies, the Environmental Protection Agency got its start on July 9, 1970 when President Nixon submitted Reorganization Plan No. 3 of 1970 for congressional approval. That reorganization plan proposed to consolidate under one roof—the EPA's—various functions being performed at that time by the departments of Interior, Health, Education and Welfare, and Agriculture, as well as by the Atomic Energy Commission, the Federal Radiation Council, and the Council on Environmental Quality. By December 1970 the plan had been approved by Congress and the EPA was in action.

Because it was created out of existing programs, the EPA was never a very small agency. In 1971, its first full year of existence, it had about 7,000 employees and a budget of $3.3 billion, $512 million of which went to operate the agency, with the remainder being passed through the EPA in grants to state and local governments. By 1980, the agency had grown to more than 12,000 employees; its budget by that time was more than $5 billion, $1.5 billion of which went to the operation of the agency itself.

The budget of the Environmental Protection Agency, like that of most federal agencies, shrank during the Reagan years. In its final budget submission, just before George Bush was sworn in as president, the Reagan administration requested $4.8 billion for the EPA for fiscal year 1989—$1.6 billion for operations, $1.5 billion for sewage treatment grants, and $1.6 billion for the Superfund (in 1989 dollars). Ignoring the latter program, which did not exist in 1980, and adjusting for inflation, the operating budget of the EPA has fallen by about 15 percent in real terms since 1980.

One must be careful not to ascribe too much importance to the budget of the Environmental Protection Agency (or any other regulatory agency). While the budget may provide some guidance as to the agency's capabilities, its spending authority is less important than the costs incurred by those subject to the agency's various regulations. The latter, called compliance costs by economists, never show up in the agency's budget, yet they can dwarf its operating costs. For example, in 1981 the EPA spent $1.8 billion on outside research, salaries, rent, and other operating expenses (expressed in 1988 dollars). Yet in that same year, as indicated above, those subject to the EPA's air and water pollution control regulations were forced to spend some $52 billion to comply with EPA requirements. These expenditures—for pollution control equipment, cleaner fuels, sludge removal, additional manpower—give a much more accurate picture of the economic importance of the EPA than does its budget. Environmental regulations may of course result in substantial benefits as well. One must look at both costs and benefits before passing judgment on the overall worth of a particular regulatory program.)

compliance costs

The rapid growth of off-budget environmental compliance costs (they were estimated to be only $29 billion in 1973) has led to periodic concern about the possible effects of pollution control spending on the overall performance of the economy. While this is not the place to review in any detail the many studies on this subject, it may be useful to summarize them briefly.[4] With a fair degree of consistency, these studies have found that pollution control spending has had a relatively minor impact on macro-economic performance. It has exacerbated inflation somewhat and slowed the rate of growth of productivity and of the GNP. On the other hand, studies have found that pollution control spending appears to have pro-vided some very modest stimulus to employment, at least during the time when spending for sewage treatment plant construction was high. None of this should be too surprising. While annual expenditures of $52 billion or more are hardly trivial, they are small in comparison to a $5 trillion GNP like that of the United States. Thus it is unlikely that environmental or other regulatory programs on the present scale will ever be found to exert a significant impact on the measured performance of the economy.[5]

FUNDAMENTAL CHOICES IN ENVIRONMENTAL REGULATION

It may prove helpful in understanding the following chapters to review some of the basic choices that must be made in environmental regulation. The first question is one whose answer is often taken for granted—whether to regulate at all. While today we tend to take the need for regulation as a given, there are several possible alternatives to environmental regula-tion. These include the private use of our legal system as well as private negotiation or mediation.

Both alternatives depend upon a clear prior definition of property rights. Imagine that it was clearly understood that any citizen had an absolute right to be compensated fully for the damages from any kind of pollution. If the smoke from a neighbor's wood stove or a factory were ruining your laundry business, you could take the offending party to court. If your damages could be accurately assessed, and if the polluter were held liable for them, the legal approach would create the right incen-tives for polluters. They would undertake certain pollution control mea-sures if the costs of these measures were less than the damages they would have to pay you. And they would continue to emit some pollution—and

[4]See Paul R. Portney, "The Macroeconomic Impacts of Federal Environmental Regula-tion," *Natural Resources Journal* vol. 21 (July 1981) pp. 459–488.

[5]If regulation is the barrier to entry and discouragement to new growth that some maintain it is, and if these effects could be quantified, this conclusion could change.

pay you for the damage it does—if it were more expensive to control than it was to reimburse you. Economists like such solutions because they minimize the total costs—control expenditures plus residual damages—associated with pollution control. Until the start of the twentieth century, all air pollution problems were handled under the nuisance and trespass provisions of common law.

Unfortunately, the real world is much more complicated than this simple example would suggest, and the added complexity makes a purely legal approach to environmental protection much less practical. First, it is not perfectly clear where property rights in clean air are, or even ought to be, vested in our society. This may seem puzzling, since most people accept the right of the citizenry to be free from pollution. Yet even this apparently sensible proposition seems strained in, say, the case of a laundry that deliberately moves from a clean location to one directly adjacent to a factory and then demands compensation for smoke damage. In other words, it seems to make some difference who was there first. Furthermore, if property values and rents are lower in polluted areas than in clean ones (as they generally are), the launderer seems to be on shakier ground still, since he would have already reaped some savings in operating costs (lower rent) by virtue of his new location. In one sense, he would be getting double benefits if the factory were forced to curtail its operations. Not surprisingly, arguments like these are often advanced in defense of firms resisting pollution control. They are not wholly without merit. [6]

Even if property rights were clearly defined, environmental protection via the legal system or private negotiation would not be without difficulties. For instance, pollution rarely occurs on a one-to-one basis as in the simple example above. There are often many polluters, thus making it difficult or impossible to know which factory, car, or wood stove is responsible for which damages. Also, there are generally many "pollutees," no one of whom may be suffering sufficient damages to merit taking legal action or initiating mediation or negotiation alone. Legal transaction costs may be so high that they inhibit the filing of class action suits, even though aggregate damages across all pollutees may be significant. Finally, some of the damaging effects of pollution may be both more subtle yet more serious than mere dirty laundry. For example, pollution may be one of the causes of cancer and other serious illnesses. Yet the long latency period between exposure and manifestation, coupled with the possibility of other causes, will make it virtually impossible to assess liability satisfactorily in a courtroom or arbitration chamber. Add to this the difficulty of valuing

[6]In the real world, in fact, property rights appear to be shared. Even under existing regulation, firms are generally permitted to emit at least some pollution without being held responsible for the damage it may cause. In addition, tax breaks are provided for some investments in pollution control equipment. In effect, this shifts some of the costs of pollution control to the taxpayer, as would be the case if the property right were initially vested in the polluter.

the pain and suffering from such illnesses, and one can quickly understand the possible shortcomings of alternative approaches to government intervention.

Having said this much, it is time to make one more quite important point. Proponents of government action, regulatory or otherwise, are quick to point as justification to the imperfections inherent in free markets. Public goods, externalities, natural monopolies, and imperfect information—these are all problems economists recognize as standing in the way of efficient resource allocation. Yet regulation seldom goes exactly as planned when it is substituted for the forces of the market.[7] It is often poorly conceived, time-consuming, arbitrary, and manipulated for political purposes completely unrelated to its original intent. Thus the real comparison one must make in contemplating a regulatory intervention is that between an admittedly imperfect market and what will inevitably be imperfect regulation. Until it is recognized that this is the dilemma before us, we will be dissatisfied with either approach.[8]

If government intervention *is* deemed desirable, one must then ask, at what level should intervention take place? While environmental protection is very important, so too are public school quality, police and fire protection, income assistance, and criminal justice. Yet the latter are all functions which, in our federal system, are entrusted largely to local or state governments. Thus it is not obvious that all or even most environmental regulation should take place at the federal level. This is a key decision that must be made in designing interventions, and one recent trend is in the direction of much more state and local regulatory activity.

Once the decision has been made to intervene at some level of government, the next choice to be made is, how should we decide how much protection to provide? There are a number of frameworks for deciding the answer, and Congress has directed different social regulatory agencies to use different frameworks in establishing levels of protection.[9] In fact, even within the EPA the approach differs depending on the regulatory program in question.

One way to select the degree of protection might be called the zero-risk

[7]See Charles Wolf, Jr., *Markets or Government* (Cambridge, Mass., MIT Press, 1988).
[8]While it is probably inappropriate in the case of environmental problems, there is a third alternative to regulation in cases involving occupational hazards or dangerous consumer products. There the government could limit its role to the provision of information about the risks inherent in different jobs and/or products. Workers and consumers could then hold out for higher wages for risky jobs (as they do now) or pay low prices for risky products. Employers or producers would then have to decide whether it was in their interest to continue to pay higher wages or receive lower prices rather than reduce the hazards. In this way, too, market forces would work toward optimal riskiness. It should be noted, of course, that this model would be successful only if workers and consumers were fully informed about such risks and only if they had a range of jobs and products from which to choose.
[9]See Lester B. Lave, *The Strategy of Social Regulation* (Washington, D.C., Brookings Institution, 1981).

or safe-levels approach. The administrator of the Environmental Protection Agency would be directed to set a particular environmental standard at a level that would ensure against any adverse health (or other kind of) effect. This approach is not uncommon. And on its face, it certainly seems reasonable. After all, would we want a standard to be set at a level that poses some recognizable threat to health? Surprisingly, perhaps, the answer is maybe.

Science and economics contribute to this unexpected response. Accumulated research in physiology, toxicology, and other health sciences suggests that for a number of environmental pollutants, particularly carcinogens, there may be no threshold concentrations below which exposures are safe. This implies that standards for these pollutants must be set at zero concentrations if the populace really is to be protected against all risks. Here economics intrudes in a jarring way. Simply put, it is impossible to eliminate all traces of environmental pollution without at the same time shutting down all economic activity, an outcome which neither the Congress nor the public would abide. Yet this is where the zero-risk framework often appears to lead if interpreted literally.

Having raised this disquieting possibility, let us push it further to consider another interesting case. Suppose a particular pollutant was harmful at ambient (or outdoor) levels to one very large group of people—the one-third of U.S. adults who choose to smoke cigarettes—but only this group. If the costs of reducing ambient concentrations of the contaminant were very large, might society not decide to forgo this health protection? It might well, in view of the role that the sensitive population has played in predisposing itself to environmental illness. Here too, then, the zero-risk approach would cause problems.

Perhaps more realistically, decisions to live with some risk might be reached even if the group at risk had done nothing to create its sensitive status. At some point, the costs of additional protection might be judged by society to be too great if the added health benefits are relatively small. Painful though such decisions may be, they are the rule rather than the exception in environmental and other policy areas. The problem with the zero-risk approach is that it prevents such tradeoffs from being made.

Another approach to the how-much-protection question is a variant of the above. It is often referred to as the technology-based approach. Under this framework, the only pollution permitted is that remaining after sources have installed "best available" or other state-of-the-art control technology. The underlying idea is simply that all technologically feasible pollution control measures will be required, and only after that will residual risks be accepted. This approach is somewhat weaker than a strict zero-risk approach since it admits the possibility of some risks. But in its strictest form, it is uncompromising with respect to trading off cost savings for less strict pollution control. For this reason, it appeals to many.

In its application, however, the technology-based approach faces several drawbacks. First, there is no unambiguously "best" technology—emissions can always be reduced further for additional control expenditures. In the limit, of course, sources could be closed down entirely—an ultimate, perhaps draconian, form of best technology. Moreover, implicit in the technology-based approach is the assumption that the control that results must be worth the cost. This might well depend upon the particulars. For instance, very strict control may be deemed essential for polluters in densely populated areas but much less important for those in remote, unpopulated regions. Yet the uniform technological approach has the liability of precluding the tradeoffs necessary to decide such questions. In addition, the technology-based approach suffers in a dynamic setting because it locks sources into specific means of control. It is unlikely that a firm required to meet this year's best technology will be told to scrap that equipment if next year's is even better. Thus this approach may deprive us of the opportunity to reduce the costs and increase the efficiency of pollution control over time.

The final framework for standard-setting discussed here formalizes the notion of balancing and incorporates it into environmental law. The relevant statutory language might direct the administrator of the EPA to set standards to protect health and other values while at the same time taking account of the costs and other adverse consequences of the regulations. The advantage of this approach is that it makes possible, indeed mandatory, the kinds of tradeoffs we have suggested might be desirable. On the other hand, it also forces the administrator to make very difficult decisions. Moreover, if all favorable and unfavorable effects are supposed to be expressed in dollars, so that precise benefit-cost ratios are required, this approach would impose a burden that economic analysis is not prepared to bear. In spite of the recent progress in valuing environmental benefits and costs,[10] the science is still far short of being able to make such comparisons in a precise way. For this reason, the balancing approach is best left in a qualitative or judgmental form.

At this point, some might reasonably chafe at the balancing framework. Why compromise citizens' health or welfare so that corporate or other polluters might remain economically healthy? This very natural question deserves a straightforward answer, one which proceeds along the following lines. While it would be nice if there were, there are no disembodied corporate entities into whose deep pockets we can reach for pollution control spending without at the same time imposing losses on ourselves or our fellow citizens. This is because corporations are merely legal creations, the financial returns to which all accrue to individuals in one

---

[10]See A. Myrick Freeman III, *The Benefits of Environmental Improvements: Theory and Practice* (Baltimore, The Johns Hopkins University Press for Resources for the Future, 1979); and Allen v. Kneese, *Measuring the Benefits of Clean Air and Water* (Washington, D.C., Resources for the Future, 1984). See also chapter 18 in this volume—Eds.

capacity or another. Thus if corporations spend more for pollution control, these costs may be passed on to others in the form of higher product prices.[11] Alternatively, if costs cannot be shifted to consumers, then stockholders, laborers, or the management of the corporations will suffer reduced earnings.

Thus far the tradeoff may still seem appealing since it is expressed in terms of dollars versus health. However, the reduced incomes of consumers, stockholders, or employees will eventually mean less spending on goods and services they value. At this point, then, the tradeoff becomes more stark. Pollution control spending can sometimes protect health, but at an eventual cost to society of forgone health, education, shelter, or other valued things.[12] The real trick in environmental policy—or any other area of government intervention—is to ensure that the value of the resulting output is greater than that which must be sacrificed. And this sacrifice will take place regardless of the framework for standard-setting that is being employed. In the economists' view of things, the balancing approach is desirable not only because it is a natural way to make decisions, but also because it brings out in the open the terms of trade, so to speak. If we dislike the compromises being made by our regulatory officials, we can demand their removal.

Once environmental standards (or ambient standards, as they are sometimes called) have been selected, the next step is deciding on the means of attainment. In other words, how do we control the sources of pollution so that the environmental goals are met? While there are other possibilities, the two most common approaches are via direct, centralized regulation, or through an incentive-based, decentralized system.[13] Under the first, individual polluters are assigned specific emissions reductions; under the latter, they are given more latitude.

Under the centralized approach, the regulating authority has considerable discretion in apportioning the emission reductions required to meet the ambient standard. Only when there is but one source of pollution is it clear where emissions must be reduced.[14] More typically, there are multiple sources; in such cases the authority has to decide how much each

[11]This is the point of such regulations, in fact, since the higher prices discourage consumers from purchasing products whose production generates pollution.

[12]Recently an effort was made to link spending like that necessitated by pollution controls to possible premature mortality via reductions in individuals' wealth. See Ralph L. Keeney, "Mortality Risks Induced by Economic Expenditures," working paper, University of Southern California, Systems Science Department (July 1988).

[13]Other possibilities are moral suasion and direct government purchase of pollution control equipment. See William J. Baumol and Wallace E. Oates, *Economics, Environmental Policy, and the Quality of Life* (Englewood Cliffs, N.J., Prentice-Hall, 1979) pp. 217–224.

[14]Even in this apparently simple case the matter is not so straightforward, because nature accounts for a share of many major pollutants. For instance, particulate matter can be blown from fields or roadways just as it can be generated by steel mills or cement plants. In such cases, the man-made share must first be assessed before required cutbacks can be determined.

Centralized approaches

source must curtail its offending activities. There are several ways this decision can be made.

If aggregate emissions must be reduced by 25 percent to meet the environmental standard, for example, each source could be required to cut back its own emissions by 25 percent. This equiproportional rule has the very attractive feature of *appearing fair.* Why only the appearance of fairness? Because of the very great diversity of sources for many environmental contaminants, ranging from neighborhood dry cleaners or car-repair shops to complex steel mills or large chemical plants. The differing characteristics and technological circumstances of these sources mean that one source may be able to reduce its emissions by 25 percent quite inexpensively, perhaps by switching to a less polluting fuel or altering slightly its manufacturing technique. Yet another source might find that it can meet its 25 percent reduction only through the installation of expensive control technology. Thus a requirement for equal-percentage reductions may mean very unequal financial burdens.

Under another approach, emission reductions might be apportioned on the basis of affordability—that is, the largest cutbacks might be required of those in the best financial shape. This, too, has some obvious appeal. Indeed, under our present individual income tax system, we ask those in higher income brackets to pay a higher percentage of their income in taxes, and this would seem to extend that principle to pollution control.

On closer inspection, however, assigning emission reductions on the basis of ability to pay also has serious drawbacks. First, it would penalize successful, well-managed firms and reward laggards that may well be largely responsible for their own poor financial state. In this sense, then, the approach gives exactly the wrong set of signals to firms and slows the replacement of failing enterprises with newer, more efficient ones. Second, there may be no relation whatsoever between a source's emissions and its financial condition. Thus a very profitable firm may have very low emissions (particularly if it has continually modernized) but under this approach would still be forced to spend heavily on further emission reductions; meanwhile, a smoke-belching firm in perilous financial condition would be let off lightly. For these reasons, an affordability criterion is less attractive than it may at first appear.

Finally, the regulatory authority could try to apportion emission reductions among sources in such a way that the required aggregate reduction was accomplished at the least total cost to society. In other words, the central regulator could look across all sources and ask where the first ton of emissions might be reduced most inexpensively, then require it to be removed there. The second ton of emissions reductions would then be assigned, again to the source that could accomplish it most cheaply. And so on until the aggregate emissions goal had been met.

This approach has the advantage of ensuring that society (through the affected sources) gives up as little as possible to get the emission reductions. But it raises the possibility of another sort of inequity. Suppose that one source, among a large number of polluters in a particular area, was always the lowest-cost abater? This is unlikely to be the case in reality, but it might hold true in certain circumstances. It would hardly seem fair to place the entire burden of emissions control on that source merely because it could reduce pollution more inexpensively than the other sources. Thus, although the cost-minimization approach has some obvious appeal, it is not ideal.

Decentralized approaches, on the other hand, do address certain of these problems, although they present difficulties of their own. Perhaps the best known of the decentralized approaches is the effluent charge or pollution tax.[15] Under this scheme, the regulatory authority imposes a tax or fee on each unit of the environmental contaminant discharged. In its purest form, the charge would be set to reflect the damage done by each unit of emissions. Rather than tell each firm how much to reduce emissions, the authority would leave it to the firm to respond to the charge however the firm best sees fit. Some sources would reduce their emissions immediately—those will be the ones that can do so at unit costs less than the amount of the charge. By doing so, they save the difference between their per-unit cost of control and the per-unit charge. Other sources will find it economical to continue discharging—they will be the ones finding it cheaper to pay the tax than to incur the required control costs.

Such an approach has several advantages. First, it ensures that the sources that do elect to take control measures are those with the lowest control costs. In other words, it mimics the least-cost approach under command-and-control, but does so without requiring the central authority to specify emission reductions for each and every source. Second, and perhaps more important, it provides a continuing incentive for firms to reduce their costs of pollution control. Since they must continue to pay the per-unit charge, it continues to be economical for them to find ways to reduce emissions for less than that charge. Third, this system requires something from all sources—either they must reduce pollution to escape the charge, or they must continue to pay the charge. No one gets off scot-free.

As might be expected, the effluent-charge route has shortcomings, which at least some economists have been slow to recognize or acknowledge.[16] For one thing, it is no picnic to determine the damage done by each

[15]For a comprehensive description and discussion, see Peter Bohm and Clifford S. Russell, "Comparative Analysis of Alternative Policy Instruments," in Allen V. Kneese and James L. Sweeney, eds., *Handbook of Natural Resource and Energy Economics,* vol. 1 (New York, North-Holland, 1985) pp. 395–460; see also Frederick R. Anderson, Allen V. Kneese, Phillip D. Reed, Serge Taylor, and Russell B. Stevenson, *Environmental Improvement Through Economic Incentives* (Washington, D.C., Resources for the Future, 1977).
[16]For an interesting analysis of attitudes toward effluent charges, see Steven Kelman, *What Price Incentives?* (Boston, Auburn House, 1981).

unit of pollution; in practice this could only be approximated at best. Some have suggested that the difficulty of apportioning damage is such a liability of the charge approach that a modified version of the approach should be used.[17] Under this variant, the central authority would first select the desired level of environmental quality and would then set the charge at a level sufficient to induce the emissions control that would achieve it. Yet even such a variant would require some trial and error, and the uncertainty this might create could make firms reluctant to come to their initial emissions control decisions. The effluent-charge route also presents one serious political problem. Under this approach, the emissions that sources are free to discharge under the current permit system would be subject to the charge. Thus many sources that presently complain about over-regulation would have a new complaint: a major effluent-tax liability.

A second variant of the incentive-based approach involves marketable pollution "rights" or permits. This approach could work in one of two basic ways. Under one version, the central authority would first decide how much total pollution was consistent with the predetermined environmental goal. It would then print up individual discharge permits, the total quantity of which added up to the maximum amount permitted. No one without a permit would be allowed to discharge the regulated pollutants. The permits could be allocated among sources in one of several ways. First, a sale might be held at which all of the permits were auctioned off to the highest bidders. Alternatively, the permits could be distributed free of charge on some predetermined (or even random) basis—perhaps on the basis of historical levels of pollution. Either way, the permits would be marketable anytime after the initial distribution.

The incentive effect under a system of marketable permits is not unlike that of the effluent charge discussed above. Those sources that currently pollute but which could reduce pollution for less than the cost of a permit would take control measures. Those sources finding it very expensive to reduce pollution would buy discharge permits instead. Thus, as if guided by the same invisible hand, the emission reductions would take place at the low-cost sources, thereby minimizing the costs associated with a given reduction in emissions. Similarly, those firms buying permits would have a continuing incentive to reduce their costs of pollution control—as soon as they could do so, they could stop buying permits and save themselves money in the process.

The permit approach has one major advantage when compared to the effluent charge: the permit approach looks more like the existing system, which involves permits issued by the EPA or state environmental authorities, than does the latter. This may sound strange, but radical change is almost always more difficult to accommodate than gradual change. Since

[17]William J. Baumol and Wallace E. Oates, "The Use of Standards and Prices for Protection of the Environment," *Swedish Journal of Economics,* vol. 73 (March 1971) pp. 42–54.

the marketable permit system is capable of accomplishing most of the same things as the effluent charge, why not advance it if it will be easier to put in place? This logic appears to have prevailed, and the inroads made by incentive-based approaches in environmental policy over the last ten years have featured permits.

Marketable permits are not without shortcomings, of course. One concern has to do with the possibility that certain sources might buy up all the permits as an anti-competitive tactic. While this ought to be rectifiable through governmental antitrust actions, in practice such actions might take time. Another question concerns the initial distribution of permits before the development of secondary markets. If all the permits are auctioned off, this approach would fall prey to the same political problems that arise under a charge approach—some sources would have to pay for emissions they are granted free under the existing system. Thus political problems could become formidable. If the initial permits are to be distributed free of charge, how should they be allocated? On the basis of previous emissions? To all citizens equally? To environmental and industry groups? This too is a potentially thorny problem, although not an insurmountable one.

A fifth and final question that arises in environmental policy is often overlooked: How do we monitor for compliance with the standards we set, and take enforcement actions against those in violation? The choice of both environmental and individual sources discharge standards ought to be (but often is not) influenced by the realities of monitoring and enforcement. Key issues involve the extent of reliance on financial penalties for noncompliance, the comparative strengths and weaknesses of civil as opposed to criminal penalties, and the choice of a monitoring strategy in a world of limited resources. All these issues and more must be addressed in designing sensible environmental policies.

U.S. ENVIRONMENTAL POLICY: A HYBRID APPROACH

As one might suspect, the fundamental questions raised above have been answered in an eclectic and hybrid way as environmental policy has evolved in the United States. With respect to the decision on intervention, federal, state, and local governments have all decided to intervene. Environmental statutes exist at all three levels—in fact, special districts have been formed in many areas around environmental problems. Thus long ago the decision was made not to entrust environmental problems and disputes solely to markets, to the courts, or to mediation services.

This observation also suggests the level at which intervention has taken place—at every level. Even federal environmental laws reserve important functions for state and local governments. For instance, under the Clean

Air and Clean Water acts, the federal government (as embodied in the EPA) sets important ambient environmental and source discharge standards, yet the monitoring and enforcement of these standards is left largely to the states and localities. In fact, some ambient and source discharge standards are themselves reserved for lower levels of government in certain important cases. In other words, even federal laws are "federalist" in nature.

As to the choice of goals (How safe should we be?), U.S. environmental laws embody a range of approaches. A number of the most important environmental laws, or parts thereof, reflect the zero-risk (or threshold) philosophy. For instance, the Clean Air Act directs that ambient standards for common air pollutants be set at levels that provide an "adequate margin of safety" against adverse health effects, while standards for the so-called hazardous air pollutants are to provide an "ample margin of safety." Under the Clean Water Act, ambient water quality standards—which are left to the states rather than the federal government to establish—are also to include a margin of safety for the protection of aquatic life.

Other environmental standards are based on the technological approach to goal-setting. This is true of the Clean Air and Clean Water acts, the Resource Conservation and Recovery Act, and the Safe Drinking Water Act. The notion of best-available technology—along with its cousins "best-conventional" and "reasonably available" technology and "lowest-achievable emissions," as well as others—plays a big role in U.S. environmental policy, even in those statutes which in other places embrace the zero-risk goal.

Even the balancing framework favored by economists is alive and well in environmental policy. For although balancing appears to be prohibited under certain sections of the Clean Air and Clean Water acts, it is *mandated* under the most important parts of the Toxic Substances Control Act and the basic pesticide law, the Federal Insecticide, Fungicide, and Rodenticide Act. One is tempted to throw up one's hands and say, You figure it out!

In one important respect, the environmental laws have been rather uniform. When it comes to the means of pursuing environmental goals, the centralized or command-and-control approach has been given precedence over incentive-based approaches. Congress has rather consistently written regulations directing the EPA to establish emissions standards (with the help of the states and localities) and to issue and enforce permits specifying those standards. However, it is certain that a variety of factors have influenced the emission reductions the EPA has required. While the agency often claims to have pursued a least-cost strategy, the uniform rollback and ability-to-pay criteria have clearly dominated in apportioning emission cutbacks.

The effluent charge approach has never really gotten off the ground in

U.S. environmental policy, although a tax on emissions of sulfur into the air was proposed by the Nixon administration in the EPA's first year and once again in 1988. Recently, however, the standards-and-permits approach to air pollution control has evolved in the direction of marketable permits, and there is talk of applying this approach more widely in air pollution control as well as in other regulatory programs. Needless to say, this is an important development.

Before concluding this chapter, some mention should be made of several generic problems that have arisen in the hybrid U.S. environmental policy since 1970. These problems have nothing to do with the well-known difficulties that arose at the Environmental Protection Agency between 1981 and early 1983;[18] rather, they have to do with the fundamental approach that Congress has taken in environmental regulation. A brief review of them may provide valuable perspective.

These problems are of four sorts.[19] The first has to do with the tremendous complexity of our environmental laws and their penchant for promising a very great deal in a very short time. For instance, the clean air and clean water laws promise "safe" air and water quality, call for the establishment of literally tens of thousands of discharge standards, mandate the creation of comprehensive monitoring networks, and impose numerous other important tasks on the administrator of the EPA. Yet the laws allotted just 180 days for completion of many of these responsibilities. Today, more than seventeen years after passage of the laws, many of those assignments have yet to be carried out.

Similarly, the Toxic Substances Control Act calls for the promulgation of separate testing rules for each new chemical. Yet although such chemicals come on the market at the rate of 1,000 per year, the EPA has issued testing rules for only a few substances. Of the more than 50,000 existing chemicals in commerce, only a small fraction have been tested for carcinogenicity or other harmful effects. Each of the environmental laws provides examples like this where Congress either misunderstood the time required to issue careful regulations, or disregarded it in the rush to get legislation on the books. The EPA has tried to run faster and faster since its creation, but has fallen farther and farther behind because of its impossible burden and an occasional lack of will.

Problems of the second sort have to do with the spotty compliance with those standards that have been issued, and our poor ability to know which standards are being violated and which sources are responsible. The rea-

[18]These difficulties were quite serious, to be sure. Never before had the competence or commitment of the EPA's top management been questioned. Ultimately all but one of the agency's highest-ranking officials resigned or were fired. See Robert W. Crandall and Paul R. Portney, "Environmental Policy," in Paul R. Portney, ed., *Natural Resources and the Environment: The Reagan Approach* (Washington, D.C., Urban Institute, 1984) pp. 47–81.
[19]Ibid., pp. 47–55.

son for these problems is two-fold, it would appear. First, monitoring both ambient environmental quality and the emissions from individual sources is much more complicated and expensive than one would imagine. Monitoring is not a straightforward matter, as might be supposed from reading the laws. Second, monitoring and enforcement have always been poor stepsisters in the eyes of Congress. Apparently it is more fashionable to write new laws and call attention to problems with existing laws than it is to engage in the dirty work of fashioning an enforceable and scientifically meaningful set of standards. Thus enforcement programs have always suffered financially at the expense of new and emerging regulatory programs.

The third sort of generic problem concerns the frequent emphasis in environmental statutes on absolutist goals. Waters are to be "fishable and swimmable" as one step toward a world of "zero discharges" into rivers, lakes, and the oceans. Conventional and hazardous air pollutants are to be at "safe" levels, as are drinking water contaminants. Such an approach has obvious political appeal—it is comforting to tell voters that they will be safe from all environmental threats. But that will simply not be the case unless standards are to be set at zero, an impossibility for most pollutants. Thus, although it is surely done, the balancing of environmental versus other important goals, economic and otherwise, is done implicitly. This has resulted in setting some standards at levels that appear to be hard to justify on any rational basis.

Finally, and perhaps inevitably, environmental statutes have become contaminated by redistributive goals which often work against the environmental programs in which they are nested—the fourth sort of problem. For example, although this policy is now under review, newly built electric power plants are still forbidden to reduce sulfur dioxide emissions by switching from high-sulfur to low-sulfur coal in order to protect the jobs of a small number of high-sulfur coal miners. This prohibition exists in spite of the tremendous cost savings that might be reaped if fuel-switching were permitted. Similarly, federal subsidies for the construction of sewage treatment plants have been continued even though the plants seem to have had a questionable impact on water quality in many areas, and although the federal subsidy has crowded out state and local spending for these same plants. Apparently, the pork-barrel aspects of the program have proved too attractive to eliminate. Water pollution from farms and other non-point sources has been overlooked altogether because of the political power of the parties that would be affected by tighter controls. Some evidence also suggests that environmental regulations have been structured to protect declining regions of the country from the effects of further economic growth in faster-growing Sunbelt states. While all these contortions of environmental policy are understandable, they also stand in the way of an effective and less costly approach to environmental protection. As such, they deserve to be starkly highlighted.

II

ECONOMIC PRINCIPLES

The first group of papers presented the environmental problem along with a foretaste of the economist's approach to it and the strategy the United States has adopted for dealing with it. This second group delves more deeply into the economic analysis. The third group will pursue some of the policy implications of this analysis.

Above all, the economist sees the environmental problem as the leading instance of resource misallocation caused by "externalities"—unintended consequences or side effects of one's actions that are borne by others. Externalities have always been with us, but as the planet grows more crowded and per-capita consumption rises, accompanied by the emergence of new polluting technologies like chlorofluorocarbons (CFCs), synthetic fertilizers, pesticides, herbicides, and plastics, externalities become more critical. In the effort to cope with these problems, causal inferences and policy recommendations must rest on a firm understanding of the economic forces at work. The objective of this second group of papers is to explore these forces.

The paper that opens this section, "Some Concepts from Welfare Economics," introduces the basic concepts that underlie both the analysis and practical conclusions of welfare economics. The key concepts are the notion of public goods or common-property resources, typified by Garrett Hardin's commons; externalities, the side effects of one person's activities on other people's welfares; and Pareto efficiency, the condition in which it is impossible to improve anyone's welfare without diminishing someone else's. From these building blocks are constructed the theoretical precepts of welfare and environmental economics, and also the criteria by which the soundness of environmental projects and policies can be judged.

The problem that was raised in Hardin's essay on the commons in Section I is subjected to careful economic analysis by H. Scott Gordon in his piece on an analogous common-property resource, an ocean fishery. The basic problem is that, because the fishermen do not own the fishery and, therefore, cannot appropriate its surplus, they have no incentive to restrict their catch to the level that would maximize its net economic yield. Instead, each increases his effort until his marginal costs are equal to the average value of the product of the fishery without regard to the external effect of his effort on the value of every other fisherman's catch. In the end, they dissipate the surplus and each fisherman earns no more than his average cost. A private owner would internalize the externality, and limit the catch to the amount at which the marginal cost of the fishing effort was equal to the marginal, rather than the average, value of product.

Undoubtedly the most widely quoted selection in this volume is Ronald Coase's "The Problem of Social Cost." Coase, like many other contributors to this volume, is concerned with what sort of social arrangement will bring about an optimal allocation of resources in the presence of externalities. A major theorem (Coase's theorem) emerges from his analysis. It states that, if it were costless for the parties imposing and those suffering from an externality to bargain with each other to their mutual advantage and to enforce the bargain, and if property rights were clearly defined, an efficient resolution of the externality would be achieved through negotiation, without social intervention. Under normal circumstances these conditions are not met and the pertinent question is: What sort of social arrangement would most nearly approximate the results of negotiations?

An important insight pointed the way to Coase's theorem. It is that sometimes the person damaged by an externality can ameliorate or avoid the damage at lower social cost than the person who creates it. Coase's colorful example of the railroad locomotive spewing its sparks over farmers' crops drives home the conclusion that the popular maxim, "Polluter pays," does not necessarily lead to the maximum net social product since it provides no incentive for the victim to reduce his exposure to damages. This example also reaffirms Coase's insistence on the importance of defining property rights clearly. For if the conditions of the theorem are met, the farmers and the railroads will be able to negotiate an arrangement by which both railroad and farmers will take all economically justifiable measures to reduce the damage from the sparks.

Ralph Turvey's paper, "On Divergences between Social Cost and Private Cost," synthesizes the major ideas of several influential articles on externalities, including Coase's. He concludes, along with Coase, that when it is not feasible for the parties to negotiate, there is no single device, such as a tax, that will induce the socially efficient use of resources in all instances. In some circumstances an effluent charge or tax may induce behavior that misallocates resources. This paper is a stern reminder of how complicated responses to externalities can be even when there is only one polluter, one pollutant, and one recipient.

By the time readers come to Alan Randall's paper, "The Problem of Market Failure," they will have encountered in this volume repeated, and not always consistent, uses of the terms "common-property resource," "public good," and "externality" to explain the market's failure to preserve and protect environmental resources. Randall recommends dropping the use of the first two terms, which he finds confusing, and focusing instead on the specific characteristics of resources that account for the market's failure to provide for their efficient use. They boil down to "nonexclusiveness," meaning the condition in which it is excessively costly, impossible, or undesirable to exclude anyone from the use of a resource, and "nonrivalry," or the condition in which one person's use of the resource does not interfere with that of anyone else. He then describes how the presence or absence of these two characteristics of resources, in various combinations, determines whether the market for them will behave efficiently. He maintains that, whereas the Coase Theorem does not apply to resources that are either nonrivalous or nonexclusive, the major environmental problems concern resources that display at least one of these properties. Therefore, he concludes, Coase's theorem is not central to environmental economics.

Each of the two lectures by Robert M. Solow that are reprinted here is concerned with the way in which our use of resources affects the distribution of well-being between present and future generations. In the first, he considers whether the optimal rate of exploitation of an exhaustable resource by private owners, which Harold Hotelling explained, is likely to approximate the rate that is socially desirable in the long run. Convergence of the two rates depends fundamentally on equality of the private discount rate and the social discount rate. If the private rate exceeds the social rate, extraction will proceed too rapidly from a long-range social point of view. But, as Solow points out, the choice of the social discount rate is itself a complicated policy decision, which depends on how much consumption we feel obligated to provide for future generations and the extent to which technological advances and resource substitution can be counted on to sustain future consumption.

In his second lecture, delivered eighteen years later, Solow broadens the question of our obligation to future generations to include the preservation of commonly owned, as well as privately owned, resources. At the same time he broadens the scope of the solution. Solow believes that the obligation is a general one to leave future generations as well off as we are, rather than to bequeath them specific resources. It may be possible to meet that obligation equally well by leaving a heritage of investment in knowledge or manmade capital as by bequeathing an unscathed environment. "Sustainability" thus is a problem that concerns choices between current consumption and investment. Protecting the environment will not necessarily help to meet obligations to future generations if it comes at the expense of investment in the capacity to produce in the future rather than at the expense of current consumption.

John V. Krutilla, in his well-known "Conservation Reconsidered," calls attention to a further reason why market forces may not lead to an allocation of resources that is optimal for present or future generations. There is a trade-off in allocating many natural resources between their uses in the production of goods and services and in the production of amenities. The private market cannot be relied on to achieve this allocation optimally because the amenities are public goods whose value could not be fully appropriated even if it were known. Assessing that value is especially troublesome when it consists of preserving a nebulous option for future use, in the case, say, of a fragile ecosystem on which the survival of a number of species depends. Had he written this paper in the 1990s, Krutilla might have cited the tropical rain forest as a leading example of a divergence between the private returns and public returns to exploitation of an environmental resource.

It is sometimes argued that the traditional concern for husbanding natural resources as a source of production for future generations is outmoded in view of advances in technology. Krutilla reminds us that the need to preserve the natural environment as a source of amenities is getting greater all the time.

5

Some Concepts from Welfare Economics

ROBERT DORFMAN

Robert Dorfman is David A. Wells Professor of Political
Economy Emeritus at Harvard University.

The essays in this collection are concerned with the economic aspects of current environmental problems. There are other aspects—moral, medical, biological, chemical, and more—but they are treated here only to the extent necessary for understanding the economic aspects. Even with this limitation, the environment remains too large a concept and word to be analyzed in a single volume. We shall therefore confine ourselves to the aspects of the environment that people seem to have in mind when they talk of "the environmental crisis." In that context the subjects of concern are the purity of air and public waters, the plentifulness and vitality of natural landscapes, fauna, and flora, the integrity of certain other natural features such as beaches and, increasingly, widespread conditions, particularly the accumulation of greenhouse gases in the atmosphere, acid precipitation, and deforestation. All of these are comprehended in the *ecosphere*—the living space shared by all living creatures (including man)—and the creatures themselves.

Man has been tampering with the ecosphere for a very long time; one might almost define man as the animal that modifies its environment consciously, not instinctively. The great transformation from primitive hunting and gathering to settled civilization occurred when man learned to convert forests and savannahs into farms and to breed domesticated varieties of plants and animals. This was the most radical change in the environment that mankind has undertaken to this very day. Over the course of time this transformation altered the ecology of entire continents, and there is good reason to believe that ancient civilizations rose and fell as a result of its progress.

79

Not all of the environmental changes engineered by man have been for the worse, though some writers presume that they have been. Quite apart from the enormous increase in food production, malarial swamps have been drained, deserts have been made habitable, and much more has been achieved. But in the last few generations mankind's propensity to change his environment has itself been transformed, as symbolized by the contrasts between the whaleboat and the radar-equipped factory ship, the waterwheel and the nuclear power plant, or the country road and the interstate highway. The power to use and adapt has become the power to destroy abruptly. Our understanding of the environment has by no means kept pace with our capacity to alter it, and our ability to control our impact has fallen far behind. Therein lies the current threat.

The visible and impending environmental impacts of our newly acquired powers have forced us to recognize that the environment consists of scarce and exhaustible resources. That is where economics enters, for economics is the science of allocating scarce resources among competing ends. What has happened, obviously, is that the economic institutions that sufficed when ecological side effects were mild and gradual have abruptly become inadequate. It is as if we were trying to control automobiles with bridles. The task for economics is to perceive just why the old institutions cannot control the new forces, and to devise new methods of control that can. The papers in this collection are addressed to that task. First we shall examine what they say about the adequacy of the economic system we have inherited.

ENVIRONMENTAL RESOURCES AND PRIVATE PROPERTY

Several of our authors [Coase in Chapter 7, J. H. Dales in Chapter 14] lay great emphasis on the concept of private property and its role in directing the use of resources. Their point is twofold. First, the institution of private property and our other economic institutions have evolved together so that our economic system is well attuned to securing the efficient use of things that are owned. This follows from the familiar argument, going back to Adam Smith's "invisible hand," that competitive markets guide resources into the uses in which they will produce the things that consumers want most. But it should be noticed, first, that this argument applies only to resources that are privately owned and to commodities that consumers buy individually and use as they see fit. Second, if any resource is not privately owned, or if a consumer wants a commodity that he cannot procure and use individually, then the invisible hand doesn't work. On the contrary, ordinary economic institutions do not provide any incentives for furnishing such resources or commodities, or for using them efficiently if they occur naturally. In particular, resources in the environment, which no private person owns, tend to be used heedlessly, with results that are all too obvious.

One explanation, then, of our environmental problems is that many vitally important resources are not owned by anyone and consequently lack the protection and guidance that a private owner normally provides. [1] This state of affairs was tolerable when the unowned resources were plentiful, and careless use had virtually no effect upon them, but now we can no longer be so relaxed about it.

It does not follow that more extensive private ownership is the solution to our environmental problems. There are good reasons that explain why it is either impractical or undesirable to confer private titles to certain resources. Since these are the underlying causes of the misuse of the environment, and any corrective measures must take them into account, it pays to examine them closely.

The resources that make up the environment are unsuitable for private ownership because they lack the "excludability property." That is, it is typically not practical to exclude people from these resources or to prevent people from benefiting from them, either because of physical impossibility, or because controlling access would be inordinately expensive or cumbersome, or because limiting access would be socially unacceptable. This is the case with the atmosphere, and with most public roads, waters, beaches, and so on. It is also the case with flood-protection works, sewage-treatment plants, and many other facilities that improve the condition of the environment. There isn't much point in owning anything from which other people cannot be excluded.

Nonexcludability is not the whole story. Another peculiarity of environmental resources is that there are likely to be enormous economies in the joint consumption or use of the resource as contrasted with individual use. This second characteristic is illustrated by the contrast between housing and streets. It is economical to divide living space into family-size lots and devote each of them to housing one or a few families, but it would be fantastically wasteful to provide each family with its private road to the central business district. Accordingly, most houses are private property, but streets are public property. Most resources in the environment are analogous to streets, not houses, and are therefore used in common by substantial numbers of people. The difficulty that this common use creates is that each user may interfere with the others, reducing the serviceability of the resource to them; moreover, he has little incentive not to do so. Typical examples are road congestion and the use of the atmosphere and waters for discharging waste products. Resources that are most economi-

---

[1] There is no implication that resources that are owned are automatically used in the public interest. For example, if the owner of a resource is a monopolist, he will not use it in a socially efficient way.

cally used in common frequently are privately owned—ski slopes, for example. But frequently also, it is deemed undesirable to have such resources controlled by private individuals. It is not much help to notice that resources from which people cannot be excluded or which it is very economical for them to share would be better managed if they were ordinary, private economic goods. Their very nature precludes that solution.

Although the characteristics of nonexcludability and of shared, mutually interfering usage are especially prevalent among environmental goods, they do occur elsewhere. For example, fire protection and scientific discoveries have the nonexcludability property. Public and university libraries are used in common, and the users interfere with each other.

In fact, the economic theory of resources that exhibit these characteristics has been worked out in other contexts. We shall outline it briefly here.

This theory rests on two concepts. One is the notion of public goods, also called common-property resources or collective goods. The defining characteristic of a public good is that no member of the community who wants its services can be excluded from them if they are available to any other member. Classic examples are public roads and lighthouses. Environmental examples include the diversity of species, the oceans, and the ozone layer.

The other basic concept of the theory of public goods is externalities. An externality is a direct and unintended side effect of an activity of one individual or firm on the welfares of other individuals or firms. The word "direct" excludes effects that are transmitted by changes in prices. The word "unintended" excludes the effects of both benevolent and aggressive activities. Examples are air pollution generated by power plants or automobiles, water pollution, depletion of fisheries by overfishing, etc. Externalities are not always harmful, though economists are most concerned about the ones that are. The beneficial effects of TV broadcasting are a case in point.

These two concepts work together to produce an important classification of public goods. If the users of a public good neither interfere with each other nor increase the good's usefulness to each other, it is a "pure public good" (or "public bad" if it is harmful, like automobile exhaust). The scenic vistas and protective ozone layer, just mentioned, are pure public goods. If the users of a public good do affect its usefulness to each other, the good is said to be a "congestible public good." Garrett Hardin's commons and the fishery analyzed by Scott Gordon (Chapter 6) are congestible.

The mutual interference of the users of a congestible public good is a special case of the general phenomenon of externalities. An externality occurs when the activities of one person have unintentional effects on the welfares or production functions of other persons.[2]

Furthermore, congestible public goods are of two types. In one type,

---

[2]Externalities are explained at greater length in Chapter 17 and elsewhere in this collection.

the users are relatively homogeneous and impose reciprocal externalities on each other, like drivers on a crowded street. In the ohter type, there are two or more different kinds of user, and the effects of each type's activities on users of other types are not reciprocated. Thus waste dischargers on a stream impose externalities on swimmers and other users downstream, but are not affected by those users.

This elaborate classification of public goods is needed because the economic analysis and practical management of public goods is different for the different types. We turn first to the simplest case, pure public goods.

Since access to pure public goods cannot be controlled, by definition, the only decision is how much of the goods to provide. If the good occurs entirely naturally, as do oceanic shipping lanes, again there is no decision to be made. But if it has to be created at some trouble and expense, as does a flood-protection dam or a lighthouse, then no single individual is likely to find providing it worthwhile, since he cannot enforce a charge for its services. [3] A social decision therefore has to be made about whether to provide the good and, if so, on what scale.

To be concrete, consider a community consisting of three individuals, A, B, and C. They are all bothered by the smoke from the power plant that supplies their electricity, and the public good in question is a smoke precipitator that might be installed in the power plant's stack. None of them is willing to pay for the precipitator singlehandedly, but each would accept a moderate increase in his electric bill (or taxes) to defray the expense. Whether a precipitator should be installed and, if so, how large a one depend on its cost and on its value to each of the beneficiaries. [4]

The relevant data are shown in Figure 1. The size of the precipitator, measured in terms of the percentage of smoke removed from the discharge plume, is plotted horizontally; the vertical scale is a scale of dollars. The line labeled P is the marginal-cost curve. The smallest conceivable precipitator costs $4 per year to own and operate, and would achieve virtually no reduction in smoke emission. If 20 percent of the smoke is being precipitated, it would cost an additional $7.50 a year to raise the proportion to 21 percent. If 80 percent is precipitated, it would cost an extra $18 a year to raise the proportion to 81 percent. The other points are interpreted similarly. This line is effectually the supply curve for smoke removal. For example, if the consumers were willing to pay $18 for each percentage point of smoke removed, and if the power company could collect smoke-

---

[3]Exception to this assertion. Wealthy people sometimes provide public goods with no expectation of reimbursement. All charities and nonprofit enterprises depend on such altruism.
[4]Alternatively, this situation could be considered an instance in which four asymmetrical users (the power plant and A, B, and C) share the local airshed. The simpler interpretation, in which the smoke precipitator is a pure public good whose benefits are shared by A, B, and C, brings out the relevant points more clearly.

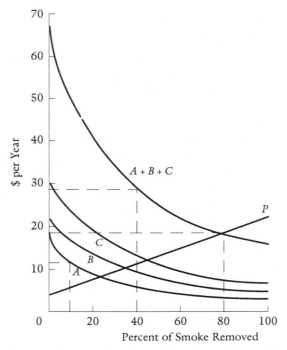

FIGURE 1. Optimal provision of a public good.

removal charges, the power plant would find it worthwhile to remove 80 percent of its smoke from its discharges.

The curves labeled *A, B,* and *C* are willingness-to-pay curves, one for each beneficiary. The curve for *A* shows how much he would be willing to pay for incremental improvements in the atmosphere. If only 10 percent of the smoke were being removed, he would be willing to pay $12 a year to have an additional 1 percent removed, and so on. If *A* were able to buy smoke removal as an individual, this would be his demand curve for it, telling how much he would buy at each stated price. Of course, he cannot do that, since the precipitator is a public good, and its services cannot be supplied to him without providing the same amount to others.

The willingness-to-pay curves for *B* and *C* are interpreted in the same way. To see how the curve labeled *A + B + C* is constructed, consider removing 40 percent of the smoke. The willingness-to-pay curves show that *A* would be willing to pay $6.00 for this improvement, *B* would be willing to pay $9.50, and *C* $13.00, for a total of $28.50 for the whole community. Since all will benefit from the improvement, its value to the community is $28.50, which is shown as the ordinate of the *A + B + C* curve at 40 percent. The entire *A + B + C* curve is constructed similarly as the *vertical* sum of the three willingness-to-pay curves, and shows the aggregate willingness, on the part of all the consumers, to pay for additional reductions in emissions.

In the light of these interpretations, it is clear that the point where P crosses $A + B + C$ has special significance. That point is the level of smoke removal where the cost of an additional percentage is just equal to the amount that all the beneficiaries taken together are willing to pay for it. If less is removed, they would be willing to pay enough to cover the cost of a higher level of removal; if more, they would be forced to pay for a purer atmosphere than they deemed worthwhile. For these data, a precipitator that will remove 80 percent of the discharge is the right size.

That simple example illustrates the essence of the theory of pure public goods: the appropriate level to provide is the one for which the vertical sum of the beneficiaries' willingness-to-pay curves crosses the marginal-cost curve. Of course, there is much more to the theory than this, dealing with such additional matters as how the cost should be distributed among the beneficiaries, but we now have the main idea.[5]

The theory of externalities, including the special kind arising from the shared use of resources, is more involved, and we shall not attempt to summarize it here. The main issues are treated by H. Scott Gordon in Chapter 6 and Ronald Coase in Chapter 7. The basic notion is that since each user imposes a cost (in terms of inconvenience or reduced productivity of the resource) on the others, no use should be allowed unless the utility or benefit of that use is great enough to counterbalance the total cost that it imposes on other users. That simple statement slides over a great many difficulties, among them the problems of measuring both the direct benefits of the use and the costs that it imposes on others. It is the essence of environmental externalities that prices are not charged for access to environmental resources, so that there are no prices to reveal how much users are willing to pay for the privilege, nor are there any prices to indicate how much each user has inconvenienced the others. The selections in Section IV deal with these problems of benefit and cost measurement.

The main point of these concepts and theories for us is that they explain why environmental resources, which partake of these difficulties, tend to be overused, misused, and abused unless special measures are taken. They show that each individual's incentives induce him to use environmental resources more heavily and to contribute less to protecting them than is socially desirable. It should be emphasized that no malign intent or even

---

[5]For brief and clear discussions of the theory of public-goods provision, see R. A. Musgrave, *The Theory of Public Finance* (New York: McGraw-Hill Book Co., 1959), pp. 74–78, or L. Johansen, *Public Economics* (Chicago: Rand-McNally, 1965), pp. 129–140. The obvious ideal would be to charge each beneficiary in accordance with his willingness to pay, but if this were the policy, the consumers could not be expected to disclose their preferences with sufficient candor. (Contrast this with behavior with respect to ordinary commodities.)

blameworthy negligence is involved. It would be silly to stay home from Coney Island just because our presence would make it even more congested. No individual can make a noticeable reduction in the prevalence of smog by putting a catalytic converter on his car. The only meaningful solutions are collective solutions.

It follows that we have to make collective decisions about the use of the environment, and that one reason for its current state is that thus far we have neglected this obligation. We have to decide both what we want to achieve and what means to use for achieving it. Many of the papers, especially those in Section III, deal with this second problem, but since there are no selections dealing systematically with the first, we shall take it up here.

CRITERIA FOR ECONOMIC PERFORMANCE

So far we have taken it as accepted ground that the environment is not being used properly at present. We are putting too many nitrates and pesticides in the water, too much nitrous oxide and lead in the air, too many chlorofluorocarbons and too much carbon dioxide in the troposphere, and much more. Although it is obvious that what we are doing is wrong, it is by no means obvious what would be right. One thing is certain: we have to go on using the environment, and using it in common. People have to live and congregate somewhere, dispose of waste products, and even use depletable resources. These activities cannot be abolished, though they can be controlled. Controlling them means finding the proper balance between the utility of these activities to the individual and the disutility they impose, via the environment, on others.

It is easier to talk about the proper balance than to define it. Economists have hammered out useful formulations—useful in the sense that they can be applied to specific problems and decisions—only after long and earnest effort. And still, these formulations are not all that might be desired, as we shall see.

The critical difficulty lies in diversity of interest. What is good for Mrs. Goose is not necessarily good for Mr. Gander, and both of them have to be weighed in the proper balance. Reconciling divergent interests is what politics is all about, but economists have something to say, too.

Because of diversity of interest, it is very hard to say which of two policies or situations is better than the other unless, perchance, everyone happens to agree. An appeal to majority rule is a cop-out. It is a way to reach a decision in concrete instances, but it will not serve as a definition of the "right" decision, and no one maintains seriously that it can be relied on to produce the "right" decisions. Our task at the moment is to define, as well as we can, what we mean by a right decision as a foundation for policies for using the environment.

After generations of hard thought, economists have arrived at five criteria for judging policies or decisions. So many criteria are needed because none of them is entirely satisfactory, for one reason or another. Four of the criteria relate to the efficiency of the economic system, the fifth to equity. We shall discuss them all, since all are invoked in the papers that follow.

The four efficiency criteria consist of two pairs. The first, and fundamental, pair relates to the success of the economy in promoting welfare or satisfaction. We shall refer to these as the "utility criteria." The second, and more tractable, pair relates to the success of the economy in producing goods or other physical results. We shall refer to these as the "productivity criteria."

Within each pair there is one fully specific criterion that purports to pick out the best single decision or situation. Unfortunately in both cases the device that singles out the *optimum optimorum* is questionable. So there is a kind of fall-back criterion that identifies the class of decisions or situations in which the best must lie, if there is a best, but does not compare or evaluate the decisions in that class. That is how the four efficiency criteria are related logically. Now we can define them, beginning with the two utility criteria.

The Broad Utility Criterion: Pareto Efficiency

The task of an economy is to produce the combination of goods and services that will promote the welfare of the members of the community as much as possible with the resources and production techniques available. The welfare of the community is some resultant of the welfares of its individual members; these welfares are generally called their utilities. [6] The level of utility of each member of the community is presumed to depend on two things: his own consumption of private goods and services, and the environmental conditions to which he is exposed. We can compare the desirability of two or more different modes of operation of the economy by noting the amount of utility that each affords to each member of the community, and drawing conclusions on the basis of these individual utilities.

A simple diagram is frequently used to illustrate these concepts and make their application more concrete. Suppose that a community consists of only two individuals, Mr. Able and Mr. Baker. The entire output of the

---

[6]Philosophers sometimes call this assertion "radical individualism." It denies that the community, or the state, or any other group has an objective or welfare distinct from the welfares of its members. Of course, to the extent that members of a group identify with it and derive satisfaction from the attainment of group objectives (e.g., Olympic medals), that attainment is reflected in the welfares of members of the group. Such attainments are a form of public good.

economy is divided between them, and both are affected by whatever public goods or environmental conditions result from the operation of the economy. Then we can depict the welfare results of the economy on a diagram like Figure 2, where Mr. Able's utility is measured along the horizontal axis and Mr. Baker's is measured vertically. Point *A,* for example, represents the results of some mode of operating the economy in which Able's utility level is 100 units and Baker's is 150 units. It may be possible to change the pattern of economic activity so as to increase Able's welfare without diminishing Baker's. For example, Baker may enjoy playing his stereo equipment at a resounding volume that interferes with the yoga meditations to which Able is devoted. Able might be willing to pay Baker as much as $100 to use earphones instead of stereo speakers (i.e., to turn over to Baker $100 worth of the private goods and services to which he would otherwise be entitled), and Baker might be willing to comply for as little as $50. In this case, if Able paid Baker $50, Baker would lose some utility by resorting to earphones but would gain it back by consuming additional private commodities and be just as well off as before, whereas Able's utility would be increased $100 worth by the reduction in the noise level and would be diminished only $50 worth by his reduced consumption of other commodities, so that he would be better off than before. This result is shown in the figure by point *B.* Point *B* is indubitably a better point for the operation of the economy as a whole: one of the members of the community has benefited and none has suffered. Coase analyzes this type of social arrangement at some length in Chapter 7.

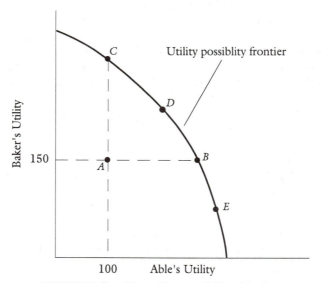

FIGURE 2.   The utility-possibility frontier.

Let us suppose that when the economy is at point B there are no further possibilities for increasing Able's utility without reducing Baker's. Then the position of the economy corresponding to point B is said to be *Pareto optimal* or *Pareto efficient*. As a formal definition, the operation of the economy is Pareto efficient if there is no way to change it that will make some member or members better off by increasing their utilities without reducing the utilities of some other members. The curve on the diagram represents all the Pareto-efficient ways of operating the economy. Each point on it shows the greatest utility that can be provided to Able in conjunction with the given level of utility for Baker, and at the same time the greatest utility than can be afforded Baker in conjunction with the given level of utility for Able. It is called the *utility-possibility frontier*. The utility-possibility frontier can be drawn only for the trivial case of a two-person community, but it can be conceived, and does exist theoretically, for every community.

Now let us go back to point A, which lies below the utility-possibility frontier. If the economy is operating there, it is not doing as good a job as possible. For, by assumption, it can be operated so as to attain points such as B or C or any point on the frontier between them that will make at least one member of the community better off without harming any one. A large part of welfare economics, and of the papers in this collection, is concerned with social arrangements that will enable a real economy to avoid points like A that are inefficient from the utility point of view, and enable it to attain points like D that are efficient.

One method for resolving the conflict of interest between Able and Baker would be for the community to require Baker to use earphones. This would increase Able's utility even more than if he had to pay for silence, but it would reduce Baker's. The result might be point E, which is drawn on the utility-possibility frontier. This assumes that this social arrangement is Pareto efficient, as we defined it. But, to illustrate a terminological pitfall, the *change* from A to E is not Pareto efficient, because a member of the community is harmed thereby.

The comparison between points such as C and E, or between social arrangements such as free bargaining between Able and Baker as against government regulation, raises important social issues. As we have drawn them, both social devices lead to Pareto-efficient points, but there is good reason to believe that free bargaining is more likely to produce this result than governmental decrees. We shall consider that issue when we take up various practical methods of control. For the moment we note that the Pareto efficiency criterion does not enable us to discriminate between such points. The next criterion does so.

The Sharp Utility Criterion: Social Welfare

We noted above that the welfare of a community is some resultant of the welfares of its individual members. Suppose we could agree on precisely what the resultant was. Then we should have a *social-welfare function,* that is, a rule by which we could evaluate the welfare of an entire community if we knew the welfares, or utilities, of its individual members. For example, the social-welfare function might state that the welfare of the community is the simple sum of the welfares of its individual members. [7] A slightly more complicated social-welfare function would be the sum of the individual utilities minus one-half the difference between the greatest utility and the lowest one. [8] This social-welfare function would pay attention not only to the aggregate of utility but to the amount of inequality in the society. Any real social-welfare function would undoubtedly be very complicated indeed, and in fact, so far as we know, no satisfactory social-welfare function dependent on individual utilities has ever been constructed. There is good reason to believe that the social-welfare function is a philosopher's stone that does not exist in any real society.

In spite of the difficulties with the concept of a social-welfare function, it is very appealing because it provides the only way to arrive at clearcut, unambiguous social evaluations. Look again at our diagram. Point D is clearly socially superior to point A, since everyone benefits from a movement from A to D. But is point E socially superior to A? It is Pareto efficient, while A is not; but, nevertheless, Baker is likely to feel definitely worse off. And Baker is half of this society. Without a social-welfare function it is not even clear that a Pareto-efficient operation of the economy is superior to an inefficient one; and it is even more impossible to compare the merits of a number of points on the utility-possibility frontier. So some people conclude that since social decisions are made in practice, some social-welfare function must be implicit in the social-decision process.

If there were a social-welfare function, we could portray it like an individual consumer's utility map. This is done in Figure 3. Figure 3 shows the same data as Figure 2 plus a number of "social-indifference curves,"

---

[7]This is an important criterion historically; it is the basis of the utilitarian philosophy of social welfare.

[8]One restriction has to be mentioned. The value of the social-welfare function has to increase whenever the utility of any individual increases, other utilities remaining the same. For example, the following is *not* a social-welfare function: the sum of individual utilities minus twice the difference between the greatest and the lowest. The reason is that this function would actually decrease if the happiest individual became still happier, all others remaining the same. Because of this restriction, it is very dangerous to try to include such considerations as altruism or envy in a social-welfare function. Strange paradoxes can result—for example, a recommended distribution of income in which the most envious people receive the highest incomes and the most generous the lowest.

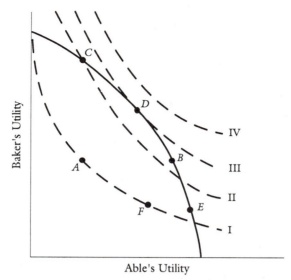

FIGURE 3. The utility-possibility frontier with social-indifference curves.

distinguished by Roman numerals. The value of the social-welfare function is constant along any social-indifference curve and is higher on higher indifference curves. Since points *A* and *F* are both on indifference curve I, they are equally desirable socially according to the social-welfare function depicted. Point *E* is above that social-indifference curve and is superior to either of them. Points *C* and *B* are better still. Point *D* is the most desirable point on the utility-possibility frontier. The points on social-indifference curve IV are even more desirable, but they are unattainable. From this diagram one can conclude that according to this social-welfare function it would be better to let Able and Baker bargain than to forbid Baker to use his loudspeakers, but that it would be better still to adopt the social arrangements that lead to point *D*, whatever they are. As the example illustrates, if there is a social-welfare function, the social desirability of any two configurations of individual utilities, Pareto efficient or not, can be compared.

But we have already cast doubt on the existence of such functions.[9] Indeed, there is even good reason to doubt that individual utilities can be defined or measured, either in principle or in practice. So criteria are needed for assessing economic performance without appeal to these dubi-

---

[9]For a more profound and decisive critique, see Kenneth J. Arrow, *Social Choice and Individual Values,* second edition (New York: John Wiley & Sons, 1963).

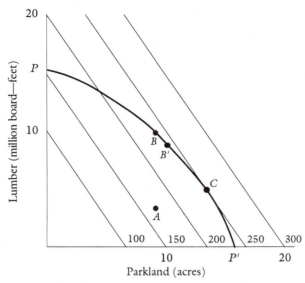

FIGURE 4. The production-possibility frontier.

ous concepts. [10] The two productive-efficiency criteria are intended to fill this need.

The Broad Productivity Criterion

The broad productivity criterion is a more hardheaded analog of Pareto efficiency. Instead of looking for that possible will-o'-the-wisp, utility, it concentrates on measurable outputs of goods and services and, possibly, environmental quality. Otherwise it is similar. Specifically, an economy is said to be productively efficient if it is producing as much of every good and service (rather than utility) as is technically possible, given the outputs of all other goods and services and the amounts of resources used.

As an example, consider an economy that produces only two commodities (there may be any number of consumers). Suppose, to be concrete, that the two commodities are lumber and parkland. The productive possibilities for the community are shown in Figure 4, with acres of parkland plotted horizontally and output of lumber vertically. The key feature of the diagram is the arc *PP'*, called the *production-possibility frontier*. It

---

[10]The Pareto-efficiency criterion does not depend on having a social-welfare function or on being able to measure individual utilities. All it requires is that we be able to tell for each individual which situation in any pair he prefers. But even this is excessively demanding for most practical purposes.

shows the greatest amount of lumber that can be produced in conjunction with a specified acreage of parkland, and the greatest amount of parkland consistent with a specified output of lumber. [11]

Any combination of lumber and parkland that is on or below the production-possibility frontier is technically feasible (for example, the combinations shown by points *A, B,* and *C*), but combinations that lie above the frontier are physically impossible. A productively efficient economy will produce an output right on the frontier (*B* or *C,* not *A*). A badly managed economy could find itself at *A,* perhaps by turning its most productive stands of timber into a park. If those stands were turned over to lumbering and an equal, less heavily wooded area turned into a park, more lumber could be produced without diminishing the amount of parkland, as at efficient point *B.* An important criterion of economic efficiency, therefore, is whether the rules and procedures of an economy tend to lead it to produce at a point like *A* or at one like *B.*

The relationship between productive efficiency (i.e., producing on the production-possibility frontier) and Pareto efficiency is clear, but not simple. On the simple assumption that everyone would like both more lumber and more parkland, an economy cannot be operating Pareto efficiently unless it is productively efficient. For if it were at a point like *A,* it would be possible to change its operations so as to produce more lumber without sacrificing any parkland. There would then be more lumber, and more utility, to distribute; and some people could be made better off without harming anyone. On the other hand, it is quite possible for an economy to be productively efficient without being Pareto efficient. This would happen if the economy were producing everything efficiently but were producing the wrong combination of things, in the sense that the outputs of some commodities could be reduced and other outputs increased in such a way that the new bundle of commodities would satisfy everybody better than the old. In terms of the graphs, an economy could produce at point *B* on the production-possibility frontier in Figure 4 and yield the utilities of a point inside the utility-possibility frontier, such as point *A* of Figure 3. To illustrate, notice that, as the figure is drawn, if the economy were producing at *B,* it could have about 1.5 acres more parkland by giving up 1.5 million board-feet of lumber (thus moving to point *B'*). [12] If there were a thousand families in the community and each would be glad to reduce its consumption of lumber by 1,000 board-feet or more in order to expand the parks by an acre, then it would not be Pareto efficient for the economy to produce at point *B;* everyone's welfare could be increased by moving to *B'.*

In short, productive efficiency is a necessary condition for Pareto effi-

---

[11]The diagonal lines will be explained below.

[12]The amount of lumber that has to be given up to obtain one more acre of parkland is known as the *marginal rate of transformation* of lumber for parkland.

ciency, but it is not sufficient. Pareto efficiency is the more fundamental and the more demanding achievement. On the other hand, it is much easier to judge the productive efficiency of any policy or economy than to assess its Pareto efficiency.

The criterion that an efficient economy produces an output that lies on its production-possibility frontier enables us to reject an output like point *A* but does not enable us to choose between outputs like *B* and *C*. The final efficiency criterion is addressed to this question.

The Sharp Productivity Criterion: GDP

The relative desirability of any two privately owned commodities is reflected in their prices. On familiar grounds, an additional dollar's worth of anything a consumer buys gives him as much satisfaction as an additional dollar's worth of anything else. Therefore, if an economy revises its operations so as to provide consumers with commodities that they value more highly, in dollars-and-cents terms, it is providing them with more utility (is moving closer to the utility-possibility frontier) even though it must reduce the quantities of some commodities to provide more of others. This remark discloses the social significance of prices as guides to economic activity, as well as suggesting how to use prices to choose among the points on the production-possibility frontier. It suggests that the best point on the frontier is the one at which the value of the goods and services produced is as great as possible, using market prices for the private goods and services. This gives rise to the criterion that the market value of goods and services produced should be as great as possible.

The market value of final goods and services produced in a country is, by definition, the gross domestic product, GDP for short. The criterion just deduced can therefore be expressed by saying that the output of marketable goods and services is optimal when the GDP is as great as possible, using the market prices corresponding to that output to evaluate the various goods and services. But that criterion is inadequate for guiding environmental decisions, or decisions about public goods in general, since it ignores the values of unmarketed public goods and services. For example, in Figure 4 the GDP account would include the value of the lumber produced but would not include the value derived from using the parkland, which is a public good for which no price is charged.

In order to bring environmental values, and the values of other public goods and services, into the picture, a more inclusive concept is needed, which we shall call augmented GDP. Augmented GDP, ADP for short, is defined as the ordinary GDP plus the value of public goods and services of all sorts, including, of course, those provided by the environment. The papers in Section IV explain at some length the issues involved in estimating the values of public goods and services, and the reasons why it is useful to have both the GDP and the AGDP concepts. For the rest of this article, when we refer to gross domestic product we shall always mean the augmented version of the concept unless we state the contrary.

This concept enables us to take due account of the values of environmental and other public goods, merely by choosing the point on the production-possibility frontier at which the AGDP is as great as possible. The use of the AGDP criterion is illustrated in Figure 4. If lumber is worth $10 per million board-feet and a year's use of parkland is worth $15 per acre, then

$$AGDP = \$10 \times \text{millions of board-feet} + \$15 \times \text{acres of park}.$$

The slanting lines in the figure are lines of constant AGDP at the prices assumed. The output at point C is worth $250, and this is greater than the value of output at any other point on or under the production-possibility frontier. According to the AGDP criterion, point C should be chosen.

When we invoke the AGDP criterion we not only extend the concept of GDP to include the value of public goods, we also extend the reasoning that justifies its use. The ordinary GDP criterion, remember, was justified by noting that consumers could be better pleased if the output of the economy could be changed so as to replace some commodities for which they would be willing to pay just $1 by others for which they would be willing to pay more than $1. But consumers do not have to pay for the use of public goods or the environment, so the AGDP criterion cannot be justified by appeal to consumers' market choices. In fact, however, when the government provides public goods it is acting as an agent for the consumers by buying something for them collectively that they cannot purchase individually. The reasoning, then, is that consumers will be better pleased by the provision of some public goods if (and only if) they value those public goods more highly than the marketable goods that have to be relinquished in order to obtain them. The best mix of public and market goods is then the one with the highest possible AGDP, where private goods are valued at market prices and public goods at what the public would be willing to pay for them if it had to.

No point below the production-possibility frontier can satisfy the AGDP criterion, since if the economy were at such a point AGDP could be increased by increasing the output of some commodity without cutting back any other output. Thus an output point that satisfies the AGDP criterion automatically satisfies the broad productivity criterion. It also satisfies the broad utility criterion—i.e., is Pareto efficient, because it leaves no room for improving anyone's consumption bundle without detracting from someone else's.

The AGDP criterion is the most practicable of the four, in fact the only one that can be applied with much assurance. We have followed a long road to arrive at it, and it is time to reconsider what we have learned. The ideal would be the sharp utility criterion, which maximizes the society's social-welfare function. But a society may not have a social-welfare function; in any event we do not know it, so that is impracticable. Next best is the broad utility criterion, but that one is indecisive among many options and, besides, depends upon unascertainable data for the most part. Next we considered the broad productivity criterion, the demand that the bundle of outputs produced by the economy should lie on the production-possibility frontier. This is a modest requirement, and one that can often be implemented, but it is insufficient as a principle of social choice because it does not instruct us which point on the production-possibility frontier is to be preferred. Finally, we arrived at the AGDP criterion, which does select a decision that satisfies both the broad utility criterion and the broad productivity criterion. These, indeed, are its main justifications.

For these reasons, the AGDP criterion is the main reliance of practical economic analysis in the environmental field. Perfection cannot be claimed for it. In particular, it pays no attention to the equity or inequity of the distribution of income or of environmental benefits. It is therefore quite possible for a society to prefer a decision that is inconsistent with this criterion to one that is consistent and, by the same token, for a proposal that will increase AGDP to be socially undesirable.

6

The Economic Theory of a Common-Property Resource: The Fishery

H. SCOTT GORDON

H. Scott Gordon is Distinguished Professor of Economics
and the History and Philosophy of Science Emeritus at
Indiana University in Bloomington.

I. INTRODUCTION

The chief aim of this paper is to examine the economic theory of natural
resource utilization as it pertains to the fishing industry. It will appear, I
hope, that most of the problems associated with the words "conservation" or "depletion" or "overexploitation" in the fishery are, in reality,
manifestations of the fact that the natural resources of the sea yield no
economic rent. Fishery resources are unusual in the fact of their common-property nature; but they are not unique, and similar problems are
encountered in other cases of common-property resource industries,
such as petroleum production, hunting and trapping, etc. Although the
theory presented in the following pages is worked out in terms of the
fishing industry, it is, I believe, applicable generally to all cases where
natural resources are owned in common and exploited under conditions
of individualistic competition.

"The Economic Theory of a Common-Property Resource: The
Fishery," by H. Scott Gordon, from *The Journal of Political
Economy,* April 1954, Sections I, II, III. Reprinted by permission.

II. BIOLOGICAL FACTORS AND THEORIES

The great bulk of the research that has been done on the primary production phase of the fishing industry has so far been in the field of biology.
Owing to the lack of theoretical economic research,[1] biologists have been
forced to extend the scope of their own thought into the economic sphere
and in some cases have penetrated quite deeply, despite the lack of the
analytical tools of economic theory.[2] Many others, who have paid no
specific attention to the economic aspects of the problem, have nevertheless recognized that the ultimate question is not the ecology of life in the
sea as such, but man's use of these resources for his own (economic)
purposes. Dr. Martin D. Burkenroad, for example, began a recent article
on fishery management with a section on "Fishery Management as Political Economy," saying that "the management of fisheries is intended for
the benefit of man, not fish; therefore effect of management upon fish-
stocks cannot be regarded as beneficial *per se.*"[3] The great Russian marine
biology theorist, T. I. Baranoff, referred to his work as "bionomics" or
"bio-economics," although he made little explicit reference to economic
factors.[4] In the same way, A. G. Huntsman, reporting in 1944 on the work
of the Fisheries Research Board of Canada, defined the problem of fisheries depletion in economic terms: "Where the take in proportion to the
effort fails to yield a satisfactory living to the fisherman";[5] and a later
paper by the same author contains, as an incidental statement, the essence
of the economic optimum solution without, apparently, any recognition of
its significance.[6] Upon the occasion of its fiftieth anniversary in 1952, the

---

[1]The single exception that I know is G. M. Gerhardsen, "Production Economics in Fisheries," *Revista de economía* (Lisbon), March, 1952.

[2]Especially remarkable efforts in this sense are Robert A. Nesbit, "Fishery Management"
("U.S. Fish and Wildlife Service, Special Scientific Reports," No. 18 [Chicago, 1943])
(mimeographed), and Harden F. Taylor, *Survey of Marine Fisheries of North Carolina*
(Chapel Hill, 1951); also R. J. H. Beverton, "Some Observations on the Principles of Fishery
Regulation," *Journal du conseil permanent international pour l'exploration de la mer* (Copen-
hagen), Vol. XIX, No. 1 (May, 1953); and M. D. Burkenroad, "Some Principles of Marine
Fishery Biology," *Publications of the Institute of Marine Science* (University of Texas), Vol.
II, No. 1 (September, 1951).

[3]"Theory and Practice of Marine Fishery Management," *Journal du conseil permanent
international pour l'exploration de la mer,* Vol. XVIII, No. 3 (January, 1953).

[4]Two of Baranoff's most important papers—"On the Question of the Biological Basis of
Fisheries" (1918) and "On the Question of the Dynamics of the Fishing Industry" (1925)—
have been translated by W. E. Ricker, now of the Fisheries Research Board of Canada
(Nanaimo, B.C.), and issued in mimeographed form.

[5]"Fishery Depletion," *Science,* XCIX (1944), 534.

[6]"The highest take is not necessarily the best. The take should be increased only as long as
the extra cost is offset by the added revenue from sales" (A. G. Huntsman, "Research on Use
and Increase of Fish Stocks," *Proceedings of the United Nations Scientific Conference on the
Conservation and Utilization of Resources* [Lake Success, 1949]).

International Council for the Exploration of the Sea published a *Rapport Jubilaire,* consisting of a series of papers summarizing progress in various fields of fisheries research. The paper by Michael Graham on "Overfishing and Optimum Fishing," by its emphatic recognition of the economic criterion, would lead one to think that the economic aspects of the question had been extensively examined during the last half century. But such is not the case. Virtually no specific research into the economics of fishery resource utilization has been undertaken. The present state of knowledge is that a great deal is known about the biology of the various commercial species but little about the economic characteristics of the fishing industry.

The most vivid thread that runs through the biological literature is the effort to determine the effect of fishing on the stock of fish in the sea. This discussion has had a very distinct practical orientation, being part of the effort to design regulative policies of a "conservation" nature.

* * *

The term "fisheries management" has been much in vogue in recent years, being taken to express a more subtle approach to the fisheries problem than the older terms "depletion" and "conservation." Briefly, it focuses attention on the quantity of fish caught, taking as the human objective of commercial fishing the derivation of the largest sustainable catch. This approach is often hailed in the biological literature as the "new theory" or the "modern formulation" of the fisheries problem.[7] Its limitations, however, are very serious, and, indeed, the new approach comes very little closer to treating the fisheries problem as one of human utilization of natural resources than did the older, more primitive, theories. Focusing attention on the maximization of the catch neglects entirely the inputs of other factors of production which are used up in fishing and must be accounted for as costs. There are many references to such ultimate economic considerations in the biological literature but no analytical integration of the economic factors. In fact, the very conception of a *net economic yield* has scarcely made any appearance at all. On the whole, biologists tend to treat the fisherman as an exogenous element in their analytical model, and the behavior of fishermen is not made into an integrated element of a general and systematic "bionomic" theory. In the case of the fishing industry the large numbers of fishermen permit valid behavioristic generalization of their activities along the lines of the standard economic theory of production. The following section attempts to apply that theory to the fishing industry and to demonstrate that the "overfishing problem" has its roots in the economic organization of the industry.

---

[7]See, e.g., R. E. Foerster, "Prospects for Managing Our Fisheries," *Bulletin of the Bingham Oceanographic Collection* (New Haven). May, 1948; E. S. Russell, "Some Theoretical Considerations on the Overfishing Problem," *Journal du conseil permanent international pour l'exploration de la mer,* 1931, and *The Overfishing Problem,* Lecture IV.

III. ECONOMIC THEORY OF THE FISHERY

In the analysis which follows, the theory of optimum utilization of fishery resources and the reasons for its frustration in practice are developed for a typical demersal fish. Demersal, or bottom-dwelling fishes, such as cod, haddock, and similar species, and the various flat-fishes, are relatively nonmigratory in character. They live and feed on shallow continental shelves where the continual mixing of cold water maintains the availability of those nutrient salts which form the fundamental basis of marine-food chains. The various feeding grounds are separated by deep-water channels which constitute barriers to the movement of these species; and in some cases the fish of different banks can be differentiated morphologically, having varying numbers of vertebrae or some such distinguishing characteristic. The significance of this fact is that each fishing ground can be treated as unique, in the same sense as can a piece of land, possessing, at the very least, one characteristic not shared by any other piece: that is, location.

(Other species, such as herring, mackerel, and similar pelagic or surface dwellers, migrate over very large distances, and it is necessary to treat the resource of an entire geographic region as one. The conclusions arrived at below are applicable to such fisheries, but the method of analysis employed is not formally applicable. The same is true of species that migrate to and from fresh water and the lake fishes proper.)

We can define the optimum degree of utilization of any particular fishing ground as that which maximizes the net economic yield, the difference between total cost, on the one hand, and total receipts (or total value production), on the other.[8] Total cost and total production can each be expressed as a function of the degree of fishing intensity or, as the biologists put it, "fishing effort," so that a simple maximization solution is possible. Total cost will be a linear function of fishing effort, if we assume no fishing-induced effects on factor prices, which is reasonable for any particular regional fishery.

The production function—the relationship between fishing effort and total value produced—requires some special attention. If we were to follow the usual presentation of economic theory, we should argue that this function would be positive but, after a point, would rise at a diminishing rate because of the law of diminishing returns. This would not mean that the fish population has been reduced, for the law refers only to the *proportions* of factors to one another, and a fixed fish population, together with an increasing intensity of effort, would be assumed to show the typical sigmoid pattern of yield. However, in what follows it will be assumed that

---

[8]Expressed in these terms, this appears to be the monopoly maximum, but it coincides with the social optimum under the conditions employed in the analysis, as will be indicated below.

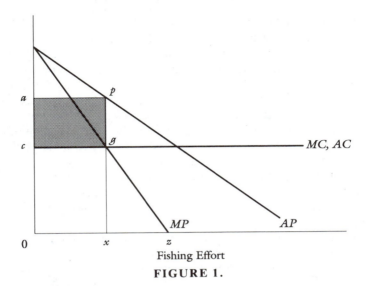

FIGURE 1.

the law of diminishing returns in this pure sense is inoperative in the fishing industry. (The reasons will be advanced at a later point in this paper.) We shall assume that, as fishing effort expands, the catch of fish increases at a diminishing rate but that it does so because of the effect of catch upon the fish population.[9] So far as the argument of the next few pages is concerned, all that is formally necessary is to assume that, as fishing intensity increases, catch will grow at a diminishing rate. Whether this reflects the pure law of diminishing returns or the reduction of population by fishing, or both, is of no particular importance.

Our analysis can be simplified if we retain the ordinary production function instead of converting it to cost curves, as is usually done in the theory of the firm. Let us further assume that the functional relationship between average production (production-per-unit-of-fishing-effort) and the quantity of fishing effort is uniformly linear. This does not distort the results unduly, and it permits the analysis to be presented more simply and in graphic terms that are already quite familiar.

In Figure 1 the optimum intensity of utilization of a particular fishing ground is shown. The curves AP and MP represent, respectively, the average productivity and marginal productivity of fishing effort. The relationship between them is the same as that between average revenue and

---

[9]Throughout this paper the conception of fish population that is employed is one of *weight* rather than *numbers*. A good deal of the biological theory has been an effort to combine growth factors and numbers factors into weight sums. The following analysis will neglect the fact that, for some species, fish of different sizes bring different unit prices.

marginal revenue in imperfect competition theory, and MP bisects any horizontal between the ordinate and AP. Since the costs of fishing supplies, etc., are assumed to be unaffected by the amount of fishing effort, marginal cost and average cost are identical and constant, as shown by the curve MC, AC.[10] These costs are assumed to include an opportunity income for the fishermen, the income that could be earned in other comparable employments. Then Ox is the optimum intensity of effort on this fishing ground, and the resource will, at this level of exploitation, provide the maximum net economic yield indicated by the shaded area $apgc$. The maximum sustained physical yield that the biologists speak of will be attained when marginal productivity of fishing effort is zero, at Oz of fishing intensity in the chart shown. Thus, as one might expect, the optimum economic fishing intensity is less than that which would produce the maximum sustained physical yield.

The area $apgc$ in Figure 1 can be regarded as the rent yielded by the fishery resource. Under the given conditions, Ox is the best rate of exploitation for the fishing ground in question, and the rent reflects the productivity of that ground, not any artificial market limitation. The rent here corresponds to the extra productivity yielded in agriculture by soils of better quality or location than those on the margin of cultivation, which may produce an opportunity income but no more. In short, Figure 1 shows the determination of the intensive margin of utilization on an intramarginal fishing ground.

We now come to the point that is of greatest theoretical importance in understanding the primary production phase of the fishing industry and in distinguishing it from agriculture. In the sea fisheries the natural resource is not private property; hence the rent it may yield is not capable of being appropriated by anyone. The individual fisherman has no legal title to a section of ocean bottom. Each fisherman is more or less free to fish wherever he pleases. The result is a pattern of competition among fishermen which culminates in the dissipation of the rent of the intramarginal grounds. This can be most clearly seen through an analysis of the relationship between the intensive margin and the extensive margin of resource exploitation in fisheries.

In Figure 2, two fishing grounds of different fertility (or location) are shown. Any given amount of fishing effort devoted to ground 2 will yield a smaller total (and therefore average) product than if devoted to 1. The maximization problem is now a question of the allocation of fishing effort between grounds 1 and 2. The optimum is, of course, where the marginal

[10]Throughout this analysis, fixed costs are neglected. The general conclusions reached would not be appreciably altered, I think, by their inclusion, though the presentation would be greatly complicated. Moreover, in the fishing industry the most substantial portion of fixed cost—wharves, harbors, etc.—is borne by government and does not enter into the cost calculations of the operators.

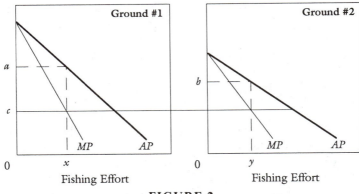

FIGURE 2.

productivities are equal on both grounds. In Figure 2, fishing effort of Ox on 1 and Oy on 2 would maximize the total net yield of $Ox + Oy$ effort if marginal cost were equal to Oc. But if under such circumstances the individual fishermen are free to fish on whichever ground they please, it is clear that this is not an equilibrium allocation of fishing effort in the sense of connoting stability. A fisherman starting from port and deciding whether to go to ground 1 or 2 does not care for *marginal* productivity but for *average* productivity, for it is the latter that indicates where the greater total yield may be obtained. If fishing effort were allocated in the optimum fashion, as shown in Figure 2, with Ox on 1, and Oy on 2, this would be a disequilibrium situation. Each fisherman could expect to get an average catch of Oa on 1 but only Ob on 2. Therefore, fishermen would shift from 2 to 1. Stable equilibrium would not be reached until the average productivity of both grounds was equal. If we now imagine a continuous gradation of fishing grounds, the extensive margin would be on that ground which yielded nothing more than outlaid costs plus opportunity income— in short, the one on which average productivity and average cost were equal. But, since average cost is the same for all grounds and the average productivity of all grounds is also brought to equality by the free and competitive nature of fishing, this means that the intramarginal grounds also yield no rent. It is entirely possible that some grounds would be exploited at a level of *negative* marginal productivity. What happens is that the rent which the intramarginal grounds are capable of yielding is dissipated through misallocation of fishing effort.

This is why fishermen are not wealthy, despite the fact that the fishery resources of the sea are the richest and most indestructible available to man. By and large, the only fisherman who becomes rich is one who makes a lucky catch or one who participates in a fishery that is put under a form of social control that turns the open resource into property rights.

Up to this point, the remuneration of fisherman has been accounted for as an opportunity-cost income comparable to earnings attainable in other industries. In point of fact, fishermen typically earn less than most others, even in much less hazardous occupations or in those requiring less skill. There is no effective reason why the competition among fishermen described above must stop at the point where opportunity incomes are yielded. It may be and is in many cases carried much further. Two factors prevent an equilibration of fishermen's incomes with those of other members of society. The first is the great immobility of fishermen. Living often in isolated communities, with little knowledge of conditions or opportunities elsewhere; educationally and often romantically tied to the sea; and lacking the savings necessary to provide a "stake," the fisherman is one of the least mobile of occupational groups. But, second, there is in the spirit of every fisherman the hope of the "lucky catch." As those who know fishermen well have often testified, they are gamblers and incurably optimistic. As a consequence, they will work for less than the going wage. [11]

The theory advanced above is substantiated by important developments in the fishing industry. For example, practically all control measures have, in the past, been designed by biologists, with sole attention paid to the production side of the problem and none to the cost side. The result has been a wide-open door for the frustration of the purposes of such measures. The Pacific halibut fishery, for example, is often hailed as a great achievement in modern fisheries management. Under international agreement between the United States and Canada, a fixed-catch limit was established during the early thirties. Since then, catch-per-unit-effort indexes, as usually interpreted, show a significant rise in the fish population. W. F. Thompson, the pioneer of the Pacific halibut management program, noted recently that "it has often been said that the halibut regulation presents the only definite case of sustained improvement of an overfished deep-sea fishery. This, I believe, is true and the fact should lend special importance to the principles which have been deliberately used to obtain this improvement." [12] Actually, careful study of the statistics indicates that the estimated recovery of halibut stocks could not have been due principally to the control measures, for the average catch was, in fact, greater during the recovery years than during the years of decline. The total amount of fish taken was only a small fraction of the estimated population reduction for the years prior to regulation. [13] Natural factors seem to be mainly responsible for the observed change in population, and the institution of control regulations almost a coincidence. Such coincidences are not

[11]"The gambling instinct of the men makes many of them work for less remuneration than they would accept as a weekly wage, because there is always the possibility of a good catch and a financial windfall" (Graham, *op. cit.,* p. 86).
[12]W. F. Thompson, "Condition of Stocks of Halibut in the Pacific," *Journal du conseil permanent international pour l'exploration de la mer,* Vol. XVIII, No. 2 (August, 1952).
[13]See M. D. Burkenroad, "Fluctuations in Abundance of Pacific Halibut," *Bulletin of the Bingham Oceanographic Collection,* May, 1948.

uncommon in the history of fisheries policy, but they may be easily explained. If a long-term cyclical fluctuation is taking place in a commercially valuable species, controls will likely be instituted when fishing yields have fallen very low and the clamor of fishermen is great; but it is then, of course, that stocks are about due to recover in any case. The "success" of conservation measures may be due fully as much to the sociological foundations of public policy as to the policy's effect on the fish. Indeed, Burkenroad argues that biological statistics in general may be called into question on these grounds. Governments sponsor biological research when the catches are disappointing. If there are long-term cyclical fluctuations in fish populations, as some think, it is hardly to be wondered why biologists frequently discover that the sea is being depleted, only to change their collective opinion a decade or so later.

Quite aside from the *biological* argument on the Pacific halibut case, there is no clear-cut evidence that halibut fishermen were made relatively more prosperous by the control measures. Whether or not the recovery of the halibut stocks was due to natural factors or to the catch limit, the potential net yield this could have meant has been dissipated through a rise in fishing costs. Since the method of control was to halt fishing when the limit had been reached, this created a great incentive on the part of each fisherman to get the fish before his competitors. During the last twenty years, fishermen have invested in more, larger, and faster boats in a competitive race for fish. In 1933 the fishing season was more than six months long. In 1952 it took just twenty-six days to catch the legal limit in the area from Willapa Harbor to Cape Spencer, and sixty days in the Alaska region. What has been happening is a rise in the average cost of fishing effort, allowing no gap between average production and average cost to appear, and hence no rent. [14]

Essentially the same phenomenon is observable in the Canadian Atlan-

---

[14]The economic significance of the reduction in season length which followed upon the catch limitation imposed in the Pacific halibut fishery has not been fully appreciated. E.g., Michael Graham said in summary of the program in 1943: "The result has been that it now takes only five months to catch the quantity of halibut that formerly needed nine. This, *of course,* has meant profit, where there was none before" (*op. cit.,* p. 156; my italics). Yet, even when biologists have grasped the economic import of the halibut program and its results, they appear reluctant to declare against it. E.g., W. E. Ricker: "This method of regulation does not necessarily make for more profitable fishing and certainly puts no effective brake on waste of effort, since an unlimited number of boats is free to join the fleet and compete during the short period that fishing is open. However, the stock is protected, and yield approximates to a maximum if quotas are wisely set; as biologists, perhaps we are not required to think any further. Some claim that any mixing into the economics of the matter might prejudice the desirable biological consequences of regulation by quotas" ("Production and Utilization of Fish Population," in a Symposium on Dynamics of Production in Aquatic Populations, Ecological Society of America, *Ecological Monographs,* XVI [October, 1946], 385). What such "desirable biological consequences" might be, is hard to conceive. Since the regulatory policies are made by man, surely it is necessary they be evaluated in terms of human, not piscatorial, objectives.

tic Coast lobster-conservation program. The method of control here is by seasonal closure. The result has been a steady growth in the number of lobster traps set by each fisherman. Virtually all available lobsters are now caught each year within the season, but at much greater cost in gear and supplies. At a fairly conservative estimate, the same quantity of lobsters could be caught with half the present number of traps. In a few places the fishermen have banded together into a local monopoly, preventing entry and controlling their own operations. By this means, the amount of fishing gear has been greatly reduced and incomes considerably improved.

That the plight of fishermen and the inefficiency of fisheries production stems from the common-property nature of the resources of the sea is further corroborated by the fact that one finds similar patterns of exploitation and similar problems in other cases of open resources. Perhaps the most obvious is hunting and trapping. Unlike fishes, the biotic potential of land animals is low enough for the species to be destroyed. Uncontrolled hunting means that animals will be killed for any short-range human reason, great or small: for food or simply for fun. Thus the buffalo of the western plains was destroyed to satisfy the most trivial desires of the white man, against which the long-term food needs of the aboriginal population counted as nothing. Even in the most civilized communities, conservation authorities have discovered that a bag-limit *per man* is necessary if complete destruction is to be avoided.

The results of anthropological investigation of modes of land tenure among primitive peoples render some further support to this thesis. In accordance with an evolutionary concept of cultural comparison, the older anthropological study was prone to regard resource tenure in common, with unrestricted exploitation, as a "lower" stage of development comparative with private and group property rights. However, more complete annals of primitive cultures reveal common tenure to be quite rare, even in hunting and gathering societies. Property rights in some form predominate by far, and, most important, their existence may be easily explained in terms of the necessity for orderly exploitation and conservation of the resource. Environmental conditions make necessary some vehicle which will prevent the resources of the community at large from being destroyed by excessive exploitation. Private or group land tenure accomplishes this end in an easily understandable fashion.[15] Significantly, land tenure is found to be "common" only in those cases where the hunting resource is migratory over such large areas that it cannot be regarded as husbandable by the society. In cases of group tenure where the numbers of the group are large, there is still the necessity of co-ordinating the practices of exploitation, in agricultural, as well as in hunting or gathering,

[15]See Frank G. Speck, "Land Ownership among Hunting Peoples in Primitive America and the World's Marginal Areas," *Proceedings of the 22nd International Congress of Americanists* (Rome, 1926), II, 323–32.

economies. Thus, for example, Malinowski reported that among the Trobriand Islanders one of the fundamental principles of land tenure is the co-ordination of the productive activities of the gardeners by the person possessing magical leadership in the group.[16] Speaking generally, we may say that stable primitive cultures appear to have discovered the dangers of common-property tenure and to have developed measures to protect their resources. Or, if a more Darwinian explanation be preferred, we may say that only those primitive cultures have survived which succeeded in developing such institutions.

Another case, from a very different industry, is that of petroleum production. Although the individual petroleum producer may acquire undisputed lease or ownership of the particular plot of land upon which his well is drilled, he shares, in most cases, a common pool of oil with other drillers. There is, consequently, set up the same kind of competitive race as is found in the fishing industry, with attending overexpansion of productive facilities and gross wastage of the resource. In the United States, efforts to regulate a chaotic situation in oil production began as early as 1915. Production practices, number of wells, and even output quotas were set by governmental authority; but it was not until the federal "Hot Oil" Act of 1935 and the development of interstate agreements that the final loophole (bootlegging) was closed through regulation of interstate commerce in oil.

Perhaps the most interesting similar case is the use of common pasture in the medieval manorial economy. Where the ownership of animals was private but the resource on which they fed was common (and limited), it was necessary to regulate the use of common pasture in order to prevent each man from competing and conflicting with his neighbors in an effort to utilize more of the pasture for his own animals. Thus the manor developed its elaborate rules regulating the use of the common pasture, or "stinting" the common: limitations on the number of animals, hours of pasturing, etc., designed to prevent the abuses of excessive individualistic competition.[17]

There appears, then, to be some truth in the conservative dictum that everybody's property is nobody's property. Wealth that is free for all is valued by none because he who is foolhardy enough to wait for its proper time of use will only find that it has been taken by another. The blade of grass that the manorial cowherd leaves behind is valueless to him, for tomorrow it may be eaten by another's animal; the oil left under the earth

[16]B. Malinowski, *Coral Gardens and Their Magic*, Vol. I, chaps. xi and xii. Malinowski sees this as further evidence of the importance of magic in the culture rather than as a means of co-ordinating productive activity; but his discussion of the practice makes it clear that the latter is, to use Malinowski's own concept, the "function" of the institution of magical leadership, at least in this connection.

[17]See P. Vinogradoff, *The Growth of the Manor* [London, 1905], chap. iv; E. Lipson, *The Economic History of England* [London, 1949], I, 72.

is valueless to the driller, for another may legally take it; the fish in the sea are valueless to the fisherman, because there is no assurance that they will be there for him tomorrow if they are left behind today. A factor of production that is valued at nothing in the business calculations of its users will yield nothing in income. Common-property natural resources are free goods for the individual and scarce goods for society. Under unregulated private exploitation, they can yield no rent; that can be accomplished only by methods which make them private property or public (government) property, in either case subject to a unified directing power.

7

The Problem of Social Cost[1]

RONALD COASE

.

Ronald Coase is Professor Emeritus at University of
Chicago Law School and a Nobel Laureate in Economics.

I. THE PROBLEM TO BE EXAMINED

This paper is concerned with those actions of business firms which have
harmful effects on others. The standard example is that of a factory the
smoke from which has harmful effects on those occupying neighbouring
properties. The economic analysis of such a situation has usually proceeded in terms of a divergence between the private and social product of
the factory, in which economists have largely followed the treatment of
Pigou in *The Economics of Welfare.* The conclusion to which this kind
of analysis seems to have led most economists is that it would be desirable
to make the owner of the factory liable for the damage caused to those
injured by the smoke, or alternatively, to place a tax on the factory owner
varying with the amount of smoke produced and equivalent in money
terms to the damage it would cause, or finally, to exclude the factory from
residential districts (and presumably from other areas in which the emis-

"The Problem of Social Cost," by Ronald Coase, from *The Journal of
Law and Economics* (October 1960). (Several passages devoted to
extended discussions of legal decisions have been omitted.) Reprinted
by permission.

[1]This article, although concerned with a technical problem of economic analysis, arose out
of the study of the Political Economy of Broadcasting which I am now conducting. The
argument of the present article was implicit in a previous article dealing with the problem of
allocating radio and television frequencies ("The Federal Communications Commission," 2
J. Law & Econ. [1959]) but comments which I have received seemed to suggest that it would
be desirable to deal with the question in a more explicit way and without reference to the
original problem for the solution of which the analysis was developed.

sion of smoke would have harmful effects on others). It is my contention that the suggested courses of action are inappropriate, in that they lead to results which are not necessarily, or even usually, desirable.

II. THE RECIPROCAL NATURE OF THE PROBLEM

The traditional approach has tended to obscure the nature of the choice that has to be made. The question is commonly thought of as one in which A inflicts harm on B and what has to be decided is: how should we restrain A? But this is wrong. We are dealing with a problem of a reciprocal nature. To avoid the harm to B would inflict harm on A. The real question that has to be decided is: should A be allowed to harm B or should B be allowed to harm A? The problem is to avoid the more serious harm. I instanced in my previous article[2] the case of a confectioner the noise and vibrations from whose machinery disturbed a doctor in his work. To avoid harming the doctor would inflict harm on the confectioner. The problem posed by this case was essentially whether it was worth while, as a result of restricting the methods of production which could be used by the confectioner, to secure more doctoring at the cost of a reduced supply of confectionery products. Another example is afforded by the problem of straying cattle which destroy crops on neighbouring land. If it is inevitable that some cattle will stray, an increase in the supply of meat can only be obtained at the expense of a decrease in the supply of crops. The nature of the choice is clear: meat or crops. What answer should be given is, of course, not clear unless we know the value of what is obtained as well as the value of what is sacrificed to obtain it. To give another example, Professor George J. Stigler instances the contamination of a stream.[3] If we assume that the harmful effect of the pollution is that it kills the fish, the question to be decided is: is the value of the fish lost greater or less than the value of the product which the contamination of the stream makes possible? It goes almost without saying that this problem has to be looked at in total *and* at the margin.

---

[2]Coase, "The Federal Communications Commission," 2 *J. Law & Econ.* 26–27 (1959).
[3]G. J. Stigler, *The Theory of Price,* 105 (1952).

III. THE PRICING SYSTEM WITH LIABILITY FOR DAMAGE

I propose to start my analysis by examining a case in which most economists would presumably agree that the problem would be solved in a completely satisfactory manner: when the damaging business has to pay for all damage caused *and* the pricing system works smoothly (strictly this means that the operation of a pricing system is without cost).

A good example of the problem under discussion is afforded by the case of straying cattle which destroy crops growing on neighbouring land. Let us suppose that a farmer and cattle-raiser are operating on neighbouring properties. Let us further suppose that, without any fencing between the properties, an increase in the size of the cattle-raiser's herd increases the total damage to the farmer's crops. What happens to the marginal damage as the size of the herd increases is another matter. This depends on whether the cattle tend to follow one another or to roam side by side, on whether they tend to be more or less restless as the size of the herd increases and on other similar factors. For my immediate purpose, it is immaterial what assumption is made about marginal damage as the size of the herd increases.

To simplify the argument, I propose to use an arithmetical example. I shall assume that the annual cost of fencing the farmer's property is $9 and the price of the crop is $1 per ton. Also, I assume that the relation between the number of cattle in the herd and the annual crop loss is as follows:

NUMBER IN HERD (STEERS)	ANNUAL CROP LOSS (TONS)	CROP LOSS PER ADDITIONAL STEER (TONS)
1	1	1
2	3	2
3	6	3
4	10	4

Given that the cattle-raiser is liable for the damage caused, the additional annual cost imposed on the cattle-raiser if he increased his herd from, say, 2 to 3 steers is $3 and in deciding on the size of the herd, he will take this into account along with his other costs. That is, he will not increase the size of the herd unless the value of the additional meat produced (assuming that the cattle-raiser slaughters the cattle) is greater than the additional costs that this will entail, including the value of the additional crops destroyed. Of course, if, by the employment of dogs, herdsmen, aeroplanes, mobile radio and other means, the amount of damage can be reduced, these means will be adopted when their cost is less than the value of the crop which they prevent being lost. Given that the annual cost of fencing is $9, the cattle-raiser who wished to have a herd with 4 steers or more would pay for fencing to be erected and maintained, assuming that other means of attaining the same end would not do so more cheaply. When the fence is erected, the marginal cost due to the liability for damage becomes zero, except to the extent that an increase in the size of the herd necessitates a stronger and therefore more expensive fence because more steers are liable to lean against it at the same time. But, of course, it may be cheaper for the cattle-raiser not to fence and to pay for the damaged crops, as in my arithmetical example, with 3 or fewer steers.

It might be thought that the fact that the cattle-raiser would pay for all crops damaged would lead the farmer to increase his planting if a cattle-raiser came to occupy the neighbouring property. But this is not so. If the crop was previously sold in conditions of perfect competition, marginal cost was equal to price for the amount of planting undertaken and any expansion would have reduced the profits of the farmer. In the new situation, the existence of crop damage would mean that the farmer would sell less on the open market but his receipts for a given production would remain the same, since the cattle-raiser would pay the market price for any crop damaged. Of course, if cattle-raising commonly involved the destruction of crops, the coming into existence of a cattle-raising industry might raise the price of the crops involved and farmers would then extend their planting. But I wish to confine my attention to the individual farmer.

I have said that the occupation of a neighbouring property by a cattle-raiser would not cause the amount of production, or perhaps more exactly the amount of planting, by the farmer to increase. In fact, if the cattle-raising has any effect, it will be to decrease the amount of planting. The reason for this is that, for any given tract of land, if the value of the crop damaged is so great that the receipts from the sale of the undamaged crop are less than the total costs of cultivating that tract of land, it will be profitable for the farmer and the cattle-raiser to make a bargain whereby that tract of land is left uncultivated. This can be made clear by means of an arithmetical example. Assume initially that the value of the crop obtained from cultivating a given tract of land is $12 and that the cost incurred in cultivating this tract of land is $10, the net gain from cultivating the land being $2. I assume for purposes of simplicity that the farmer owns the land. Now assume that the cattle-raiser starts operations on the neighbouring property and that the value of the crops damaged is $1. In this case $11 is obtained by the farmer from sale on the market and $1 is obtained from the cattle-raiser for damage suffered and the net gain remains $2. Now suppose that the cattle-raiser finds it profitable to increase the size of his herd, even though the amount of damage rises to $3; which means that the value of the additional meat production is greater than the additional costs, including the additional $2 payment for damage. But the total payment for damage is now $3. The net gain to the farmer from cultivating the land is still $2. The cattle-raiser would be better off if the farmer would agree not to cultivate his land for any payment less than $3. The farmer would be agreeable to not cultivating the land for any payment greater than $2. There is clearly room for a mutually satisfactory bargain which would lead to the abandonment of cultivation.[4] But the same argument applies not only to the whole tract cultivated by the farmer but also to any subdivision of it. Suppose, for example, that the cattle have a well-defined route, say, to a brook or to a shady area. In these circumstances, the amount of damage to the crop along the route may well be great and if so, it could be that the farmer and the cattle-raiser would find it profitable to make a bargain whereby the farmer would agree not to cultivate this strip of land.

But this raises a further possibility. Suppose that there is such a well-defined route. Suppose further that the value of the crop that would be obtained by cultivating this strip of land is $10 but that the cost of cultivation is $11. In the absence of the cattle-raiser, the land would not be cultivated. However, given the presence of the cattle-raiser, it could well be that if the strip was cultivated, the whole crop would be destroyed by the cattle. In which case, the cattle-raiser would be forced to pay $10 to the farmer. It is true that the farmer would lose $1. But the cattle-raiser would lose $10. Clearly this is a situation which is not likely to last indefinitely since neither party would want this to happen. The aim of the farmer would be to induce the cattle-raiser to make a payment in return for an agreement to leave this land uncultivated. The farmer would not be able to obtain a payment greater than the cost of fencing off this piece of land nor so high as to lead the cattle-raiser to abandon the use of the neighbouring property. What payment would in fact be made would depend on the shrewdness of the farmer and the cattle-raiser as bargainers. But as the payment would not be so high as to cause the cattle-raiser to abandon this location and as it would not vary with the size of the herd, such an agreement would not affect the allocation of resources but would merely alter the distribution of income and wealth as between the cattle-raiser and the farmer.

I think it is clear that if the cattle-raiser is liable for damage caused and

---

[4]The argument in the text has proceeded on the assumption that the alternative to cultivation of the crop is abandonment of cultivation altogether. But this need not be so. There may be crops which are less liable to damage by cattle but which would not be as profitable as the crop grown in the absence of damage. Thus, if the cultivation of a new crop would yield a return to the farmer of $1 instead of $2, and the size of the herd which would cause $3 damage with the old crop would cause $1 damage with the new crop, it would be profitable to the cattle-raiser to pay any sum less than $2 to induce the farmer to change his crop (since this would reduce damage liability from $3 to $1) and it would be profitable for the farmer to do so if the amount received was more than $1 (the reduction in his return caused by switching crops). In fact, there would be room for a mutually satisfactory bargain in all cases in which change of crop would reduce the amount of damage by more than it reduces the value of the crop (excluding damage)—in all cases, that is, in which a change in the crop cultivated would lead to an increase in the value of production.

the pricing system works smoothly, the reduction in the value of production elsewhere will be taken into account in computing the additional cost involved in increasing the size of the herd. This cost will be weighed against the value of the additional meat production and, given perfect competition in the cattle industry, the allocation of resources in cattle-raising will be optimal. What needs to be emphasized is that the fall in the value of production elsewhere which would be taken into account in the costs of the cattle-raiser may well be less than the damage which the cattle would cause to the crops in the ordinary course of events. This is because it is possible, as a result of market transactions, to discontinue cultivation of the land. This is desirable in all cases in which the damage that the cattle would cause, and for which the cattle-raiser would be willing to pay, exceeds the amount which the farmer would pay for use of the land. In conditions of perfect competition, the amount which the farmer would pay for the use of the land is equal to the difference between the value of the total production when the factors are employed on this land and the value of the additional product yielded in their next best use (which would be what the farmer would have to pay for the factors). If damage exceeds the amount the farmer would pay for the use of the land, the value of the additional product of the factors employed elsewhere would exceed the value of the total product in this use after damage is taken into account. It follows that it would be desirable to abandon cultivation of the land and to release the factors employed for production elsewhere. A procedure which merely provided for payment for damage to the crop caused by the cattle but which did not allow for the possibility of cultivation being discontinued would result in too small an employment of factors of production in cattle-raising and too large an employment of factors in cultivation of the crop. But given the possibility of market transactions, a situation in which damage to crops exceeded the rent of the land would not endure. Whether the cattle-raiser pays the farmer to leave the land uncultivated or himself rents the land by paying the land-owner an amount slightly greater than the farmer would pay (if the farmer was himself renting the land), the final result would be the same and would maximise the value of production. Even when the farmer is induced to plant crops which it would not be profitable to cultivate for sale on the market, this will be a purely short-term phenomenon and may be expected to lead to an agreement under which the planting will cease. The cattle-raiser will remain in that location and the marginal cost of meat production will be the same as before, thus having no long-run effect on the allocation of resources.

IV. THE PRICING SYSTEM WITH
NO LIABILITY FOR DAMAGE

I now turn to the case in which, although the pricing system is assumed to work smoothly (that is, costlessly), the damaging business is not liable for any of the damage which it causes. This business does not have to make a payment to those damaged by its actions. I propose to show that the allocation of resources will be the same in this case as it was when the damaging business was liable for damage caused. As I showed in the previous case that the allocation of resources was optimal, it will not be necessary to repeat this part of the argument.

I return to the case of the farmer and the cattle-raiser. The farmer would suffer increased damage to his crop as the size of the herd increased. Suppose that the size of the cattle-raiser's herd is 3 steers (and that this is the size of the herd that would be maintained if crop damage was not taken into account). Then the farmer would be willing to pay up to $3 if the cattle-raiser would reduce his herd to 2 steers, up to $5 if the herd were reduced to 1 steer and would pay up to $6 if cattle-raising was abandoned. The cattle-raiser would therefore receive $3 from the farmer if he kept 2 steers instead of 3. This $3 foregone is therefore part of the cost incurred in keeping the third steer. Whether the $3 is a payment which the cattle-raiser has to make if he adds the third steer to his herd (which it would be if the cattle-raiser was liable to the farmer for damage caused to the crop) or whether it is a sum of money which he would have received if he did not keep a third steer (which it would be if the cattle-raiser was not liable to the farmer for damage caused to the crop) does not affect the final result. In both cases $3 is part of the cost of adding a third steer, to be included along with the other costs. If the increase in the value of production in cattle-raising through increasing the size of the herd from 2 to 3 is greater than the additional costs that have to be incurred (including the $3 damage to crops), the size of the herd will be increased. Otherwise, it will not. The size of the herd will be the same whether the cattle-raiser is liable for damage caused to the crop or not.

It may be argued that the assumed starting point—a herd of 3 steers—was arbitrary. And this is true. But the farmer would not wish to pay to avoid crop damage which the cattle-raiser would not be able to cause. For example, the maximum annual payment which the farmer could be induced to pay could not exceed $9, the annual cost of fencing. And the farmer would only be willing to pay this sum if it did not reduce his earnings to a level that would cause him to abandon cultivation of this particular tract of land. Furthermore, the farmer would only be willing to pay this amount if he believed that, in the absence of any payment by him, the size of the herd maintained by the cattle-raiser would be 4 or more steers. Let us assume that this is the case. Then the farmer would be willing to pay up to $3 if the cattle-raiser would reduce his herd to 3 steers, up to $6 if the herd were reduced to 2 steers, up to $8 if one steer only were kept and up to $9 if cattle-raising were abandoned. It will be noticed that the change in the starting point has not altered the amount which would accrue to the cattle-raiser if he reduced the size of his herd by any given amount. It is still true that the cattle-raiser could receive an additional $3 from the farmer if he agreed to reduce his herd from 3 steers to 2 and that the $3 represents the value of the crop that would be destroyed by adding the third steer to the herd. Although a different belief on the part of the farmer (whether justified or not) about the size of the herd that the cattle-raiser would maintain in the absence of payments from him may affect the total payment he can be induced to pay, it is not true that this different belief would have any effect on the size of the herd that the cattle-raiser will actually keep. This will be the same as it would be if the cattle-raiser had to pay for damage caused by his cattle, since a receipt foregone of a given amount is the equivalent of a payment of the same amount.

It might be thought that it would pay the cattle-raiser to increase his herd above the size that he would wish to maintain once a bargain had been made, in order to induce the farmer to make a larger total payment. And this may be true. It is similar in nature to the action of the farmer (when the cattle-raiser was liable for damage) in cultivating land on which, as a result of an agreement with the cattle-raiser, planting would subsequently be abandoned (including land which would not be cultivated at all in the absence of cattle-raising). But such manoeuvres are preliminaries to an agreement and do not affect the long-run equilibrium position, which is the same whether or not the cattle-raiser is held responsible for the crop damage brought about by his cattle.

It is necessary to know whether the damaging business is liable or not for damage caused since without the establishment of this initial delimitation of rights there can be no market transactions to transfer and recombine them. But the ultimate result (which maximises the value of production) is independent of the legal position if the pricing system is assumed to work without cost.

V. THE PROBLEM ILLUSTRATED ANEW

The harmful effects of the activities of a business can assume a wide variety of forms. An early English case concerned a building which, by obstructing currents of air, hindered the operation of a windmill.[5] A recent case in Florida concerned a building which cast a shadow on the cabana,

[5]See Gale on *Easements* 237–39 (13th ed. M. Bowles 1959).

swimming pool and sunbathing areas of a neighbouring hotel.[6] The problem of straying cattle and the damaging of crops which was the subject of detailed examination in the two preceding sections, although it may have appeared to be rather a special case, is in fact but one example of a problem which arises in many different guises. To clarify the nature of my argument and to demonstrate its general applicability, I propose to illustrate it anew by reference to four actual cases.

Let us first reconsider the case of *Sturges v. Bridgman*[7] which I used as an illustration of the general problem in my article on "The Federal Communications Commission." In this case, a confectioner (in Wigmore Street) used two mortars and pestles in connection with his business (one had been in operation in the same position for more than 60 years and the other for more than 26 years). A doctor then came to occupy neighbouring premises (in Wimpole Street). The confectioner's machinery caused the doctor no harm until, eight years after he had first occupied the premises, he built a consulting room at the end of his garden right against the confectioner's kitchen. It was then found that the noise and vibration caused by the confectioner's machinery made it difficult for the doctor to use his new consulting room. "In particular . . . the noise prevented him from examining his patients by auscultation[8] for diseases of the chest. He also found it impossible to engage with effect in any occupation which required thought and attention." The doctor therefore brought a legal action to force the confectioner to stop using his machinery. The courts had little difficulty in granting the doctor the injunction he sought. "Individual cases of hardship may occur in the strict carrying out of the principle upon which we found our judgment, but the negation of the principle would lead even more to individual hardship, and would at the same time produce a prejudicial effect upon the development of land for residential purposes."

The court's decision established that the doctor had the right to prevent the confectioner from using his machinery. But, of course, it would have been possible to modify the arrangements envisaged in the legal ruling by means of a bargain between the parties. The doctor would have been willing to waive his right and allow the machinery to continue in operation if the confectioner would have paid him a sum of money which was greater than the loss of income which he would suffer from having to move to a more costly or less convenient location or from having to curtail his activities at this location or, as was suggested as a possibility, from having to build a separate wall which would deaden the noise and vibration. The confectioner would have been willing to do this if the amount he would

---

[6]See *Fontainebleu Hotel Corp. v. Forty-Five Twenty-Five, Inc.*, 114 So. 2d 357 (1959).
[7]11 Ch. D. 852 (1879).
[8]Auscultation is the act of listening by ear or stethoscope in order to judge by sound the condition of the body.

have to pay the doctor was less than the fall in income he would suffer if he had to change his mode of operation at this location, abandon his operation or move his confectionery business to some other location. The solution of the problem depends essentially on whether the continued use of the machinery adds more to the confectioner's income than it subtracts from the doctor's.[9] But now consider the situation if the confectioner had won the case. The confectioner would then have had the right to continue operating his noise and vibration-generating machinery without having to pay anything to the doctor. The boot would have been on the other foot: the doctor would have had to pay the confectioner to induce him to stop using the machinery. If the doctor's income would have fallen more through continuance of the use of this machinery than it added to the income of the confectioner, there would clearly be room for a bargain whereby the doctor paid the confectioner to stop using the machinery. That is to say, the circumstances in which it would not pay the confectioner to continue to use the machinery and to compensate the doctor for the losses that this would bring (if the doctor had the right to prevent the confectioner's using his machinery) would be those in which it would be in the interest of the doctor to make a payment to the confectioner which would induce him to discontinue the use of the machinery (if the confectioner had the right to operate the machinery). The basic conditions are exactly the same in this case as they were in the example of the cattle which destroyed crops. With costless market transactions, the decision of the courts concerning liability for damage would be without effect on the allocation of resources. It was of course the view of the judges that they were affecting the working of the economic system—and in a desirable direction. Any other decision would have had "a prejudicial effect upon the development of land for residential purposes," an argument which was elaborated by examining the example of a forge operating on a barren moor, which was later developed for residential purposes. The judges' view that they were settling how the land was to be used would be true only in the case in which the costs of carrying out the necessary market transactions exceeded the gain which might be achieved by any rearrangement of rights. And it would be desirable to preserve the areas (Wimpole Street or the moor) for residential or professional use (by giving non-industrial users the right to stop the noise, vibration, smoke, etc., by injunction) only if the value of the additional residential facilities obtained was greater than the value of cakes or iron lost. But of this the judges seem to have been unaware.

*   *   *

The reasoning employed by the courts in determining legal rights will often seem strange to an economist because many of the factors on which

---

[9]Note that what is taken into account is the change in income after allowing for alterations in methods of production, location, character of product, etc.

the decision turns are, to an economist, irrelevant. Because of this, situations which are, from an economic point of view, identical will be treated quite differently by the courts. The economic problem in all cases of harmful effects is how to maximise the value of production. In the case of *Bass v. Gregory* fresh air was drawn in through the well which facilitated the production of beer but foul air was expelled through the well which made life in the adjoining houses less pleasant. The economic problem was to decide which to choose: a lower cost of beer and worsened amenities in adjoining houses or a higher cost of beer and improved amenities. In deciding this question, the "doctrine of lost grant" is about as relevant as the colour of the judge's eyes. But it has to be remembered that the immediate question faced by the courts is *not* what shall be done by whom *but* who has the legal right to do what. It is always possible to modify by transactions on the market the initial legal delimitation of rights. And, of course, if such market transactions are costless, such a rearrangement of rights will always take place if it would lead to an increase in the value of production.

VI. THE COST OF MARKET TRANSACTIONS TAKEN INTO ACCOUNT

The argument has proceeded up to this point on the assumption (explicit in Sections III and IV and tacit in Section V) that there were no costs involved in carrying out market transactions. This is, of course, a very unrealistic assumption. In order to carry out a market transaction it is necessary to discover who it is that one wishes to deal with, to inform people that one wishes to deal and on what terms, to conduct negotiations leading up to a bargain, to draw up the contract, to undertake the inspection needed to make sure that the terms of the contract are being observed and so on. These operations are often extremely costly, sufficiently costly at any rate to prevent many transactions that would be carried out in a world in which the pricing system worked without cost.

In earlier sections, when dealing with the problem of the rearrangement of legal rights through the market, it was argued that such a rearrangement would be made through the market whenever this would lead to an increase in the value of production. But this assumed costless market transactions. Once the costs of carrying out market transactions are taken into account it is clear that such a rearrangement of rights will only be undertaken when the increase in the value of production consequent upon the rearrangement is greater than the costs which would be involved in bringing it about. When it is less, the granting of an injunction (or the knowledge that it would be granted) or the liability to pay damages may result in an activity being discontinued (or may prevent its being started) which would be undertaken if market transactions were costless. In these conditions the initial delimitation of legal rights does have an effect on the efficiency with which the economic system operates. One arrangement of rights may bring about a greater value of production than any other. But unless this is the arrangement of rights established by the legal system, the costs of reaching the same result by altering and combining rights through the market may be so great that this optimal arrangement of rights, and the greater value of production which it would bring, may never be achieved. The part played by economic considerations in the process of delimiting legal rights will be discussed in the next section. In this section, I will take the initial delimitation of rights and the costs of carrying out market transactions as given.

It is clear that an alternative form of economic organisation which could achieve the same result at less cost than would be incurred by using the market would enable the value of production to be raised. As I explained many years ago, the firm represents such an alternative to organising production through market transactions. [10] Within the firm individual bargains between the various cooperating factors of production are eliminated and for a market transaction is substituted an administrative decision. The rearrangement of production then takes place without the need for bargains between the owners of the factors of production. A landowner who has control of a large tract of land may devote his land to various uses taking into account the effect that the interrelations of the various activities will have on the net return of the land, thus rendering unnecessary bargains between those undertaking the various activities. Owners of a large building or of several adjoining properties in a given area may act in much the same way. In effect, using our earlier terminology, the firm would acquire the legal rights of all the parties and the rearrangement of activities would not follow on a rearrangement of rights by contract, but as a result of an administrative decision as to how the rights should be used.

It does not, of course, follow that the administrative costs of organising a transaction through a firm are inevitably less than the costs of the market transactions which are superseded. But where contracts are peculiarly difficult to draw up and an attempt to describe what the parties have agreed to do or not to do (e.g. the amount and kind of a smell or noise that they may make or will not make) would necessitate a lengthy and highly involved document, and, where, as is probable, a long-term contract would be desirable, [11] it would be hardly surprising if the emergence of a firm or the extension of the activities of an existing firm was not the solution adopted on many occasions to deal with the problem of harmful

---

[10]See Coase, "The Nature of the Firm," 4 *Economica,* New Series, 386 (1937). Reprinted in *Readings in Price Theory,* 331 (1952).

[11]For reasons explained in my earlier article, see *Readings in Price Theory,* n. 14 at 337.

effects. This solution would be adopted whenever the administrative costs of the firm were less than the costs of the market transactions that it supersedes and the gains which would result from the rearrangement of activities greater than the firm's costs of organising them. I do not need to examine in great detail the character of this solution since I have explained what is involved in my earlier article.

But the firm is not the only possible answer to this problem. The administrative costs of organising transactions within the firm may also be high, and particularly so when many diverse activities are brought within the control of a single organisation. In the standard case of a smoke nuisance, which may affect a vast number of people engaged in a wide variety of activities, the administrative costs might well be so high as to make any attempt to deal with the problem within the confines of a single firm impossible. An alternative solution is direct government regulation. Instead of instituting a legal system of rights which can be modified by transactions on the market, the government may impose regulations which state what people must or must not do and which have to be obeyed. Thus, the government (by statute or perhaps more likely through an administrative agency) may, to deal with the problem of smoke nuisance, decree that certain methods of production should or should not be used (e.g. that smoke preventing devices should be installed or that coal or oil should not be burned) or may confine certain types of business to certain districts (zoning regulations).

The government is, in a sense, a superfirm (but of a very special kind) since it is able to influence the use of factors of production by administrative decision. But the ordinary firm is subject to checks in its operations because of the competition of other firms, which might administer the same activities at lower cost and also because there is always the alternative of market transactions as against organisation within the firm if the administrative costs become too great. The government is able, if it wishes, to avoid the market altogether, which a firm can never do. The firm has to make market agreements with the owners of the factors of production that it uses. Just as the government can conscript or seize property, so it can decree that factors of production should only be used in such-and-such a way. Such authoritarian methods save a lot of trouble (for those doing the organising). Furthermore, the government has at its disposal the police and the other law enforcement agencies to make sure that its regulations are carried out.

It is clear that the government has powers which might enable it to get some things done at a lower cost than could a private organisation (or at any rate one without special governmental powers). But the governmental administrative machine is not itself costless. It can, in fact, on occasion be extremely costly. Furthermore, there is no reason to suppose that the restrictive and zoning regulations, made by a fallible administration subject to political pressures and operating without any competitive check, will necessarily always be those which increase the efficiency with which the economic system operates. Furthermore, such general regulations which must apply to a wide variety of cases will be enforced in some cases in which they are clearly inappropriate. From these considerations it follows that direct governmental regulation will not necessarily give better results than leaving the problem to be solved by the market or the firm. But equally there is no reason why, on occasion, such governmental administrative regulation should not lead to an improvement in economic efficiency. This would seem particularly likely when, as is normally the case with the smoke nuisance, a large number of people are involved and in which therefore the costs of handling the problem through the market or the firm may be high.

There is, of course, a further alternative which is to do nothing about the problem at all. And given that the costs involved in solving the problem by regulations issued by the governmental administrative machine will often be heavy (particularly if the costs are interpreted to include all the consequences which follow from the government engaging in this kind of activity), it will no doubt be commonly the case that the gain which would come from regulating the actions which give rise to the harmful effects will be less than the costs involved in government regulation.

The discussion of the problem of harmful effects in this section (when the costs of market transactions are taken into account) is extremely inadequate. But at least it has made clear that the problem is one of choosing the appropriate social arrangement for dealing with the harmful effects. All solutions have costs and there is no reason to suppose that government regulation is called for simply because the problem is not well handled by the market or the firm. Satisfactory views on policy can only come from a patient study of how, in practice, the market, firms and governments handle the problem of harmful effects. Economists need to study the work of the broker in bringing parties together, the effectiveness of restrictive covenants, the problems of the large-scale real-estate development company, the operation of government zoning and other regulating activities. It is my belief that economists, and policy-makers generally, have tended to over-estimate the advantages which come from governmental regulation. But this belief, even if justified, does not do more than suggest that government regulation should be curtailed. It does not tell us where the boundary line should be drawn. This, it seems to me, has to come from a detailed investigation of the actual results of handling the problem in different ways. But it would be unfortunate if this investigation were undertaken with the aid of a faulty economic analysis. The aim of this article is to indicate what the economic approach to the problem should be.

VII. THE LEGAL DELIMITATION OF RIGHTS AND THE ECONOMIC PROBLEM

The discussion in Section V not only served to illustrate the argument but also afforded a glimpse at the legal approach to the problem of harmful effects. The cases considered were all English but a similar selection of American cases could easily be made and the character of the reasoning would have been the same. Of course, if market transactions were costless, all that matters (questions of equity apart) is that the rights of the various parties should be well-defined and the results of legal actions easy to forecast. But as we have seen, the situation is quite different when market transactions are so costly as to make it difficult to change the arrangement of rights established by the law. In such cases, the courts directly influence economic activity. It would therefore seem desirable that the courts should understand the economic consequences of their decisions and should, insofar as this is possible without creating too much uncertainty about the legal position itself, take these consequences into account when making their decisions. Even when it is possible to change the legal delimitation of rights through market transactions, it is obviously desirable to reduce the need for such transactions and thus reduce the employment of resources in carrying them out.

A thorough examination of the presuppositions of the courts in trying such cases would be of great interest but I have not been able to attempt it. Nevertheless it is clear from a cursory study that the courts have often recognized the economic implications of their decisions and are aware (as many economists are not) of the reciprocal nature of the problem. Furthermore, from time to time, they take these economic implications into account, along with other factors, in arriving at their decisions. The American writers on this subject refer to the question in a more explicit fashion than do the British. Thus, to quote Prosser on Torts, a person may

> make use of his own property or . . . conduct his own affairs at the expense of some harm to his neighbors. He may operate a factory whose noise and smoke casue some discomfort to others, so long as he keeps within reasonable bounds. It is only when his conduct is unreasonable, *in the light of its utility and the harm which results* [italics added], that it becomes a nuisance. . . . As it was said in an ancient case in regard to candle-making in a town, "Le utility del chose excusera le noisomeness del stink."
>
> The world must have factories, smelters, oil refineries, noisy machinery and blasting, even at the expense of some inconvenience to those in the vicinity and the plaintiff may be required to accept some not unreasonable discomfort for the general good. [12]

[12]See W. L. Prosser, *The Law of Torts* 398–99, 412 (2d ed. 1955). The quotation about the ancient case concerning candle-making is taken from Sir James Fitzjames Stephen, *A General*

The standard British writers do not state as explicitly as this that a comparison between the utility and harm produced is an element in deciding whether a harmful effect should be considered a nuisance. But similar views, if less strongly expressed, are to be found.[13] The doctrine that the harmful effect must be substantial before the court will act is, no doubt, in part a reflection of the fact that there will almost always be some gain to offset the harm. And in the reports of individual cases, it is clear that the judges have had in mind what would be lost as well as what would be gained in deciding whether to grant an injunction or award damages. Thus, in refusing to prevent the destruction of a prospect by a new building, the judge stated:

> I know no general rule of common law, which . . . says, that building so as to stop another's prospect is a nuisance. Was that the case, there could be no great towns; and I must grant injunctions to all the new buildings in this town. . . .[14]

*   *   *

. . . The problem which we face in dealing with actions which have harmful effects is not simply one of restraining those responsible for them. What has to be decided is whether the gain from preventing the harm is greater than the loss which would be suffered elsewhere as a result of stopping the action which produces the harm. In a world in which there are costs of rearranging the rights established by the legal system, the courts, in cases relating to nuisance, are, in effect, making a decision on the economic problem and determining how resources are to be employed. It was argued that the courts are conscious of this and that they often make, although not always in a very explicit fashion, a comparison between what would be gained and what lost by preventing actions which have harmful effects. But the delimitation of rights is also the result of statutory enactments. Here we also find evidence of an appreciation of the reciprocal nature of the problem. While statutory enactments add to the list of nuisances, action is also taken to legalize what would otherwise be nuisances under the common law. The kind of situation which economists are prone to

---

View of the Criminal Law of England 106 (1890). Sir James Stephen gives no reference. He perhaps had in mind *Rex. v. Ronkett,* included in Seavey, Keeton and Thurston, *Cases on Torts* 604 (1950). A similar view to that expressed by Prosser is to be found in F. V. Harper and F. James, *The Law of Torts* 67–74 (1956); *Restatement, Torts* §§826, 827 and 828.

[13]See Winfield on *Torts* 541–48 (6th ed. T. E. Lewis 1954); Salmond on the *Law of Torts* 181–90 (12th ed. R. F. V. Heuston 1957); H. Street, *The Law of Torts* 221–29 (1959).

[14]*Attorney General v. Doughty,* 2 Ves. Sen. 453, 28 Eng. Rep. 290 (Ch. 1752). Compare in this connection the statement of an American judge, quoted in Prosser, *op. cit. supra* n. 16 at 413 n. 54: "Without smoke, Pittsburgh would have remained a very pretty village," Musmanno, J., in *Versailles Borough v. McKeesport Coal & Coke Co.,* 1935, 83 Pitts. Leg. J. 379, 385.

consider as requiring corrective government action is, in fact, often the result of government action. Such action is not necessarily unwise. But there is a real danger that extensive government intervention in the economic system may lead to the protection of those responsible for harmful effects being carried too far.

VIII. PIGOU'S TREATMENT IN "THE ECONOMICS OF WELFARE"

The fountainhead for the modern economic analysis of the problem discussed in this article is Pigou's *Economics of Welfare* and, in particular, that section of Part II which deals with divergences between social and private net products which come about because

> one person A, in the course of rendering some service, for which payment is made, to a second person B, incidentally also renders services or disservices to other persons (not producers of like services), or such a sort that payment cannot be exacted from the benefited parties or compensation enforced on behalf of the injured parties.[15]

Pigou tells us that his aim in Part II of *The Economics of Welfare* is

> to ascertain how far the free play of self-interest, acting under the existing legal system, tends to distribute the country's resources in the way most favorable to the production of a large national dividend, and how far it is feasible for State action to improve upon 'natural' tendencies.[16]

To judge from the first part of this statement, Pigou's purpose is to discover whether any improvements could be made in the existing arrangements which determine the use of resources. Since Pigou's conclusion is that improvements could be made, one might have expected him to continue by saying that he proposed to set out the changes required to bring them about. Instead, Pigou adds a phrase which contrasts "natural" tendencies with State action, which seems in some sense to equate the present arrangements with "natural" tendencies and to imply that what is required to bring about these improvements is State action (if feasible). That this is more or less Pigou's position is evident from Chapter I of Part II.[17] Pigou starts by referring to "optimistic followers of the classical

[15]A. C. Pigou, *The Economics of Welfare* 183 (4th ed. 1932). My references will all be to the fourth edition but the argument and examples examined in this article remained substantially unchanged from the first edition in 1920 to the fourth in 1932. A large part (but not all) of this analysis had appeared previously in *Wealth and Welfare* (1912).

[16]*Id.* at xii.

[17]*Id.* at 127–30.

economists"[18] who have argued that the value of production would be maximised if the government refrained from any interference in the economic system and the economic arrangements were those which came about "naturally." Pigou goes on to say that if self-interest does promote economic welfare, it is because human institutions have been devised to make it so. (This part of Pigou's argument, which he develops with the aid of a quotation from Cannan, seems to me to be essentially correct.) Pigou concludes:

> But even in the most advanced States there are failures and imperfections. . . . there are many obstacles that prevent a community's resources from being distributed . . . in the most efficient way. The study of these constitutes our present problem. . . . its purpose is essentially practical. It seeks to bring into clearer light some of the ways in which it now is, or eventually may become, feasible for governments to control the play of economic forces in such wise as to promote the economic welfare, and through that, the total welfare, of their citizens as a whole.[19]

Pigou's underlying thought would appear to be: Some have argued that no State action is needed. But the system has performed as well as it has because of State action. Nonetheless, there are still imperfections. What additional State action is required?

If this is a correct summary of Pigou's position, its inadequacy can be demonstrated by examining the first example he gives of a divergence between private and social products.

> It might happen . . . that costs are thrown upon people not directly concerned, through, say, uncompensated damage done to surrounding woods by sparks from railway engines. All such effects must be included—some of them will be positive, others negative elements—in reckoning up the social net product of the marginal increment of any volume of resources turned into any use or place.[20]

The example used by Pigou refers to a real situation. In Britain, a railway does not normally have to compensate those who suffer damage by fire caused by sparks from an engine. Taken in conjunction with what he says in Chapter 9 of Part II, I take Pigou's policy recommendations to be, first, that there should be State action to correct this "natural" situation and, second, that the railways should be forced to compensate those whose

---

[18]In *Wealth and Welfare,* Pigou attributes the "optimism" to Adam Smith himself and not to his followers. He there refers to the "highly optimistic theory of Adam Smith that the national dividend, in given circumstances of demand and supply, tends 'naturally' to a maximum" (p. 104).

[19]Pigou, *op. cit. supra* n. 35 at 129–30.

[20]*Id.* at 134.

woods are burnt. If this is a correct interpretation of Pigou's position, I would argue that the first recommendation is based on a misapprehension of the facts and that the second is not necessarily desirable.

Let us consider the legal position. Under the heading "Sparks from engines," we find the following in Halsbury's *Laws of England:*

> If railway undertakers use steam engines on their railway without express statutory authority to do so, they are liable, irrespective of any negligence on their part, for fires caused by sparks from engines. Railway undertakers are, however, generally given statutory authority to use steam engines on their railway; accordingly, if an engine is constructed with the precautions which science suggests against fire and is used without negligence, they are not responsible at common law for any damage which may be done by sparks. . . . In the construction of an engine the undertaker is bound to use all the discoveries which science has put within its reach in order to avoid doing harm, provided they are such as it is reasonable to require the company to adopt, having proper regard to the likelihood of the damage and to the cost and convenience of the remedy; but it is not negligence on the part of an undertaker if it refuses to use an apparatus the efficiency of which is open to bona fide doubt.

To this general rule, there is a statutory exception arising from the Railway (Fires) Act, 1905, as amended in 1923. This concerns agricultural land or agricultural crops.

> In such a case the fact that the engine was used under statutory powers does not affect the liability of the company in an action for the damage. . . . These provisions, however, only apply where the claim for damage . . . does not exceed £200 [£100 in the 1905 Act], and where written notice of the occurrence of the fire and the intention to claim has been sent to the company within seven days of the occurrence of the damage and particulars of the damage in writing showing the amount of the claim in money not exceeding £200 have been sent to the company within twenty-one days.

Agricultural land does not include moorland or buildings and agricultural crops do not include those led away or stacked. [21] I have not made a close study of the parliamentary history of this statutory exception, but to judge from debates in the House of Commons in 1922 and 1923, this exception was probably designed to help the smallholder. [22]

Let us return to Pigou's example of uncompensated damage to surrounding woods caused by sparks from railway engines. This is presumably intended to show how it is possible "for State action to improve on 'natural' tendencies." If we treat Pigou's example as referring to the posi-

---

[21] See 31 Halsbury, *Laws of England* 474–75 (3d ed. 1960), Article on Railways and Canals, from which this summary of the legal position, and all quotations, are taken.
[22] See 152 H.C. Deb. 2622–63 (1922); 161 H.C. Deb. 2935–55 (1923).

tion before 1905, or as being an arbitrary example (in that he might just as well have written "surrounding buildings" instead of "surrounding woods"), then it is clear that the reason why compensation was not paid must have been that the railway had statutory authority to run steam engines (which relieved it of liability for fires caused by sparks). That this was the legal position was established in 1860, in a case, oddly enough, which concerned the burning of surrounding woods by a railway,[23] and the law on this point has not been changed (apart from the one exception) by a century of railway legislation, including nationalisation. If we treat Pigou's example of "uncompensated damage done to surrounding woods by sparks from railway engines" literally, and assume that it refers to the period after 1905, then it is clear that the reason why compensation was not paid must have been that the damage was more than £100 (in the first edition of *The Economics of Welfare*) or more than £200 (in later editions) or that the owner of the wood failed to notify the railway in writing within seven days of the fire or did not send particulars of the damage, in writing, within twenty-one days. In the real world, Pigou's example could only exist as a result of a deliberate choice of the legislature. It is not, of course, easy to imagine the construction of a railway in a state of nature. The nearest one can get to this is presumably a railway which uses steam engines "without express statutory authority." However, in this case the railway would be obliged to compensate those whose woods it burnt down. That is to say, compensation would be paid in the absence of Government action. The only circumstances in which compensation would not be paid would be those in which there had been Government action. It is strange that Pigou, who clearly thought it desirable that compensation should be paid, should have chosen this particular example to demonstrate how it is possible "for State action to improve on 'natural' tendencies."

Pigou seems to have had a faulty view of the facts of the situation. But it also seems likely that he was mistaken in his economic analysis. It is not necessarily desirable that the railway should be required to compensate those who suffer damage by fires caused by railway engines. I need not show here that, if the railway could make a bargain with everyone having property adjoining the railway line and there were no costs involved in making such bargains, it would not matter whether the railway was liable for damage caused by fires or not. This question has been treated at length in earlier sections. The problem is whether it would be desirable to make the railway liable in conditions in which it is too expensive for such bargains to be made. Pigou clearly thought it was desirable to force the railway to pay compensation and it is easy to see the kind of argument that would have led him to this conclusion. Suppose a railway is considering whether to run an additional train or to increase the speed of an existing

[23] *Vaughan v. Taff Railway Co.*, 3 H. and N. 743 (Ex. 1858) and 5 H. and N. 679 (Ex. 1860).

train or to install spark-preventing devices on its engines. If the railway were not liable for fire damage, then, when making these decisions, it would not take into account as a cost the increase in damage resulting from the additional train or the faster train or the failure to install spark-preventing devices. This is the source of the divergence between private and social net products. It results in the railway performing acts which will lower the value of total production—and which it would not do if it were liable for the damage. This can be shown by means of an arithmetical example.

Consider a railway, which is *not* liable for damage by fires caused by sparks from its engines, which runs two trains per day on a certain line. Suppose that running one train per day would enable the railway to perform services worth $150 per annum and running two trains a day would enable the railway to perform services worth $250 per annum. Suppose further that the cost of running one train is $50 per annum and two trains $100 per annum. Assuming perfect competition, the cost equals the fall in the value of production elsewhere due to the employment of additional factors of production by the railway. Clearly the railway would find it profitable to run two trains per day. But suppose that running one train per day would destroy by fire crops worth (on an average over the year) $60 and two trains a day would result in the destruction of crops worth $120. In these circumstances running one train per day would raise the value of total production but the running of a second train would reduce the value of total production. The second train would enable additional railway services worth $100 per annum to be performed. But the fall in the value of production elsewhere would be $110 per annum; $50 as a result of the employment of additional factors of production and $60 as a result of the destruction of crops. Since it would be better if the second train were not run and since it would not run if the railway were liable for damage caused to crops, the conclusion that the railway should be made liable for the damage seems irresistible. Undoubtedly it is this kind of reasoning which underlies the Pigovian position.

The conclusion that it would be better if the second train did not run is correct. The conclusion that it is desirable that the railway should be made liable for the damage it causes is wrong. Let us change our assumption concerning the rule of liability. Suppose that the railway is liable for damage from fires caused by sparks from the engine. A farmer on lands adjoining the railway is then in the position that, if his crop is destroyed by fires caused by the railway, he will receive the market price from the railway; but if his crop is not damaged, he will receive the market price by sale. It therefore becomes a matter of indifference to him whether his crop is damaged by fire or not. The position is very different when the railway is *not* liable. Any crop destruction through railway-caused fires would then reduce the receipts of the farmer. He would therefore take out of cultivation any land for which the damage is likely to be greater than the net return of the land (for reasons explained at length in Section III). A change from a regime in which the railway is *not* liable for damage to one in which it *is* liable is likely therefore to lead to an increase in the amount of cultivation on lands adjoining the railway. It will also, of course, lead to an increase in the amount of crop destruction due to railway-caused fires.

Let us return to our arithmetical example. Assume that, with the changed rule of liability, there is a doubling in the amount of crop destruction due to railway-caused fires. With one train per day, crops worth $120 would be destroyed each year and two trains per day would lead to the destruction of crops worth $240. We saw previously that it would not be profitable to run the second train if the railway had to pay $60 per annum as compensation for damage. With damage at $120 per annum the loss from running the second train would be $60 greater. But now let us consider the first train. The value of the transport services furnished by the first train is $150. The cost of running the train is $50. The amount that the railway would have to pay out as compensation for damage is $120. If follows that it would not be profitable to run any trains. With the figures in our example we reach the following result: if the railway is not liable for fire-damage, two trains per day would be run; if the railway is liable for fire-damage, it would cease operations altogether. Does this mean that it is better that there should be no railway? This question can be resolved by considering what would happen to the value of total production if it were decided to exempt the railway from liability for fire-damage, thus bringing it into operation (with two trains per day).

The operation of the railway would enable transport services worth $250 to be performed. It would also mean the employment of factors of production which would reduce the value of production elsewhere by $100. Furthermore it would mean the destruction of crops worth $120. The coming of the railway will also have led to the abandonment of cultivation of some land. Since we know that, had this land been cultivated, the value of the crops destroyed by fire would have been $120, and since it is unlikely that the total crop on this land would have been destroyed, it seems reasonable to suppose that the value of the crop yield on this land would have been higher than this. Assume it would have been $160. But the abandonment of cultivation would have released factors of production for employment elsewhere. All we know is that the amount by which the value of production elsewhere will increase will be less than $160. Suppose that it is $150. Then the gain from operating the railway would be $250 (the value of the transport services) minus $100 (the cost of the factors of production) minus $120 (the value of crops destroyed by fire) minus $160 (the fall in the value of crop production due to the abandonment of cultivation) plus $150 (the value of production elsewhere of the released factors of production). Overall, operating the railway will increase the value of total production by $20. With these figures it is clear that it is better that the railway should not be liable for the damage it causes, thus enabling it to operate profitably. Of course, by altering the figures, it could be shown that there are other cases in which it would be desirable that the railway should be liable for the damage it causes. It is enough for my purpose to show that, from an economic point of view, a situation in which there is "uncompensated damage done to surrounding woods by sparks from railway engines" is not necessarily undesirable. Whether it is desirable or not depends on the particular circumstances.

How is it that the Pigovian analysis seems to give the wrong answer? The reason is that Pigou does not seem to have noticed that his analysis is dealing with an entirely different question. The analysis as such is correct. But it is quite illegitimate for Pigou to draw the particular conclusion he does. The question at issue is not whether it is desirable to run an additional train or a faster train or to install smoke-preventing devices; the question at issue is whether it is desirable to have a system in which the railway has to compensate those who suffer damage from the fires which it causes or one in which the railway does not have to compensate them. When an economist is comparing alternative social arrangements, the proper procedure is to compare the total social product yielded by these different arrangements. The comparison of private and social products is neither here nor there. A simple example will demonstrate this. Imagine a town in which there are traffic lights. A motorist approaches an intersection and stops because the light is red. There are no cars approaching the intersection on the other street. If the motorist ignored the red signal, no accident would occur and the total product would increase because the motorist would arrive earlier at his destination. Why does he not do this? The reason is that if he ignored the light he would be fined. The private product from crossing the street is less than the social product. Should we conclude from this that the total product would be greater if there were no fines for failing to obey traffic signals? The Pigovian analysis shows us that it is possible to conceive of better worlds than the one in which we live. But the problem is to devise practical arrangements which will correct defects in one part of the system without causing more serious harm in other parts.

I have examined in considerable detail one example of a divergence between private and social products and I do not propose to make any further examination of Pigou's analytical system. But the main discussion of the problem considered in this article is to be found in that part of Chapter 9 in Part II which deals with Pigou's second class of divergence and it is of interest to see how Pigou develops his argument. Pigou's own description of this second class of divergence was quoted at the beginning of this section. Pigou distinguishes between the case in which a person renders services for which he receives no payment and the case in which a person renders disservices and compensation is not given to the injured parties. Our main attention has, of course, centred on this second case. It is therefore rather astonishing to find, as was pointed out to me by Professor Francesco Forte, that the problem of the smoking chimney—the "stock instance"[24] or "classroom example"[25] of the second case—is used by Pigou as an example of the first case (services rendered without payment) and is never mentioned, at any rate explicitly, in connection with the second case.[26] Pigou points out that factory owners who devote resources to preventing their chimneys from smoking render services for which they receive no payment. The implication, in the light of Pigou's discussion later in the chapter, is that a factory owner with a smokey chimney should be given a bounty to induce him to install smoke-preventing devices. Most modern economists would suggest that the owner of the factory with the smokey chimney should be taxed. It seems a pity that economists (apart from Professor Forte) do not seem to have noticed this feature of Pigou's treatment since a realisation that the problem could be tackled in either of these two ways would probably have led to an explicit recognition of its reciprocal nature.

In discussing the second case (disservices without compensation to those damaged), Pigou says that they are rendered "when the owner of a site in a residential quarter of a city builds a factory there and so destroys a great part of the amenities of neighbouring sites; or, in a less degree, when he uses his site in such a way as to spoil the lighting of the house opposite; or when he invests resources in erecting buildings in a crowded centre, which by contracting the air-space and the playing room of the neighbourhood, tend to injure the health and efficiency of the families living there."[27] Pigou is, of course, quite right to describe such actions as "uncharged disservices." But he is wrong when he describes these actions as "anti-social."[28] They may or may not be. It is necessary to weigh the harm against the good that will result. Nothing could be more "anti-social" than to oppose any action which causes any harm to anyone.

* * *

Indeed, Pigou's treatment of the problems considered in this article is extremely elusive and the discussion of his views raises almost insuperable difficulties of interpretation. Consequently it is impossible to be sure that one has understood what Pigou really meant. Nevertheless, it is difficult to resist the conclusion, extraordinary though this may be in an economist of Pigou's stature, that the main source of this obscurity is that Pigou had not thought his position through.

[24]Sir Dennis Robertson, I *Lectures on Economic Principles* 162 (1957).
[25]E. J. Mishan, "The Meaning of Efficiency in Economics," 189, *The Bankers' Magazine* 482 (June 1960).
[26]Pigou, *op, cit. supra* n. 35 at 184.
[27]*Id.* at 185–86.
[28]*Id.* at 186 n. 1. For similar unqualified statements see Pigou's lecture "Some Aspects of the Housing Problem" in B. S. Rowntree and A. C. Pigou, "Lectures on Housing," in 18 *Manchester Univ. Lectures* (1914).

IX. THE PIGOVIAN TRADITION

It is strange that a doctrine as faulty as that developed by Pigou should have been so influential, although part of its success has probably been due to the lack of clarity in the exposition. Not being clear, it was never clearly wrong. Curiously enough, this obscurity in the source has not prevented the emergence of a fairly well-defined oral tradition. What economists think they learn from Pigou, and what they tell their students, which I term the Pigovian tradition, is reasonably clear. I propose to show the inadequacy of this Pigovian tradition by demonstrating that both the analysis and the policy conclusions which it supports are incorrect.

I do not propose to justify my view as to the prevailing opinion by copious references to the literature. I do this partly because the treatment in the literature is usually so fragmentary, often involving little more than a reference to Pigou plus some explanatory comment, that detailed examination would be inappropriate. But the main reason for this lack of reference is that the doctrine, although based on Pigou, must have been largely the product of an oral tradition. Certainly economists with whom I have discussed these problems have shown a unanimity of opinion which is quite remarkable considering the meagre treatment accorded this subject in the literature. No doubt there are some economists who do not share the usual view but they must represent a small minority of the profession.

The approach to the problems under discussion is through an examination of the value of physical production. The private product is the value of the additional product resulting from a particular activity of a business. The social product equals the private product minus the fall in the value of production elsewhere for which no compensation is paid by the business. Thus, if 10 units of a factor (and no other factors) are used by a business to make a certain product with a value of $105; and the owner of this factor is not compensated for their use, which he is unable to prevent; and these 10 units of the factor would yield products in their best alternative use worth $100; then, the social product is $105 minus $100 or $5. If the business now pays for one unit of the factor and its price equals the value of its marginal product, then the social product rises to $15. If two units are paid for, the social product rises to $25 and so on until it reaches $105 when all units of the factor are paid for. It is not difficult to see why economists have so readily accepted this rather odd procedure. The analysis focusses on the individual business decision and since the use of certain resources is not allowed for in costs, receipts are reduced by the same amount. But, of course, this means that the value of the social product has no social significance whatsoever. It seems to me preferable to use the opportunity cost concept and to approach these problems by comparing the value of the product yielded by factors in alternative uses or by alternative arrangements. The main advantage of a pricing system is that it leads to the employment of factors in places where the value of the product yielded is greatest and does so at less cost than alternative systems (I leave aside that a pricing system also eases the problem of the redistribution of income). But if through some God-given natural harmony factors flowed to the places where the value of the product yielded was greatest without any use of the pricing system and consequently there was no compensation, I would find it a source of surprise rather than a cause for dismay.

The definition of the social product is queer but this does not mean that the conclusions for policy drawn from the analysis are necessarily wrong. However, there are bound to be dangers in an approach which diverts attention from the basic issues and there can be little doubt that it has been responsible for some of the errors in current doctrine. The belief that it is desirable that the business which causes harmful effects should be forced to compensate those who suffer damage (which was exhaustively discussed in Section VIII in connection with Pigou's railway sparks example) is undoubtedly the result of not comparing the total product obtainable with alternative social arrangements.

The same fault is to be found in proposals for solving the problem of harmful effects by the use of taxes or bounties. Pigou lays considerable stress on this solution although he is, as usual, lacking in detail and qualified in his support. [29] Modern economists tend to think exclusively in terms of taxes and in a very precise way. The tax should be equal to the damage done and should therefore vary with the amount of the harmful effect. As it is not proposed that the proceeds of the tax should be paid to those suffering the damage, this solution is not the same as that which would force a business to pay compensation to those damaged by its actions, although economists generally do not seem to have noticed this and tend to treat the two solutions as being identical.

Assume that a factory which emits smoke is set up in a district previously free from smoke pollution, causing damage valued at $100 per annum. Assume that the taxation solution is adopted and that the factory-owner is taxed $100 per annum as long as the factory emits the smoke. Assume further that a smoke-preventing device costing $90 per annum to run is available. In these circumstances, the smoke-preventing device would be installed. Damage of $100 would have been avoided at an expenditure of $90 and the factory-owner would be better off by $10 per annum. Yet the position achieved may not be optimal. Suppose that those who suffer the damage could avoid it by moving to other locations or by taking various precautions which would cost them, or be equivalent to a loss in income of, $40 per annum. Then there would be a gain in the value of production of $50 if the factory continued to emit its smoke and those

[29] *Id.* 192–4, 381 and *Public Finance* 94–100 (3d ed. 1947).

now in the district moved elsewhere or made other adjustments to avoid the damage. If the factory owner is to be made to pay a tax equal to the damage caused, it would clearly be desirable to institute a double tax system and to make residents of the district pay an amount equal to the additional cost incurred by the factory owner (or the consumers of his products) in order to avoid the damage. In these conditions, people would not stay in the district or would take other measures to prevent the damage from occurring, when the costs of doing so were less than the costs that would be incurred by the producer to reduce the damage (the producer's object, of course, being not so much to reduce the damage as to reduce the tax payments). A tax system which was confined to a tax on the producer for damage caused would tend to lead to unduly high costs being incurred for the prevention of damage. Of course this could be avoided if it were possible to base the tax, not on the damage caused, but on the fall in the value of production (in its widest sense) resulting from the emission of smoke. But to do so would require a detailed knowledge of individual preferences and I am unable to imagine how the data needed for such a taxation system could be assembled. Indeed, the proposal to solve the smoke pollution and similar problems by the use of taxes bristles with difficulties: the problem of calculation, the difference between average and marginal damage, the interrelations between the damage suffered on different properties, etc. But it is unnecessary to examine these problems here. It is enough for my purpose to show that, even if the tax is exactly adjusted to equal the damage that would be done to neighbouring properties as a result of the emission of each additional puff of smoke, the tax would not necessarily bring about optimal conditions. An increase in the number of people living or of businesses operating in the vicinity of the smoke-emitting factory will increase the amount of harm produced by a given emission of smoke. The tax that would be imposed would therefore increase with an increase in the number of those in the vicinity. This will tend to lead to a decrease in the value of production of the factors employed by the factory, either because a reduction in production due to the tax will result in factors being used elsewhere in ways which are less valuable, or because factors will be diverted to produce means for reducing the amount of smoke emitted. But people deciding to establish themselves in the vicinity of the factory will not take into account this fall in the value of production which results from their presence. This failure to take into account costs imposed on others is comparable to the action of a factory owner in not taking into account the harm resulting from his emission of smoke. Without the tax, there may be too much smoke and too few people in the vicinity of the factory; but with the tax there may be too little smoke and too many people in the vicinity of the factory. There is no reason to suppose that one of these results is necessarily preferable.

I need not devote much space to discussing the similar error involved in the suggestion that smoke-producing factories should, by means of zoning regulations, be removed from the districts in which the smoke causes harmful effects. When the change in the location of the factory results in a reduction in production, this obviously needs to be taken into account and weighed against the harm which would result from the factory remaining in that location. The aim of such regulation should not be to eliminate smoke pollution but rather to secure the optimum amount of smoke pollution, this being the amount which will maximise the value of production.

X. A CHANGE OF APPROACH

It is my belief that the failure of economists to reach correct conclusions about the treatment of harmful effects cannot be ascribed simply to a few slips in analysis. It stems from basic defects in the current approach to problems of welfare economics. What is needed is a change of approach.

Analysis in terms of divergencies between private and social products concentrates attention on particular deficiencies in the system and tends to nourish the belief that any measure which will remove the deficiency is necessarily desirable. It diverts attention from those other changes in the system which are inevitably associated with the corrective measure, changes which may well produce more harm than the original deficiency. In the preceding sections of this article, we have seen many examples of this. But it is not necessary to approach the problem in this way. Economists who study problems of the firm habitually use an opportunity cost approach and compare the receipts obtained from a given combination of factors with alternative business arrangements. It would seem desirable to use a similar approach when dealing with questions of economic policy and to compare the total product yielded by alternative social arrangements. In this article, the analysis has been confined, as is usual in this part of economics, to comparisons of the value of production, as measured by the market. But it is, of course, desirable that the choice between different social arrangements for the solution of economic problems should be carried out in broader terms than this and that the total effect of these arrangements in all spheres of life should be taken into account. As Frank H. Knight has so often emphasized, problems of welfare economics must ultimately dissolve into a study of aesthetics and morals.

A second feature of the usual treatment of the problems discussed in this article is that the analysis proceeds in terms of a comparison between a state of laissez faire and some kind of ideal world. This approach inevitably leads to a looseness of thought since the nature of the alternatives being compared is never clear. In a state of laissez faire, is there a monetary, a legal or a political system and if so, what are they? In an ideal world, would there be a monetary, a legal or a political system and if so, what would they be? The answers to all these questions are shrouded in mystery and every man is free to draw whatever conclusions he likes. Actually very little analysis is required to show that an ideal world is better than a state of laissez faire, unless the definitions of a state of laissez faire and an ideal world happen to be the same. But the whole discussion is largely irrelevant for questions of economic policy since whatever we may have in mind as our ideal world, it is clear that we have not yet discovered how to get to it from where we are. A better approach would seem to be to start our analysis with a situation approximating that which actually exists, to examine the effects of a proposed policy change and to attempt to decide whether the new situation would be, in total, better or worse than the original one. In this way, conclusions for policy would have some relevance to the actual situation.

A final reason for the failure to develop a theory adequate to handle the problem of harmful effects stems from a faulty concept of a factor of production. This is usually thought of as a physical entity which the businessman acquires and uses (an acre of land, a ton of fertiliser) instead of as a right to perform certain (physical) actions. We may speak of a person owning land and using it as a factor of production but what the land-owner in fact possesses is the right to carry out a circumscribed list of actions. The rights of a land-owner are not unlimited. It is not even always possible for him to remove the land to another place, for instance, by quarrying it. And although it may be possible for him to exclude some people from using "his" land, this may not be true of others. For example, some people may have the right to cross the land. Furthermore, it may or may not be possible to erect certain types of buildings or to grow certain crops or to use particular drainage systems on the land. This does not come about simply because of Government regulation. It would be equally true under the common law. In fact it would be true under any system of law. A system in which the rights of individuals were unlimited would be one in which there were no rights to acquire.

If factors of production are thought of as rights, it becomes easier to understand that the right to do something which has a harmful effect (such as the creation of smoke, noise, smells, etc.) is also a factor of production. Just as we may use a piece of land in such a way as to prevent someone else from crossing it, or parking his car, or building his house upon it, so we may use it in such a way as to deny him a view or quiet or unpolluted air. The cost of exercising a right (of using a factor of production) is always the loss which is suffered elsewhere in consequence of the exercise of that right—the inability to cross land, to park a car, to build a house, to enjoy a view, to have peace and quiet or to breathe clean air.

It would clearly be desirable if the only actions performed were those in which what was gained was worth more than what was lost. But in choosing between social arrangements within the context of which individual decisions are made, we have to bear in mind that a change in the existing system which will lead to an improvement in some decisions may well lead to a worsening of others. Furthermore we have to take into account the costs involved in operating the various social arrangements (whether it be the working of a market or of a government department), as well as the costs involved in moving to a new system. In devising and choosing between social arrangements we should have regard for the total effect. This, above all, is the change in approach which I am advocating.

8

On Divergences between Social Cost and Private Cost

RALPH TURVEY

.

Ralph Turvey is Visiting Professor of Economics at the
London School of Economics.

The notion that the resource-allocation effects of divergences between
marginal social and private costs can be dealt with by imposing a tax or
granting a subsidy equal to the difference now seems too simple a notion.
Three recent articles have shown us this. First came Professor Coase's
"The Problem of Social Cost", then Davis and Whinston's "Externalities,
Welfare and the Theory of Games" appeared, and, finally, Buchanan and
Stubblebine have published their paper "Externality".[1] These articles
have an aggregate length of eighty pages and are by no means easy to read.
The following attempt to synthesise and summarise the main ideas may
therefore be useful. It is couched in terms of external diseconomies, i.e. an
excess of social over private costs, and the reader is left to invert the
analysis himself should he be interested in external economies.

The scope of the following argument can usefully be indicated by
starting with a brief statement of its main conclusions. The first is that if
the party imposing external diseconomies and the party suffering them are
able and willing to negotiate to their mutual advantage, state intervention
is unnecessary to secure optimum resource allocation. The second is that
the imposition of a tax upon the party imposing external diseconomies can

"On Divergences between Social Cost and Private Cost," by Ralph
Turvey, from *Economica* (August 1963). Reprinted by permission.

[1]*Journal of Law and Economics,* Vol. III, October, 1960, *Journal of Political Economy,* June,
1962, and *Economica,* November, 1962, respectively. [Professor Coase's article is reprinted
in this volume, Selection 7.]

be a very complicated matter, even in principle, so that the *a priori* prescription of such a tax is unwise.

To develop these and other points, let us begin by calling *A* the person, firm or group (of persons or firms) which imposes a diseconomy, and *B* the person, firm or group which suffers it. How much *B* suffers will in many cases depend not only upon the *scale* of *A*'s diseconomy-creating activity, but also upon the precise *nature* of *A*'s activity and upon *B*'s reaction to it. If *A* emits smoke, for example, *B*'s loss will depend not only upon the quantity emitted but also upon the height of *A*'s chimney and upon the cost to *B* of installing air-conditioning, indoor clothes-dryers or other means of reducing the effect of the smoke. Thus to ascertain the optimum resource allocation will frequently require an investigation of the nature and costs both of alternative activities open to *A* and of the devices by which *B* can reduce the impact of each activity. The optimum involves that kind and scale of *A*'s activity and that adjustment to it by *B* which maximises the algebraic sum of *A*'s gain and *B*'s loss as against the situation where *A* pursues no diseconomy-creating activity. Note that the optimum will frequently involve *B* suffering a loss, both in total and at the margin.[2]

If *A* and *B* are firms, gain and loss can be measured in money terms as profit differences. (In considering a social optimum, allowance has of course to be made for market imperfections.) Now assuming that they both seek to maximise profits, that they know about the available alternatives and adjustments and that they are able and willing to negotiate, they will achieve the optimum without any government interference. They will internalize the externality by merger,[3] or they will make an agreement whereby *B* pays *A* to modify the nature or scale of its activity.[4] Alternatively,[5] if the law gives *B* rights against *A*, *A* will pay *B* to accept the optimal amount of loss imposed by *A*.

If *A* and *B* are people, their gain and loss must be measured as the amount of money they respectively would pay to indulge in and prevent *A*'s activity. It could also be measured as the amount of money they respectively would require to refrain from and to endure *A*'s activity, which will be different unless the marginal utility of income is constant. We shall assume that it is constant for both *A* and *B*, which is reasonable when the payments do not bulk large in relation to their incomes.[6] Under this assumption, it makes no difference whether *B* pays *A* or, if the law gives *B* rights against *A*, *A* compensates *B*.

---

[2]Buchanan-Stubblebine, pp. 380–1.

[3]Davis-Whinston, pp. 244, 252, 256; Coase, pp. 16–17.

[4]Coase, p. 6; Buchanan-Stubblebine agree, p. 383.

[5]See previous references.

[6]Dr. Mishan has examined the welfare criterion for the case where the only variable is the scale of *A*'s activity, but where neither *A* nor *B* has a constant marginal utility of income. Cf. his paper "Welfare Criteria for External Effects," *American Economic Review*, September, 1961.

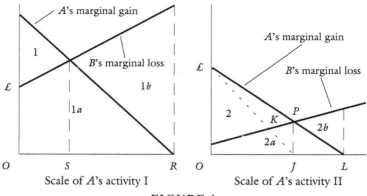

FIGURE 1.

Whether A and B are persons or firms, to levy a tax on A which is *not* received as damages or compensation by B may prevent optimal resource allocation from being achieved—still assuming that they can and do negotiate.[7] The reason is that the resource allocation which maximises A's *gain less B's loss* may differ from that which maximises A's *gain less A's tax less B's loss.*

The points made so far can usefully be presented diagrammatically (Figure 1). We assume that A has only two alternative activities, I and II, and that their scales and B's losses are all continuously variable. Let us temporarily disregard the dotted curve in the right-hand part of the diagram. The area under A's curves then gives the total gain to A. The area under B's curves gives the total loss to B after he has made the best adjustment possible to A's activity. This is thus the direct loss as reduced by adjustment, plus the cost of making that adjustment.

If A and B could not negotiate and if A were unhampered by restrictions of any sort, A would choose activity I at a scale of OR. A scale of OS would obviously give a larger social product, but the optimum is clearly activity II at scale OJ, since area 2 is greater than area 1. Now B will be prepared to pay up to $(1a + 1b - 2a)$ to secure this result, while A will be prepared to accept down to $(1 + 1a - 2 - 2a)$ to assure it. The difference is $(1b - 1 + 2)$, the maximum gain to be shared between them, and this is clearly positive.

If A is liable to compensate B for actual damages caused by either activity I or II, he will choose activity II at scale OJ (i.e. the optimum allocation), pay $2a$ to B and retain a net gain of 2. The result is the same as when there is no such liability, though the distribution of the gain is very different: B will pay A up to $(1a + 1b - 2a)$ to secure this result. Hence whether or not we should advocate the imposition of a liability on A for

[7]Buchanan-Stubblebine, pp. 381–3.

damages caused is a matter of fairness, not of resource allocation. Our judgment will presumably depend on such factors as who got there first, whether one of them is a non-conforming user (e.g. an establishment for the breeding of maggots on putrescible vegetable matter in a residential district), who is richer, and so on. Efficient resource allocation requires the imposition of a liability upon *A* only if we can show that inertia, obstinacy, etc. inhibit *A* and *B* from reaching a voluntary agreement.[8]

We can now make the point implicit in Buchanan-Stubblebine's argument, namely that there is a necessity for any impost levied on *A* to be paid to *B* when *A* and *B* are able to negotiate. Suppose that *A* is charged an amount equal to the loss he imposes on *B;* subtracting this from his marginal gain curve in the right-hand part of the diagram gives us the dotted line as his marginal net gain. If *A* moves to point *J* it will then pay *B* to induce him to move back to position *K* (which is sub-optimal) as it is this position which maximises the *joint* net gain to *A* and *B* together.

There is a final point to be made about the case where *A* and *B* can negotiate. This is that if the external diseconomies are reciprocal, so that each imposes a loss upon the other, the problem is still more complicated.[9]

We now turn to the case where *A* and *B* cannot negotiate, which in most cases will result from *A* and / or *B* being too large a group for the members to get together. Here there are certain benefits to be had from resource re-allocation which are not privately appropriable. Just as with collective goods,[10] therefore, there is thus a case for collective action to achieve optimum allocation. But all this means is that *if* the state can ascertain and enforce a move to the optimum position at a cost less than the gain to be had, and *if* it can do this in a way which does not have unfavourable effects upon income distribution, then it should take action.

These two "ifs" are very important. The second is obvious and requires no elaboration. The first, however, deserves a few words. In order to ascertain the optimum type and scale of *A*'s activity, the authorities must estimate all of the curves in the diagrams. They must, in other words, list and evaluate all the alternatives open to *A* and examine their effects upon *B* and the adjustments *B* could make to reduce the loss suffered. When this is done, if it can be done, it is necessary to consider how to reach the optimum. Now, where the nature as well as the scale of *A*'s activity is variable, it may be necessary to control both, and this may require two controls, not one. Suppose, for instance, that in the diagram, both activities are the emission of smoke: I from a low chimney and II from a tall chimney. To induce *A* to shift from emitting *OR* smoke from the low chimney to emitting *OJ* smoke from the tall chimney, it will not suffice to

[8]Cf. the comparable argument on pp. 94–8 of my *The Economics of Real Property,* 1957, about the external economy to landlords of tenants' improvements.
[9]Davis-Whinston devote several pages of game theory to this problem.
[10]Buchanan-Stubblebine, p. 383.

levy a tax of *PJ* per unit of smoke.[11] If this alone were done, *A* would continue to use a low chimney, emitting slightly less than *OR* smoke. It will also be necessary to regulate chimney heights. A tax would do the trick alone only if it were proportioned to losses imposed rather than to smoke emitted, and that would be very difficult.

These complications show that in many cases the cost of achieving optimum resource allocation may outweigh the gain. If this is the case, a second-best solution may be appropriate. Thus a prohibition of all smoke emission would be better than *OR* smoke from a low chimney (since 1 is less than 1*b*) and a requirement that all chimneys be tall would be better still (giving a net gain of 2 less 2*b*). Whether these requirements should be imposed on existing chimney-owners as well as on new ones then introduces further complications relating to the short run and the long run.

There is no need to carry the example any further. It is now abundantly clear that any general prescription of a tax to deal with external diseconomies is useless. Each case must be considered on its own and there is no *a priori* reason to suppose that the imposition of a tax is better than alternative measures or indeed, that any measures at all are desirable unless we assume that information and administration are both costless.[12]

To sum up, then: when negotiation is possible, the case for government intervention is one of justice not of economic efficiency; when it is not, the theorist should be silent and call in the applied economist.

---

[11]Note how different *PJ* is from *RT,* the initial observable marginal external diseconomy.
[12]Coase, pp. 18, 44.

9

The Problem of Market Failure*

ALAN RANDALL

Alan Randall is Professor of Agricultural Economics at
Ohio State University in Columbus.

I

The concept of market failure seems entrenched in the conventional wisdom of the economics discipline, if the conventional wisdom is most
clearly revealed by what respected economists tell undergraduate students
and government policy makers. The typical treatment proceeds as follows:
the concept of Pareto-optimality is explained; the idea that competitive
markets tend to allocate resources efficiently is developed; the notion that,
under certain conditions prevalent in the real world, markets fail to perform efficiently is introduced; and the search for ameliorative measures,
involving government as law-maker, tax collector, and/or regulator, is
undertaken. This approach pervades economic discussions of public finance, the provision of collective goods, management of natural resources, and environmental quality.[1]

"The Problem of Market Failure," by Alan Randall, from *Natural
Resources Journal*, v. 23, pp. 131–48. Reprinted by permission.

*This paper was drafted while the author was visiting at the University of New England,
Armidale N.S.W., Australia. An earlier version was presented in the 1980 Reuben A. Gustav-
son lecture series at the University of Chicago.

[1]A few examples from the environmental quality area should suffice: See the editors' introduction to R. Dorfman and N. Dorfman, *Economics of the Environment, Selected Readings*
(1977); Ruff, *"The Economic Common Sense of Pollution,"* 19 *Pub. Interest* 69 (1970); Davis
and Kamien, *"Externalities, Information, and Alternative Collective Action,"* 1 *The PPB*

The conventional wisdom, almost by definition, is an amalgam of new ideas from prior seasons and persistent ideas from earlier times. Since 1960, a vocal group (but still perhaps a minority) of economists who reject the market failure approach has arisen. [2] This group's critique has led the better economists who use the market failure paradigm toward a sharper, more precise and more sophisticated analysis. [3] Nevertheless, considerable confusion remains concerning the nature of market failure, its significance in theory and as an observable phenomenon, and the appropriate policy response to it.

Contemporary discussions of market failure usually list four distinct phenomena, although two or more of these may occur together and, if so, their effects are mutually reinforcing. The phenomena are externality, "public goods," "common property resources," and monopoly. To further complicate matters, two kinds of monopoly are recognized: (1) market concentration in the extreme, and (2) "natural monopoly," which is treated as endemic to decreasing-cost industries. This paper focuses on externality, "public goods," "common property resources," and "natural monopoly." A massive literature on market concentration already exists; thus this paper will not explore that topic.

Section II argues that "common property resources" and "public goods" are confusing terms which refer to confused concepts. That section offers an alternative and more precise terminology, based on notions of nonexclusiveness and nonrivalry. Section III discusses externality in a post-Coasian context and examines the "Coase Theorem" in its "weak" and "strong" versions. The discussion shows that the strong version is a useful pedagogical device with little policy relevance, while the weak version is seen as a general theorem on the existence of markets, rather than a theorem central to environmental economics.

The paper goes on to examine the conditions under which an externality may *persist* and finds that, in general, externality cannot persist. Inefficiency, however, may surely persist, but for reasons more closely related to nonexclusiveness and nonrivalry than to externality. Externality is, then, a vacuous and entirely unhelpful term, and can be replaced by the more general term *inefficiency* with no loss of content.

On the other hand, an intrusion or invasion may occur—or, in the terminology this article seeks to eliminate, "an externality may *arise*," resulting in (1) a non-consensual change in the product mix of society and (2) a welfare shock for some. Assuming, as most authors (including Coase

---

System 67 (1969); and A. Freeman, R. Haveman and A. Kneese, *The Economics of Environmental Policy* (1973).

[2]The seminal article is Coase, *"The Problem of Social Cost,"* 3 *J.L. & Econ.* 1 (1960). However, others make a more vehement attack on the market failure paradigm: for example, Demsetz, *"Information and Efficiency: Another Viewpoint,"* 12 *J.L. & Econ.* 1 (1969).

[3]Including most of those cited above, note 1.

and Buchanan) do, that citizens may appropriately use the powers of government to deal with these kinds of events, what should be done? In the literature that addresses this question, one can identify two post-Coasian traditions: a Coase–Buchanan tradition and a Coase–Posner tradition. Section IV evaluates these two traditions, develops an argument which finds more merit in the Coase–Buchanan tradition, and sketches some institutional reforms in that tradition.

Section V presents concluding comments that suggest what, if anything, can be salvaged from the "market failure" baggage of the economist's conventional wisdom.

II

For the current generation of economists, Gordon popularized the term "common property resource."[4] The analysis of Gordon and most subsequent authors has focused on the unowned resource, *res nullius,* and is basically correct in that context. The problem is that rights to the resource are *nonexclusive.*

Considerable confusion arises because the now standard "common property resource" analysis is not applicable to res communis, property held in common.[5] While it is unlikely that *res communis* rights will be strictly nonattenuated,[6] many of the solutions proposed for "common property resources" (i.e., res nullius) problems actually fall within the *res communis* classification. Ownership is vested in some kind of collective and rules of access (usually exclusive and enforceable to a considerable degree, and often transferable under stated conditions) are established to minimize abuse of the resource and overinvestment of factors of production in its exploitation.

One may ask why the *res nullius,* or non-exclusiveness, problem is so often handled in basically free enterprise economies by establishing some form of *res communis* rather than nonattenuated property rights. The answer may lie in traditional beliefs that private ownership is inappropriate for certain kinds of resources. However, the author suspects that it lies more often in the high cost of exclusion. For example, for many species of commercial fish, the costs of specifying and enforcing exclusive property

---

[4]Gordon, *"The Economic Theory of a Common-Property Resource: The Fishery,"* 62 *J. Pol. Econ.* 124 (1954).

[5]Ciriacy-Wantrup & Bishop. *" 'Common property' as a Concept in Natural Resources Policy,"* 15 *Nat. Res. J.* 713 (1975).

[6]S. N. S. Cheung defines nonattenuated rights as (1) exclusive, (2) transferable, (3) enforced and (4) in no way inconsistent with the marginal conditions for Pareto optimality. *See* Cheung, *"The Structure of a Contract and the Theory of a Nonexclusive Resource,"* 13 *J.L. & Econ.* 49 (1970).

rights in individual fish are prohibitive, in that they exceed any possible gains which could arise from the voluntary exchange thus permitted. Similar difficulties may apply to individual units of ambient air and water, and oil and groundwater in large pools.

In these cases, transactions costs (broadly defined)[7] are not merely so large as to prohibit the kinds of trade which would lead to Pareto-efficiency, but so large as to prohibit the establishment of nonattenuated property rights, a necessary precondition for such trade. Further, it is not "the large numbers problem," the reason usually offered to explain high transactions costs, but some peculiarities in the physical nature of the resource itself that are to blame.[8] For example, fencing the open sea is technologically more demanding and thus vastly more expensive than fencing the open range.

"Public goods" (or "collective goods") is another term which generates confusion. In most definitions, at least one of two phenomena is involved: nonexclusiveness, discussed above, and nonrivalry. The latter refers to Samuelson's notion of a good which may be enjoyed ("consumed") by some without diminution of the amount effectively available for others.[9]

Confusion arises as to whether both of these phenomena are necessary to make a good "public" or, if one is enough, which one?[10] However, nonexclusiveness and nonrivalry may occur together or separately, and the economic analyses of the two phenomena are quite different. Accordingly, a focus on questions of exclusiveness and rivalry permits precise analysis, while the term "public good" only introduces confusion. Debate about what is required for a good to be called "public" seems not only unhelpful but misdirected.

The economics profession could abolish the terms "common property resources," and "public goods" with no loss of information and considerable gain in clarity. The terms nonexclusiveness and nonrivalry represent vast improvements, useful in all contexts and relevant for both goods and resources. For the "non-pure public good," the term congestible good is entirely adequate. It describes a good which is nonrival for some number

[7]As for example, by Coase, note 2, at 6; Crocker, *"Externalities, Property Rights and Transactions Costs: An Empirical Study,* 14 *J.L. & Econ.* 451, at 462-4 (1971); and Dahlman, *"The Problem of Externality,"* 22 *J.L. & Econ.* 141, 144 (1979).

[8]"The Large Numbers Problem," alone, will never cause prohibitive transactions costs. The market for bread, with myriad buyers and sellers, works as well as any, and much better than the market for clear air which has a similar number of potential buyers and sellers.

[9]Samuelson, *"The Pure Theory of Public Expenditure,"* 36 *Rev. Econ. & Stat.* 387 (1954).

[10]Samuelson, *ibid.,* focused on nonrivalry, but also made mention of the free-rider problem, a manifestation of nonexclusiveness. Davis and Whinston, *"On the Distinction Between Private and Public Goods,"* 57 *Am. Econ. Rev.* 360 (1967) focused on nonrivalry, and discussed the economic properties of nonexclusive public goods, exclusive public goods, and non-pure (*i.e.* congestible) public goods, both exclusive and nonexclusive. Head, *"Public Goods: The Polar Case Reconsidered* 53 *Econ. Rec.* 227 (1977), argues that a "public good" must have both characteristics, nonexclusiveness and nonrivalry.

of users, while rivalry sets in as that number is increased and becomes intense as the number of users approaches the capacity constraint. For such goods, initial capital costs tend to be high, while the marginal cost of adding an additional user remains low until the capacity constraint is approached. Thus, average cost per user steadily declines until the capacity constraint is approached. Examples include almost all services provided in a confined space, and all services provided in capital-intensive delivery systems of constrained capacity, *e.g.,* roads, bridges, railbeds, canals, transmission lines and pipelines.

The point of these last observations is that the so-called "natural monopoly" problem can be adequately analyzed under the rubric of congestible goods.

What are the economic consequences of nonrivalry? In principle, the efficient amount of nonrival good may be provided. One may express individual preferences across the range of possible levels of provision and identify individual total and marginal "willingness to pay" (WTP) schedules. These schedules are aggregated vertically, across individuals. The efficient level of provision is identified (in the Samuelson[11] solution) as that level at which aggregate marginal WTP just equals the marginal cost of provision, given that aggregate total WTP exceeds total cost. The problem, for efficiency, is that no non-discriminatory pricing scheme can achieve this solution. A price high enough to generate revenue covering costs of provision would inefficiently exclude some potential users who value the good positively and would result in provision of less than the efficient amount of the good. As Davis and Whinston[12] point out, such a solution would not be Pareto-efficient but may be a second-best solution. With an adequate exclusionary device, the private sector could provide nonrival goods in this manner. In the absence of exclusion, the best hope is for public sector provision of the efficient quantity, but financing procedures permitting this outcome would necessarily violate the pricing conditions for Pareto-efficiency.

Discriminatory pricing would permit the Lindahl solution, which provides the efficient quantity (*i.e.,* the same quantity as the Samuelson solution) at Pareto-efficient prices. The Lindahl-price is, in general, different for each individual. Private (or public) sector provision in a Pareto-efficient manner would therefore require not just exclusion of non-payers, but exclusion of each individual who does not pay his individual Lindahl-price. This latter kind of exclusion is much more technologically demanding than the exclusion adequate for ordinary (*i.e.,* rival) goods, and is required only in the case of nonrival goods. For the want of a better name, let us call it hyperexclusion.

Now, a classification system based on concepts of exclusion and ri-

[11]Samuelson, above, note 9, at 387, 389.
[12]David and Whinston, above, note 10, at 363.

TABLE 1. A Classification of Goods Based on Concepts of Rivalry and Exclusion

	NONEXCLUSIVE	EXCLUSIVE	HYPEREXCLUSIVE
Nonrival	1	4	7
Congestible	2	5	8
Rival	3	6	9

valry—and designed to replace the confusing and often misleading notions of "common property resources" and "public goods"—can be spelled out.

Conceptually, all of the nine types of goods identified in Table 1 can exist. Examples of types 1 through 6 occur quite frequently. The economic characteristics of each type are summarized below. Each of the nine types has characteristics which distinguish it from the others and, in each case, the economic analysis of the possibilities for efficient pricing and the provision of efficient quantities have distinguishing features.

Goods of types 1 through 3 cannot be reliably provided by the private sector, or by the public sector financing them with user charges. Public sector provision, financed from general revenues, is possible. An all-wise public sector could, in concept, provide goods of type 1 in efficient quantities but not, of course, at efficient prices. For types 2 and 3, the lack of exclusion would result in overutilization; thus, both price and quantity aspects of Pareto-efficiency would be violated. There are subtle but important differences in the analyses appropriate for determining the efficient quantity in each case.

Goods of types 4 and 5 could be provided by the private sector, or by the public sector, financing them with user charges. Second-best solutions may be achieved, but Pareto-efficiency is unattainable. Davis and Whinston[13] speculate on the nature of the second-best solutions for each of these two distinct types of goods.

Type 6, the "private goods" which neoclassical microeconomic theory treats as typical, may be provided by the private sector in a Pareto-efficient manner, if all the conditions for Pareto-efficiency are satisfied.

Type 7 represents nonrival goods provided, by the private or public sector, at Lindahl-prices. Pareto-efficiency is achieved under these conditions.

Goods of types 8 and 9 may, in principle, be provided in efficient quantities by private or public sector. In these cases, hyperexclusion requires, among other things, that the provider enjoy monopoly status. Such a perfectly discriminating monopolist would extract, via Lindahl-pricing, all of the economic surplus which in the case of typical private goods (type

---

[13]*Ibid.* at 363.

6) is divided among producers and consumers. Pure profits may therefore arise, in violation of the conditions for Pareto-efficiency.

Note that Pareto-efficient provision through the market is conceivable only for goods of types 6 and 7, and that in case 7, hyperexclusion—which remains technologically elusive—is required.

What are the prospects of shifting goods among the categories defined in Table 1? Since the three rivalry-related concepts refer to fundamental physical characteristics of the goods involved, there are very few prospects of shifting goods vertically without transforming the nature of the good itself.

The exclusion-related concepts represent the interaction of institutional and technological factors. To achieve exclusion requires a structure of laws and institutions to establish and enforce exclusive property rights, but the effectiveness of enforcement—and thus of the rights themselves—depends upon the available technology of exclusion and the costs of implementing that technology relative to the gains from trade thereby permitted. For some goods—*e.g.,* grocery items, consumer durables, residences, and farmland—these costs are tolerably low, and enterprise-oriented societies have generally shifted such goods horizontally from the nonexclusive to the exclusive categories. For other goods—*e.g.,* fish in the ocean and ambient air—strict exclusion is technologically difficult and costly relative to the potential gains. Therefore, one finds, at best, *res communis* types of rules of access which are surely attenuated, but which may permit some improvements over the *res nullius* situation.

For reference, we identify the place of the terms "common property resource" and "public good" in the nine category classification of Table 1. "Common property resource" is usually the misnomer applied to factors of production in category 3, but on occasion has also been applied to items in categories 1 and 2. "Public good" has been applied, by various authors, to commodities in some or all of categories 1, 4, and 7. Similarly, "non-pure public good" has been applied to some or all of categories 2, 5, and 8. Here, we rest our case for substituting the classification of Table 1 for the customary, but imprecise and confusing, terminology of "common property resources" and "public goods."

What do nonexclusiveness and nonrivalry imply about market failure? First, what does market failure mean? Some have defined it as the failure of markets to exist. However, it has long been clear that the non-existence of certain markets is a rational market response to transactions costs in excess of potential gains from trade.[14] If market failure means inefficiency, it must refer in concept to goods in all categories except 6 and 7, and in practice to all goods. However, such a concept is not helpful, as it identifies market failure in cases where other institutional devices will also predicta-

---

[14]Demsetz, *"The Exchange and Enforcement of Property Rights,"* 7 *J.L. & Econ.* 11 (1964).

bly fail, to greater or lesser degrees. If market failure means the failure of markets to do as well as some other institutional device(s), the concept comes to grief on the absence of unanimity about what is meant by "to do as well." Some progress can be made if we confine ourselves to one of the several aspects of efficiency: the provision of goods, services and amenities in efficient quantities. For nonrival goods the possibility exists that government, by direct provision, may outperform the market. Where nonexclusiveness is the problem, citizens working through government may institute changes in property rights which cause the market to provide more nearly the efficient amount. But these improvements are possibilities, not certainties, and their validity must be demonstrated on a case-by-case basis recognizing all the costs and imperfections of both market and governmental institutions, rather than established by mere appeal to market failure notions. Finally, the focus on quantity provided without reference to pricing and the division of surpluses among producers and between producers and consumers will be unsatisfactory to some.

III

Externality is usually defined as a situation in which the utility of an affected party is influenced by a vector of activities under his control but also by one or more activities under the control of another (or others). Since the writings of Coase[15] and Buchanan and Stubblebine,[16] most authors have focused on Pareto-relevant externalities: those which are inefficient. Many categories of interactions which satisfy the definition of externality are efficiently handled in markets, and no possibility of Pareto-relevance exists for these categories when markets function well. For other kinds of interactions—air and water pollution are commonly cited examples—intense debate has occurred about whether Pareto-relevant externality may persist. It is generally accepted that some externality—*e.g.,* some positive level of air or water pollution may persist even in a Pareto-efficient situation.

The Coase Theorem comes in two versions, a strong and a weak version. The strong version states: given a structure of property rights which is completely specified and exclusive, costlessly transferable, and costlessly enforced, voluntary exchange will eliminate all Pareto-relevant externality, *and* the resultant allocation of resources will be independent of the specific assignment of property rights. This theorem relies upon a number of restrictive assumptions, notably that income effects are zero, nonat-

[15]Coase, above, note 2.
[16]Buchanan and Stubblebine, *"Externality,"* 29 *Economica* 371 (1962).

tenuated property rights may be costlessly established and maintained, and markets in goods and rights are frictionless.[17] For these reasons, the strong theorem must be regarded more as a pedagogical device than a source of policy prescriptions.

The weak version of the Coase Theorem casts much more light on the problem of externality. It states: given a structure of property rights consistent with Pareto-efficiency, voluntary exchange will eliminate Pareto-relevant externality and thereby establish an efficient allocation of resources. The weak version may, of course, be developed from the strong by relaxing the above-mentioned assumptions. Our understanding of the concepts of externality and Pareto-relevance is enhanced by working through this process.

Under the strong theorem assumptions, for any initial assignment of rights, voluntary exchange results in the same equilibrium level of abatement of an annoyance (*i.e.,* external diseconomy). Efficiency is achieved, the allocative dimensions of the efficient solution are unaffected by the specific assignment of rights, and—by definition—the Pareto-relevant externality is eliminated while some Pareto-irrelevant (and therefore efficient) annoyance remains.

Recognizing income effects (and assuming them to be positive), the specific assignment of rights does make a difference. With rights favoring the receptor, the equilibrium solution involves more abatement. In other words, when rights favor the receptor, more of the externality is Pareto-relevant than when rights favor the emitter. Positive transactions costs have a similar effect, and the difference between the equilibrium levels of abatement under opposite assignments of rights grows as the level of transactions costs increases. With transactions costs sufficiently large to preclude trade, the resource allocation implied by the initial rights structure is retained as the equilibrium solution. Under receptor rights, no externality remains; under emitter rights, no abatement occurs. Yet, assuming the transactions industry itself is efficient, each of these solutions is efficient given the assignment of rights which underlie it. This is the import of the weak Coase Theorem. Under receptor rights, the externality is Pareto-relevant in its entirety; under emitter rights, none of it is Pareto-relevant. Given the magnitude of transactions costs (and, usually to a lesser extent, income effects[18]), the initial assignment of rights determines what is Pareto-relevant.

---

[17]Coase, above, note 2, recognizes the crucial role of transactions costs in the strong theorem, and devotes considerable space to exploring its implications.

[18]For some formulae permitting rigorous calculation of the size of income effects, *see* Willig, *"Consumer Surplus Without Apology,"* 66 *Am. Econ. Rev.* 587 (1976); and Randall and Stoll, *"Consumer's Surplus in Commodity Space,"* 70 *Am. Econ. Rev.* 449 (1980). For many ordinary situations, income effects are quite small. However, where rights involving income-elastic goods and amenities which command a large share of the individual's budget are concerned, income effects are empirically important.

What, then, is Pareto-relevance? It is a general term, with no special relevance to externality, meaning the existence of unrealized gains from trade. Similarly, in an economy with a rights structure conducive to trade, Pareto-irrelevance describes any situation—without reference to its desirability, or lack thereof, in normative terms—which cannot be changed through voluntary exchange. In such an economy, the hunger of the undernourished is Pareto-irrelevant.

With respect to an external diseconomy, the specific assignment of rights does two things. First, it determines the directional flow of payments (if any) resulting from trade (if any). In other words, it determines which party faces a "pay or suffer" situation. Second, given the magnitude of transactions costs and income effects, the assignment of rights determines how much of the annoyance persists at the completion of trade, i.e., how much of it is *ipso facto* declared Pareto-irrelevant. Assuming income effects to be small, consider transactions costs. With low transactions costs, the initial assignment of rights has a relatively small influence on how much of the annoyance persists. With high transactions costs the influence is large. With prohibitive transactions costs, rights determine whether the annoyance persists in its entirety or is eliminated completely. Note that, for any initial assignment of rights, both parties, emitter and receptor, would prefer—given a choice—the low transactions costs situation in which some but not all of the annoyance persists.

Under what circumstances, given an externality which results in significant annoyance, would transactions costs be high or prohibitive? Only two possibilities exist. One, institutions are designed to impede trade in rights. The discussion thus far has, through its focus on nonattenuated structures of rights, eliminated this possibility by assumption. Two, exclusion and transfer are expensive for technological reasons. Something about the physical nature of the good itself and/or the technical processes required to delineate and enforce exclusive property rights therein renders exclusion expensive, thus making nonexclusiveness the norm for that kind of good. One particular aspect of the physical nature of some goods is nonrivalry which makes hyperexclusion, not just exclusion, necessary for Pareto-efficiency.

The import of all this is that, in economies which maintain institutions conducive to trade and efficiency, those things called externality cannot persist in inefficient quantities unless accompanied by nonexclusiveness and/or nonrivalry. Externality, by itself, is simply not persistent. In this sense, as Cheung[19] and Dahlman[20] have already pointed out, externality is not a useful term. Externality can refer only to temporary disequilibria indistinguishable from any other form of inefficiency which results from failure to realize potential gains from trade.

[19]Cheung, above, note 6.
[20]Dahlman, above, note 7.

In the absence of nonexclusiveness and nonrivalry, the specification of rights influences the distribution of income and wealth and (given costs and income effects) resource allocation. This influence is important, but not special to the concept of externality. There are (at least in theory) an infinite number of Pareto-efficient solutions, each differing in its allocative and distributional implications, and each associated with some unique initial distribution of endowments including rights.

Since externality is nothing special, it follows that the weak Coase Theorem is not specifically about externality. The weak Coase Theorem is a general theorem (or perhaps a tautology as is common with economic "theorems") about the existence of markets. It says: unless trade is impeded, trade will eliminate unrealized gains from trade, ensuring efficiency *by definition*. While this proposition, thus stated, is obvious with the benefit of a generation of post-Coasian hindsight, the Coasian analysis made an extremely valuable contribution to economics. By focusing so mercilessly on the logic of markets, it led economists to look for markets where none had previously been suspected and to ask the right questions about those cases in which no observable markets were found.

The weak Coase Theorem ensures that "externality" will be efficiently abated, but not that nonexclusive and/or nonrival goods will be be provided in efficient quantities. Since the major environmental problems—for example, air and water pollution—concern nonexclusive and/or nonrival goods, it follows that the Coase Theorem is not a theorem central to environmental economics, in spite of some early interpretations to that effect.[21]

To summarize, the weak Coase Theorem is seen not as a theorem about externality and environmental quality, but as a general theorem about the existence of markets. It draws attention to the imperatives of trade and the instability of situations characterized by unrealized gains from trade. It performs for economics a service similar to that performed for physics by the dictum: nature abhors a vacuum.

IV

While externality alone cannot persist, events may occur which shock a previously stable system. The invasion of the wheat field by cattle, in Coase's famous example,[22] was just such an event. Otherwise, the expecta-

[21]For example, Bramhall and Mills, *"A Note on the Asymmetry Between Fees and Payments,"* 2 *Water Resources Research* 615 (1966); Kamien, Schwartz and Dolbear, *"Asymmetry Between Bribes and Charges,"* 2 *Water Resources Research* 147 (1966); and A. Kneese and B. Bower, *Managing Water Quality: Economics Technology, Institutions* (1968).
[22]Coase, above, note 2.

tion of trampling damage would have precluded, one way or another, the planting of wheat. The event was (1) unexpected, and (2) an invasion, an attack on the wheat grower's property generically different from, for example, a drastic fall in the world price of wheat.

While the wheat-cattle example involved only exclusive and rival goods, conceivably events may occur to shock a previously stable system that includes some nonexclusive and/or nonrival goods. For both types of events, the specification of property rights will determine whether or not the welfare positions of those who had established themselves in equilibrium positions with respect to the previously stable environment will be protected. Will they be permitted to choose between maintaining their previous positions and trading to preferred positions, or will they be placed in a "pay or suffer" situation offering at best the hope of trading to a slightly better position in which they pay some and suffer some?

For events which involve only exclusive rival goods, the specification of rights will influence the eventual post-shock resource allocation. Where nonexclusive and/or nonrival goods are involved, the effect on post-shock resource allocation may be quite drastic, and the result may diverge substantially from a Samuelson (i.e., quantity efficient) solution.

The possibility of these kinds of events raises two different questions. First, what protections should be provided for individuals in the face of externally imposed (or threatened) welfare shock? Second, where the post-shock situation is inefficient, what institutional devices should be used to restore a reasonable degree of efficiency in the aggregate product mix and consumption bundle? While much of the existing literature attempts to handle these questions together, some advantages may be gained by considering them sequentially.

The first question is one of security of individual expectations. Non-attenuated property rights encourage (and, in the best of all worlds, guarantee) efficiency, but by what mechanisms do they perform such desirable services? The answer is by encouraging resource reallocation, as needed, in an environment of secure rights.

The individual enjoys protection for his person and his property from physical attack, dispossession, invasion, trespass and nuisance. Anyone who is sure he could use that individual's resources more profitably must buy those resources in voluntary exchange, thereby compensating the individual to the full extent of his perceived loss. The individual, having made prudent investments in securing his property—i.e., bearing his share of the transactions costs in a non-frictionless world—is free to use his resources as he sees best, substantially secure in the expectation that the rewards from so doing will accrue to him.

The individual's welfare position is not guaranteed. Changes in the pattern of relative scarcity and technological innovations may threaten the individual. Successful adjustments are then required of those who seek to maintain their welfare positions, and those who fail will be disadvantaged.

In this way incentives for continual adjustment and resource reallocation are maintained in a world where scarcity and technology are dynamic.

The role of technology bears closer examination. New technology may render the individual's skills and resources less valuable in the competitive arena, but may not violate his personal or property rights through attack, invasion or trespass.

One can make a sound case that property rights should be used to protect individuals from attack and dispossession and their property from invasion, trespass and nuisance, and that this protection should be broad-based and pervasive. The argument is most fully developed in the writings of James M. Buchanan, who took his cue from the emphasis in Coase's seminal article on the importance of secure property rights in providing a basis for conflict-resolving trade.[23]

In its most complete elaboration, the Coase–Buchanan tradition proposes a two-stage, constitutional-contractarian, approach. Following initial establishment of rights in an admittedly idealized constitutional stage, all subsequent reassignment would be through voluntary exchange (which implements the *strict* Pareto-improvement criterion). Stage 2 has the decided advantages of providing a secure basis for rights, promoting social stability, and ensuring that those affected by invasive shock events would be protected. Stagnation would not, however, follow, since desirable invasions would proceed, with full compensation of those threatened. Stage 1 is idealized and unrealistic in its "veil of ignorance" aspects, but is included because Stage 2, alone, would enshrine a *status quo* which could not command universal consent.

The idea of once-and-for-all assignment (or reassignment) of rights followed by consensual change thereafter has much to recommend it. However, its application to nonexclusive and/or nonrival goods is limited. In societies which seek to establish and maintain nonattentuated structures of rights, inefficiency persists in cases where it is prohibitively expensive to achieve the necessary exclusion or hyperexclusion at the individual level.

An alternative post-Coasian tradition, here called the Coase–Posner tradition, takes its cue from the discussion (in the later sections of Coase's seminal article) which, recognizing the asymmetry introduced by positive transactions costs, suggests assigning liability so as to minimize total costs or maximize the aggregate net value of product.[24] This approach is, presumably, adaptable to situations where nonexclusiveness and/or nonrivalry pose problems for pure Coase–Buchanan approaches.

The Coase–Posner tradition is hospitable to, and in some versions

[23] *See* J. Buchanan, *Freedom in Constitutional Contract* (1977), and Buchanan, *The Limits of Liberty: Between Anarchy and Leviathan* (1975).
[24] *See* R. Posner, *Economic Analysis of Law* (1972); note that Dahlman, above, note 7, places himself firmly in the Coase–Posner tradition, in the final two sections of his paper.

actively promotes, a case by case post-shock determination of property rights on benefit cost grounds. Where nonexclusiveness and/or nonrivalry are involved, the Coase–Posner solution would involve some combination of property rights reassignment and direct government actions, aimed at promoting the result which most nearly satisfies the objectives of providing such goods in Samuelson-efficient amounts while minimizing transactions costs. One can be sympathetic with the desire to see law promote efficiency, which obviously underlies Coase–Posner thinking. However, the problem is that expectations in a Coase–Posner world are no longer secured by (explicit or implicit) property rights. Implicit property rights— those assumed to exist securely because the threat has yet to be introduced—would be entirely insecure. Explicit property rights would also be insecure, since they could be changed whenever changes in technology and/or relative scarcity tilted the benefit cost ratio in favor of some alternative rights assignment.

This insecurity of property rights is undesirable for several reasons. Security of rights provides a sound basis for economic decisions, especially those with longer time profiles, such as saving and investment. More generally, it is an essential component of the whole legitimizing process. Rights that are "right" (i.e., recognized as legitimate) are not merely easier to enforce; they provide a sound basis for long-term social stability.[25] Rights which shift with the benefit cost numbers are unlikely to enjoy the aura of legitimacy. Further, stability of rights discourages self-interested investment in institution-changing behaviors while unstable rights encourage it. Voluntary exchange is an unimpeachable method of conflict resolution. However, for the individual who wants a right he does not have, voluntary exchange is the method of last resort.[26] Again, rights which shift with the benefit cost numbers would tend to discourage voluntary exchange, while encouraging efforts to generate and gain recognition for the kind of benefit cost data which would ensure reassignment of rights in a Coase–Posner world.[27]

These arguments suggest that Coase–Posner solutions are unacceptable in cases where nonattenuated property rights can be maintained. Where nonexclusiveness and nonrivalry persist, Coase–Posner approaches could

---

[25]The fundamental importance of this point is recognized by writers as diverse as J. Buchanan, *Freedom in Constitutional Contract* 93 (1977) and J. Commons, *The Legal Foundations of Capitalism* 325, 330 (1924).

[26]The rational individual who wants a right he does not have may, in an environment of unstable rights, rank his alternatives as follows: (1) ask an executive agency to give the right to him; (2) ask a judge to give it to him; (3) ask a legislature to give it to him; and (4) if all the above fail, attempt to buy it. In an environment of stable and secure rights, the voluntary exchange option becomes immeasurably more attractive.

[27]Note Dahlman's note 7, at 161, argument that benefit cost analysts should regard the Coasian (or, more accurately, the Coase–Posner–Dahlman) argument as entirely encouraging in its implications for their employment prospects.

be accepted only if the Coase–Buchanan tradition is clearly unadaptable to such cases. However, a pragmatic acceptance of the fact that prohibitive costs of exclusion and hyperexclusion will continue to relegate some important resources, goods and amenities to the nonexclusive and nonrival categories does not, *per se,* invalidate the concept that secure rights should provide protection from invasion and attack.

The concept of *res communis* offers an imperfect but by no means unthinkable solution. Where individualized exclusion is infeasible, why not provide for exclusion at the community level? Surely that would be better than either abandoning the objectives of efficient resource allocation and protection from invasion where nonexclusive and/or nonrival goods are involved, or pursuing efficiency at the cost of Coase–Posner instability of rights.

Proposals in this spirit would involve some form of property rights assigned at the collective rather than at the individual level. Yet, to exploit fully Coase–Buchanan concepts, individuals and small communities of interest would need considerable veto power over the collective decision process in order to protect themselves from majority-imposed welfare shock.

Of course, the current regulatory approach to, for example, air and water pollution control and the mainstream economists' most frequently proposed alternative, the Pigovian tax,[28] involve the assertion of rights to nonexclusive and nonrival environmental resources at the collective level. Yet, these approaches have their problems. Vesting of collective rights at the national level provides no guarantee of protection for communities invaded by greater-than-typical pollution loads, and thus violates the Coase–Buchanan spirit. More generally, the prospect that emissions standards or Pigovian tax rates may change, perhaps frequently, as the cognizant administrator's perceptions of the marginal benefits and costs of pollution abatement change, has more in common with the Coase–Posner tradition than the Coase–Buchanan tradition.

A recent proposal for rights in various nonexclusive and nonrival "quality of life" goods, vested at the community level, deserves consideration, since it is firmly in the Coase–Buchanan tradition.[29] While recognizing that action (or inaction) must be at the community level, this proposal preserves the Coase–Buchanan notion of secure explicit and implicit property rights and protection from externally-imposed shock invasions. Given the "need" to site a locally obnoxious facility somewhere (or, equally

---

[28]Note that this discussion of the Pigovian tax does not resurrect the concept of externality. The state-of-the-art proof of the "optimality" (the Samuelson-efficiency) of such taxes relies not on the concept of externality but on the notion of pollution as a nonrival discommodity. *See* W. Baumol and W. Oates, *The Theory of Environmental Policy* 33 (1975).

[29]*See* O'Hare, *"Not on My Block, You Don't: Facility Sitting and the Strategic Importance of Compensation,"* 25 *Pub. Pol.* 407 (1977).

plausibly, the "need" to impose a heavy pollution load somewhere), a compensation auction would be held and the low bidding community would become the fully compensated—and, therefore, happy—host of an invading force which would otherwise be resented.[30] If no community submitted a "low enough" bid, the "need" for the facility would be reevaluated. Thus, a market would be established which would tend to achieve efficiency in determining the total number of such facilities and their locations.

Two crucial details have yet to be worked out: (1) the collective decision process leading to submission of the community bid, and (2) the mechanism for determination of intra-community compensation in cases where various sections of the community were differently impacted. While the vesting of rights at the community, rather than the national, level represents a move which is firmly in the Coase–Buchanan tradition, the proper relationships between the community and its component individuals and groups have yet to be elucidated.

Nevertheless, the idea of rights to protection from externally imposed invasions threatening welfare shock, vested at the community level, has much to recommend it. It seems to have the potential (more so than most of the alternatives) to provide solutions to the two problems identified early in this section: (1) to provide protection for individuals who face externally threatened welfare shock, and (2) to approach a Samuelson solution which provides nonexclusive and nonrival goods in efficient amounts.

V

This article has examined the conventional wisdom notions of market failure, and found them wanting. The idea that individualistic markets do not provide certain kinds of goods efficiently, or even passably well, is not rejected. "So," you might say, "market failure lives." Not so. If the conventional diagnoses and analyses of market failure cannot withstand the rigors of deductive logic, they do disservices of at least two kinds: (1) they provide an inviting target for those who tend to overrate, absolutely and perhaps relatively, the capacities of individualistic markets, giving them an unnecessary advantage in debate, and (2) they tend to misdirect the efforts of those who seek workable solutions to the problems posed by nonexclusive and nonrival goods.

The notions of "common property resources" and "public goods" are

---

[30]Japan has introduced a program in which owners of power plants and some kinds of chemical industries compensate communities disadvantaged by their location, See OECD, *Environmental Policies in Japan* (1977).

rejected as imprecise, confusing and, in some applications, downright misleading. The concepts of nonexclusiveness and nonrivalry, phenomena which may occur separately or together, are precise and lead to correct analyses. Congestible goods, which are nonrival for users fewer than some threshold number but intensely rival as the capacity constraint is approached, are recognized. The congestible goods model may also serve as a means for diagnosing and analyzing problems attributed to "natural monopoly."

Externality is found to be nothing special: merely an inefficient disequilibrium situation which cannot persist alone. The inefficiency can persist if nonexclusiveness and nonrivalry are involved but, in that case, it is attributable to nonexclusiveness and nonrivalry, not to externality.

Nonexclusiveness is attributable to institutional or technological conditions. Where the technology of exclusion is tolerably inexpensive, market logic would suggest that efficiency can be promoted by establishing exclusive institutions. However, the physical properties of some goods are such that exclusion is prohibitively expensive with existing technology. In addition, for nonrival goods simple exclusion is insufficient to ensure efficiency: hyperexclusion, which is by and large technically infeasible, is needed. For these kinds of goods, institutional reforms permitting Pareto-efficient provision via the market are completely elusive. Yet, it is possible that progress toward provision of Samuelson-efficient quantities can be achieved via the establishment of *res communis* rights vested in appropriate collectives. Pareto-efficiency would remain unattainable, but to reject such solutions for that reason would be to fall prey to the "grass is greener" fallacy by comparing imperfect collective solutions with perfect, but unattainable individualistic solutions.[31]

Thus, the only thing which can be salvaged from the conventional wisdom idea of market failure is the possibility that collective institutions might be able to provide nonexclusive and/or nonrival goods in quantities approaching the efficient amount (but probably not at near-efficient prices) while individualistic markets cannot. This possibility is not an especially robust nor attractive survivor, since (1) the collective alternative to individual markets is imperfect, (2) its superior performance, in terms of providing near-efficient quantities, cannot be assumed but must be established on a case by case basis, and (3) its coercive aspects are unlikely to be entirely eliminated. Nevertheless, the collective alternative is too important to be ignored.

Finally, events that threaten attacks on the person, dispossesion of property, trespass and nuisance may occur. Such events may impose a welfare shock on affected individuals. Resources may be reallocated, at least to some degree, and where nonexclusive and nonrival goods are involved, the resultant resource allocation may deviate very substantially

---

[31] *See* Demsetz, above, note 2.

from the quantity-efficient. Given such possibilities, one may ask (1) what protections should be provided for individuals thus threatened, and (2) what mechanism should be used to restore the aggregate consumption bundle to a reasonably quantity-efficient condition? The post-Coasian literature has produced two distinct approaches, which the author has labelled the Coase–Posner and Coase–Buchanan approaches. For reasons having to do with the promotion of social stability and the security of individual expectations, as well as the promotion of voluntary exchange as a conflict resolution device, the Coase–Buchanan tradition has more merit. While more customary approaches such as regulation and taxation of discommodities can be explained (and, to a limited degree, justified) as reasonable attempts to solve problems in a fundamentally imperfect world, the search for institutional devices more firmly in the Coase–Buchanan tradition is to be encouraged.

10

The Economics of Resources
or the Resources of Economics

ROBERT M. SOLOW
.

Robert M. Solow is Institute Professor of Economics at the
Massachusetts Institute of Technology, and a Nobel
Laureate in Economics.

It is easy to choose a subject for a distinguished lecture like this, before a
large and critical audience with a wide range of interests. You need a topic
that is absolutely contemporary, but somehow perennial. It should survey
a broad field, without being superficial or vague. It should probably bear
some relation to economic policy, but of course it must have some serious
analytical foundations. It is nice if the topic has an important literature in
the past of our subject—a literature which you can summarize brilliantly
in about eleven minutes—but it better be something in which economists
are interested today, and it should appropriately be a subject you have
worked on yourself. The lecture should have some technical interest,
because you can't waffle for a whole hour to a room full of professionals,
but it is hardly the occasion to use a blackboard.

I said that it is easy to choose a subject for the Ely Lecture. It has to
be, because twelve people, counting me, have done it.

I am going to begin with a quotation that could have come from
yesterday's newspaper, or the most recent issue of the *American Economic
Review*.

> Contemplation of the world's disappearing supplies of minerals, forests, and
> other exhaustible assets has led to demands for regulation of their exploitation.

"The Economics of Resources or the Resources of Economics," by
Robert M. Solow, from *American Economic Review*, vol. 64, no. 2
(May 1974), pp. 1–14. Reprinted by permission.

The feeling that these products are now too cheap for the good of future generations, that they are being selfishly exploited at too rapid a rate, and that in consequence of their excessive cheapness they are being produced and consumed wastefully has given rise to the conservation movement.

The author of those sentences is not Dennis Meadows and associates, not Ralph Nader and associates, not the President of the Sierra Club; it is a very eminent economic theorist, a Distinguished Fellow of this Association, Harold Hotelling, who died at the age of seventy-eight, just a few days ago. Like all economic theorists, I am much in his debt, and I would be happy to have this lecture stand as a tribute to him. These sentences appeared at the beginning of his article "The Economics of Exhaustible Resources," not in the most recent *Review,* but in the *Journal of Political Economy* for April 1931. So I think I have found something that is both contemporary and perennial. The world has been exhausting its exhaustible resources since the first cave-man chipped a flint, and I imagine the process will go on for a long, long time.

Mr. Dooley noticed that "th' Supreme Court follows the iliction returns." He would be glad to know that economic theorists read the newspapers. About a year ago, having seen several of those respectable committee reports on the advancing scarcity of materials in the United States and the world, and having, like everyone else, been suckered into reading the *Limits to Growth,* I decided I ought to find out what economic theory has to say about the problems connected with exhaustible resources. I read some of the literature, including Hotelling's classic article—the theoretical literature on exhaustible resources is, fortunately, not very large—and began doing some work of my own on the problem of optimal social management of a stock of a nonrenewable but essential resource. I will be mentioning some of the results later. About the time I finished a first draft of my own paper and was patting myself on the back for having been clever enough to realize that there was in fact something still to be said on this important, contemporary but somehow perennial topic—just about then it seemed that every time the mail came it contained another paper by another economic theorist on the economics of exhaustible resources. It was a little like trotting down to the sea, minding your own business like any nice independent rat, and then looking around and suddenly discovering that you're a lemming. Anyhow, I now have a nice collection of papers on the theory of exhaustible resources; and most of them are still unpublished, which is just the advantage I need over the rest of you.

A pool of oil or vein of iron or deposit of copper in the ground is a capital asset to society and to its owner (in the kind of society in which such things have private owners) much like a printing press or a building or any other reproducible capital asset. The only difference is that that natural resource is not reproducible, so the size of the existing stock can never increase through time. It can only decrease (or, if none is mined for a while, stay the same). This is true even of recyclable materials; the laws of thermodynamics and life guarantee that we will never recover a whole pound of secondary copper from a pound of primary copper in use, or a whole pound of tertiary copper from a pound of secondary copper in use. There is leakage at every round; and a formula just like the ordinary multiplier formula tells us how much copper use can be built on the world's initial endowment of copper, in terms of the recycling or recovery ratio. There is always less ultimate copper use left than there was last year, less by the amount dissipated beyond recovery during the year. So copper remains an exhaustible resource, despite the possibility of partial recycling.

A resource deposit draws its market value, ultimately, from the prospect of extraction and sale. In the meanwhile, its owner, like the owner of every capital asset, is asking: What have you done for me lately? The only way that a resource deposit in the ground and left in the ground can produce a current return for its owner is by appreciating in value. Asset markets can be in equilibrium only when all assets in a given risk class earn the same rate of return, partly as current dividend and partly as capital gain. The common rate of return is the interest rate for that risk class. Since resource deposits have the peculiar property that they yield no dividend so long as they stay in the ground, in equilibrium the value of a resource deposit must be growing at a rate equal to the rate of interest. Since the value of a deposit is also the present value of future sales from it, after deduction of extraction costs, resource owners must expect the net price of the ore to be increasing exponentially at a rate equal to the rate of interest. If the mining industry is competitive, net price stands for market price minus marginal extraction cost for a ton of ore. If the industry operates under constant costs, that is just market price net of unit extraction costs, or the profit margin. If the industry is more or less monopolistic, as is frequently the case in extractive industry, it is the marginal profit—marginal revenue less marginal cost—that has to be growing, and expected to grow, proportionally like the rate of interest.

This is the fundamental principle of the economics of exhaustible resources. It was the basis of Hotelling's classic article. I have deduced it as a condition of stock equilibrium in the asset market. Hotelling thought of it mainly as a condition of flow equilibrium in the market for ore: if net price is increasing like compound interest, owners of operating mines will be indifferent at the margin between extracting and holding at every instant of time. So one can imagine production just equal to demand at the current price, and the ore market clears. No other time profile for prices can elicit positive production in every period of time.

It is hard to overemphasize the importance of this tilt in the time profile for net price. If the net price were to rise too slowly, production would be pushed nearer in time and the resource would be exhausted quickly, precisely because no one would wish to hold resources in the ground and earn less than the going rate of return. If the net price were to rise too fast, resource deposits would be an excellent way to hold wealth, and owners would delay production while they enjoyed supernormal capital gains.

According to the fundamental principle, if we observe the market for an exhaustible resource near equilibrium, we should see the net price—or marginal profit—rising exponentially. That is not quite the same thing as seeing the market price to users of the resource rising exponentially. The price to consumers is the net price plus extraction costs, or the obvious analogy for monopoly. The market price can fall or stay constant while the net price is rising if extraction costs are falling through time, and if the net price or scarcity rent is not too large a proportion of the market price. That is presumably what has been happening in the market for most exhaustible resources in the past. (It is odd that there are not some econometric studies designed to find out just this. Maybe econometricians don't follow the election returns.) Eventually, as the extraction cost falls and the net price rises, the scarcity rent must come to dominate the movement of market price, so the market price will eventually rise, although that may take a very long time to happen. Whatever the pattern, the market price and the rate of extraction are connected by the demand curve for the resource. So, ultimately, when the market price rises, the current rate of production must fall along the demand curve. Sooner or later, the market price will get high enough to choke off the demand entirely. At that moment production falls to zero. If flows and stocks have been beautifully coordinated through the operations of futures markets or a planning board, the last ton produced will also be the last ton in the ground. The resource will be exhausted at the instant that it has priced itself out of the market. The Age of Oil or Zinc or Whatever It Is will have come to an end. (There is a limiting case, of course, in which demand goes asymptotically to zero as the price rises to infinity, and the resource is exhausted only asymptotically. But it is neither believable nor important.)

Now let us do an exercise with this apparatus. Suppose there are two sources of the same ore, one high-cost and the other low-cost. The cost difference may reflect geographical accessibility and transportation costs, or some geological or chemical difference that makes extraction cheap at one site and dear at the other. The important thing is that there are cost differences, though the final mineral product is identical from both sources.

It is easy to see that production from both sources cannot coexist in the market for any interval of time. For both sources to produce, net price for each of them must be growing like compound interest at the market rate. But they must market their ore at the same price, because the product is identical. That is arithmetically impossible, if their extraction costs differ.

So the story has to go like this. First one source operates and supplies the whole market. Its net price rises exponentially, and the market price moves correspondingly. At a certain moment, the first source is exhausted. At just that moment and not before, it must become economical for the second source to come into production. From then on, the world is in the single-source situation: the net price calculated with current extraction costs must rise exponentially until all production is choked off and the second source is exhausted. (If there are many sources, you can see how it will work.)

Which source will be used first? Your instinct tells you that the low-cost deposit will be the first one worked, and your instinct is right. You can see why, in terms of the fundamental principle. At the beginning, if the high-cost producer is serving the market, the market price must cover high extraction costs plus a scarcity rent that is growing exponentially. The low-cost producer would refrain from undercutting the price and entering the market only if his capital gains justify holding off and entering the market later. But just the reverse will be true. Any price high enough to keep the high-cost producer in business will tempt the low-cost producer to sell ore while the selling is good and invest the proceeds in any asset paying the market rate of interest. So it must be that the low-cost producer is the first to enter. Price rises and output falls. Eventually, at precisely the moment when the low-cost supply is exhausted, the price has reached a level at which it pays the high-cost producer to enter. From then on, *his* net price rises exponentially and production continues to fall. When cumulative production has exhausted the high-cost deposit, the market price must be such as to choke the demand off to zero—or else just high enough to tempt a still higher-cost source into production. And so it goes. Apart from market processes, it is actually socially rational to use the lower-cost deposits before the higher-cost ones.

You can take this story even further, as William Nordhaus has done in connection with the energy industry. Suppose that, somewhere in the background, there is a technology capable of producing or substituting for a mineral resource at relatively high cost, but on an effectively inexhaustible resource base. Nordhaus calls this a "backstop technology." (The nearest we now have to such a thing is the breeder reactor using U^{238} as fuel. World reserves of U^{238} are thought to be enough to provide energy for over a million years at current rates of consumption. If that is not a backstop technology, it is at least a catcher who will not allow a lot of passed balls. For a better approximation, we must wait for controlled nuclear fusion or direct use of solar energy. The sun will not last forever, but it will last at least as long as we do, more or less by definition.) Since there is no scarcity rent to grow exponentially, the backstop technology can operate as soon as the market price rises enough to cover its extraction costs (including, of course, profit on the capital equipment involved in production). And as soon as that happens, the market price of the ore or its substitute stops rising. The "backstop technology" provides a ceiling for the market price of the natural resource.

The story in the early stages is as I have told it. In the beginning, the successive grades of the resource are mined. The last and highest-cost source gives out just when the market price has risen to the point where the backstop technology becomes competitive. During the earlier phases, one imagines that resource companies keep a careful eye on the prospective costs associated with the backstop technology. Any laboratory success or failure that changes those prospective costs has instantaneous effects on the capital value of existing resource deposits, and on the most profitable rate of current production. In actual fact, those future costs have to be regarded as uncertain. A correct theory of market behavior and a correct theory of optimal social policy will have to take account of technological uncertainty (and perhaps also uncertainty about the true size of mineral reserves).

Here is a mildly concrete illustration of these principles. There is now a workable technology for liquefying coal—that is, for producing synthetic crude oil from coal.[1] Nordhaus puts the extraction-and-preparation cost at the equivalent of seven or eight 1970 dollars per barrel of crude oil, including amortization and interest at 10 percent on the plant; I have heard higher and lower figures quoted. If coal were available in unlimited amounts, that would be all. But, of course, coal is a scarce resource, though more abundant than drillable petroleum, so a scarcity rent has to be added to that figure, and the rent has to be increasing like the rate of interest during the period when coal is being used for this purpose.

In the meanwhile, the extraction and production cost for this technology is large compared with the scarcity rent on the coal input, so the market price at which the liquefied-coal-synthetic-crude activity would now be economic is rising more slowly than the rate of interest. It may even fall if there are cost-reducing technological improvements; and that is not unlikely, given that research on coal has not been splashed as liberally with funds as research on nuclear energy. In any case, political shenanigans and monopoly profits aside, scarcity rents on oil form a larger fraction of the market price of oil, precisely because it is a lower cost fuel. The price of a barrel of oil should therefore be rising faster than the implicit price at which synthetic crude from coal could compete. One day those curves will intersect, and that day the synthetic-crude technology will replace the drilled-petroleum technology.

Even before that day, the possibility of coal liquefaction provides a kind of ceiling for the price of oil. I say "kind of" to remind you that coal-mining and moving capacity and synthetic-crude plant cannot be created overnight. One might hope that the ceiling might also limit the consuming world's vulnerability to political shenanigans and monopoly profits. I suppose it does in some ultimate sense, but one must not slide

[1]As best one can tell at the moment, shale oil is a more likely successor to oil and natural gas than either gasified or liquefied coal. The relevant costs are bound to be uncertain until more research and development has been done. I tell the story in terms of liquefied coal only because it is more picturesque that way.

over the difficulties: for example, who would want to make a large investment in coal liquefaction or coal gasification in the knowledge that the current price of oil contains a large monopoly element that could be cut, at least temporarily, if something like a price war should develop?

The fundamental principle of the economics of exhaustible resources is, as I have said, simultaneously a condition of flow equilibrium in the market for the ore and of asset equilibrium in the market for deposits. When it holds, it says quite a lot about the probable pattern of exploitation of a resource. But there are more than the usual reasons for wondering whether the equilibrium conditions have any explanatory value. For instance, the flow market that has to be cleared is not just one market; it is the sequence of markets for resource products from now until the date of exhaustion. It is, in other words, a sequence of futures markets, perhaps a long sequence. If the futures markets actually existed, we could perhaps accept the notion that their equilibrium configuration is stable; that might not be true, but it is at least the sort of working hypothesis we frequently accept as a way of getting on with business. But there clearly is not a full set of futures markets; natural-resource markets work with a combination of myopic flow transactions and rather more farsighted asset transactions. It is legitimate to ask whether observed resource prices are to be interpreted as approximations to equilibrium prices, or whether the equilibrium is so unstable that momentary prices are not only a bad indicator of equilibrium relationships, but also a bad guide to resource allocation.

That turns out not to be an easy question to answer. Flow considerations and stock considerations work in opposite directions. The flow markets by themselves could easily be unstable; but the asset markets provide a corrective force. Let me try to explain why.

The flow equilibrium condition is that the net price grow like compound interest at the prevailing rate. Suppose net prices are expected by producers to be rising too slowly. Then resource deposits are a bad way to hold wealth. Mine owners will try to pull out; and if they think only in flow terms, the way to get out of the resource business is to increase current production and convert ore into money. If current production increases, for this or any other reason, the current price must move down along the demand curve. So initially pessimistic price expectations on the part of producers have led to more pressure on the current price. If expectations about future price changes are responsive to current events, the consequence can only be that pessimism is reinforced and deepened. The initial disequilibrium is worsened, not eliminated, by this chain of events. In other words, the market mechanism I have just described is unstable. Symmetrical reasoning leads to the conclusion that if prices are initially expected to be rising too fast, the withholding of supplies will lead to a speculative run-up of prices which is self-reinforcing. Depending on which way we start, initial disequilibrium is magnified, and production is tilted either toward excessive current dumping or toward speculative withholding of supply. (Still other assumptions are possible and may lead to qualitatively different results. For instance, one could imagine that expectations focus on the price level rather than its rate of change. There is much more work to be done on this question.)

Such things have happened in resource markets; but they do not seem always to be happening. I think that this story of instability in spot markets needs amendment; it is implausible because it leaves the asset market entirely out of account. The longer run prospect is not allowed to have any influence on current happenings. Suppose that producers do have some notion that the resource they own has a value anchored somewhere in the future, a value determined by technological and demand considerations, not by pure and simple speculation. Then if prices are now rising toward that rendezvous at too slow a rate, that is indeed evidence that owning resource deposits is bad business. But that will lead not to wholesale dumping of current production, but to capital losses on existing stocks. When existing stocks have been written down in value, the net price can rise toward its future rendezvous at more or less the right rate. As well as being destabilized by flow reactions, the market can be stabilized by capitalization reactions. In fact the two stories can be made to merge: the reduction in flow price coming from increased current production can be read as a signal and capitalized into losses on asset values, after which near-equilibrium is reestablished.

I think the correct conclusion to be drawn from this discussion is not that either of the stories is more likely to be true. It is more complex: that in tranquil conditions, resource markets are likely to track their equilibrium paths moderately well, or at least not likely to rush away from them. But resource markets may be rather vulnerable to surprises. They may respond to shocks about the volume of reserves, or about competition from new materials, or about the costs of competing technologies, or even about near-term political events, by drastic movements of current price and production. It may be quite a while before the transvaluation of values—I never thought I could quote Nietzsche in an economics paper—settles down under the control of sober future prospects. In between, it may be a cold winter.

So far, I have discussed the economic theory of exhaustible resources as a partial-equilibrium market theory. The interest rate that more or less controls the whole process was taken as given to the mining industry by the rest of the economy. So was the demand curve for the resource itself. And when the market price of the resource has ridden up the demand curve to the point where the quantity demanded falls to zero, the theory says that the resource in question will have been exhausted.

There is clearly a more cosmic aspect to the question than this; and I do not mean to suggest that it is unimportant, just because it is cosmic. In particular, there remains an important question about the social interest in the pace of exploitation of the world's endowment of exhaustible natural resources. This aspect has been brought to a head recently, as everyone knows, by the various Doomsday forecasts that combine a positive finding that the world is already close to irreversible collapse from shortage of natural resources and other causes with the normative judgment that civilization is much too young to die. I do not intend to discuss those forecasts and judgments now, but I do want to talk about the economic issues of principle involved.

First, there is a proposition that will be second nature to everyone in this room. What I have called the fundamental principle of the economics of exhaustible resources is, among other things, a condition of competitive equilibrium in the sequence of futures markets for deliveries of the natural resource. This sequence extends out to infinity, even if the competitive equilibrium calls for the resource to be exhausted in finite time. Beyond the time of exhaustion there is also equilibrium: supply equals demand equals zero at a price simultaneously so high that demand is choked off and so low that it is worth no one's while to lose interest by holding some of the resource that long. Like any other competitive equilibrium with the right background assumptions, this one has some optimality properties. In particular, as Hotelling pointed out, the competitive equilibrium maximizes the sum of the discounted consumer-plus-producer surpluses from the natural resource, *provided* that society wishes to discount future consumer surpluses at the same rate that mine owners choose to discount their own future profits.

Hotelling was not so naive as to leap from this conclusion to the belief that *laissez-faire* would be an adequate policy for the resource industries. He pointed to several ways in which the background assumptions might be expected to fail: the presence of externalities when several owners can exploit the same underground pool of gas or oil; the considerable uncertainty surrounding the process of exploration with the consequent likelihood of wasteful rushes to stake claims and exploit, and the creation of socially useless windfall profits; and, finally, the existence of large monopolistic or oligopolistic firms in the extractive industries.

There is an amusing sidelight here. It is not hard to show that, generally speaking, a monopolist will exhaust a mine more slowly than a competitive industry facing the same demand curve would do. (Hotelling did not explore this point in detail, though he clearly knew it. He did mention the possibility of an extreme case in which competition will exhaust a resource in finite time and a monopolist only asymptotically.) The amusing thing is that if a conversationist is someone who would like to see resources conserved *beyond* the pace that competition would adopt, then the monopolist is the conservationist's friend. No doubt they would both be surprised to know it.

Hotelling mentions, but rather pooh-poohs, the notion that market rates of interest might exceed the rate at which society would wish to discount future utilities or consumer surpluses. I think a modern economist would take that possibility more seriously. It is certainly a potentially important question, because the discount rate determines the whole tilt of the equilibrium production schedule. If it is true that the market rate of interest exceeds the social rate of time preference, then scarcity rents and market prices will rise faster than they "ought to" and production will have to fall correspondingly faster along the demand curve. Thus the resource will be exploited too fast and exhausted too soon.

The literature has several reasons for expecting that private discount rates might be systematically higher than the correct social rate of discount. They fall into two classes. The first class takes it more or less for granted that society ought to discount utility and consumption at the same rates as reflective individuals would discount their own future utility and consumption. This line of thought then goes on to suggest that there are reasons why this might not happen. One standard example is the fact that individuals can be expected to discount for the riskiness of the future, and some of the risks for which they will discount are not risks to society but merely the danger of transfers within the society. Since there is not a complete enough set of insurance markets to permit all these risks to be spread properly, market interest rates will be too high. Insecurity of tenure, as William Vickrey has pointed out, is a special form of uncertainty with particular relevance to natural resources.

A second standard example is the existence of various taxes on income from capital; since individuals care about the after-tax return on investment and society about the before-tax return, if investment is carried to the point where the after-tax yield is properly related to the rate of time preference, the before-tax profitability of investment will be too high. I have nothing to add to this discussion.

The other class of reasons for expecting that private discount rates are too high and will thus distort intertemporal decisions away from social optimality denies that private time preference is the right basis for intertemporal decisions. Frank Ramsey, for instance, argued that it was ethically indefensible for society to discount future utilities. Individuals might do so, either because they lack imagination (Böhm-Bawerk's "defective telescopic faculty") or because they are all too conscious that life is short. In social decision-making, however, there is no excuse for treating generations unequally, and the time-horizon is, or should be, very long. In solemn conclave assembled, so to speak, we ought to act as if the social rate of time preference were zero (though we would simultaneously discount future *consumption* if we expect the future to be richer than the present). I confess I find that reasoning persuasive, and it provides another reason for expecting that the market will exhaust resources too fast.

This point need not be divorced so completely from individual time preference. If the whole infinite sequence of futures markets for resource products could actually take place and find equilibrium, I might be inclined to accept the result (though I would like to know who decides the initial endowments within and between generations). But of course they cannot take place. There is no way to collect bids and offers from everyone who will ever live. In the markets that actually do take place, future generations are represented only by us, their eventual ancestors. Now generations overlap, so that I worry about my children, and they about theirs, and so on. But it does seem fundamentally implausible that there should be anything *ex post* right about the weight that is actually given to the welfare of those who will not live for another thousand years. We have actually done quite well at the hands of *our* ancestors. Given how poor they were and how rich we are, they might properly have saved less and consumed more. No doubt they never expected the rise in income per head that has made us so much richer than they ever dreamed was possible. But that only reinforces the point that the future may be too important to be left to the accident of mistaken expectations and the ups and downs of the Protestant ethic.

Several writers have studied directly the problem of defining and characterizing a socially-optimal path for the exploitation of a given pool of exhaustible resources. The idea is familiar enough: instead of worrying about market responses, one imagines an idealized planned economy, constrained only by its initial endowment, the size of the labor force, the available technology, and the laws of arithmetic. The planning board then has to find the best feasible development for the economy. To do so, it needs a precise criterion for comparing different paths, and that is where the social rate of time preference plays a role.

It turns out that the choice of a rate of time preference is even more critical in this situation than it is in the older literature on optimal capital accumulation without any exhaustible resources. In that theory, the criterion usually adopted is the maximization of a discounted sum of one-period social welfare indicators, depending on consumption per head, and summed over all time from now to the infinite future. The typical result, depending somewhat on the particular assumptions made, is that consumption per head rises through time to a constant plateau defined by the "modified Golden Rule." In that ultimate steady state, consumption per head is lower the higher is the social rate of discount; and, correspondingly, the path to the steady state is characterized by less saving and more interim consumption, the higher the social rate of discount. That is as it should be: the main beneficiaries of a high level of ultimate steady-state consumption are the inhabitants of the distant future, and so, if the planning board discounts the future very strongly, it will choose a path that favors the near future over the distant future.

When one adds exhaustible resources to the picture, the social rate of time preference can play a similar, but even more critical, role. As a paper by Geoffrey Heal and Partha Dasgupta and one of my own show, it is possible that the optimal path with a positive discount rate should lead to consumption per head going asymptotically to zero, whereas a zero discount rate leads to perpetually rising consumption per head. In other words, even when the technology and the resource base could permit a plateau level of consumption per head, or even a rising standard of living, positive social time preference might in effect lead society to prefer eventual extinction, given the drag exercised by exhaustible resources. Of course, it is part of the point that it is the planning board in the present that plans for future extinction: nobody has asked the about-to-become-defunct last generation whether *it* approved of weighting its satisfactions less than those of its ancestors.

Good theory is usually trying to tell you something, even if it is not the literal truth. In this context, it is not hard to interpret the general tenor of the theoretical indications. We know in general that even well-functioning competitive markets may fail to allocate resources properly over time. The reason, I have suggested, is because, in the nature of the case, the future brings no endowment of its own to whatever markets actually exist. The intergenerational distribution of income or welfare depends on the provision that each generation makes for its successors. The choice of a social discount rate is, in effect, a policy decision about that intergenerational distribution. What happens in the planning parable depends very much—perhaps dramatically—on that choice; and one's evaluation of what happens in the market parable depends very much on whether private choices are made with a discount rate much larger than the one a deliberate policy decision would select. The pure theory of exhaustible resources is trying to tell us that, if exhaustible resources really matter, then the balance between present and future is more delicate than we are accustomed to think; and then the choice of a discount rate can be pretty important and one ought not to be too casual about it.

In my own work on this question, I have sometimes used a rather special criterion that embodies sharp assumptions about intergenerational equity: I have imposed the requirement that consumption per head be constant through time, so that no generation is favored over any other, and asked for the largest steady consumption per head that can be maintained forever, given all the constraints including the finiteness of resources. This criterion, like any other, has its pluses and its minuses and I am not committed to it by any means. Like the standard criterion—the discounted sum of one-period utilities—this one will always pick out an *efficient* path, so one at least gets the efficiency conditions out of the analysis. The highest-constant-consumption criterion also has the advantage of highlighting the crucial importance of certain technological assumptions.

It is clear without any technical apparatus that the seriousness of the resource-exhaustion problem must depend in an important way on two aspects of the technology: first, the likelihood of technical progress, especially natural-resource-saving technical progress, and, second, the ease with which other factors of production, especially labor and reproducible capital, can be substituted for exhaustible resources in production.

My own practice, in working on this problem, has been to treat as the central case (though not the only case) the assumption of zero technological progress. This is not because I think resource-saving inventions are unlikely or that their capacity to save resources is fundamentally limited. Quite the contrary—if the future is anything like the past, there will be prolonged and substantial reductions in natural-resource requirements per unit of real output. It is true, as pessimists say, that it is just an assumption and one cannot be sure; but to assume the contrary is also an assumption, and a much less plausible one. I think there is virtue in analyzing the zero-technical-progress case because it is easy to see how technical progress can relieve and perhaps eliminate the drag on economic welfare exercised by natural-resource scarcity. The more important task for theory is to try to understand what happens or can happen in the opposite case.

As you would expect, the degree of substitutability is also a key factor. If it is very easy to substitute other factors for natural resources, then there is in principle no "problem." The world can, in effect, get along without natural resources, so exhaustion is just an event, not a catastrophe. Nordhaus's notion of a "backstop technology" is just a dramatic way of putting this case; at some finite cost, production can be freed of dependence on exhaustible resources altogether.

If, on the other hand, real output per unit of resources is effectively bounded—cannot exceed some upper limit of productivity which is in turn not too far from where we are now—then catastrophe is unavoidable. In-between there is a wide range of cases in which the problem is real, interesting, and not foreclosed. Fortunately, what little evidence there is suggests that there is quite a lot of substitutability between exhaustible resources and renewable or reproducible resources, though this is an empirical question that could absorb a lot more work than it has had so far.

Perhaps the most dramatic way to illustrate the importance of substitutability, and its connection with Doomsday, is in terms of the permanent sustainability of a constant level of consumption. In the simplest, most aggregative, model of a resource-using economy one can prove something like the following: if the elasticity of substitution between exhaustible resources and other inputs is unity or bigger, and if the elasticity of output with respect to reproducible capital exceeds the elasticity of output with respect to natural resources, then a constant population can maintain a positive constant level of consumption per head forever. This permanently maintainable standard of living is an increasing, concave, and unbounded function of the initial stock of capital. So the drag of a given resource pool can be overcome *to any extent* if only the initial stock of capital is large enough. On the other hand, if the elasticity of substitution between natural resources and other inputs is less than one, or if the elasticity of output with respect to resources exceeds the elasticity of output with respect to reproducible capital, then the largest constant level of consumption sustainable forever with constant population is—zero. We know much too little about which side of that boundary the world is on—technological progress aside—but at least the few entrails that have been read seem favorable.[2]

Perhaps I should mention that when I say "forever" in this connection, I mean "for a very long time." The mathematical reasoning does deal with infinite histories, but actually life in the solar system will only last for a finite time, though a very long finite time, much longer than this lecture, for instance. That is why I think it takes economics as well as the entropy law to answer our question.

I began this lecture by talking of the conditions for competitive equilibrium in the market for natural resources. Now I have been talking of centralized planning optima. As you would expect, it turns out that under the standard assumptions, the Hotelling rule, the fundamental principle of natural-resource economics, is a necessary condition for efficiency and therefore for social optimality. So there is at least a prayer that a market-guided system might manage fairly well. But more than the Hotelling condition is needed.

I have already mentioned one of the extra requirements for the intertemporal optimality of market allocations: it is that the market discount future profits at the same rate as the society would wish to discount the welfare of future inhabitants of the planet. This condition is often given as an argument for public intervention in resource allocation because—as I have also mentioned—there are reasons to expect market interest rates to exceed the social rate of time preference, or at least what philosophers like us think it ought to be. If the analysis is right, then the market will tend to consume exhaustible resources too fast, and corrective public intervention should be aimed at slowing down and stretching out the exploitation of the resource pool. There are several ways that could be done, in principle, through conservation subsidies or a system of graduated severance taxes, falling through time.

Realistically speaking, however, when we say "public intervention" we mean rough and ready political action. An only moderately cynical observer will see a problem here: it is far from clear that the political process can be relied on to be more future-oriented than your average corporation. The conventional pay-out period for business is of the same order of magnitude as the time to the next election, and transferring a given individual from the industrial to the government bureaucracy does not transform him into a guardian of the far future's interests. I have no ready solution to this problem. At a minimum, it suggests that one ought to be as suspicious of uncritical centralization as of uncritical free-marketeering. Maybe the safest course is to favor specific policies—like graduated severance taxes—rather than blanket institutional solutions.

[2]See pp. 60–70 in William D. Nordhaus and James Tobin.

There is another, more subtle, extra requirement for the optimality of the competitive market solution to the natural-resource problem. Many patterns of exploitation of the exhaustible-resource pool obey Hotelling's fundamental principle myopically, from moment to moment, but are wrong from a very long-run point of view. Such mistaken paths may even stay very near the right path for a long time, but eventually they veer off and become bizarre in one way or another. If a market-guided system is to perform well over the long haul, it must be more than myopic. Someone—it could be the Department of the Interior, or the mining companies, or their major customers, or speculators—must always be taking the long view. They must somehow notice in advance that the resource economy is moving along a path that is bound to end in disequilibrium of some extreme kind. If they do notice it, and take defensive actions, they will help steer the economy from the wrong path toward the right one.[3] Usually the "wrong" path is one that leads to exhaustion at a date either too late or too soon; anyone who perceives this will be motivated to arbitrage between present and future in ways that will push the current price toward the "right" path.[4]

It is interesting that this need for someone to take the long view emerged also when the question at hand was the potential instability of the market for natural resources if it concentrates too heavily on spot or flow decisions, and not enough on future or stock decisions. In that context too, a reasonably accurate view of the long-term prospects turns out to be a useful, maybe indispensable, thing for the resource market to have.

This lecture has been—as Kenneth Burke once said about the novel—words, all words. Nevertheless, it has been a discourse on economic theory, not on current policy. If some of you have been daydreaming about oil and the coming winter, I assure you that I have been thinking about shadow prices and transversality conditions at infinity. If I turn briefly to policy at the end, it is not with concrete current problems in mind. After all, nothing I have been able to say takes account of the international oil cartel, the political and economic ambitions of the Middle Eastern potentates, the speeds of adjustment to surprises in the supply of oil, or the doings of our own friendly local oligopolists. The only remarks I feel entitled to make are about the long-run pursuit of a general policy toward exhaustible resources.

---

[3]This sort of process has been studied in a different context by Frank Hahn and by Karl Shell and Joseph Stiglitz.

[4]For example, suppose the current price is too low, in the sense that, if it rises according to the current principle, the demand path will be enough to exhaust the resource before the price has risen high enough to choke demand to zero. A clever speculator would see that there will be money to be made just after the date of exhaustion, because anyone with a bit of the resource to sell could make a discrete jump in the price and still find buyers. Such a speculator would wish to buy now and hold for sale then. But that action would tend to raise the current price (and, by the fundamental principle, the whole price path) and reduce demand, so that the life of the resource would be prolonged. The speculation is thus corrective.

Many discussions of economic policy—macroeconomics aside—boil down to a tension between market allocation and public intervention. Marketeers keep thinking about the doughnut of allocative efficiency and informational economy and *dirigistes* are impressed with the size of the hole containing externalities, imperfections, and distributional issues. So it is with exhaustible resources. One is impressed with what a system of ideal markets, including futures markets, can accomplish in this complicated situation; and one can hardly miss seeing that our actual oligopolistic, politically involved, pollution-producing industry is not exactly what the textbook ordered. I have nothing new to add to all that. The unusual factor that the theory of exhaustible resources brings to the fore is the importance of the long view, and the value of reasonable information about reserves, technology, and demand in the fairly far future.

This being so, one is led to wonder whether public policy can contribute to stability and efficiency along those lines. One possibility is the encouragement of organized futures trading in natural resource products. To be useful, futures contracts would have to be much longer-term than is usual in the futures markets that now exist, mostly for agricultural products. I simply do not know enough to have an opinion about the feasibility of large scale futures trading, or about the ultimate contribution that such a reform would make to the stability and efficiency of the market for resource products. But in principle it would seem to be a good idea.

The same considerations suggest that the market for exhaustible resources might be one of the places in the economy where some sort of organized indicative planning could play a constructive role. This is not an endorsement of centralized decision-making, which is likely to have imperfections and externalities of its own. Indeed it might be enough to have the government engaged in a continuous program of information-gathering and dissemination covering trends in technology, reserves and demand. One could at least hope to have professional standards govern such an exercise. I take it that the underlying logic of indicative planning is that some comparison and coordination of the main participants in the market, including the government, could eliminate major errors and resolve much uncertainty. In the case of exhaustible resources, it could have the additional purpose of generating a set of consistent expectations about the distant future. In this effort, the pooling of information and intentions from both sides of the market could be useful, with the effect of inducing behavior that will steer the economy away from ultimately inferior paths. It is also likely, as Adam Smith would have warned, that a certain amount of conspiracy against the public interest might occur in such sessions, so perhaps they ought to be recorded and the tapes turned over to Judge Sirica, who will know what to do with them.

REFERENCES

P. Dasgupta and G. Heal, "The Optimal Depletion of Exhaustible Resources," *Rev. Econ. Stud.,* 1974.

F. H. Hahn, "Equilibrium Dynamics with Heterogeneous Capital Goods," *Quart. J. Econ.,* Nov. 1966, *80,* 633–646.

H. Hotelling, "The Economics of Exhaustible Resources," *J. Polit, Econ.,* April 1931, *39,* 137–175.

W. D. Nordhaus, "The Allocation of Energy Resources, *Brookings Papers on Econ. Activ.*

————— and J. Tobin, "Is Economic Growth Obsolete?" in National Bureau of Economic Research, *Economic Growth,* 50th Anniversary Colloq. V, New York 1972.

K. Shell and J. E. Stiglitz, "The Allocation of Investment in a Dynamic Economy," *Quart. J. Econ.,* Nov. 1967, *81.*

R. M. Solow, "Intergenerational Equity and Exhaustible Resources," *Rev. Econ. Stud.,* 1974.

J. E. Stiglitz, "Growth with Exhaustible Natural Resources," *Rev. Econ. Stud.,* 1974.

M. Weinstein and R. Zeckhauser, "Use Patterns for Depletable and Recyclable Resources," *Rev. Econ. Stud.,* 1974.

11

Sustainability: An Economist's Perspective

ROBERT M. SOLOW

.

Robert M. Solow is Institute Professor of Economics at the
Massachusetts Institute of Technology and a Nobel
Laureate in Economics.

This talk is different from anything else anyone has heard at Woods Hole;
certainly for the last two days. Three people have asked me, "Do you plan
to use any transparencies or slides?" Three times I said, "No," and three
times I was met with this blank stare of disbelief. I actually have some
beautiful aerial photographs of Prince William Sound that I could have
brought along to show you, and I also have a spectacular picture of
Michael Jordan in full flight that you would have liked to have seen. But
in fact I don't need or want any slides or transparencies. I want to talk to
you about an idea. The notion of sustainability or sustainable growth
(although, as you will see, it has nothing necessarily to do with growth) has
infiltrated discussions of long-run economic policy in the last few years. It
is very hard to be against sustainability. In fact, the less you know about
it, the better it sounds. That is true of lots of ideas. The questions that
come to be connected with sustainable development or sustainable growth
or just sustainability are genuine and deeply felt and very complex. The
combination of deep feeling and complexity breeds buzzwords, and sustainability has certainly become a buzzword. What I thought I might do,
when I was invited to talk to a group like this, was to try to talk out loud
about how one might think straight about the concept of sustainability,

This paper was presented as the Eighteenth J. Seward Johnson
Lecture to the Marine Policy Center, Woods Hole Oceanographic
Institution, at Woods Hole, Massachusetts, on June 14, 1991.

179

what it might mean and what its implications (not for daily life but for your annual vote or your concern for economic policy) might be.

Definitions are usually boring. That is probably true here too. But here it matters a lot. Some people say they don't know what sustainability means, but it sounds good. I've seen things on restaurant menus that strike me the same way. I took these two parts of a definition from a UNESCO document: ". . . every generation should leave water, air and soil resources as pure and unpolluted as when it came on earth." Alternatively, it was suggested that "each generation should leave undiminished all the species of animals it found existing on earth." I suppose that sounds good, as it is meant to. But I believe that kind of thought is fundamentally the wrong way to go in thinking about this issue. I must also say that there are some much more carefully thought out definitions and discussions, say by the U.N. Environment Programme and the World Conservation Union. They all turn out to be vague; in a way, the message I want to leave with you today is that sustainability is an essentially vague concept, and it would be wrong to think of it as being precise, or even capable of being made precise. It is therefore probably not in any clear way an exact guide to policy. Nevertheless, it is not at all useless.

Pretty clearly the notion of sustainability is about our obligation to the future. It says something about a moral obligation that we are supposed to have for future generations. I think it is very important to keep in mind—I'm talking like a philosopher for the next few sentences and I don't really know how to do that—that you can't be morally obligated to do something that is not feasible. Could I be morally obligated to be like Peter Pan and flap my wings and fly around the room? The answer is clearly not. I can't have a moral obligation like that because I am not capable of flapping my arms and flying around the room. If I fail to carry out a moral obligation, you must be entitled to blame me. You could properly say unkind things about me. But you couldn't possibly say unkind things about me for not flying around the room like Peter Pan because you know, as well as I do, that I can't do it.

If you define sustainability as an obligation to leave the world as we found it in detail, I think that's glib but essentially unfeasible. It is, when you think about it, not even desirable. To carry out literally the injunction of UNESCO would mean to make no use of mineral resources; it would mean to do no permanent construction or semi-permanent construction; build no roads; build no dams; build no piers. A mooring would be all right but not a pier. Apart from being essentially an injunction to do something that is not feasible, it asks us to do something that is not, on reflection, desirable. I doubt that I would feel myself better off if I had found the world exactly as the Iroquois left it. It is not clear that one would really want to do that.

To make something reasonable and useful out of the idea of sustainability, I think you have to try a different kind of definition. The best thing

I could think of is to say that it is an obligation to conduct ourselves so that we leave to the future the option or the capacity to be as well off as we are. It is not clear to me that one can be more precise than that. Sustainability is an injunction not to satisfy ourselves by impoverishing our successors. That sounds good too, but I want you to realize how problematic it is—how hard it is to make anything precise or checkable out of that thought. If we try to look far ahead, as presumably we ought to if we are trying to obey the injunction to sustainability, we realize that the tastes, the preferences, of future generations are something that we don't know about. Nor do we know anything very much about the technology that will be available to people 100 years from now. Put yourself in the position of someone in 1880 trying to imagine what life would be like in 1980 and you will see how wrong you would be. I think all we can do in this respect is to imagine people in the future being much like ourselves and attributing to them, imputing to them, whatever technology we can "reasonably" extrapolate—whatever that means. I am trying to emphasize the vagueness but not the meaningless of that concept. It is not meaningless, it is just inevitably vague.

We are entitled to please ourselves, according to this definition, so long as it is not at the expense (in the sense that I stated) of future well-being. You have to take into account, in thinking about sustainability, the resources that we use up and the resources that we leave behind, but also the sort of environment we leave behind including the built environment, including productive capacity (plant and equipment) and including technological knowledge. *To talk about sustainability in that way is not at all empty.* It attracts your attention, first, to what history tells us is an important fact, namely, that goods and services can be substituted for one another. If you don't eat one species of fish, you can eat another species of fish. Resources are, to use a favorite word of economists, fungible in a certain sense. They can take the place of each other. That is extremely important because it suggests that we do not owe to the future any particular thing. There is no specific object that the goal of sustainability, the obligation of sustainability, requires us to leave untouched.

What about nature? What about wilderness or unspoiled nature? I think that we ought, in our policy choices, to embody our desire for unspoiled nature as a component of well-being. But we have to recognize that different amenities really are, to some extent, substitutable for one another, and we should be as inclusive as possible in our calculations. It is perfectly okay, it is perfectly logical and rational, to argue for the preservation of a particular species or the preservation of a particular landscape. But that has to be done on its own, for its own sake, because this landscape is intrinsically what we want or this species is intrinsically important to preserve, not under the heading of sustainability. Sustainability doesn't require that any *particular* species of owl or any *particular* species of fish or any *particular* tract of forest be preserved. Substitutability is also important on the production side. We know that one kind of input can be substituted for another in production. There is no reason for our society to feel guilty about using up aluminum as long as we leave behind a capacity to perform the same or analogous functions using other kinds of materials—plastics or other natural or artificial materials. In making policy decisions we can take advantage of the principle of substitutability, remembering that what we are obligated to leave behind is a generalized capacity to create well-being, not any particular thing or any particular natural resource.

If you approach the problem that way in trying to make plans and make policies, it is certain that there will be mistakes. We will impute to the future tastes that they don't have or we will impute to them technological capacities that they won't have or we will fail to impute to them tastes and technological capacities that they do have. The set of possible mistakes is usually pretty symmetric.

That suggests to me the importance of choosing robust policies whenever we can. We should choose policies that will be appropriate over as wide a range of possible circumstances as we can imagine. But it would be wrong for policy to be paralyzed by the notion that one can make mistakes. Liability to error is the law of life. And, as most people around Woods Hole know, you choose policies to avoid potentially catastrophic errors, if you can. You insure wherever you can, but that's it.

The way I have put this, and I meant to do so, emphasizes that sustainability is about distributional equity. It is about who gets what. It is about the sharing of well-being between present people and future people. I have also emphasized the need to keep in mind, in making plans, that we don't know what they will do, what they will like, what they will want. And, to be honest, it is none of our business.

It is often asked whether, at this level, the goal or obligation of sustainability can be left entirely to the market. It seems to me that there is no reason to believe in a doctrinaire way that it can. The future is not adequately represented in the market, at least not the far future. If you remember that our societies live with real interest rates of the order of 5 or 6 percent, you will realize that that means that the dollar a generation from now, thirty years from now, is worth 25 cents today. That kind of discount seems to me to be much sharper than we would seriously propose in our public capacity, as citizens thinking about our obligation to the future. It seems to me to be a stronger discount than most of us would like to make. It is fair to say that those people a few generations hence are not adequately represented in today's market. They don't participate in it, and therefore there is no doctrinaire reason for saying, "Oh well, ordinary supply and demand, ordinary market behavior, will take care of whatever obligation we have to the future."

Now, in principle, government could serve as a trustee, as a representative for future interests. Policy actions, taxes, subsidies, regulations could, in principle, correct for the excessive present-mindedness of ordinary people like ourselves in our daily business. Of course, we are not sure that government will do a good job. It often seems that the rate at which governments discount the future is rather sharper than that at which the bond market does. So we can't be sure that public policy will do a good job. That is why we talk about it in a democracy. We are trying to think about collective decisions for the future, and discussions like this, not with just me talking, are the way in which policies of that kind ought to be thrashed out.

Just to give you some idea of how uncertain both private and public behavior can be in an issue like this, let me ask you to think about the past, not about the future. You could make a good case that our ancestors, who were considerably poorer than we are, whose standard of living was considerably less than our own, were probably excessively generous in providing for us. They cut down a lot of trees, but they saved a lot and they built a lot of railroad rights-of-way. Both privately and publicly they probably did better by us than a sort of fair-minded judge in thinking about the equity (whether they got their share and we got our share or whether we profited at their expense) would have required. It would have been okay for them to save a little less, to enjoy a little more and given us a little less of a start than our generation has had. I don't think there is any simple generalization that will serve to guide policy about these issues. There is every reason to discuss economic policy and social policy from this point of view, and anything else is likely to be ideology rather than analysis.

Once you take the point of view that I have been urging on you in thinking about sustainability as a matter of distributional equity between the present and the future, you can see that it becomes a problem about saving and investment. It becomes a problem about the choice between current consumption and providing for the future.

There is a sort of dual connection—a connection that need not be intrinsic but is there—between environmental issues and sustainability issues. The environment needs protection by public policy because each of us knows that by burdening the environment, by damaging it, we can profit and have some of the cost, perhaps most of the cost, borne by others. Sustainability is a problem precisely because each of us knows or realizes that we can profit at the expense of the future rather than at the expense of our contemporaries and the environment. We free-ride on each other and we free-ride on the future.

Environmental policy is important for both reasons. One of the ways we free-ride on the future is by burdening the environment. And so current environmental protection—this is what I meant by a dual connection— will almost certainly contribute quite a lot to sustainability. Although, I want to warn you, not automatically. Current environmental protection contributes to sustainability if it comes at the expense of current consumption. Not if it comes at the expense of investment, of additions to future capacity. So, there are no absolutes. There is nothing precise about this notion but there are perhaps approximate guides to public policy that come out of this way of reasoning about the idea of sustainability. A correct principle, a correct general guide is that when we use up something—and by we I mean our society, our country, our civilization, however broadly you want to think—when we use up something that is irreplaceable, whether it is minerals or a fish species, or an environmental amenity, then we should be thinking about providing a substitute of equal value, and the vagueness comes in the notion of value. The something that we provide in exchange could be knowledge, could be technology. It needn't even be a physical object.

Let me give you an excellent example from the recent past of a case of good thought along these lines and also a case of bad thought along these lines. Commercially usable volumes of oil were discovered in the North Sea some years ago. The two main beneficiaries of North Sea oil were the United Kingdom and Norway. It is only right to say that the United Kingdom dissipated North Sea oil, wasted it, used it up in consumption and on employment. If I meet Mrs. Thatcher in heaven, since that is where I intend to go, the biggest thing I will tax her with is that she blew North Sea oil. Here was an asset that by happenstance the U.K. acquired. If the sort of general approach to sustainability that I have been suggesting to you had been taken by the Thatcher government, someone would have said, "It's okay we are going to use up the oil, that's what it is for, but we will make sure that we provide something else in exchange, that we guide those resources, at least in large part, into investment in capacity in the future." That did not happen. As I said, if you ask where (and by the way the curve of production from the North Sea fields is already on the way down; that asset is on its way to exhaustion) it went, it went into maintaining consumption in the United Kingdom and, at the same time, into unemployment.

Norway, on the other hand, went about it in the typical sober way you expect of good Scandinavians. The Norwegians said, here is a wasting asset. Here is an asset that we are going to use up. Scandinavians are also slightly masochistic, as you know. They said the one thing we must avoid is blowing this; the one thing we must avoid is a binge. They tried very hard to convert a large fraction of the revenues, of the rentals, of the royalties from North Sea oil into investment. I confess I don't know how well they succeeded but I am willing to bet that they did a better job of it than the United Kingdom.

This brings me to the one piece of technical economics that I want to mention. There is a neat analytical result in economics (mainly done by John Hartwick of Queen's University in Canada) which studies an economy that takes what we call the rentals, the pure return to a non-renewable resource, and invests those rentals.[1] That is, it uses up a natural asset like the North Sea oil field, but makes a point of investing whatever revenues intrinsically inhere to the oil itself. That policy can be shown to have neat sustainability properties. In a simple sort of economy, it will guarantee a perpetually constant capacity to consume. By the way, it is a very simple rule, and it is really true only for very simple economies; but it has the advantage, first of all, of sounding right, of sounding like justice, and secondly, of being practical. It is a calculation that could be made. It is a calculation that we don't make and I am going to suggest in a minute that we should be making it. You might want to do better. You might feel so good about your great-grandchildren that you would like to do better than invest the rents on the non-renewable resources that you use up. But in any case, it is, at a minimum, a policy that one could pursue for the sake of sustainability. I want to remind you again that most environmental protection can be regarded as an act of investment. If we were to think that our obligation to the future is in principle discharged by seeing that the return to non-renewable resources is funnelled into capital formation, any kind of capital formation—plant and equipment, research and development, physical oceanography, economics or environmental investment—we could have some feeling that we were about on the right track.

Now I want to mention what strikes me as sort of a paradox—as a difficulty with a concept of sustainability. I said, I kind of insisted, that you should think about it as a matter of equity, as a matter of distributional equity, as a matter of choice of how productive capacity should be shared between us and them, them being the future. Once you think about it that way you are almost forced logically to think about equity not between periods of time but equity right now. There is something inconsistent about people who profess to be terribly concerned about the welfare of future generations but do not seem to be terribly concerned about the welfare of poor people today. You will see in a way why this comes to be a paradox. The only reason for thinking that sustainability is a problem is that you think that some people are likely to be shortchanged, namely, in the future. Then I think you really are obligated to ask, "Well, is anybody being shortchanged right now?"

The paradox arises because if you are concerned about people who are currently poor, it will turn out that your concern for them will translate into an increase in current consumption, not into an increase in investment. The logic of sustainability says, "You ought to be thinking about poor people today, and thinking about poor people today will be disadvantageous from the point of view of sustainability." Intellectually, there is no difficulty in resolving that paradox, but practically there is every

---

[1]John M. Hartwick, "Substitution among exhaustible resources and intergenerational equity," *Review of Economic Studies* 45(2): 347–543 (June 1978).

difficulty in the world in resolving that paradox. And I don't have the vaguest notion of how it can be done in practice.

The most dramatic way in which I can remind you of the nature of that paradox is to think about what it will mean for, say, CO_2 discharge when the Chinese start to burn their coal in a very large way; and, then, while you are interested in moral obligation, I think you should invent for yourself how you are going to explain to the Chinese that they shouldn't burn the coal, even living at their standard of living they shouldn't burn the coal, because the CO_2 might conceivably damage somebody in 50 or 100 years.

Actually the record of the U.S. is not very good on either the inter-generational equity or the intra-generational equity front. We tolerate, for a rich society, quite a lot of poverty, and at the same time we don't save or invest a lot. I've just spent some time in West Germany, and there is considerably less apparent poverty in the former Federal Republic than there is here; and at the same time they are investing a larger fraction of their GNP than we are by a large margin.

It would not be very hard for us to do better. One thing we might do, for starters, is to make a comprehensive accounting of rents on non-renewable resources. It is something that we do not do. There is nothing in the national accounts of the U.S. which will tell you what fraction of the national income is the return to the using up of non-renewable resources. If we were to make that accounting, then we would have a better idea than we have now as to whether we are at least meeting that minimal obligation to channel those rents into saving and investment. And I also suggested that careful attention to current environmental protection is another way that is very likely to slip in some advantage in the way of sustainability, provided it is at the expense of current consumption and not at the expense of other forms of investment.

I have left out of this talk, as some of you may have noticed until now, any mention of population growth; and I did that on purpose, although it might be the natural first order concern if you are thinking about sustainability issues. Control of population growth would probably be the best available policy on behalf of sustainability. You know that, I know that, and I have no particular competence to discuss it any further; so I won't, except to remind you that rapid population growth is fundamentally a Third World phenomenon, not a developed country phenomenon. So once again, you are up against the paradox that people in poor countries have children as insurance policies for their own old age. It is very hard to preach to them not to do that. On the other hand, if they continue to do that, then you have probably the largest, single danger to sustainability of the world economy.

All that remains for me is to summarize. What I have been trying to say goes roughly as follows. Sustainability as a moral obligation is a general obligation not a specific one. It is not an obligation to preserve this or preserve that. It is an obligation, if you want to make sense out of it, to preserve the capacity to be well off, to be as well off as we. That does not preclude preserving specific resources, if they have an independent value and no good substitutes. But we shouldn't kid ourselves, that is part of the value of specific resources. It is not a consequence of any interest in sustainability. Secondly, an interest in sustainability speaks for investment generally. I mentioned that directing the rents on non-renewable resources into investment is a good rule of thumb, a reasonable and dependable starting point. But what sustainability speaks for is investment, investment of any kind. In particular, environmental investment seems to me to correlate well with concerns about sustainability and so, of course, does reliance on renewable resources as a substitute for non-renewable ones. Third, there is something faintly phony about deep concern for the future combined with callousness about the state of the world today. The catch is that today's poor want consumption not investment. So the conflict is pretty deep and there is unlikely to be any easy way to resolve it. Fourth, research is a good thing. Knowledge on the whole is an environmentally neutral asset that we can contribute to the future. I said that in thinking about sustainability you want to be as inclusive as you can. Investment in the broader sense and investment in knowledge, especially technological and scientific knowledge, is as environmentally clean an asset as we know. And the last thing I want to say is, don't forget that sustainability is a vague concept. It is intrinsically inexact. It is not something that can be measured out in coffee spoons. It is not something that you could be numerically accurate about. It is, at best, a general guide to policies that have to do with investment, conservation and resource use. And we shouldn't pretend that it is anything other than that.

Thank you very much.

REFERENCES

World Commission on Environment and Development, *Our Common Future* (The Brundtland Report). Oxford: Oxford University Press, 1987.

World Conservation Union, *Caring for the Earth.* Gland, Switzerland, 1991; see especially p. 10.

World Resources Institute, *World Resources 1992–93: Toward Sustainable Development.* New York: Oxford University Press, 1992. See especially Ch. 1.

12

Conservation Reconsidered

JOHN V. KRUTILLA*

.

John V. Krutilla, now an independent consultant, was a
Senior Fellow at Resources for the Future.

> "It is the clear duty of Government, which is the trustee for unborn generations as
> well as for its present citizens, to watch over, and if need be, by legislative enactment,
> to defend, the exhaustible natural resources of the country from rash and reckless
> spoliation. How far it should itself, either out of taxes, or out of State loans, or by
> the device of guaranteed interest, press resources into undertakings from which the
> business community, if left to itself, would hold aloof, is a more difficult problem.
> Plainly, if we assume adequate competence on the part of governments, there is a valid
> case for *some* artificial encouragement to investment, particularly to investments the
> return from which will only begin to appear after the lapse of many years."
>
> A. C. Pigou

Conservation of natural resources has meant different things to different
people. But to the economist from the time of Pigou, who first took notice
of the economics of conservation [10, p. 27ff], until quite recently, the
central concerns have been associated with the question of the optimal
intertemporal utilization of the fixed natural resource stocks. The gnawing
anxiety provoked by the Malthusian thesis of natural resource scarcity
was in no way allayed by the rates of consumption of natural resource
stocks during two world wars occurring between the first and fourth
editions of Pigou's famous work. In the United States, a presidential
commission, reviewing the materials situation following World War II,

"Conservation Reconsidered," by John V. Krutilla, from
American Economic Review, vol. 57, (1967), pp. 777–86.
Reprinted by permission.

*The author is indebted to all of his colleagues at Resources for the Future and to Harold
Barnett, Paul Davidson, Otto Davis, Chandler Morse, Peter Pearse, and Ralph Turvey for
many helpful suggestions on an earlier draft of this paper.

concluded that an end had come to the historic decline in the cost of natural resource commodities [12, pp. 13–14]. This conclusion reinforced the concern of many that the resource base ultimately would be depleted.

More recently, on the other hand, a systematic analysis of the trends in prices of natural resource commodities did not reveal any permanent interruption in the decline relative to commodities and services in general [11]. Moreover, a rather ambitious attempt to test rigorously the thesis of natural resource scarcity suggested instead that technological progress had compensated quite adequately for the depletion of the higher quality natural resource stocks [1]. Further, given the present state of the arts, future advances need not be fortuitous occurrences; rather the rate of advance can be influenced by investment in research and development. Indeed, those who take an optimistic view would hold that the modern industrial economy is winning its independence from the traditional natural resources sector to a remarkable degree. Ultimately, the raw material inputs to industrial production may be only mass and energy [1, p. 238]. [1]

While such optimistic conclusions were being reached, they were nevertheless accompanied by a caveat that, while we may expect production of goods and services to increase without interruption, the level of living may not necessarily be improved. More specifically, Barnett and Morse concluded that the quality of the physical environment—the landscape, water, and atmospheric quality—was deteriorating.

These conclusions suggest that on the one hand the traditional concerns of conservation economics—the husbanding of natural resource stocks for the use of future generations—may now be outmoded by advances in technology. On the other hand, the central issue seems to be the problem of providing for the present and future the amenities associated with unspoiled natural environments, for which the market fails to make adequate provision. While this appears to be the implication of recent research, [2] and is certainly consistent with recent public policy in regard to preserving natural environments, the traditional economic rationale for conservation does not address itself to this issue directly. [3] The use of Pigou's social time preference may serve only to hasten the conversion of natural environments into low-yield capital investments. [4] On what basis,

---

[1]The conclusions were based on data relevant to the U.S. economy. While they may be pertinent to Western Europe also, all of my subsequent observations are restricted to the United States.

[2]For example, see [7].

[3]It must be acknowledged that with sufficient patience and perception nearly all of the argument for preserving unique phenomena of nature can be found in the classic on conservation economics by Ciriacy-Wantrup [3].

[4]An example of this was the recent threat to the Grand Canyon by the proposed Bridge and Marble Canyon dams. Scott makes a similar point with reference to natural resource commodities [13].

then, can we make decisions when we confront a choice entailing action which will have an irreversible adverse consequence for rare phenomena of nature? I investigate this question below.

Let us consider an area with some unique attribute of nature—a geomorphologic feature such as the Grand Canyon, a threatened species, or an entire ecosystem or biotic community essential to the survival of the threatened species.[5] Let us assume further that the area can be used for certain recreation and/or scientific research activities which would be compatible with the preservation of the natural environment, or for extractive activities such as logging or hydraulic mining, which would have adverse consequences for scenic landscapes and wildlife habitat.

A private resource owner would consider the discounted net income stream from the alternative uses and select the use which would hold prospects for the highest present net value. If the use which promises the highest present net value is incompatible with preserving the environment in its natural state, does it necessarily follow that the market will allocate the resources efficiently? There are several reasons why private and social returns in this case are likely to diverge significantly.

Consider the problem first in its static aspects. By assumption, the resources used in a manner compatible with preserving the natural environment have no close substitutes; on the other hand, alternative sources of supply of natural resource commodities are available.[6] Under the circumstances and given the practical obstacles to perfectly discriminating pricing, the private resource owner would not be able to appropriate in gate receipts the entire social value of the resources when used in a manner compatible with preserving the natural state. Thus the present values of his expected net revenues are not comparable as between the competing uses in evaluating the efficiency of the resource allocation.

Aside from the practical problem of implementing a perfectly discriminating pricing policy, it is not clear even on theoretic grounds that a comparison of the total area under the demand curve on the one hand and market receipts on the other will yield an unambiguous answer to the allocative question. When the existence of a grand scenic wonder or a unique and fragile ecosystem is involved, its preservation and continued availability are a significant part of the real income of many individ-

---

[5]Uniqueness need not be absolute for the following arguments to hold. It may be, like Dupuit's bridge, a good with no adequate substitutes in the "natural" market area of its principal clientele, while possibly being replicated in other market areas to which the clientele in question has no access for all practical purposes.

[6]The asymmetry in the relation posited is realistic. The historic decline in cost of natural resource commodities relative to commodities in general suggests that the production and exchange of the former occur under fairly competitive conditions. On the other hand, increasing congestion at parks, such as Yellowstone, Yosemite, and Grand Canyon, suggests there are no adequate substitutes for these rare natural environments.

uals.[7] Under the conditions postulated, the area under the demand curve, which represents a maximum willingness to pay, may be significantly less than the minimum which would be required to compensate such individuals were they to be deprived in perpetuity of the opportunity to continue enjoying the natural phenomenon in question. Accordingly, it is conceivable that the potential losers cannot influence the decision in their favor by their aggregate willingness to pay, yet the resource owner may not be able to compensate the losers out of the receipts from the alternative use of the resource. In such cases—and they are more likely encountered in this area—it is impossible to determine whether the market allocation is efficient or inefficient.

Another reason for questioning the allocative efficiency of the market for the case in hand has been recognized only more recently. This involves the notion of *option demand* [14]. This demand is characterized as a willingness to pay for retaining an option to use an area or facility that would be difficult or impossible to replace and for which no close substitute is available. Moreover, such a demand may exist even though there is no current intention to use the area or facility in question and the option may never be exercised. If an option value exists for rare or unique occurrences of nature, but there is no means by which a private resource owner can appropriate this value, the resulting resource allocation may be questioned.

Because options are traded on the market in connection with other economic values, one may ask why no market has developed where option value exists for the preservation of natural environments.[8] We need to consider briefly the nature of the value in question and the marketability of the option.

From a purely scientific viewpoint, much is yet to be learned in the earth and life sciences; preservation of the objects of study may be defended on these grounds, given the serendipity value of basic research. We know also that the natural biota represents our reservoir of germ plasm, which has economic value. For example, modern agriculture in advanced countries represents cultivation figuratively in a hot-house environment in which crops are protected against disease, pests, and drought by a variety of agricultural practices. The energy released from some of the genetic characteristics no longer required for survival under cultivated conditions is redirected toward greater productivity. Yet because of the instability introduced with progressive reduction of biological diversity, a need occa-

[7]These would be the spiritual descendants of John Muir, the present members of the Sierra Club, the Wilderness Society, National Wildlife Federation, Audubon Society and others to whom the loss of a species or the disfigurement of a scenic area causes acute distress and a sense of genuine relative impoverishment.

[8]For a somewhat differently developed argument, see [6].

sionally arises for the reintroduction of some genetic characteristics lost in the past from domestic strains. It is from the natural biota that these can be obtained.

The value of botanical specimens for medicinal purposes also has been long, if not widely, recognized. Approximately half of the new drugs currently being developed are obtained from botanical specimens.[9] There is a traffic in medicinal plants which approximates a third of a billion dollars annually. Cortisone, digitalis, and heparin are among the better known of the myriad drugs which are derived from natural vegetation or zoological sources. Since only a relatively small part of the potential medicinal value of biological specimens has yet been realized, preserving the opportunity to examine all species among the natural biota for this purpose is a matter of considerable importance.

The option value may have only a sentimental basis in some instances. Consider the rallying to preserve the historical relic, "Old Ironsides."[10] There are many persons who obtain satisfaction from mere knowledge that part of wilderness North America remains even though they would be appalled by the prospect of being exposed to it. Subscriptions to World Wildlife Fund are of the same character. The funds are employed predominantly in an effort to save exotic species in remote areas of the world which few subscribers to the Fund ever hope to see. An option demand may exist therefore not only among persons currently and prospectively active in the market for the object of the demand, but among others who place a value on the mere existence of biological and/or geomorphological variety and its widespread distribution.[11]

If a genuine value for retaining an option in these respects exists, why has not a market developed? To some extent, and for certain purposes, it has. Where a small natural area in some locality in the United States is threatened, the property is often purchased by Nature Conservancy,[12] a private organization which raises funds through voluntary subscriptions.[13] But this market is grossly imperfect. First, the risk for private investors associated with absence of knowledge as to whether a particular ecosystem has special characteristics not widely shared by others is enor-

---

[9]For an interesting account of the use of plants for medicinal purposes, see [8].

[10]The presumption in favor of option value is applicable also to historic and cultural features; rare works of art, perhaps, being the most prominent of this class.

[11]The phenomenon discussed may have an exclusive sentimental basis, but if we consider the "bequest motivation" in economic behavior, discussed below, it may be explained by an interest in preserving an option for one's heirs to view or use the object in question.

[12]Not to be confused with a public agency of the same name in the United Kingdom.

[13]Subscriptions to World Wildlife Fund, the Wilderness Society, National Parks Association, etc. may be similar, but, of course, much of the effect these organizations have on the preservation of natural areas stems not from purchasing options, but from influencing public programs.

mous. [14] Moreover, to the extent that the natural environment will support basic scientific research which often has unanticipated practical results, the serendipity value may not be appropriable by those paying to preserve the options. But perhaps of greatest significance is that the preservation of the grand scenic wonders, threatened species, and the like involves comparatively large land tracts which are not of merely local interest. Thus, all of the problems of organizing a market for public goods arise. Potential purchasers of options may be expected to bide time in the expectation that others will meet the necessary cost, thus eliminating cost to themselves. Since the mere existence or preservation of the natural environment in question satisfies the demand, those who do not subscribe cannot be excluded except by the failure to enroll sufficient subscribers for its preservation.

Perhaps of equal significance to the presumption of market failure are some dynamic characteristics of the problem suggested by recent research. First, consider the consumption aspects of the problem. Davidson, Adams, and Seneca have recently advanced some interesting notions regarding the formation of demand that may be particularly relevant to our problem [5, p. 186].

> When facilities are not readily available, skills will not be developed and, consequently, there may be little desire to participate in these activities. If facilities are made available, opportunities to acquire skill increase, and user demand tends to rise rapidly over time as individuals learn to enjoy these activities. Thus, participation in and enjoyment of water recreational activities by the present generation will stimulate future demand without diminishing the supply presently available. Learning-by-doing, to the extent it increases future demand, suggests an interaction between present and future demand functions, which will result in a public good externality, as present demand enters into the utility function of future users.

While this quotation refers to water-based recreation, it is likely to be more persuasive in connection with some other resource-based recreation activity. Its relevance for wilderness preservation is obvious. When we consider the remote backcountry landscape, or the wilderness scene as the object of experience and enjoyment, we recognize that utility from the experience depends predominantly upon the prior acquisition of technical skill and specialized knowledge. This, of course, must come from experience initially with less arduous or demanding activities. The more the present population is initiated into activities requiring similar but less advanced skills (e.g., car camping), the better prepared the future popula-

[14]The problem here is in part like a national lottery in which there exists a very small chance for a very large gain. Unlike a lottery, rather large sums at very large risk typically would be required.

tion will be to participate in the more exacting activities. Given the phenomenal rise of car camping, if this activity will spawn a disproportionate number of future back-packers, canoe cruisers, cross-country skiers, etc., the greater will be the induced demand for wild, primitive, and wilderness-related opportunities for indulging such interest. Admittedly, we know little about the demand for outdoor experiences which depend on unique phenomena of nature—its formation, stability, and probable course of development. These are important questions for research, results of which will have significant policy implications.

In regard to the production aspects of the "new conservation," we need to examine the implications of technological progress a little further. Earlier I suggested that the advances of technology have compensated for the depletion of the richer mineral deposits and, in a sense, for the superior stands of timber and tracts of arable land. On the other hand, there is likely to be an asymmetry in the implications of technological progress for the production of goods and services from the natural resource base, and the production of natural phenomena which give rise to utility without undergoing fabrication or other processing.[15] In fact, it is improbable that technology will advance to the point at which the grand geomorphologic wonders could be replicated, or extinct species resurrected. Nor is it obvious that fabricated replicas, were they even possible, would have a value equivalent to that of the originals. To a lesser extent, the landscape can be manufactured in a pleasing way with artistry and the larger earth-moving equipment of today's construction technology. Open pit mines may be refilled and the surroundings rehabilitated in a way to approximate the original conditions. But even here the undertaking cannot be accomplished without the cooperation of nature over a substantial period of time depending on the growth rate of the vegetal cover and the requirements of the native habitat.[16] Accordingly, while the supply of fabricated goods and commercial services may be capable of continuous expansion from a given resource base by reason of scientific discovery and mastery of technique, the supply of natural phenomena is virtually inelastic. That is, we may preserve the natural environment which remains to provide amenities of this sort for the future, but there are significant limitations on reproducing it in the future should we fail to preserve it.

If we consider the asymmetric implications of technology, we can conceive of a transformation function having along its vertical axis amenities derived directly from association with the natural environment and fabricated goods along the horizontal axis. Advances in technology would stretch the transformation function's terminus along the horizontal axis but not appreciably along the vertical. Accordingly, if we simply take the

[15]I owe this point to a related observation, to my knowledge first made by Ciriacy-Wantrup [3, p. 47].

[16]That is, giving rise to option value for members of the present population.

effect of technological progress over time, considering tastes as constant, the marginal trade-off between manufactured and natural amenities will progressively favor the latter. Natural environments will represent irreplaceable assets of appreciating value with the passage of time.

If we consider technology as constant, but consider a change in tastes progressively favoring amenities of the natural environment due to the learn-by-doing phenomenon, natural environments will similarly for this reason represent assets of appreciating value. If both influences are operative (changes in technology with asymmetric implications, and tastes), the appreciating value of natural environments will be compounded.

This leads to a final point which, while a static consideration, tends to have its real significance in conjunction with the effects of parametric shifts in tastes and technology. We are coming to realize that consumption-saving behavior is motivated by a desire to leave one's heirs an estate as well as by the utility to be obtained from consumption. [17] A bequest of maximum value would require an appropriate mix of public and private assets, and, equally, the appropriate mix of opportunities to enjoy amenities experienced directly from association with the natural environment along with readily producible goods. But the option to enjoy the grand scenic wonders for the bulk of the population depends upon their provision as public goods.

Several observations have been made which may now be summarized. The first is that, unlike resource allocation questions dealt with in conventional economic problems, there is a family of problems associated with the natural environment which involves the irreproducibility of unique phenomena of nature—or the irreversibility of some consequence inimical to human welfare. Second, it appears that the utility to individuals of direct association with natural environments may be increasing while the supply is not readily subject to enlargement by man. Third, the real cost of refraining from converting our remaining rare natural environments may not be very great. Moreover, with the continued advance in technology, more substitutes for conventional natural resources will be found for the industrial and agricultural sectors, liberating production from dependence on conventional sources of raw materials. Finally, if consumption-saving behavior is motivated also by the desire to leave an estate, some portion of the estate would need to be in assets which yield collective consumption goods of appreciating future value. For all of these reasons we are confronted with a problem not conventionally met in resource economics. The problem is of the following nature.

At any point in time characterized by a level of technology which is less advanced than at some future date, the conversion of the natural environment into industrially produced private goods has proceeded further than it would have with the more advanced future technology. Moreover, with

---

[17]See [2]; also [9].

the apparent increasing appreciation of direct contact with natural environments, the conversion will have proceeded further, for this reason as well, than it would have were the future composition of tastes to have prevailed. Given the irreversibility of converted natural environments, however, it will not be possible to achieve a level of well-being in the future that would have been possible had the conversion of natural environments been retarded. That this should be of concern to members of the present generation may be attributable to the bequest motivation in private economic behavior as much as to a sense of public responsibility.[18]

Accordingly, our problem is akin to the dynamic programming problem which requires a present action (which may violate conventional benefit-cost criteria) to be compatible with the attainment of future states of affairs. But we know little about the value that the instrumental variables may take. We have virtually no knowledge about the possible magnitude of the option demand. And we still have much to learn about the determinants of the growth in demand for outdoor recreation and about the quantitative significance of the assymmetry in the implications of technological advances for producing industrial goods on the one hand and natural environments on the other. Obviously, a great deal of research in these areas is necessary before we can hope to apply formal decision criteria comparable to current benefit-cost criteria. Fully useful results may be very long in coming; what then is a sensible way to proceed in the interim?

First, we need to consider what we need as a minimum reserve to avoid potentially grossly adverse consequences for human welfare. We may regard this as our scientific preserve of research materials required for advances in the life and earth sciences. While no careful evaluation of the size of this reserve has been undertaken by scientists, an educated guess has put the need in connection with terrestrial communities at about ten million acres for North America [4, p. 128]. Reservation of this amount of land—but a small fraction of one per cent of the total relevant area—is not likely to affect appreciably the supply or costs of material inputs to the manufacturing or agricultural sectors.

The size of the scientific preserve required for aquatic environments is still unknown. Only after there is developed an adequate system of classification of aquatic communities will it be possible to identify distinct environments, recognize the needed reservations, and, then, estimate the opportunity costs. Classification and identification of aquatic environments demand early research attention by natural scientists.

---

[18]The rationale above differs from that of Stephen Marglin which is perhaps the most rigorous one relying on a sense of public responsibility and externalities to justify explicit provision for future generations. In this case also, my concern is with providing *collective consumption goods for the present and future*, whereas the traditional concern in conservation economics has been with provision of *private intermediate goods for the future*.

Finally, one might hope that the reservations for scientific purposes would also support the bulk of the outdoor recreation demands, or that substantial additional reservations for recreational purposes could be justified by the demand and implicit opportunity costs. Reservations for recreation, as well as for biotic communities, should include special or rare environments which can support esoteric tastes as well as the more common ones. This is a matter of some importance because outdoor recreation opportunities will be provided in large part by public bodies, and within the public sector there is a tendency to provide a homogenized recreation commodity oriented toward a common denominator. There is need to recognize, and make provision for, the widest range of outdoor recreation tastes, just as a well-functioning market would do. We need a policy and a mechanism to ensure that all natural areas peculiarly suited for specialized recreation uses receive consideration for such uses. A policy of this kind would be consistent both with maintaining the greatest biological diversity for scientific research and educational purposes and with providing the widest choice for consumers of outdoor recreation.

REFERENCES

1. H. J. Barnett and C. Morse, *Scarcity and Growth: The Economics of Natural Resource Availability.* Baltimore 1963.
2. S. B. Chase, Jr., *Asset Prices in Economic Analysis.* Berkeley 1963.
3. S. V. Ciriacy-Wantrup, *Resources Conservation.* Berkeley 1952.
4. F. Darling and J. P. Milton, ed., *Future Environments of North America, Transformation of a Continent.* Garden City, N.Y. 1966.
5. P. Davidson, F. G. Adams, and J. Seneca, "The Social Value of Water Recreation Facilities Resulting from an Improvement in Water Quality: The Delaware Estuary," in A. V. Kneese and S. C. Smith, eds., *Water Research,* Baltimore 1966.
6. A. E. Kahn, "The Tyranny of Small Decisions: Market Failures, Imperfections, and the Limits of Economics," *Kyklos,* 1966, 19 (1), 23–47.
7. A. V. Kneese, *The Economics of Regional Water Quality Management.* Baltimore 1964.
8. M. B. Kreig, *Green Medicine: The Search for Plants That Heal.* New York 1964.
9. F. Modigliani and R. Brumberg, "Utility Analysis and the Consumption Function: An Interpretation of Cross-Section Data," in K. K. Kurihara, ed., *Post-Keynesian Economics,* New Brunswick 1954.
10. A. C. Pigou, *The Economics of Welfare,* 4th ed., London 1952.
11. N. Potter and F. T. Christy, Jr., *Trends in Natural Resources Commodities: Statistics of Prices, Output, Consumption, Foreign Trade, and Employment in the United States, 1870–1957,* Baltimore 1962.

12. The President's Materials Policy Commission, *Resources for Freedom, Foundation for Growth and Security,* Vol. I. Washington DC 1952.
13. A. D. Scott, *Natural Resources: The Economics of Conservation.* Toronto 1955.
14. B. A. Weisbrod, "Collective Consumption Services of Individual Consumption Goods," *Quart. Jour. Econ.,* Aug. 1964, 77, 71–77.

III

DESIGNING AND IMPLEMENTING ENVIRONMENTAL POLICIES

The preceding section set the economic framework for analyzing environmental policies. We turn now to a less formal examination of specific policies and to the challenges of implementing them. There are two fundamentally different approaches to policy. The one that has been almost uniformly applied in practice in the United States is the so-called command and control policy, which mandates specific standards for emissions, or for the abatement technology, for all dischargers. The alternative preferred by most economists relies on price incentives that aim to induce dischargers to limit their discharges to levels that equate their marginal costs of control to the marginal social damages from emissions.

The main trouble with centralized command-and-control strategies, as A. Michael Spence and Martin L. Weitzman point out in "Regulatory Strategies for Pollution Control," is that different dischargers are required to conform to the same emissions standards regardless of their individual costs of reducing emissions. In practice, different dischargers of the same waste materials often have widely varying abatement costs. The total cost could usually be reduced greatly if emissions were cut where that could be done most economically. That is why economists, for most purposes, tend to favor decentralized incentive systems that allow each discharger flexibility in reducing emissions depending upon its individual costs. Two strategies have received the most attention. The first is based on emission charges and the second on marketable emission permits.

Traditionally, economists have argued, like Larry E. Ruff in Section I, for a system of effluent charges (or in some instances abatement subsidies). Faced with a choice between paying a fee for discharging waste or bearing the cost of reducing emissions, a firm will limit its emissions to the level at which the marginal cost of controlling emissions is equal to the effluent charge (or abatement subsidy). The marginal cost of control will, in this manner, be equalized among all dischargers, and the total cost of abatement minimized. (This conclusion applies, by the way, to individuals as well as to firms. Faced with a charge for disposing of household waste or a carbon tax on fuel, individuals will adjust the amount they discharge by weighing their personal gain from additional consumption against the price of generating additional wastes.)

The efficiency of effluent charges can, in principle, be achieved in a different way with tradeable emissions permits, a concept that is introduced here by J. H. Dales and elaborated by Thomas H. Tietenberg. Instead of setting a price for emissions, the regulatory authority decides on the total quantity of emissions to be allowed and issues corresponding permits. Whether the initial distribution is by public auction or some other allocation system, the purchase and sale of permits in the market eventually sets a price for permits that performs the same function as an emission charge. Firms will limit their emissions to the level at which the marginal cost of control is equal to the price at which they can buy or sell permits, and marginal costs will be equalized across dischargers.

Conceptually, the most important difference between these two incentive systems is that in the case of effluent charges the regulatory authority sets the price of emissions and allows the total quantity to adjust, whereas with emission permits it sets the total quantity and lets the price adjust. Whether it sets the price or the quantity, if the authority wants to achieve the socially optimal level of pollution, it must aim for the the one at which the marginal social damage function and the marginal total abatement cost function intersect. If emissions are controlled beyond that point, the cost will be excessive, given the social damages. If they are controlled below that point, emissions will be excessive, given the cost of further control.

If the regulator had perfect information about the cost and damage functions, either effluent charges or permits could be made to yield the optimal level of emissions at minimum cost. Spence and Weitzman emphasize that knowledge regarding either of these functions is far from perfect and that the social cost of correcting policy errors by trial and error can be very high. They propose that, in order to minimize the chance of serious error in setting policy, one should choose between a strategy that fixes the total quantity of effluent and one that fixes the price depending on the particular circumstances. For instance, if there is a range in the damage function where the marginal social cost of emissions is thought to rise rapidly, it would be important to keep emissions below that range. A policy that fixed the permissible level of emissions, such as marketable permits or standards, could assure this, whereas an effluent charge would not. On the other hand, if marginal control costs are believed to rise rapidly within some range, a fixed limitation on emissions could result in unexpectedly high control costs and it would be preferable to fix the price of emissions.

Readers will note that the comparison in Spence and Weitzman is between effluent charges and standards, rather than permits, but the conclusions regarding when to choose standards apply as well to the choice of permits. The authors propose a hybrid control policy that contains some elements of standard setting and charges, without raising the possibility of marketable permits, which were not regarded as a major option when their article was written in 1978.

J. H. Dales's "Land, Water, and Ownership" is a seminal article on the possibilities for using tradeable emission permits. It also provides an illuminating discussion of the role that the absence of property rights plays in the misallocation of water resources between amenity and waste disposal uses. Although Dales despairs of the economist's ability to decide on the optimal division between these two uses, he believes economics provides a sound basis for implementing the division once it is decided upon and he stresses the advantages of tradeable permits for this purpose.

The bulk of Thomas H. Tietenberg's "Transferable Discharge Permits and the control of Stationary Source Air Pollution" is devoted to a survey of the formidable technical, legal, and administrative challenges that would have to be faced in moving to an unrestricted program of tradeable emissions permits. (Equally difficult decisions would have to be dealt with in implementing a system of charges.) One of the advantages of permit systems is that the initial distribution of permits can be tailored to greatly reduce the cost of the program to dischargers compared with a program of pollution charges. As Tietenberg shows, this possibility presents the policy maker with a host of options that need to be evaluated. A potential drawback of permits is the danger that one or a few relatively large dischargers might dominate and control a market for permits.

None of the incentive-based systems for regulating pollution discussed in these three articles has been fully implemented in its textbook form. However, nontradeable permits have been a staple of environmental policy for implementing emissions standards, and under the Clean Air Act of 1990 the Environmental Protection Agency is called upon to implement a system of marketable permits for limiting SO_2 emissions of major utilities to control acid rain deposition. Tietenberg describes the tentative steps in the direction of tradeable permits that were introduced in the late 1970s. Various devices to permit "internal" balancing of increases in emissions with reductions within a single firm were implemented, and an "offset" policy was introduced to allow new sources of air pollution to locate in so-called nonattainment areas. New entrants were required to offset their emissions by arranging for even larger reductions in emissions from existing sources, introducing a bias against new sources. Each transaction required regulatory approval, and markets have been relatively thin. As a result, transactions costs have been high and the number of transactions few. The experiment, thus far, has been too restricted to provide a basis for evaluating how well a true marketable permit system might work. Under the 1990 act, a more genuine experiment has now begun.

In his article "Economic Incentives in the Management of Hazardous Wastes," Clifford S. Russell deals primarily with the problem of enforcement and administration of regulations. Evidence points to widespread and serious violations of permits in the case of conventional (meaning nonhazardous) wastes, due, Russell says, not to lack of technical ability or information but to failure to allocate sufficient resources to monitoring and enforcement. The design and enforcement of hazardous waste regulations pose more serious problems.

The special problem in managing hazardous waste arises from the presumption that even in minute concentrations such substances can pose unacceptable risks to human health and to the fact that the quantity discharged by a single source is often relatively small. How does one compel compliance with regulations for wastes that are generated in such small quantities that they need not be discharged or treated where and when they are generated but can be containerized, concealed, and discharged anywhere?

It is generally agreed that there is no room in the case of hazardous waste for conventional incentive-based systems that allow individual dischargers discretion in deciding how much to discharge. Instead, regulations must be written to assure that no discharger exceeds a level that is deemed critical. The problem of how to assure compliance remains. Russell proposes a deposit-refund system in which generators of hazardous wastes would be required to make deposits that would be refunded when they delivered the wastes to approved disposal facilities. The system could be applied, as well, to controlling the disposal of some nonhazardous wastes such as worn-out tires and automobiles.

Peter Passell, in his *New York Times* article, considers the consequences of the nation's failure to control the disposal of hazardous wastes in the past and questions the current policy for cleaning them up. The United States is contemplating spending hundreds of billions of dollars to restore thousands of hazardous waste dumps to their pristine condition, some of them abandoned sites and others federally owned. Passell reminds us that the salient question is not whether to clean up the environment, but how much to clean it up. He calls for a weighing of benefits and costs. Rather than attempting to return all sites to pristine condition, he would ask in each case, what is an acceptable level of risk in the light of the cost of reducing it further? Beyond a certain point, he says, far greater reductions in health risks could be achieved for the same or less cost by applying the resources to reducing exposure to other hazardous substances, such as radon gas.

13

Regulatory Strategies for Pollution Control*

A. MICHAEL SPENCE

MARTIN L. WEITZMAN

A. Michael Spence is Dean of the Business School at
Stanford University and Martin L. Weitzman is Ernest E.
Monrad Professor of Russian Studies in the Department of
Economics at Harvard University.

INTRODUCTION

The debate over regulatory policies to control pollution has been dominated by advocates of both effluent standards and effluent charges. This
paper argues for a combination of the two that achieves the objectives of
each but is more finely attuned than either alone to the practical problems

"Regulatory Strategies for Pollution Control," by A. M. Spence and
M. L. Weitzman, from *Approaches to Controlling Air Pollution* (MIT
Press), A. E. Friedlander, ed., pp. 199–219. Reprinted by permission.

*This paper is based on some recent work by the authors and others on the subject of
controlling via quantities and prices. Weitzman was the first to study the advantages of
quantity controls as a response to uncertainty. Spence and Roberts used the device of
transferable licenses in the pollution context to examine the merits of effluent standards and
of mixtures of fees and standards. The purpose of transferable licenses to employ an overall
quantity control. The authors would like to thank the members of an early workshop on the
Clean Air Act for their encouragement and their aid in helping us to understand various
facets of the problems. We want particularly to thank Douglas Allen for his help. Robert
Dorfman, Peter Diamond, and Robert Solow pointed to some problems with which we have
tried to deal. There remain some issues of a theoretical kind that are best confronted in the
more technical papers listed in the references.

that are often encountered in regulation. The conventional case for effluent charges makes some good sense, but it is based on an unrealistic assessment of the amount of information actually available to regulators, information about the sources, extent, and costs of pollution. It therefore misrepresents the practical options actually open to government regulatory bodies. On the face of it, the case for effluent standards might appear to be stronger. Yet, as we shall argue, the strategy for setting standards can also be improved.

The regulatory problem, in general terms, is to constrain sectors of the economy to achieve the "right" reductions in pollution levels. At the same time, however, the regulatory authority must hedge against the major sources of uncertainty that arise when limited information about polluters and effluents is available. Dealing effectively with this uncertainty involves building into the regulatory process sufficient flexibility to avoid the worst consequences of mistakes. This paper argues that a combination of effluent charges and standards will be the most effective policy.

The problem of uncertainty is not inconsequential. Pollution entails substantial adverse effects for society. But the costs of cleanup or the costs to the industrial sector of equipment to reduce effluents at their source are also substantial. Any mistake about these relative costs and the implied appropriate regulatory policy can cost the economy as a whole many millions of dollars. Because of the magnitude of resources involved, it is important to devote a sufficient amount of effort to ensure that the incentives faced by the regulated sector are consistent with social objectives.

Economists tend to think of pollution as a problem that results from a market failure. That failure is the complete absence of a market for the effects that polluters have on those adversely affected by pollution. (A market, in this context, should be thought of as an arrangement in which people pay for the things they do that affect others.) The fact that markets in the effects of effluents do not and are not likely to exist implies that the costs of pollution are *not* necessarily a factor in the decisions of the polluters. That implication, in turn, tends to result in excessive effluents.

Economists do not just identify the problem as one of a missing market. With a certain rationality, we also incline toward fixing the problem with something that looks like and acts like a market, that is, a system in which prices are attached to effluents. It is usually called an effluent charge or an effluent fee system. The effluent charges are imposed. The regulated sector is then allowed to adjust to them by some combination of paying the charges and reducing effluents.

To most people, the identification of the pollution problem as one of a missing market or a market failure seems odd, to say the least. There never was a market in pollution, and there is not likely to be one in the near future. Pollution, for many people, is a problem like that of preserving wilderness areas or keeping the streets clean. It has little to do with markets, potential or actual. Opponents of prostitution do not react favorably to proposals to control that industry by imposing (perhaps sizeable) excise taxes that approximate the social and moral costs of the activity. There has been a similar feeling about effluent charges as a response to the degradation of the atmosphere. Recently, opposition to effluent fees by regulators and environmental groups seems to have declined somewhat.

To the extent that the presumption in favor of effluent charges is based upon the view that the underlying problem is a missing market, the case has not been convincing to the majority of the public or to noneconomist professionals such as engineers and lawyers. Moreover, public acceptance is not entirely an academic matter. Many of the major regulatory efforts in the United States have been fashioned and modified in the Congress. The general form of the regulations does not usually emanate from a technically oriented agency like the EPA.

Those who mistrust the effluent charge approach typically favor effluent standards. These are maximum levels of effluents that are deemed acceptable and consistent with the maintenance of the quality of the environment in which people live. They are set by the political process and met by the regulated sector. This approach to the regulation of pollution corresponds much more closely to most people's perception of the problem. There is a collective decision about what is and is not acceptable conduct, and the government's task is simply to enforce the laws. The setting of standards is the major perceived alternative to effluent charges. Thus far, standards have been winning the battle in the political arena.

The economists' case for effluent charges is not entirely based upon the missing market hypothesis. To suggest that it is would be to do an injustice to its proponents. The case is often amplified and buttressed by other arguments. Standards are rigid, at least in principle, and therefore are insensitive to costs. Of course, if costs turn out to be much higher than expected, the standards can be relaxed. But the effluent charge advocate would argue that this type of relaxation occurs automatically and in a controlled way with an effluent charge. Moreover, the knowledge that a standard may be relaxed can create an incentive for the regulated to create the impression that costs are or will be high.

It is argued further that standards distribute the cleanup among polluters in ways that are potentially inefficient. What one would like is a system that causes the sources with the lowest cleanup costs to do most of the cleaning up. Effluent charges do distribute cleanup activity among sources efficiently.[1] Whatever the level of effluents actually achieved, it is achieved at least cost.

There is a final argument. It is that the government should take the

---

[1]The reason is that the sources of pollution clean up to the point at which the marginal costs of cleanup equals the effluent fee. But then the marginal costs of cleanup are the same for all sources, the implication being that cleanup cannot be shifted among sources in a way that reduces costs.

position of standing in for the public, whose interest is not represented in the absence of regulation. The public interest is properly represented when the additional benefits of reduced effluents are commensurate with the additional costs that result from effluent reduction. The public, after all, pays these costs in the end, in the form of higher prices, displacement of jobs, and so on. Effluent fees, it is sometimes argued, are a reasonable way of putting the public benefit into the equation. The fees "represent" the benefits to the public of reduced effluents or, equivalently, the costs of pollution to the public.[2] By contrast, it is said, rigid standards seem to imply that the social costs of exceeding the standard are high enough to make it unreasonable to contemplate emissions in excess of the standards. They implicitly misrepresent the damages from pollution. No one really believes that the social cost of exceeding the standard is infinite. Then, the argument goes, one is left wondering about the rationale for adopting what must inevitably be seen as a somewhat arbitrary standard in the first place.

There are two basic problems in regulating pollution, setting aside enforcement problems for the moment. One is distributing cleanup among effluent sources in an efficient manner. The second is trading off costs and benefits and adjusting effluent levels until costs and benefits are commensurate. Effluent charges accomplish the first objective. If the initial effluent charge is not the correct one, it can be adjusted and, if necessary, readjusted until the appropriate trade-off between costs and benefits is achieved. Thus, the second objective is achieved by a process of trial and error. In the course of this process, the regulators need only adjust one number, the effluent charge. It is adjusted when the incremental benefits of effluent reduction are perceived to differ significantly from the effluent charge, for it is the effluent charge that is supposed to approximate these benefits. The idea, then, is that the regulators go through a cycle: setting charges, obtaining a response from the regulated sector, and resetting the charges. A similar argument could be made for effluent standards, but the economist would argue that it is more difficult to iterate in this case because there are more variables to control.

The underlying presumption is that the regulated sector can and will effortlessly, costlessly, and frictionlessly adjust to changes in the regulations until the hypothetical optimum is reached. This is not a very accurate description of our world. The image of a frictionless and responsive regulated sector is misleading for several reasons. First, the limited information about costs and cleanup technologies that the government possesses is also a problem for effluent sources. The real expenditures required to meet particular standards or to respond to particular regulations cannot be taken back. They are sunk costs. Each time the regulations are changed there are additional costs.

---

[2]The benefits, of course, are extremely difficult to measure and to agree on. And no regulatory approach is excused from undertaking this task.

Second, the cleanup technology is usually highly capital intensive. The investments required to respond to a particular set of standards cannot easily be reversed. In fact, they may not be reversible at all.

Third, it takes time to mobilize any organization to engage in a new activity. Once in motion, most business organizations do not easily change direction. Frequent changes in regulations may create serious implementation problems for a well-intentioned business management.

Fourth, the organizational inertia just described applies also to the regulatory organization involved in implementation and enforcement. Learning on the job is rendered significantly complicated by frequent changes in the rules that are being enforced.

Fifth, if the regulated sector anticipates regulatory changes that, in turn, are responses to costs in the regulated sector, then the simple model of the regulated sector responding myopically to each new regulation, be it a standard or a charge, is not realistic.[3] If the regulated sector's behavior affects the rules and the polluters know it, they are unlikely to take each new set of rules at face value.

These factors conspire to make any regulatory process that involves repeated adjustment to new information costly, time-consuming, and perhaps infeasible. These remarks apply to any fine-tuning regulatory process and not just to the effluent charge approach. The same comments would apply to a system of continuously adjusted standards.

If it is costly or impossible to deal with uncertainty by adjusting until the relevant aspects of costs and benefits are known, then it is important to think of the regulatory problem as one of imposing rules based on the best available information, however limited. The combination of the rules and the regulated sector's responses will produce results (effluent levels, cleanup costs, and price changes for many products) that must be endured for some extended period of time. This is not to say that that the rules are immutable but rather that the initial rules are important because their effects will last for an extended period.

One cannot help feeling that the occasional hostility of the debate between proponents of standards and effluent charges is in part the result of a failure of communication among the interested parties about what the practical constraints on regulatory activity may be. Perhaps the two sides have been operating with essentially different models. Economists, using the previously described frictionless model, have labored with some success to explain the merits of the price system and appear to have difficulty understanding why there are so many recalcitrants among the policy makers. The proponents of standards have felt that their approach is safer and more practical, although the case for standards as a control strategy has been less effectively defended on formal grounds.

---

[3]It is not easy to predict how the regulated sector will respond to a situation in which regulations are anticipated to change in response to costs. It seems reasonable to surmise that the regulated sector's efforts might be sluggish under these conditions.

In the literature and in discussions of control problems with those who have had to frame, implement, and respond to regulation, certain facts emerge upon which most would agree. They are facts that can act as guides in developing regulatory strategies.

First and most important, both the benefits and costs of effluent reduction are uncertain at the time the regulations are imposed. The uncertainty can be reduced through the expenditure of time and resources, but it cannot be eliminated.[4]

Second, adjustments in levels of effluents cannot be made easily or costlessly, for the reasons cited earlier.

Third, any control system requires some form of monitoring. The costs of and available technologies for monitoring vary from one pollution problem to the next.[5]

Fourth, the link between ambient air quality standards and emission levels at effluent sources is imperfectly understood. The diffusion models required to predict ambient air quality from emissions are complicated. They have not been available long, and those that are available are not necessarily understandable to state enforcement agencies.

Fifth, regulation not only affects the regulated sector's incentives to reduce emissions, but it also affects the incentives for research and development in the area of new technologies for cleanup, a subject to which we shall return.

Regulation places restrictions and imposes costs on the regulated sector. There is a large variety of different kinds of restrictions that can be imposed. Perhaps unfortunately, two have attracted almost all of the attention. They are standards and effluent charges. After looking at the important characteristics of costs and benefits in the second section, in the third we examine standards and charges. The reason is essentially two-fold. First is to argue that, between the two, the preferred option depends on some important features of pollution damages and cleanup costs, especially the uncertainty about costs. Second, an examination of the relative merits of each alternative as a response to limited information is essential to understand what we think may be a better practical alternative: combining standards with effluent penalties for emissions in excess of the standards.

Therefore, in the fourth section we outline a regulatory strategy based on standards supplemented by penalties that look and act like effluent

---

[4]The decision about how far to go in collecting information is one of the more important policy decisions that arises. It is rarely made with a view to what difference further information will actually make.

[5]Moreover, monitoring technology is capable of changing over time. It appears, for example, that the monitoring technology for particulates from stationary sources is insufficiently advanced to permit the use of effluent charges in that context. On the other hand, it is also true that this situation may change if the government devotes more resources to the advancement of monitoring technology. At present, this essentially public R and D investment seems to be deficient.

charges. This strategy combines the better features of both of the currently debated alternatives. We believe it will appeal to the practically oriented as a reasonable and useful modification of current control systems based on standards.

THE NATURE OF BENEFITS, COSTS, AND UNCERTAINTY

Although we can and must quantify the benefits of clean air to make intelligent decisions about desirable levels of air quality, it would be a mistake to think that it is easy to determine the damages from pollution. Ideally, we would like what the economist calls a damages function—a dollar measure of the harm or disutility caused by various levels of pollution. But such numbers are hard to come by. For one thing, there is a large psychic component—polluted air is undesirable, in part, because it is unpleasant. Attaching numbers to people's preferences is always difficult, especially when they disagree. And even when less subjective elements are involved, many of the more tangible damages from pollution, including health effects, are hard to measure.

So a crucial aspect of pollution damages is uncertainty. This acknowledgment does not mean we know nothing about the benefits of clean air. We may know upper and lower limits but be somewhat fuzzy about the area in between.

The uncertainty in benefits does not really affect pollution control strategy per se (the choice between standards and fees, for example). Whatever course is actually tried out may not make us less uncertain about benefits. There is a slight qualification because changes in ambient air quality attributable to regulation provide new data that can be used to improve estimates of the health effects and other impacts of pollution. But the "value" of cleaner air would still remain largely uncertain. In contrast, the way that the regulated sector responds to a particular set of regulations will, over time, tend to reduce our uncertainty about cleanup costs. Uncertainty in benefits can be narrowed only by more research, carried out presumably by some branch of the public sector. Under the circumstances, the economist tends to suppress the uncertainty in benefits by working with the reasonable compromise of an expected damages function. The expected damages function may be higher or lower depending on whether damages are anticipated to be higher or lower. It represents our best single estimate of damages at the time when a regulatory decision must be reached. Henceforth when we speak of damages, we will implicitly be speaking of expected damages.

There remains a problem of translating the various consequences of pollution into an index that is commensurable with cleanup or abatement costs. In discussing this problem with a variety of people who have been involved in the framing and implementation of the clean air act, we have discovered a way of phrasing the issue that seems to have some appeal. Imagine that pollutants are currently at some fixed levels. One can ask what maximum additional abatement costs would be tolerated to achieve a further 10 percent reduction in pollutants. The answer to this question translates into a statement about marginal damages measured in dollars. But it seems easier to confront the issue by comparing the consequences of effluent reduction and abatement costs directly in this way than to attempt to attach a dollar value to effluent reduction abstracted from abatement costs.

Perhaps the most important single property of pollution damage is that the extra damages of an additional unit of effluent often increase (or at least do not decrease) with the overall level of pollution. This is sometimes called the "principle of increasing marginal damages." Although it is not universal, the principle seems to have general validity. When the air is fairly clean, an extra unit of effluent does less damage than when the air is already heavily polluted.

As we shall see presently, the form of an optimal pollution strategy very much depends on the shape or curvature of the damages function. Two extreme cases merit special attention.

A relatively straight damages function means that marginal damages do not increase very much with pollution levels. This function would be characteristic of a situation in which increased pollution leads to a steady, even deterioration without any dramatic changes.

When the damages function is highly curved at some level of pollution, marginal damages are increasing rapidly around that point. This might be a fair description of a "threshold effect" in pollution. In such situations, marginal pollution damages rise precipitously as pollution starts to become dangerous or uncomfortable.

The principle of nonconstant marginal damages means that it is difficult to place a single, unambiguous price tag on pollution. Unfortunately, marginal damages depend on the level of pollution. To price effluents correctly, we would need to know what the level of pollution is or will be. The fact that we do not know the pollution level in advance makes it difficult for a fee system to function well. Naturally this problem is going to be more acute when marginal damages are changing rapidly than when they are relatively constant.

Turning now to costs, our starting point is the cleanup cost function. This is simply a schedule giving the dollar outlays necessary to obtain a certain reduction in emissions.

Most analysts believe that the incremental cost required to eliminate an extra unit of pollution goes up as the effluent level declines.[6] That is, it is less costly to eliminate the first 5 percent of effluents than it is to eliminate

---

[6]See, for example, the costs derived by Donald N. DeWees in Chapter 7 of this volume [i.e., Friedlauder, ed., *Approaches to Controlling Air Pollution—Eds.*].

the second 5 percent, and so forth. Economists call this phenomenon the "principle of increasing marginal costs."

The shape of the cleanup cost function will have some bearing on the form of an optimal pollution strategy, just as does the shape of the damages function. Sharply increasing marginal costs give rise to more highly curved cost functions, whereas slightly increasing marginal costs are associated with relatively straighter cost functions. There are no general principles for determining the curvature of cost functions. It depends on the situation and varies from case to case. In short, it is an empirical matter.

It seems to be a fact of life that the regulators don't know cleanup costs to a high degree of accuracy at the time the regulations are imposed. It is especially true when a new or unproven technology is involved, as with auto emission controls. There is no way of knowing beforehand exactly what it will cost to achieve a certain cleanup level. Estimates can be made, but the final costs will not be known until mass-produced equipment is in place, if then.

The uncertainty in cleanup costs is essentially due to lack of experience. It can be reduced by research but not altogether eliminated. No one knows precisely what it will cost to achieve some cleanup level because it has never before been tried. Once a full-scale effort has been launched, the relevant costs will eventually become known but not before.

There is another important feature of cleanup costs that goes along with the uncertainty. Not only are costs unknown, but it is also difficult and expensive to find out what they are. Sometimes economists and others share a tendency to conceptualize regulation as a process of continual fine-tuning. A certain strategy is adopted, and marginal costs and marginal benefits are observed; if they are not equal, the fees, standards, or other parameters are smoothly adjusted until an optimum is obtained.

As argued earlier, this may be an inappropriate way of viewing the problem. In order to have a chance to work, a regulatory strategy must be left in place for an extended period after it has been adopted. As we see it, analysis of regulatory strategy should start from the following point of departure: *the regulators are forced to make decisions in an uncertain environment and they must live with the consequences for some time.* Among these consequences is the possibility that costs will turn out to be higher or lower than was expected. The above principle will provide a framework for analyzing certain important issues that are outside the scope of the fine-tuning model. With it in mind, we turn to the relative merits and demerits of effluent fees and standards.

FEES VS. STANDARDS

The two best-known regulatory strategies for controlling pollution are effluent *standards* and effluent *fees*. They are easily comprehended and are frequently contrasted. In this section we propose to analyze carefully the comparative advantage of each of these control strategies.[7]

It is useful to analyze standards and fees for at least two reasons. For one thing, this issue is of interest in itself because there is a long-standing policy debate about the comparative merits of these two control modes. For this reason alone it is important to understand how fees and standards work and to be able to identify situations in which each one is likely to outperform the other. A second motivating factor is our own interest in promoting a mixed standard-fee system, which we, and others, feel may be superior to either standards or fees alone. To understand how the proposed mixed system works and just exactly why it is better requires a thorough acquaintance with the basic subcomponents out of which it is constructed.

By far the easiest pollution strategy for the public to comprehend is the one based on standards. Some branch of the government acting on behalf of society's interests establishes upper limits on emissions for each polluting firm. It might appear at first glance that with standards cleanup costs are borne by the polluter, but in fact they will eventually be passed on to the consumer in the form of increased prices, reduced employment, and so on. The standards approach is popular in large part because it represents a direct assault on the problem that fits comfortably with legal, moral, and historical traditions. If the problem is that some identifiable group is overpolluting, the obvious remedy is to force it to clean up to a level more in keeping with society's needs as a whole. What could be simpler than decreeing that pollution be cut back to some level that approximates the public's interest?

An alternative strategy, one frequently favored by economists, is the effluent fee system. Effluent fees are prices that attach to effluents. A polluter pays for his effluent an amount proportional to his discharge volume. In controlling the fee, a regulatory agency indirectly controls

---

[7]In the interest of focusing sharply on the essential economic differences between fees and standards, we are abstracting away a host of "noneconomic" factors. This treatment does not mean that we feel they are unimportant. On the contrary, there may be significant legal, administrative, or historical reasons for favoring standards or fees in one situation or another. Even such factors as the capability of the monitoring system play no small role. At the present time we do not seem to have an in-place technology capable of continually monitoring stationary particulate sources to the degree of accuracy required for making an effluent fee system work effectively. The desirability of having the government encourage research in the monitoring area is something that cannot be stressed enough. Certainly the regulated sector has little direct incentive to develop or improve the monitoring system.

effluent discharge by manipulating the incentive to engage in cleanup. A higher fee encourages polluters to clean up more whereas a lower fee elicits more pollution.

To many people a fee system is an unfamiliar and peculiar method of controlling pollution. To make a fair evaluation, it is important to understand exactly how it works. When a fee is imposed on emissions, it indirectly controls pollution in the following manner. A polluter, in order to maximize profits or minimize costs, will fix emissions at that level at which the incremental cost of cleaning up an extra unit of pollution equals the fee. If the fee exceeds the marginal cleanup cost, money could be saved by making the extra investment needed to cut back pollution slightly, and vice versa when the marginal cleanup cost is greater than the fee.

Effluent fees and standards differ in how the burden of cleanup costs is shared. A full analysis of the distributional implications of either system would constitute an excessively lengthy aside. However, it is worth noting one point: an effluent fee system generates government revenues, which can be used for public expenditures or to reduce the burden of taxes collected in other ways. How they are used will in large part determine the distributional impact of the control strategy. Economists regard these distributional issues as impossible to decide on purely economic grounds, but this difficulty does not mean they are unimportant. In our experience, many practitioners react negatively to the potentially large payments that must be made under a fee system.

In an uncertain world in which cleanup costs are not precisely known to the regulators at the time a decision must be made, the comparative advantages of standards or fees derive from the following basic observation: *standards fix pollution levels but leave cleanup costs uncertain;* in contrast, *fees fix (incremental) cleanup costs but leave pollution levels uncertain.* Which of these features is more desirable depends on the underlying economic situation. As we shall see, sometimes one feature is more important, sometimes the other is. We propose to examine a few extreme cases to illustrate the general principles that are involved.

The fact that effluents are fixed under a standards system tends to make that approach relatively more desirable as the damages function is more highly curved. When marginal damages rise rapidly around some threshold level, it would probably be foolish to use effluent fees to control pollution because the pollution level remains uncertain. If the marginal social benefit of clean air is low in some range but increases precipitously as pollution starts to become dangerous or uncomfortable, then for a wide range of costs the effluents should be at or near the threshold level. And under standards they will be. But if effluent fees are used, pollution levels will vary with costs. Should cleanup costs turn out to be higher than anticipated, profit-maximizing polluters could elevate pollution levels into the danger zone. If costs are lower than expected, polluters may be motivated to clean up well beyond the threshold level, to an extent that is not socially justified because marginal damages at that point are insignificant. In a world of cost uncertainty in which regulators must work with fixed fees or standards, a threshold effect in pollution damages makes a strong case for standards.

The opposite kind of conclusion holds with respect to the curvature of the cost function. The straighter the cleanup cost function, the stronger the case for standards. If incremental costs increase only slightly with cleanup levels, it is very difficult to control pollution by fees. Suppose a fee is named. A polluter will set emissions at the level at which marginal cleanup costs equal the fee. Now, when marginal costs vary little as pollution changes, it means that pollution varies greatly as marginal costs change. Because polluters set marginal cleanup costs equal to the effluent fee, even slight fee changes will be translated into large swings in the pollution level. If the fee is set correctly in the first place, everything will be fine. But, as we have tried to emphasize, there is a large amount of uncertainty in any real-world regulatory environment. With relatively straight cleanup costs, the slightest miscalculation of the fee will result in either much more or much less than the desired pollution level. In such a situation, standards tend to be better because a high premium is placed on the rigid output controllability that only they can provide under uncertainty.

Just as a more highly curved damages function favors standards, so a relatively straight damages function is more conducive to fees. If the damages function is close to being linear, it would be foolish to name standards. If the marginal social damage is approximately constant in the relevant range, a superior policy is to confront the effluent sources with an effluent charge equal to the marginal damages. Then the polluters will automatically bring themselves close to a social optimum by picking the emission level that equates marginal cleanup costs to the fee.[8] With a straight damages function it is much better to have the polluters find their own desired emission level on the basis of a fee than to have the regulators determine it for them by setting a rigid emission standard at a time when costs are uncertain. In this case, the fee system is more attractive because it gives the ability to fix marginal costs in an uncertain world.

Perhaps the main reason that some economists traditionally favor fees over standards is that fees automatically induce an efficient distribution of cleanup effort among different sources. The economist tends to view standards as piecemeal regulation that offers no guarantee that the overall level of pollution will be attained at least cost.

Here is an example. Suppose it has been decreed on the basis of crude cost calculations that mobile sources should cut back sulfur dioxide emissions by 30 percent and stationary sources by 50 percent. Because the

---

[8]Of course, effluent sources may first have to expend resources to learn about the costs and control technologies.

regulators don't really know what the actual costs of pollution abatement will turn out to be (and maybe even if they do know), there is no guarantee that the incremental costs of a small further reduction will be the same for both sources. It would be better to require the source with the smaller incremental cleanup costs to pollute a little less and to permit the other source to pollute a little more. The same overall sulfur dioxide level would be attained but at less total cost.

It is important to understand that imposing a uniform fee on all emitters of a specific pollutant automatically guarantees that the overall cleanup level will be obtained at least total cost. Each polluter, in order to maximize profits or minimize costs, will set emissions at that level at which the marginal cost of cleaning up an extra unit equals the fee. Because the marginal cleanup costs of each polluter are equal to the same fee, they are equal to each other. This is the hallmark of a least-cost allocation of cleanup activity. Only when marginal cleanup costs differ would it be possible to obtain the same overall pollution level at less cost. As in the previous example, this effect would be accomplished by allowing less pollution from the source with low marginal cleanup costs and more from the high-cost polluter.

The automatic efficiency of a fee system is a definite point in its favor because it results in cost savings. If we lived in an infinitely flexible control environment where the regulators could continuously and costlessly adjust the fee, the efficiency argument might be overwhelming. But, in practice, adjustments are usually very costly. The consequences of any regulatory action are going to be with us for a while. And then the uncertainty about pollution levels that is inherent in a fee system can become troublesome for the reasons just discussed. Standards may be preferable to fees in a multiple-source setting even though standards are inefficient. It all depends. If it is important to hold overall pollution to some prescribed level, that need may take precedence over having a cost-minimizing way of achieving an uncertain level of pollution.

The cost-saving or efficiency argument for fees becomes more significant as the variety of polluters increases. A fee system enables the regulators automatically to screen out the low-cost polluters by encouraging them to clean up more relative to the high-cost polluters. This screening effect is more significant if there are many different types of polluters because the possible cost savings are greater. If there are three distinctly different types of sulfur dioxide emitters that have independent cleanup technologies instead of one large pollution source that yields the same aggregate effect, the case for fees is strengthened, other things being equal.

Note that the desirable cost-screening effect of fees doesn't work unless the different pollution sources really are different. A fee system for automobiles is not likely to permit much cost screening because cleanup costs are not likely to differ much from one auto to another. If costs of several polluters are highly correlated, as with automobiles, it is best to lump the units together and view them as one *type* of pollution source. The more different types of polluters there are, the greater the potential cost savings of a fee system due to the screening effect.

It is perhaps useful to summarize at this stage. The comparative advantage of fees and standards for controlling pollution depends on the shapes of the damages and cost functions, on the magnitude of the uncertainty, and on the number of effluent sources with relatively independent cleanup costs. Standards are favored as damages are curved or costs are straight. Fees are favored as damages are straight or there is a larger number of independent polluters.

The purpose of this section has been to give the reader a feeling for the way fees and standards work and when each one is likely to work better than the other. In the next section we are going to propose a mixed standards-fee system that outperforms either pure system.

THE PRESSURE-VALVE APPROACH TO REGULATION: STANDARDS WITH EFFLUENT PENALTIES

In the presence of limited information, standards are often preferable to effluent charges because they prevent the levels of effluents from running up when costs turn out to be higher than the initial estimates. This feature of standards is particularly important when the incremental damages increase rapidly with the level of effluents. On the other hand, standards are rigid and therefore unresponsive to situations in which cleanup costs turn out to be high. Under these conditions, of course, it is possible to relax the standards. Relaxation, in the form of a delay, has occurred in the case of automobile emissions. And it is done elsewhere when the need arises. But the conditions under which the standard is relaxed and the way it is relaxed are of extreme importance in determining the effectiveness of the control program.

The problem, in a nutshell, is to devise a mechanism that responds to high cleanup costs for individual sources and at the same time does not create an incentive for noncompliance with the standard. What is needed is a penalty, specified in advance and paid by the source in case its emissions exceed the standard set for that source. Moreover, to maintain the incentive to clean up, the penalty should increase with the amount by which emissions exceed the standard. A practical way to achieve this goal is to establish a penalty *per unit* of emissions in excess of the standard.[9] That penalty will act like a high effluent charge for the sources that turn

---

[9]There are other possibilities. The penalty per *unit* of emissions in excess of the standard could increase with the amount of the excess.

out to have high costs. Those sources that have costs close to prior expectations will find it desirable to meet the standards. For them, and they will be in the majority, the system will function as if there were simple standards. For the very high cost polluters, the system will function as if there were fees.

The proposal is to add effluent penalties to the standards that take effect only when the standards have been exceeded. It is a system in which a set of standards is supplemented by *pressure valves,* which release only when the costs for an individual source exceed the estimated costs by a significant amount. Each individual polluter decides for himself whether his cleanup costs are high enough to justify paying the fee for exceeding the standard. Moreover, when the escape valve releases, it is not a complete release. The standard for a high-cost polluter is *not* reset at no cost to the source. Rather, the polluter pays a penalty, which can be avoided at a future date, in the event of a reduction in the costs of cleanup. Therefore the incentive to maintain the effort to reduce the cleanup costs is retained.

This approach has several attractive features. Most sources will have costs in the neighborhood of the average of prior expectations. They will therefore meet the standards. Only those sources whose costs are significantly higher than estimated will choose to exceed the standards, and they will pay the penalties. High-cost sources will clean up less than those with lower costs. This is one of the more important attractive properties of the effluent charge approach, but it is one that can be overridden by other considerations. The protection of standards against high pollution levels is substantially maintained. So long as the penalty is set above expected marginal damages, the escape valve will operate only when costs and benefits differ by a sizeable amount.

The penalty that the individual polluter faces more closely approximates the social costs of his contribution to pollution than under pure standards or effluent charges. The penalty is neither prohibitive, as in effect would be the case under rigid standards, nor is it too lenient, as when it is equal to the incremental damages at the anticipated outcome. In the latter case, the system would function like an effluent charge system and would therefore have the problem of suboptimally high levels of pollution that high costs would cause.

The setting of the optimal penalty is a matter of some complexity.[10] For practical purposes, however, there is only one important principle. The penalty should be related to the damages or social costs of pollution. As a rough approximation, the penalty per unit of effluent should be some-

[10]For a technical discussion of related issues, see Martin L. Weitzman, "Prices vs. Quantities," *Review of Economic Studies,* October 1974; and Marc J. Roberts and Michael Spence, "Effluent Charges and Licenses Under Uncertainty," *Journal of Public Economics,* 5 (1976): 193–208.

what higher than the marginal damages at the levels of effluents that would obtain if all the standards were met. [11] It is important to remember that the incremental damages increase with the level of effluents. Therefore, as a first approximation, one could do worse than setting the penalty equal to the marginal damages at the level of effluents prior to the imposition of regulation. Or to put it in the form of a decision rule, assess the marginal damages of pollution at existing (precontrol) levels of effluents and set the unit effluent penalty equal to that number. If that is done, then the escape valve will function only for those sources whose costs are so high as to exceed the benefits of cleanup at relatively high precontrol levels of pollution. If marginal damages increase dramatically with effluents, then that penalty will be high, as it should be. On the other hand, if marginal damages do not increase rapidly, then the penalty will be lower, and that relationship also is desirable. The setting of the escape valve can and should be tied to the curvature of the damage functions. Setting the penalty at the precontrol marginal damages is one relatively simple way of accomplishing this end, although there are others.

Analysts have worried that the effluent charge system, even when it is desirable (for the reasons discussed earlier), possesses some embarrassing features on the financial side. It generates a lot of revenue that then has to be disbursed. And effluent sources pay double: once for cleanup and once for the effluent fees on the emissions after cleaning up. That system may impose a rather heavy financial burden on some sources. The standards and penalties approach does not have this pair of problems. Most sources do not pay anything to the government. They meet the standards. Some sources pay some penalties. But the penalties are only on the emissions in excess of the standards. The payments are therefore smaller by orders of magnitude than the fees that would be paid under a simple effluent charge system. Indeed, another way of describing this control strategy is as an effluent charge that doesn't take effect until certain targets or standards (one for each source) have been exceeded.

In most of the pollution problems with which we are familiar, damages increase at an increasing rate with the levels of pollution. These are the circumstances under which the use of standards is preferred, when the choice is between standards and simple effluent charges. We are suggesting that one can improve upon standards by instituting an automatic escape valve that applies to individual sources and prevents the worst dislocations in the event of unexpectedly difficult cleanup problems. The escape valve

[11]There is currently a movement to employ penalties as part of the enforcement programs in some states. Connecticut is one of them. These penalties are set to equal the cleanup costs avoided by those not in compliance. It is important to stress that this is a technique *for enforcing standards*. It should not be confused with the proposal put forward here and elsewhere. The penalties in the escape-valve approach are related to damages and may be translated into maximum tolerable abatement costs, as discussed earlier. But maximum tolerable costs and actual costs are very different.

takes the form of a high (but not prohibitively high) penalty that is proportional to the amount by which effluents exceed the standard.

The basic argument for this modification is that it makes the penalties for the individual polluter more closely approximate the actual damages he is causing than is the case under either rigid standards or effluent fees. The modified system is a form of flexible standard approach, but it is preferable to systems in which standards are simply relaxed without penalty (then no one would comply). The costs of exceeding the standards are nontrivial and nonzero. These costs are also specified in advance. The system therefore maintains the incentives to attempt to meet the standard and to reduce levels of effluents. It also removes the incentive to try to appear to have high cleanup costs and thereby impress the regulators to relax standards. In addition, the penalty-modified standards will not result in the collection of enormous volumes of revenue from the private sector. And implementation of the system does not require a major change in the direction of policy. All that is required is a redefinition of how to assess penalties for noncompliance. It is a practical and, we think, useful addition to the current system of control via standards.

There is another rather important set of issues that have emerged in discussions with regulators and interested observers. They concern the incentives for research and development in pollution control technology that the regulations create. It is clear that the regulations not only affect cleanup activity in the private sector, but they also determine the way in which technological development proceeds in this relatively new industry. The effects of regulation on the patterns of technological development are potentially among the most important long-run effects that regulations can have.

This subject can be rather difficult and complicated, but there are a few observations that can be made in support of the penalty-supplemented standards that are being proposed here. Both standards and effluent charges have potentially distortionary effects on the pattern of research and development. From a social standpoint, the R and D problem is one of investing in technologies that are good "gambles." Good gambles are technologies that have a significant chance of reducing the costs of effluent reduction and at the same time do not run significant risks of being ineffective in reducing emissions. Society wants resources and effort expended in attractive but nonspeculative ventures.

We can think of an R and D investment program as having a cost (the initial investment) and a distribution of outcomes defined in terms of the levels of effluent reduction that can be achieved at the conclusion of the program. It is useful to think of that distribution of outcomes as having a mean and a variance. What we want is a program with relatively low costs that has a high mean and a low variance. The low variance is desirable because it means that the chances of a disaster (no significant reductions in pollution) are reduced.

With these objectives in mind, what can be said about effluent charges and standards? Consider effluent charges first. Because the penalties facing the regulated sector are proportional to emissions under effluent charges, that sector will respond by trading off the costs of the R and D program on the one hand and the mean of the distribution of effluent reductions on the other. The variance is likely to be ignored as a factor in the choice of an R and D program, but the variance is irrelevant only when the damages function is linear. Therefore, the effluent charge approach in the context of the R and D problem has the potentially fatal flaw that it fails to provide the regulated sector with an incentive to respond negatively to the variance of the outcomes. [12] It therefore fails to provide protection against the risk of potentially high pollution levels. This problem is the analogue of the tendency of effluent charges to generate excessive levels of effluents when costs turn out to be high. Both problems result from the fact that a curved damage function is approximated by a straight line.

Standards pose a different problem. Under standards, provided they are not expected to be relaxed, the private sector will respond by selecting a program that minimizes the probability of failing to meet the standards. That may be a very costly program. And thus, in an important sense, standards may cause the R and D program to be excessively costly and conservative. A substantial reduction in costs with a small increase in the probability of noncompliance will be rejected even if it might appear to be a rather good decision from a social point of view. Costs figure in the R and D program under standards only if there are programs that assure the meeting of the standards. The private sector will then select the least-cost program that meets the standard.

As a rough approximation, standards respond first to the mean and variance, especially the variance, and then to costs if the standards can be met with certainty. This pattern also produces distortions. Costs and benefits may not be properly traded off.

Because the penalty-modified standards more closely approximate the damages, these sorts of problems will not arise in as severe a form. The private sector will hedge against uncertainty but not to the exclusion of cost considerations. Thus the case for the mixed approach is based in part on the need to structure the incentives for research and development in an appropriate way. The automobile air-pollution case is an example of the desirability of having a somewhat flexible standard in a situation in which the development of the control technology is a central feature of the

---

[12]It may be that the regulated sector behaves as if it were risk averse. If it does, then it will respond negatively to the variance even though the penalties are linear in effluents. This behavior will mitigate, to some extent, the tendency of effluent fees to cause risk to be ignored in R and D programs. A full analysis of this complex set of problems is unfortunately beyond the scope of this paper.

problem. It is a complicated case, which Mills and White treat in detail elsewhere in this volume. [13]

THE NUMBER OF EFFLUENT SOURCES

The preceding analysis has assumed that firms are the source of pollution and that their number is fixed or varies only slowly over time. It further assumes that the number of sources is not particularly sensitive to the regulations, whatever form they take. There are, however, important problems in which this assumption may not be true. The best-known and perhaps most important is the automobile case. When the number of sources is fixed, the issue is confined to determining the optimal effluent level per source. When the number of sources varies there are two issues. One concerns the effluents per source; the other is the number of sources. In the automobile case, effective pollution policy involves controlling both the effluents per vehicle and the number of vehicles.

In the short space available to us, we cannot deal completely with this problem. However, some comments are in order. First, many of the considerations discussed previously are applicable to this situation and point to the merits of a mixed approach to controlling effluents per vehicle. One difference is that a strong case can be made that the control program directed at the effluents per vehicle should be supplemented with a tax on the vehicles themselves. Without such a tax, the pollution cost of an additional vehicle is not borne by the purchaser; therefore, unless the demand is inelastic, there will be too many vehicles on the road.

Even in the case of stationary sources, there is a long-run decision as to their number and locations. For these decisions to be made correctly, something more than the correct marginal incentives is required. The absolute magnitude of the penalties paid by an effluent source should equal the estimated incremental social cost of that source. Otherwise the long-run entry and exit decisions may not be the correct ones. This possibility suggests that the mixed system proposed earlier should have appended to it an effluent charge equal to or slightly below the marginal damages at the pollution levels implied by the standards. This charge has two properties. It is paid on all units of effluent, so that not only the marginal social cost but also the total cost of the source is internalized. And the effluent fee is below the supplementary penalty that takes effect only when the standard has been exceeded.

Thus, the appropriate regulatory response to cleanup cost uncertainty would include an effluent charge of the conventional kind and a standard

[13]See Edwin Mills and Lawrence J. White, Chapter 8 of Friedlander, *op. cit.*

accompanied by a penalty for emissions in excess of the standard. This system would act like a nonrigid standard system. It would also provide an incentive to clean up beyond the standard for those sources that turn out to have low abatement costs. And it would approximate the appropriate long-run incentives in the private sector for the correct entry and exit of effluent sources.

14

Land, Water, and Ownership

J. H. DALES*

J. H. Dales is Professor of Economics Emeritus at the
University of Toronto.

I

Increasing public concern about the pollution of natural water systems in
North America has confronted governments with a new problem in resource administration, and challenged economists to devise an artificial
pricing system for water that will itself promote wise use of the resource,
thereby greatly simplifying the lives of water administrators. The pricing
problem turns out, not unexpectedly, to be a deliciously complex tangle of
joint uses, externalities, and peak-load problems. The administrative
problem of approximating optimum shadow prices by actual user charges
promises to be a nightmare.

The economic and administrative complexity of water problems is
commonly explained as being inherent in the nature of a fluid resource.
Because of the self-mixing quality of a fluid, one use of water at a given
point may affect other uses at the same point; and because water flows
through space, use at one point may also affect uses at other points.
Opportunity cost pricing is accordingly very complex because of the number of alternative opportunities that may be affected by any one use at any
one point, not to mention the complications introduced by time of use,

"Land, Water, and Ownership," by J. H. Dales, from the *Canadian
Journal of Economics,* November 1968. Reprinted by permission.

*I am grateful to colleagues for comments on earlier versions of this paper. My main debt,
however, is to Mr. J. C. McManus, who has spent much time discussing the paper with me
and has done his best to prevent me from making errors. The paper has been written during
a sabbatical year financed in part by the Canada Council.

varying stream flow, different rates of self-regeneration of different stretches of water, and the chemical interactions of different types of waste after they have been discharged into a natural water system.

Before we submit to this incubus of complexity, however, we might seek comfort in the reflection that the great virtue of a pricing system is that it solves, avoids, mediates, or somehow manages to dispel, all sorts of complexities, particularly those that arise from various interdependencies between uses and users of goods. Yet the existence of a natural pricing system depends crucially on the institution of ownership. What is not owned cannot be priced since prices are payments for property rights or rights to the use of an asset.[1] In the course of allocating property rights to assets among different owners, the price system in fact transforms most potential "technological externalities" into "pecuniary externalities," a synonym for prices. Thus we hear very little about externalities of land use, precisely because property rights to land use are well established and allocated by the price system. It is quite otherwise where water is concerned.

We can now re-formulate the water problem and blame its complexity not on nature and the laws of fluids, but on man and his failure to devise property rights to the use of natural water systems. Economists tend to assume implicitly that it is impossible to own water and therefore seek to devise artificial price systems that are identical to what prices "would be" if ownership were possible. The alternative strategy is to devise an ownership system and then let a price system develop. The purpose of this article is to suggest that there are very considerable advantages to attacking our water problems by means of a system of explicit ownership rather than by a system of shadow prices.

A geographical reflection is also in order. Despite the large numbers of people who live on the St. Lawrence, the Fraser, and the St. John, most Canadians live on lakes, or on rivers that flow into lakes, rather than into oceans, whereas most Americans, despite the large population around the southern rim of the Great Lakes, live on river systems that discharge into salt water. Lakes are much less "mobile" and much less "self-mixing" than rivers. Most of the water in lakes *stays* there for prolonged periods, and recent research has shown that during much of the year the shallow, inshore waters of large lakes are effectively isolated from the very large

---

[1]One never owns physical assets, but only the rights to use physical assets. Professor Ronald Coase writes that a factor of production "is usually thought of as a physical entity which the businessman acquires and uses (an acre of land, a ton of fertilizer) instead of as a right to perform certain (physical) actions. We may speak of a person owning land and using it as a factor of production but what the land-owner in fact possesses is the right to carry out a circumscribed list of actions. The rights of a land-owner are not unlimited." See his "The Problem of Social Cost," *Journal of Law and Economics,* Oct. 1960, 1–44. [Reprinted in this volume, Selection 6.]

volumes of water in their deep centres. [2] (Similar propositions apparently apply to oceans; otherwise the serious pollution problems of coastal cities such as San Francisco, New York, and Vancouver would not exist.)

People who live on river systems, as most Americans do, tend to pass on their pollution to the next downstream community, thereby creating vexing externality problems. American literature on pollution has been strongly influenced by the uni-directional flow of rivers, which makes it relatively easy to solve the identification problem of who pollutes whom. River pollution therefore lends itself to economic analysis in terms of externalities, and shadow-pricing schemes to offset them. People who live on lake systems (most Canadians) tend to pollute themselves; because inshore lake water, far from being uni-directional, tends to slosh up and down the shoreline, lake pollution tends to be a sort of Hobbesean war of all against all. It therefore requires economic analysis in terms of social decision-making and social welfare functions rather than in terms of the effects of autonomous upstream communities on autonomous downstream communities. The economics of Canadian water pollution is therefore quite different from the economics of American water pollution.

II

Rent theory has been the traditional vehicle for studying the economics of natural resources, and a review of some of the effects of the ownership-rental system as applied to land highlights the opposite effects induced by the absence of an ownership-rental system as applied to water. Following normal practice, we shall speak of the supply of land (or water) available to any society as fixed by nature. Though fixed in supply when measured in natural units (acres or gallons), the *quality* of land and water can be changed by human action. (Were we to measure the quantity of land in efficiency units of a given quality, its supply would be variable; we conduct the present argument, however, in terms of natural units, fixed supplies, and variable qualities.) Imagine a society where all land is being used and even the poorest of it commands a positive rent (this assumption allows us to avoid those not very illuminating discussions about no-rent land and the relationships between the extensive and intensive margins) and sup-

[2]See G. K. Rodgers and D. V. Anderson, "The Thermal Structure of Lake Ontario," *Proceedings Sixth Conference Great Lakes Research,* 1963, University of Michigan publication 10, 59–69; P. F. Hamblin and G. K. Rodgers, *The Currents in the Toronto Region of Lake Ontario,* publication (PR 29) of the Great Lakes Institute of the University of Toronto, 1967; and G. K. Rodgers, "Thermal Regime and Circulation in the Great Lakes," in Claude E. Dolman, ed., *Water Resources of Canada,* Royal Society of Canada, studia varia 11 (Toronto, 1967), 87–95.

pose that an initial state of equilibrium exists, and in particular that at existing land values and rents there is neither investment nor disinvestment in the quality of the soil. Population growth when superimposed on this initial state will lead to increases in land values and rents; these increases in turn will lead to economies in the use of soil by means of the substitution of manufactured fertilizers and other intensive farming practices against inputs of natural soil fertility. The process may be described as one of investing in soil fertility, and the equilibrium stock of soil fertility will accordingly rise (or its rate of decline will fall). In a general form, the conclusion is that *the level of rent determines the quality of the soil that it is economic to maintain.* It is also clear, as Ricardo showed, that when man-made inputs are substituted for natural inputs in the food-producing industry the real cost of food increases and the standard of living in terms of food falls. Rising rents, therefore, tend to slow down population growth and lessen the population pressure that produces them.

The working out of these processes in the historical development of the United States has been brilliantly described by Bunce.[3] In the early history of the country there was a high ratio of soil fertility to the human demands on it, and rent was accordingly low; "high farming" practices on the European model were therefore rejected, and economic use necessitated soil-depleting practices. In the course of time the man-soil fertility ratio rose, as a result both of population growth and of soil depletion; rents rose; more intensive farming practices became economic, and the rate of soil depletion was thereby reduced. A slowing down in the rate of population growth after 1900 further reduced pressure on the land; it is possible that in general soil depletion has now been brought to a halt in the United States and Canada, and that soil erosion and soil-depleting practices in some areas are balanced by soil-building investments in others.

The contrast between the history of land and of water use on this continent is eloquent. Property rights were established in land, with rent being the payment for the right to use the soil fertility; there were no water rents because property rights to water use were not established. Rising land rents led to more intensive land use and after 350 years there is no problem of population pressure on the land in North America. Water rents were zero; over-use of the water led to continuous reduction in water quality; and there is now a growing problem of population pressure on North American water resources. If we accept a simple dynamic extension of rent theory and assume a direct relationship between the level of rent and the development of improved technologies, the land-water comparison is again suggestive. Rising land rents have been associated with phenomenal improvements in land-use technology; zero rents for water have been associated with virtually zero improvement in water-use technology so far as quality-depleting uses are concerned.

[3] A. C. Bunce, *The Economics of Soil Conservation* (Ames, Iowa, 1945).

The short-run function of land rent, of course, is to allocate parcels of land among different users and different uses. What is interesting is that potential externalities of land use, for example, the operation of a pig farm in the centre of a choice residential area, seldom materialize. Land being immobile, the ownership-rental system seems to work in such a way as to produce "natural zoning" in land use. Differential rents provide the mechanism for such zoning, and the result is that potential technological externalities are continuously transformed into pecuniary externalities, or prices. It should also be noted that a formal economic description of this process depends on a recognition of space, and particularly of the socially "insulating" quality of space. So long as space exists—and we must remember that in most economic analysis it does not—"zoning" solutions to externalities, or what Mishan has called "separate facilities" solutions,[4] are possibilities. Given space, there is no need for pig farmers and business executives to live as neighbours, and therefore no need to devise a system of bribes to compensate one or the other party for damages suffered.

The absence of an ownership-rental system for water has meant that water use has in fact been determined by such things as historical priority, gall, and force and fraud; it cannot be otherwise when property rights do not exist and when the price for the use of a valuable asset is zero. When no pricing process exists, there is no mechanism to transform technological externalities into pecuniary externalities. Accordingly we *do* observe striking examples of externalities in water use; stinking streams flow through choice residential areas, and anglers experience a mixture of rage and resignation as their favourite streams are polluted by industrial wastes. And then there are the externalities of all against all—householders help to destroy swimming beaches by their use of detergents (which promote algal growth), motorists pollute the air they breathe, and we all promote municipal and industrial pollution by insisting on cheap products and low taxes.

These considerations suggest the enormous social benefits that have resulted from applying an ownership-rental system to land, and, by contrast, the enormous social friction and economic waste that result from not applying an ownership-rental system to water. It has, of course, been relatively "easy" to apply property rights to land because land is both divisible and immobile. The awkward problem remains: is it *possible* to apply an ownership-rental system to the use of our water resources?

[4]E. J. Mishan, *The Costs of Economic Growth* (New York, 1967), chap.8.

III

To speak of owning an asset is to use a convenient abbreviation for a complex interaction between a legal concept and an economic concept. An asset may be thought of as "a bundle of potential utility-yielding services that can be used in alternative ways." In the same vein, ownership consists of "a bundle of legally defined user rights to an asset." As Coase has pointed out, it is rights, never objects, that are owned, and the rights themselves are always limited by law; "outright" ownership can never, by definition, extend to the use of an asset for illegal purposes.[5]

From the whole spectrum of possible ownership arrangements, we shall pick four major types for brief comment. What we shall call *common-property* ownership is, from an economic point of view, virtually non-ownership. A common-property asset is one that can be used by everyone, for almost any purpose, at zero cost. Examples are the medieval commons, the high seas, wild game, freeways, and (until recently in this country) air and water. Common-property ownership is justified economically *only* when the costs of enforcing a more restricted form of property-rights would be greater than the benefits of doing so. H. S. Gordon has shown that, neglecting enforcement costs, common-property ownership of an asset is economically inefficient in that the asset will be over-used by comparison with assets that are subject to more restrictive property rights.[6]

Empirically it is clear that if the asset is depletable it will be continuously depleted on the grounds that "everybody's property is nobody's property": medieval commons were overstocked; modern freeways (but not toll-ways) quickly become congested; wild animals (but never domestic animals) become scarce or extinct; and the deteriorating quality of our air and water resources has become a matter of widespread concern. The concept of a free good has always been a contradiction in terms; it is time we appreciated its sardonic overtones, for anything that is treated as a free good is indeed likely to become a valueless thing.

In general common-property assets are nominally owned by some public body, usually a government, and the owner may restrict use of the property in a variety of ways. Some roads may be used by motorists but not by cyclists or pedestrians; some wild animals may be photographed, but not shot; on some lakes canoes and sailboats may be used, but not motor-boats. It seems reasonable to refer to such property as *restricted common-property;* though the type of use is restricted, it is still common-property in the sense that everyone can use it for designated purposes at zero cost. If uses that deplete the asset in a physical sense are banned, the

---

[5]See Coase, "The Problem of Social Cost." Reprinted in this volume, pp. 109–38.
[6]H. S. Gordon, "The Economic Theory of a Common-Property Resource: The Fishery," *Journal of Political Economy,* April 1954, 124–42. Reprinted in this volume, pp. 97–108.

quality of the asset can be maintained, though "congestion" problems may reduce its value to other users.

When the use of an asset is restricted by law to particular persons, or a particular person, we have what can conveniently be called *status-tenure* or *fixed-tenure* ownership. Such ownership guarantees exclusivity of use to the parties authorized to use the property, but these user rights are not transferable. Though secure right of access to an asset by a limited group of people is valuable, the absence of the right of transferability prevents an explicit price system from developing. Nevertheless implicit prices are likely to appear. If the right to send one's children to a particularly good school is limited to those who live in a particular area, the value of the rights is likely to become reflected in the value of real estate in the area concerned. The "regulatory" branches of modern governments create an enormous variety of valuable property rights that are imperfectly transferable, and that tend to be capitalized and monetized in ways that are usually unsuspected by their creators. The value of tariff protection, a quota to grow tobacco, a licence to transport milk or to operate a taxicab, are reflected in the values of tariff-protected businesses, tobacco farms, milk routes, and taxi fleets.[7] Though the indirect monetization of such rights is seldom illegal, contemporary populations choose to be as hypocritical about the process as medieval populations were about the evasions of prohibitions on the payment of interest; social inhibitions about a rational approach to property and prices have outlived social inhibitions about rational approaches to astronomy and sex.

From status-tenure to full *ownership,* in the usual contemporary sense of the term, is but a short step. Once the property right is separated from the person, it becomes transferable, and transfers of assets (rights) then take place at explicit prices. *Transferable* property rights stand in a one-to-one relationship to prices; everything that is owned is priced, and everything that is priced is owned—which is to say nothing about either the form of ownership (transferable property rights to assets may be owned by individuals, corporations, or governments) or about the precise functional relationship between ownership and prices. Ideological hang-ups on concepts of property rights and ownership are understandable because such concepts touch the very roots of society. We have not yet learned to discuss such matters unemotionally. Though we are inclined to take a condescending view of medieval man's distrust of full property rights to land, we tend to become quite agitated when valuable government-granted rights (licences to import, for example) are traded in the market place, or when suggestions to extend property rights to air and water are put forward for discussion. Property and prices still raise ancient fears that "the rich will eat out the poor."

---

[7]On this general question, see Charles A. Reich, "The New Property," *Yale Law Journal,* 73, no. 5 (April 1964), 733–87.

IV

Since the right to use water is valuable, and since ownership consists of user rights, it should in principle be possible to devise an ownership-rental system for water. As is well known, however, certain characteristics of a natural water system create special problems in ownership.

The characteristics of an ownership system reflect in part the "divisibility" of the asset to which it is applied. Let us define an *asset-unit* as the smallest physical amount of the asset to which it is practicable to apply property rights, i.e., for which it is practicable to enforce exclusivity of use. In land, the asset-unit is very small, perhaps a few square yards; when the asset-unit is small compared to the quantity of the asset available, the asset can be held by a large number of individual owners. In such cases a "private property" form of ownership is likely to work well; decisions about the use of the asset will be decentralized among many owners, and a reasonably competitive market in asset-units will emerge.

In water, the asset-unit is very large. If water were completely "self-mixing," no one would pay anything to own Lake Ontario unless he could also own the whole Great Lakes drainage basin above the St. Lawrence River. As we have seen, however, water, especially in large lakes, mixes only slowly and imperfectly; because of this, and because of the self-purifying characteristic of water, the quality of water in the eastern end of Lake Ontario may be effectively independent of the uses made of the water at the western end of the lake. Even so, it is clear that the asset-unit is very large. It might be possible to divide the Great Lakes water system into, say, a dozen "regions" each of which would be self-contained for practical purposes, but it would certainly be impossible to divide them into a thousand such regions. In a democratic society it would be unacceptable to allow as few as a dozen, or even a score, of owners to control such an immense property as the Great Lakes drainage system. The only sensible alternative is the one actually adopted, namely, monopoly ownership by government. The reverse side of this coin is that the government must decide how its property is to be used and must enforce its decisions—assuming that it wishes to avoid the horrors of the common-property approach to resource management.

The decision about how water shall be used must be an arbitrary one from the standpoint of economics. Let me argue this point on the basis of a simple (but seemingly realistic) classification of water uses.

If we ignore such uses as navigation and the generation of hydroelectric power, which have insignificant effects on water quality, it seems reasonable to classify other uses into two categories: waste disposal and "all other" uses, which we shall call amenity use. These two uses are competitive. Though it is not true that fishermen, swimmers, industries that use water for processing purposes, and municipal authorities responsible for residential water supplies all have the same quality demands, it is true that some of these users would be benefited, and none would be harmed, by an improvement in water quality. Waste disposers, on the other hand, would be harmed by such an improvement since it could only occur if less waste were discharged into the water. We thus reduce the many uses of water to two: amenity use and waste disposal. The social problem is then to decide on the division of water services between these two conflicting uses. In principle, the division should be made in such a way that the value of a marginal increment in the one good is equal to the value of a marginal decrement in the other. But since the value of a marginal change in amenity use cannot be measured, the optimum amount of waste disposal cannot be identified. In practice, the decision is made on a political rather than an economic calculus. Once there is a political demand for "pollution control," anti-pollution measures tend to be instituted incrementally until complaints about their cost outweigh complaints about pollution! That sort of solution, applied also to such things as education, road systems, and various social welfare schemes, seems to me to be eminently sensible, *faute de mieux*.

In water quality problems, however, it is important to keep in mind that, within limits, water can be "regionalized" for practical purposes, and that "zoning" solutions to quality problems are therefore possible in some cases. In practice it would probably be wise to provide for different ratios of amenity use to waste disposal use in different water "regions"; the socially insulating quality of space should be utilized wherever possible. People are mobile, and if they can consume the amenity services of water in the upper reaches of a river and the waste disposal services of the same river in its lower reaches, there is no need to force them to decide on the optimum division between amenity uses and waste disposal uses of the water in both the upper and lower parts of the river. But again, alas, economics has little to say about a feasible or desirable delimitation of water "regions"; a sensible "mapping" of water must be left to the good judgment of physical scientists and politicians.

The contention that there exists no economically optimum division between amenity and pollution uses of water will be resisted by exponents of damage-cost pricing.[8] In the classic example of an upstream community polluting a downstream community, an allegation of damage to the downstream user seems to rest on three assumptions: that the downstream community owns its water and in particular owns the right not to have its water polluted by others; that the downstream community gains no advantage from the upstream pollution, i.e., that its residents buy no goods from their upstream neighbours at prices that are lower than they would be if the upstream community were forced to reduce its pollution; and that

[8]A good exposition of damage-cost pricing is to be found in Allen V. Kneese, *The Economics of Regional Water Quality Management* (Baltimore, 1964).

the upstream residents suffer no disadvantages from the downstream pollution because they never visit the downstream area for fishing, swimming, or other recreational purposes. The property rights assumption has not generally been true in the past in North America, and even to-day it is far from clear that a downstream community has any more right to use the river water for swimming and drinking than the upstream community has to use it for waste disposal purposes. The other two assumptions about the inter-community immobility of goods and people are, in general, untenable. The "polluter-polutee" view of the problem that underlies the recommendation of damage-cost pricing derives from the apparently easy identification of the two parties on a river. Once the mobility of goods and people up and down a river is taken into account, however, identification becomes much more difficult and the problem appears much like the "war of all against all" that is characteristic of lake and ocean pollution.

Even if everyone is at once a polluter and a pollutee, however, the optimum amount of pollution could be achieved if the value of a marginal dose of pollution could be measured. But it cannot be measured, because its value is the value of the amenity use forgone, which cannot be measured. Attempts have been made to measure the recreational value of particular land and water areas, but all such measurements are made on the partial equilibrium assumption that the recreational use of neighbouring areas is held constant. In general, however, a reduction in the amenity capacity of one river or lake will result in increased pressure on the amenity capacities of other rivers and lakes in the same general area. So far as I know, no one has been able to measure the amenity value of an acre-foot of water under general equilibrium assumptions. All we can be reasonably sure about is that the recreational value of water rises as population grows and the standard of living increases.

In brief, it seems to me that it is unrealistic to view water management as a problem in externalities, and that the question of how water should be used is purely a matter of collective decision-making. Economics cannot be of any significant help in making this decision. Even the principle that property rights should be set so as to maximize social products is of no use in the case of water because the values of amenity uses of water—recreation, and the simple aesthetic satisfaction that most of us gain from looking at, or even merely contemplating the existence of, clean water—cannot be measured, though such values are certainly part of any society's gross national welfare. Social welfare functions, community indifference curves, and benefit-cost analysis are ways of visualizing the social decision-making problem, but not of solving it.

What is special about the ownership of water, therefore, is that the owner must decide, without the benefit of economics, how his asset is to be divided among different uses. (When asset-units are small relative to the amount of the asset available, as in land, decentralized ownership is possible and the amount of the asset devoted to different uses is, for practical purposes, determined by market forces.) But this special quality of water (and air) ownership does not make it impossible to apply a rental system to water management.

V

If economics has nothing useful to say about the ownership decision of how water should be used, it has a great deal to say about how the decision, once made, should be implemented. What the government-owner of a natural water system must decide is how many equivalent tons of wastes may be discharged into the waters of each water region. The decision has at least three arbitrary components. Since in given circumstances a ton of one waste is likely to be more injurious than a ton of another, some equivalence must be established between different waste products, and since circumstances differ widely I assume that some average equivalence is chosen for each region in order to simplify the problem and reduce administrative difficulties. The other two sources of arbitrariness from the economic point of view, the mapping of regions and the choice of the amount of pollution in each region, have already been discussed. Let us now suppose that the owner has decided that during the next five years no more than x equivalent tons of waste per year are to be dumped into the waters of region A, and that x represents a 10 per cent reduction from the amount of waste that is currently being discharged into the region's waters. How can the government-owner enforce this decision?

The government can enforce its decision in one of six main ways. It can *regulate:* (1) a waste quota can be assigned to each waste discharger and set so that the sum of the quotas does not exceed x; or (2) an across-the-board regulation that each discharger must reduce his waste discharge by 10 per cent may be promulgated. It can *subsidize:* (3) dischargers can be subsidized to reduce their wastes, either individually or (4) on an across-the-board basis of so much per ton of waste discharge reduced. It can *charge:* (5) an effluent charge can be levied on dischargers, either individually, or (6) on an across-the-board basis of so much per ton of waste discharged.

I suggest that it is intuitively obvious that the *individual,* or point-by-point, procedures would involve staggering administrative costs. Yet it should be noted that politicians and civil servants seem to favour point-by-point *regulation,* and that economists who recommend damage-cost pricing favour point-by-point *charging* schemes. It seems intuitively obvious that in practice no point-by-point procedure could distribute the cost of reducing pollution among polluters in an economically optimal way, i.e., in a way that would minimize the total cost of reducing pollution by 10 per cent. To suppose that optimality in this sense is possible is to suppose that the administrative authority is able to solve a set of thousands of simultaneous equations, when the information required to write the equations in numerical form is not only not available, but also often unobtainable. It is also obvious that an across-the-board regulation to the effect that all dischargers must reduce their wastes by 10 per cent would result in a non-optimal distribution of the cost burden.[9]

Let us then examine the across-the-board schemes of subsidization and charging. Both possess the advantages of low administrative costs relative to the point-by-point schemes, and both would result in an optimum distribution of costs among dischargers; all dischargers would reduce their wastes up to the point where the marginal cost of doing so equalled the subsidy provided, or the charge levied. Both schemes have two disadvantages: a certain amount of experimentation would be necessary to establish the level of subsidy, or charge, that would produce a 10 per cent reduction in waste discharge; and the levels would have to be varied annually to take account of industrial and demographic growth (or decline) in the region in order to keep to the target of x equivalent tons of waste discharge. The subsidy scheme, however, has two disadvantages that the charging scheme does not have. First, if a subsidy of so much per ton of waste reduced is set, extra profits will accrue to those firms that can reduce their wastes at a cost per ton that is less than the subsidy provided, and no change in relative prices of goods is necessary. In the charging scheme excess profits will not be generated, and there will necessarily be a change in relative prices of goods, which in turn will result in a socially desirable adjustment of consumption patterns to reflect the differential costs of waste disposal as between different goods. Second, the subsidization scheme provides no incentive to choose production methods that reduce the amount of waste generated (and may indeed have the opposite effect!), whereas the charging scheme provides incentives both to reduce waste and improve the technology of treating waste before it is discharged. The across-the-board charging scheme is therefore clearly the best of the six possible ways of implementing the government's decision.

Its victory is made decisive by the fact that it lends itself easily to a market mechanism, whereas the subsidy scheme does not. The government's decision is, let us say, that for the next five years no more than x equivalent tons of waste per year are to be discharged into the waters of region A. Let it therefore issue x pollution rights and put them up for sale, simultaneously passing a law that everyone who discharges one equivalent

---

[9]Paul A. Bradley, "Producers' Decisions and Water Quality Control" (Background Paper D 29-3 in *Pollution and Our Environment,* papers presented at a conference held in Montreal, Oct. 31 to Nov. 4, 1966, by the Canadian Council of Resource Ministers), discusses various possible reactions of individual firms to the regulation of effluent standards and to a system of effluent charges. Standards, charges, and subsidies are discussed extensively in Kneese, *Economics of Regional Water Quality Management,* chaps. 4 and 8.

ton of waste into the natural water system during a year must hold one pollution right throughout the year. Since x is less than the number of equivalent tons of waste being discharged at present, the rights will command a positive price—a price sufficient to result in a 10 per cent reduction in waste discharge. The market in rights would be continuous. Firms that found that their actual production was likely to be less than their initial estimate of production would have rights to sell, and those in the contrary situation would be in the market as buyers. Anyone should be able to buy rights; clean-water groups would be able to buy rights and not exercise them. A forward market in rights might be established. The rights should be for one year only, the price of one right for one year representing the annual rental value of the water for waste disposal purposes. [10] (There is no reason, though, why speculators should not gamble in one year on the price of rights in later years.) The virtues of the market mechanism are that no person, or agency, has to *set* the price—it is set by the competition among buyers and sellers of rights—and that the price in the market automatically "allows for" the regional growth (or decline) factor. If the region experiences demographic or industrial growth the price of rights will automatically rise and induce existing dischargers to reduce their wastes in order to make room for the newcomers. The government should make it clear that it reserves the right to alter the allowable level of pollution (the number of rights it issues) at stated time intervals (say, every five or ten years). All that is required to make the market work is the inflexible resolve of the government not to change the rights issue during the interval, no matter what the political pressures to do so may be, and to enforce rigidly the requirement that a ton-year of waste discharge *must* be paid for by the holding of one pollution right for one year. Pollution rights are fully transferable property rights, and any welching on the enforcement of the right would be a breach of trust.

The automaticity of the market mechanism reduces administrative costs by relieving administrators of the necessity of setting the charge for rights and changing it periodically to reflect economic growth or decline. The administrative costs of enforcement would remain, of course, but they would be no greater than the costs of enforcing any of the other implementation schemes that we have considered. Technological change in the form of automatic monitoring devices to measure the volume of effluents from discharge points promises to reduce the costs of policing for all anti-pollution schemes.

Compliance with any point-by-point regulatory or subsidization scheme of pollution control establishes a sort of *status-tenure* property right. The right inheres to the discharger that earns it, and is only transfer-

---

[10]Professor Neufeld has suggested that it would be desirable to issue rights of different durations; more complicated schemes than the one outlined in the text could easily be arranged.

able (at the capitalized value of its implicit price) when the property to which it applies is sold. The market mechanism of the across-the-board charging scheme separates the property right to water use from the other assets of the discharger, and thereby makes the property right *fully transferable*. Full transferability and explicit prices are, as has been noted, considered preferable to status tenure and implicit prices by contemporary populations in Western democratic societies.

VI

Having puffed the merits of the across-the-board *cum* market mechanism scheme of pollution control, I must now take note of its deficiencies. There are four arbitrary elements: the mapping of water regions; the setting of waste equivalents; the choice of the allowable amount of waste discharge; and the choice of time interval during which the number of pollution rights is fixed. By comparison with some ideal, Pareto-optimal scheme laid up in Heaven, each of these decisions is bound to introduce elements of non-optimality into the arrangements I have proposed. In each case, however, I suggest that the saving in administrative costs is likely to outweigh the loss in terms of resource misallocation, measured from some theoretical optimum that ignores administrative and other transactions costs—notably the cost of acquiring enough information to administer an optimal pricing system. [11]

The question of the possible effects of pollution charges on the location of industries (and population) requires special comment. It is often suggested in the literature that waste discharged into a large, lightly populated river system does less damage than if it is discharged into a small, thickly populated river system, and that accordingly pollution charges for use of the former ought to be lower than for use of the latter. This reasoning assumes that the only costs of waste discharge are the objective, measurable, costs to residents in the area, or more generally—if people are allowed to live in one area and vacation in another—that the damage done to amenities by an extra ton of waste is everywhere the same. A system of charging that equalized marginal measurable costs as between water systems would then minimize the objective costs of disposing of a given tonnage of wastes over all the water systems in an area. But this argument does *not* hold if the goal is to minimize *total* costs of disposing of a given tonnage of wastes.

[11]A referee for this paper wrote that my scheme requires "that the questions of how much pollution, where pollution is to be allowed, how it is to be measured . . . etc., are all answered beforehand. But *these* are the really big questions." I agree. I don't think that economic analysis can answer these questions; it can, however, point to the best means of implementing the given answers.

In general, as one river system (or one part of a lake) becomes more polluted, the amenity value of neighbouring unpolluted waters rises. Moreover, when pollution reaches a level that is inconsistent with all recreational uses, added waste discharge has no recreational cost, while added pollution (that destroys swimming even if not, say, boating) in a popular vacation area probably has a very high recreational cost. The demand for amenity uses of water is certainly not a continuous function of water quality. Not enough is known, or perhaps knowable, about the demand for amenity uses of water to devise a fully optimal use of water in an *"n*-region system." In general, though, when congestion problems arise—when people begin to realize the existence of a spatially generalized pollution problem—it is clear that as pollution levels in one area rise the amenity value of relatively clear water in neighbouring areas rises; thus the opportunity cost of using such waters for waste disposal purposes also rises. This consideration by itself, therefore, suggests that pollution charges should be *higher* in areas where pollution is currently at low levels than in areas where it is at high levels—the reverse of the pricing system usually recommended. The system of low pollution charges for a low pollution level tends to spread pollution evenly over the countryside. I prefer the opposite system of high pollution charges for a low pollution level; it tends to create the separate facilities recommended by Mishan.

In any event, in the present state of knowledge about amenity values of water, it is obvious that the spatial pattern of pollution, or the price differentials between regions for pollution rights, will reflect an arbitrary decision by government. In the scheme outlined in this paper initial differentials in the prices of pollution rights would probably not be large if waste disposal were to be reduced by 10 per cent in each region. As time goes on, however, price differentials will tend to change as other forces lead to the centralization or decentralization of industries and populations. These tendencies can be offset, or encouraged, by the government's decision about the absolute and relative numbers of rights made available for sale in different regions. Thus the government-owner of a water (or air) system must decide not only the over-all quality of his asset, but also the quality of the asset in each region.

It should be noted, finally, that the market in pollution rights is not a "true" or "natural" market. In natural markets price creates two-way communication between sources of supply and demand and affects amounts supplied as well as amounts demanded. (Where supply is fixed in natural units, as in the land market, price affects the equilibrium quality of the asset, and mediates between the users of land on the one hand and the users of the products of land on the other.) My market provides only for one-way communication. It transmits the government-owner's decisions about the use of water to the users of the asset, but there is no feedback from the users to the owner. A rise in the price of a pollution right signals that the waste disposal use of water is becoming more valuable; but this does *not* mean that the supply of allowable waste disposal capacity should be increased, for the value of the competing amenity use of water is also likely to be increasing under the impact of the same growth forces that make the waste disposal use more valuable. The price signals that the government gets from the market are "false," in the sense that they are largely echoes of its own arbitrary decision about the supply rights. The market proposed in this paper is therefore nothing more than an administrative tool. But administrative tools that have some *prima facie* claim to efficiency should not be ignored in an increasingly administered society.

15

Transferable Discharge Permits and the Control of Stationary Source Air Pollution*

THOMAS H. TIETENBERG
.

Thomas H. Tietenberg is Christian A. Johnson
Distinguished Teaching Professor of Economics at Colby
College.

I. INTRODUCTION

Background

Recently, in the United States, there has been an increasing interest in the use of economic incentives to achieve environmental quality goals, particularly for air pollution.[1] Though there is a range of such possible eco-

"Transferable Discharge Permits and the Control of Stationary
Source Air Pollution," by Thomas H. Tietenberg, from *Land
Economics,* v. 5 (1980), pp. 391–416. Reprinted by permission.

The author wishes to acknowledge the benefit of helpful comments from Barbara Ingle, William Baumol, Scott Atkinson, Cliff Russell and two anonymous referees as well as to acknowledge the research assistance provided by Alison Jones and Cynthia Keating. An earlier version of this paper was delivered at the International Institute for Environment and Society, Berlin, West Germany in September, 1979.
*We have omitted Section IV, "Empirical Results"—*Eds.*
[1]See, for example, the examples provided in Yandle (1978).

nomic incentives, two have received the most analytical attention—effluent or emission charges and transferable discharge permits. In this paper attention is focused on the latter approach.[2]

Although the central concept involved can be traced at least back to Alfred Marshall and Henry George, it seems fair to attribute much of the revival of interest to Dales (1968a†; 1968b) for water and Crocker (1966) for air. The concept is a disarmingly simple and powerful one. Its intellectual genesis is to be found in the realization that the behavioral sources of the pollution problem could be traced to an ill-defined set of property rights. The right to discharge pollutants had historically, in the absence of government intervention, been allocated at a zero price to all potential users. While this allocation process *can* be cost justified when the costs of controlling pollution (including the administrative and enforcement costs) exceed the costs of pollution damage incurred by the absence of control, it *cannot* be justified when the level of pollution damage indicates that some form of control is appropriate. One natural outcome of using this perspective to characterize the nature of the pollution problem is that it suggests a particular policy approach—the establishment of a correctly defined right to discharge.

The current appeal of this concept for air pollution in the United States is based upon several considerations. First, and probably foremost, is the realization that a movement from the current purely regulatory system of control to a system based on transferable discharge permits promises the potential for achieving a better quality of air than currently enjoyed with a substantially lower commitment of resources to pollution control. Secondly, for air pollution, the transferable discharge permit (henceforth TDP) system would be compatible with the existing legislation. It would represent a modification of, rather than a radical departure from, the current regulatory approach. In addition, the transferable discharge permit system is administratively flexible, yet feasible. It avoids many of the most costly rigidities in the current approach and, yet, the implementation requirements are realistic. Finally, unlike the current system the TDP system provides an incentive for emitters to adopt new control techniques which can clean up more emissions at lower cost (since they can sell the resulting excess permits), which in turn stimulates the development of these techniques.

---

[2]This approach is characterized by a variety of different names. The policy instrument has been referred to as a marketable emission permit, a pollution right and a transferable discharge permit. I have chosen in this paper to use the last of these exclusively.
†Selection number 14 in this collection—*Eds.*

The Policy Context

Although the existing regulatory approach to air pollution control is exceedingly complex, it is possible to capture briefly the essence both of the basic approach and the recent reforms which have moved it further in the direction of an increasing reliance on economic incentives. This synopsis can then serve as a point of departure and benchmark for the discussion of the possibilities for moving toward a less restrictive transferable discharge permit system.

The existing system of air pollution control relies on ambient standards (or concentration targets) defined by the Environmental Protection Agency (EPA) and implementation plans submitted by the states (SIPs) to detail the procedures by which the states intend to meet those standards.[3] In addition, for several classes of new and modified polluting sources, the EPA has issued emission standards directly.[4]

This basic approach does little to control the degradation of the air quality in regions where the air is cleaner than the ambient standards. Therefore, prodded initially by a court suit[5] and, subsequently, the amendments to the Clean Air Act of 1977[6] the EPA moved to define a policy (referred to as the PSD policy) to prevent the significant deterioration of the air in areas not currently in violation of the air quality standards. The approach taken was to designate allowable pollution increments and ceilings (either of which could serve as the binding constraint) for three separate types of regional classifications. Those increments are currently allocated on a first come, first served basis although a market approach, such as discussed in this paper, is clearly an option for the future.

The other two major reforms that took place in the late 1970s and had the effect of moving policy closer to a market approach are the alternative emission reduction approach[7] (known popularly as the "bubble" concept) and the emissions offset policy[8] applied to nonattainment areas[9] and to certain sources in attainment areas if they could be expected to contribute

---

[3]The procedures to be followed by states are detailed in 40 CFR 51 and the EPA approval procedures are spelled out in 40 CFR 52.

[4]See 40 CFR 60 for criteria pollutants and 40 CFR 61 for hazardous pollutants.

[5]Sierra Club v. Ruckelshaus, 334 Supp. 253 (DDC 1972) which was later affirmed by the Supreme Court, 412 U.S. 541.

[6]91 Stat 731. The details of the policy can be found in 40 CFR 51.24 and 40 CFR 52.21. Proposed amendments to that policy resulting from a successful court suit against EPA (Alabama Power Company v. Costle, 13 ERC 1225) are located in 44 FR 41924 (5 September 1979).

[7]See 44 FR 71780 (11 December 1979).

[8]See 40 CFR 51 Appendix S, originally presented in 44 FR 3274 (16 January 1979).

[9]Nonattainment areas are designated geographic areas which have not yet attained air as clean as required by the ambient standards. See 40 CFR 81.300 et seq.

pollution to nonattainment areas[10] or to cause an attainment area to exceed the standard.[11] The "bubble" concept specifically allows existing emitters to propose modifications of their emission standards based on the substitution of a more relaxed degree of control one source for a more stringent degree of control of another source of the same pollutant. These substitutions can, under certain circumstances, occur between plants or even between firms. This design feature carries the bubble policy a long way toward a fully transferable permit system. The object, of course, is to allow a firm to meet its emission reduction goal as flexibly and cheaply as possible while insuring that air quality is not degraded by the substitution. This concept replaced an established procedure based upon levying a separate emission standard on each source. Both the business community[12] and EPA[13] expect the costs of pollution control to be considerably reduced by this method.

The emission offset policy was originally designed as a means for allowing economic growth in nonattainment areas while insuring no further degradation of their air quality. It allows potential new entrants to a nonattainment area to procure sufficient reductions from existing firms (over and above their previous legal requirements) so as to offset the increases in pollution which would otherwise occur upon their initiation of production in the area without the compensating reduction. The significance of this program is that it allows the transfer of credit for emission reductions from existing sources to new sources, whereas the bubble concept allows transfers only among existing emitters.

These reforms are important because they introduce into the basic system two additional degrees of transferability that can serve to reduce the costs of compliance significantly. They both allow limited interfirm transfers to be approved on a case by case basis as long as total emissions do not increase and air quality is not adversely affected. In addition, the bubble concept allows an existing emitter to actually exceed the SIP emission standards for some of its sources as long as it compensates with sufficient emission reductions from the other sources in the plant. Thus, the bubble concept establishes the very important principle (for existing sources, but not for new sources) that the SIP standards are not inviolable, a principle which expands the set of trading possibilities (and, hence, the cost reduction possibilities) considerably.

The key difference between the existing system and a TDP system with full transferability lies in the fact that in the latter the control authority allows all sources to participate in the trades and allows all emission reductions to be traded in a regularized market. In contrast the bubble and offset policies have restrictions on what emission reductions can be traded

---

[10]See 44 FR 3283, §IIE (16 January 1979).
[11]See 44 FR 3884, §III (18 January 1979).
[12]See Nulty (1979).
[13]40 CFR 71780.

(e.g., only those additional reductions above the standard in the offset policy) and on what sources can participate in trades (e.g., only existing sources in areas demonstrating attainment in the bubble policy).

An Overview of the Paper

On a purely ideological level the move to this more general system is clearly merited. On a more pragmatic level, however, the implementation of an unrestricted TDP market faces a number of technical, legal and administrative challenges.

The purpose of this paper is to examine the prospects for moving toward this less restrictive TDP system by surveying and synthesizing the most prominent published English language literature on the subject. While the focus will be on air pollution the relevant water pollution literature will be included insofar as it yields insights which are pertinent to controlling air pollution. The objective is to draw together what is known and is now known about this form of control and to assess the implications of this research for the design of an operating TDP market.

The paper opens with a discussion of design criteria and the implications of these criteria for the design of the permit itself and for the design of the market within which the TDPs can be transferred. This is followed by a brief survey of the requirements for such a system to be enforceable.

II. ENVIRONMENTAL GOALS AND DESIGN OF THE MARKET

Rudiments of a TDP Approach

The current regulatory approach to air pollution control involves the specification of emission standards for each emitter. Any emissions above these standards are in noncompliance and therefore the responsible emitter is subject to some kind of sanction. The problem with this approach is that there is no guarantee that the particular allocation of emission standards chosen by the control authority will achieve the air pollution goals at anything approaching minimum cost. Indeed much available empirical evidence suggests that the typical allocations are significantly more expensive than the minimum cost allocation.[14]

The bubble concept and the offset strategies are partial responses to this deficiency, but they fall somewhat short of a marketable permit system. They rely on a case by case approach to transferability and impose severe

[14]See, for example, the studies cited in Kneese (1977, pp. 195–202), Atkinson and Lewis (1974), and Anderson et al. (1979).

restrictions on the conditions under which transfers can take place. In essence they are the administratively easiest modifications to make from the current system but the existing restrictions on trades prevent the system from approaching a minimum cost system.

The transferable discharge permit system responds to these deficiencies in the current system by establishing a regular market for discharge permits with few restrictions on transferability. This approach is both legally and philosophically compatible with the existing legislative framework. The allocation problem is handled by the remarkably simple device of making the emission standards (legitimized by the discharge permit) completely transferable. In this way those emitters facing very steep control costs can purchase permits from emitters having less costly options, thereby subsidizing the more intensive control of emissions by these low cost emitters. Remarkably, as the discussion of the literature below reveals, under competitive conditions the reallocation of permits which takes place by virtue of making them transferable can cause substantial reductions in the amount of resources committed to pollution control while meeting the air quality standards.

Specifying the Objective

The first step in designing a less restrictive TDP system is specifying the objective the system is supposed to accomplish. There are two objectives commonly discussed in the economics literature—efficiency and cost-effectiveness.

Ignoring implementation costs the efficiency criterion clearly dominates. Intuitively, the efficient allocation balances, at the margin, the damage cost incurred from remaining uncontrolled pollution with the costs of avoiding the damage. It is the allocation which minimizes the sum of damage costs and avoidance costs.[15]

The problem with this objective from a policy point of view is that it is difficult to achieve. For the control authority to achieve this objective it would have to know the control costs and the damage costs associated with each emitter. Given the large number of emitters, that is a tall order and one not likely to be fulfilled. Various conceptual schemes are currently being developed to overcome some of the theoretical hurdles (for example, the public good problems associated with the incorrect revelation of preferences) but at this stage it seems clear that these schemes are not yet sufficiently developed to permit them to be used as a basis for policy.[16]

---

[15]A formal mathematical characterization of this allocation in the context of a general equilibrium model can be found in Tietenberg (1973b).

[16]See, for example, Bohm (1972), Clarke (1971), Randall et al. (1974), Scherr and Babb (1975), and Barnett and Yandle (1973).

For these pragmatic reasons there has been a reluctant acceptance of a cost-effectiveness criterion.[17] This criterion is based upon a dichotomization of the control problem—the specification of a policy target and the establishment of a system to assure that target is met. Cost-effectiveness deals only with the second component. It provides no guidance for choosing the appropriate policy target or the appropriate level of that target, but it provides a good deal of guidance in selecting among the various ways of meeting that target. It suggests that the "best" allocation is the one which achieves the target at minimum cost. If the target happens to be efficient, then a cost-effective allocation will also be efficient. There is nothing in the cost-effectiveness criterion, however, which would guarantee that result.

Choice of Policy Targets

The key to applying the cost-effectiveness criterion involves choosing the target which is to be met at minimum cost. Two candidate targets have received prominence in the literature, either explicitly or implicitly. The first, an aggregate emissions target, focuses on the total weight of emissions placed into the receiving medium (air or water) in a particular geographic area during some period of time. The aggregate emissions cost-effectiveness (ECE) criterion envisions the establishment of some legal ceiling (the standard) on the allowable weight of emissions and then allocating the responsibility for meeting that standard among the emitters in such a fashion as to cause it to be met with the smallest possible commitment of control resources. The chief virtue of this criterion is its administrative simplicity. It is relatively easy to monitor and the contribution of each emitter to the policy target is not difficult to define.

This policy target also has a substantial disadvantage for local pollutants—it is not uniquely related to the level of damages caused.[18] The reason is quite clear. For local pollutants the damage they cause is related to their ground level concentration in the air. Their concentration, in turn, is based upon the proximity of the emitters to each other, and the process of accumulation within the environmental medium, as well as upon the amount emitted. Thus while the aggregate level of emissions is important, many other factors are as well.

To compensate for these problems a second policy target, the ambient standard, has emerged and it is this target which is mandated by the Clean

---

[17]See, for example, Baumol and Oates (1975).

[18]Local pollutants can be contrasted with global pollutants for which the damage depends solely on the aggregate volume of residuals emitted. For a discussion of the distinction between these two types of pollutants in terms of their mathematical specification and the policy implications of those specifications, see Tietenberg (1978a).

Air Act. The ambient standard represents a target concentration level measured at a specific location for specific averaging times (for example, one hour, daily, annually). This standard has the virtue that it is closely related to damage caused, but it has the disadvantage that the relationship of each emitter to this standard is no longer as clear cut. Because this target involves a threshold concentration level, if each emitter in an area were to hypothetically and sequentially reduce its emission rate by a constant amount, the effect on air quality in those areas where the thresholds have been exceeded would be much more dramatic for some emitters than for others. The contribution of any single emitter to air quality at a particular receptor location depends upon the location of the emitter vis à vis the monitoring site, and the flow characteristics of the environmental medium between the monitoring site and the emission site, as well as upon the level of emissions. The ambient cost-effectiveness (ACE) criterion requires that the responsibility for cutting back emissions be allocated among emitters so as to minimize the cost of meeting the ambient standard as measured at specific monitoring stations. [19]

Geographic Extent of the Market

Geographic considerations enter the design of a TDP system in two main ways: the geographic domain of the policy targets and the area of applicability of the permits. [20] The most natural domain for the policy targets is the local airshed while the most desirable area of applicability for the permits includes all territory with emitters contributing pollutants to that airshed.

Unfortunately, a local airshed is rarely a well-defined physical entity. Any sub-planet partitioning of the air space must be plagued by a certain degree of arbitrariness. Yet it is equally clear that some local partitioning is absolutely necessary. To illustrate this consider what would happen if the aggregate emissions target were defined on a national domain. Permits would then be issued or sold to emitters. Since there are no geographic constraints, it is quite possible that a disproportionate share of those permits might end up in urban areas (e.g., Los Angeles) with particularly harsh existing pollution problems. This would cause the damage in these areas to be unacceptably high. The basic point is that an aggregate emissions target has a weak connection with damage caused under any circumstance, but the weakness is exacerbated as the geographic area covered by the definition of this target is expanded.

---

[19]The Clean Air Act requires that the standards be met everywhere, which means that the monitoring stations have to be sufficiently numerous and sufficiently dispersed to provide adequate coverage of the geographic area.

[20]Geographic considerations also affect the design of the permit itself. This is discussed in the next section.

The air quality targets are, by their very nature, defined in terms of concentration measurements at specific locations. The issue for this set of targets is the number of such locations needed to adequately cover a particular airshed. It is not possible to come up with a specific suggestion without reference to the characteristics of the particular airshed, but it is possible to suggest some of the considerations which would affect such a choice. The danger of an inadequate number of such locations is that the permit system will be designed to insure compliance with the measured concentrations only, leaving the distinct possibility of air quality deterioration in other unmonitored parts of the airshed. In practice, because of the flow of pollutants within the airshed, it is probably possible to gain an adequate description of the state of an airshed with a relatively few monitoring stations. To my knowledge there have been no empirical studies devoted to this critical design question.

Thus the notion of a local airshed, while lacking a precise, unambiguous means of defining its boundaries, appears both necessary and workable as a basis for defining the policy targets.[21] It would be tempting and administratively simple to project the boundaries of this airshed, however defined, on the land area underlying it and call this the area of applicability of the permit market.[22]

This definition of the area of applicability would present problems for two reasons: it would not, in all likelihood, correspond to existing political boundaries, and it would omit from consideration many emitters that contribute to the degradation of the air in that airshed, but are located some distance up wind from it. Each of these places important constraints on the ability of any control authority to achieve the desired policy targets at minimum cost.

An example of the former problem is provided by the Northeast corridor of the United States. The New York metropolitan airshed, under any reasonable definition, overlies parts of at least three states (New Jersey, New York and Connecticut). For a coordinated TDP market to be established in these regions a regional political authority has to be established with sufficient authority to implement the system.[23]

To make matters even worse there is accumulating evidence that some pollutants are transported rather long distances.[24] As a result the initial sources of the pollutants may be quite far removed from the airshed and

---

[21]The Clean Air Act follows this route with the establishment of air quality control regions. See 42 USC 7407.

[22]An area of applicability is defined as the jurisdictional range of the permit market. Emitters within the area of applicability have to have permits issued by the control authority governing that area while those emitters outside the area do not.

[23]The United States has made a start in this direction with the establishment of multistate air quality control regions, but these entities have not, as yet, been endowed with sufficient authority to implement a TDP system. See 42 USC 7407.

[24]See, for example, Cleveland and Graedel (1979).

well beyond the jurisdiction of the control authority unless the jurisdiction of that authority extends well beyond the perimeter established by the airshed.

The long distance transport of pollutants also makes quite likely the possibility that any given emitter will contribute to pollution problems in more than one airshed. Conceptually this presents no problem as long as that emitter is forced to have permits from all control authorities whose airsheds that emitter affects. Legally, however, this notion of overlapping jurisdictions may well be troublesome. The current approach involves a procedure for source states to identify multistate emissions and to notify recipient states, and a procedure for recipient states to petition the administrator for disapproval of the implementation plan involving any such source.[25] This is a far cry from what the theoretical optimum would suggest.

The considerations do not eliminate the desirability of a TDP system (they plague all other local control systems as well), but they do suggest caution in relying exclusively on this approach. For example, a TDP system should be complemented by legislation which rules out various control options that solve the local pollution problem by exporting the pollution to other airsheds. The most obvious current example of such a system is the use of a tall stack to inject the polluting substance into the atmosphere at such a penetrating height and velocity that there is no effect on ground level concentrations in the immediate area, but the effect is pronounced at more distant locations.[26]

Coverage of Emitters

Once the area of applicability of the permit system is established it remains to determine whether or not all emitters within that area are required to have permits. The major concern is whether emitters who individually represent very small contributions to the pollution (e.g., residences or automobile owners) in an airshed ought to be forced to purchase rights.

Macintosh (1974, p. 65) suggests that they should and devises a special branch of the control authority to deal with the purchase and sale of permits denominated in especially small units. The analogy he draws is the odd-lot trading system employed by the major stock exchanges.

Tietenberg (1974, p. 283), on the other hand, argues that some emitters

[25]See 91 Stat 724.
[26]Some steps in this direction have already been taken. The law already rules out stacks which are taller than "good engineering practice" as a means of meeting air quality standards. Yet the law specifically prohibits the administrator from regulating stack height. See 91 Stat 721.

should be excluded on the grounds that the costs of monitoring and enforcing the permits on each small emitter may far outweigh the benefits derived from the reduced pollution which results from their inclusion.

One possibility is to allow exemptions when the *collective* reduction in air pollution from the class of emitters being considered does not justify the additional expenditures on monitoring and enforcement. In addition this discussion suggests that different market designs may be appropriate for these different types of sources.[27] Small emitters will be included when collectively they are an important source of pollution in that airshed and the enforcement costs of inclusion under the most favorable market design do not seem to outweigh those benefits. The current reforms focus almost exclusively on large sources.[28]

Initial Distribution of Permits

It is useful to distinguish between the initial distribution of the permits and some ultimate distribution at a later date. The control authority has a good deal of control over the initial allocation while the market will handle the ultimate allocation.

It is one of the desirable properties of an appropriately designed[29] transferable discharge permit system that the ultimate allocation of these permits among emitters will be cost-effective regardless of their initial allocation as long as the permit market is competitive. This is formally proven by Montgomery (1972). Intuitively, this follows from the transferability of the right. Emitters that can control their pollution most cheaply will have an incentive to sell to emitters that have higher clean-up costs. This does not imply that the initial allocation of these permits is unimportant; on the contrary, it is quite important because it affects the level of costs faced by the emitters and the amount of revenue generated by the control authority.

Conceptually the initial distribution possibilities can be arrayed on a spectrum where on one end of the spectrum the entitlement to discharge is reserved for the control authority (and granted to emitters only upon receipt of their payment) while on the other it is reserved for the emitters

[27]These alternative market designs are explored in what follows on p. 407.

[28]In the Alabama Power decision the court provided that EPA may exempt from review those situations determined to be *de minimis*. The administrator exercises this authority by establishing emission or air quality thresholds and exempting or limiting the review of sources falling below these thresholds. Interestingly, the court limited the administrator's discretion in setting these thresholds by ruling out the use of a cost-effectiveness rationale. See 44 FR 41937, §10 (5 September 1979).

[29]The question of what permit designs are compatible with cost-effectiveness is treated in section III.

(and granted to the control authority only upon receipt of adequate compensation). Each of these implies different behavior on the part of the control authority and different allocations of the cost burden for meeting the community air quality goals.

On one end of the spectrum is the "government pays" principle. If the entitlements are considered to be currently held by the existing set of emitters, then in order to improve the quality of air the control authority would have to buy back some of the permits.* As long as they were bought at market price (as opposed to a government dictated price) the local community would end up paying for the improvement in air quality through the tax system.

A rather different cost incidence is obtained when the entitlements are considered vested in the state. In this "polluter pays" case the emitters must not only pay for the pollution control equipment, they must also purchase the permits from the control authority. In this case, the cost falls on the consumers of the product, the owners of the emitting firms and the employees of the emitting firms. While members of the local community may end up paying the costs of increasing the quality of air under this scheme as well, they will do so in their role as employee, employer or consumer, not as taxpayer.

There is a middle ground of this spectrum which does not seem to have been explored systematically. In one possibility entitlements could be vested in the state, but the control authority could choose to give them at no cost to the existing emitters on some rational allocative basis. This makes the public sector budgetary impact rather small while holding down the costs to the emitters. While they must pay for the costs of the pollution control equipment, the emitters do not have to pay for their initial allocation of permits (although they do have to purchase from other dischargers any additional permits beyond their initial allocation).

It should be realized, however, that this middle part of the spectrum requires that the control authority define the initial allocation. With the "polluter pays" case the initial allocation will be handled by the market. In the "government pays" case it is based on historic emissions. When the permits are given away, some basis for allocating them must be determined. Some criterion for allocation must therefore be developed.

One of the key determinants of whether the expanded market system is politically feasible is whether the various participants (for example, employers, employees, taxpayers, etc.) view it as "better" than the existing system. One traditional measure of whether a policy change is "better" is whether the new policy is Pareto superior to the old policy. [30] It is a rather remarkable characteristic of this transferable discharge permit system that

*The discharger's receipts from the sale would more than cover his abatement costs—*Eds.*
[30] A Pareto superior reallocation is one in which no one is made worse off and at least one of the participants is made better off. [See Chapter 5 in this volume—*Eds.*]

it provides an opportunity to implement a Pareto superior policy change.[31]

The existing system has assigned emission control responsibility via the state implementation plans. Presumably this assignment is consistent with achieving the ambient standards. If the permits in an expanded system are initially allocated to the emitters so as to allow each polluter the right to pollute as much as was previously allowed by the implementation plans, then any transfers among emitters which take place after the initial allocation, because they are voluntary, will result in mutual gains for those trading. Since this choice guarantees that no emitter faces higher costs than it otherwise would, this implies that consumers and employees should be as well off. Thus this particular allocation has the potential to be a minimum disruption means of moving from the current system to a full market system, should such a move be deemed desirable.[32]

If there were perfect foresight on the part of all emitters and a competitive market structure for the TDPs, the control authority would not have to worry about the intertemporal allocation of TDPs. All current emitters would foresee the future demands for the permits as well as their effect on prices and choose their control investments accordingly.[33] New emitters would simply bid the TDPs away from existing emitters prior to their entry into the market.

Both a lack of perfect foresight, however, and a noncompetitive market structure can create problems, although probably not serious problems, with this harmonious view. The lack of perfect foresight could cause existing firms to underinvest in pollution control if they underestimate the future prices of the permits or overinvest if they overestimate future prices. This is due to the fact that the permit can subsequently be sold for a profit in the right circumstances. Underinvestment could arise because many of the methods of emission reduction involve durable equipment purchases. Prior to the purchase emitters have a choice of any point on their long run marginal cost of control function. If they expect a heavy demand for permits from new emitters, they might rationally choose to purchase some

[31]The careful reader might argue that existing emitters would be better off but potential new emitters would be worse off. Since the current system forces new emitters in nonattainment areas to purchase offsets (as would the market system), this initial reaction is not valid for those areas. It is valid for attainment areas, however, since the allowable increments are currently given, free of charge, to the first requestors.

[32]The one caveat to this proposition that has to be kept in mind is that it ignores the change in administrative costs which would accompany the transition to an expanded market system. If administrative costs rise by a sufficient amount the new policy will not be Pareto superior. One study which has examined this question concludes that a TDP system would probably result in *lower* administrative costs than the current system. See Anderson, et al. (1979, chap. 6, p. 71; chap. 7, p. 38).

[33]The EPA currently allows a firm to bank excess emission reductions for later use (i.e., during an expansion or for transfer as on offset). See 44 FR 51935 (5 September 1979) and 44 FR 3280 (16 January 1979).

temporarily idle capacity. However, once the equipment is purchased an emitter must operate on its associated short run marginal cost function. This built-in lack of *ex post* flexibility could create problems for growth when all existing emitters systematically underinvest. This is, however, a problem new firms face in many markets whenever a resource is in fixed supply. Since they seem to handle it in these other markets, there is reason to believe they can handle it in this one too.

Noncompetitive markets can also potentially create problems. If the holders of the permits are few in number, they might, through collusion, charge higher than competitive prices for the transfer of a permit to a new emitter. This could stifle growth and create winfall profits for the sellers. This is a likely problem, however, only when the number of sellers is small and there are no alternative locations the buyer could choose. Therefore this is a problem to be confronted in few (if any) markets and can be dealt with in those few with additional constraints on the transferability of the permits.

If the control authority concludes that the existing incentives do not provide for an adequate reserve of permits for new emitters, two distinct approaches are available. The first assigns a very large role to the control authority while the latter assigns more of the responsibility to the private sector. The public sector approach envisions the analytical derivation of the annual optimal stockpile of permits to be held by the control authority to accommodate future growth. To operationalize this notion the public sector would attempt to simulate the operation of the market under various growth scenarios. Then decision theory could be used to derive the supply of permits for each year which maximizes the present value of expected social benefits. While this is a conceptually easy model to construct, the data requirements are somewhat formidable.

An alternative approach, suggested by Roberts and Spence (1976) for a slightly different problem than the one considered here, could be used to shift some of the uncertainty onto the emitters while keeping their incentives compatible with cost-minimization for the system as a whole. Their system envisions complementing a TDP system with a subsidy for unused TDPs and a penalty for exceeding the emission level permitted by the TDP. For our purposes here the interesting facet of this approach is the subsidy for holding unused permits. This subsidy creates an additional incentive for the emitter (above the normal incentive of holding the permit for higher future prices) to retain some capability to trade permits to other emitters in the future. This capability can be achieved by constructing a pollution control capacity larger than current needs and stockpiling the permits or by constructing control facilities which provide rather more *ex post* flexibility than would be achieved with competing, presumably cheaper, alternative facilities. [34]

---

[34]In this context *ex post* flexibility refers to situations where some increases in the degree of emissions control can be accomplished without sharply increasing marginal cost.

This subsidy approach has the virtue that it will encourage emitters to retain the capability to sell permits in the future when the cost of doing so is relatively low, but not when it is high because the subsidy in this latter case will not be sufficient to make these actions profitable. It has the undesirable feature that it requires larger expenditures by the control authority.

Enforcing the Permit System

The enforcement of this kind of permit system depends on the technical ability to detect violations and the legal ability to deal with the violations once detected. The ability to monitor emissions is a key aspect of the system because without it emitters do not have to worry whether they have the appropriate number of permits or not and the incentive properties of the system are lost. Similarly, violators should be sufficiently penalized to make violations under normal circumstances an unattractive option, but not so harsh as to make them lack credibility.

The state of the art in monitoring pollution from a policy perspective has been surveyed by Anderson et al. (1977, pp. 90–106). A transferable discharge permit system designed to achieve ambient standards requires both ambient concentration monitoring and emission monitoring. Since emissions monitoring is the most difficult component of an enforcement policy, we shall concentrate on it. Anderson et al. (1977) summarize:

> Direct, continuous monitoring may be feasible for larger installations where suitable instruments are available, and their cost is small relative to the overall cost of operating the plant. But . . . instrument technology does not provide solutions for every monitoring problem (p. 101).

The article goes on to conclude, however, that direct continuous monitoring of emissions is not essential to the success of the program. There are a number of techniques available for sampling or estimating the emissions flow. These may lack the precision and completeness of direct continuous monitoring, but in the absence of that alternative they may be acceptable.

One such approach would use a production function relationship to relate various input and output combinations to the emission rate. These could be estimated econometrically on the basis of initial sampling. Once these relationships were validated they could be used instead of direct continuous monitoring. The records on inputs and outputs would be used to estimate emission rates. These empirical estimating relationships could be subjected to periodic validation. When new pollution control measures are adopted by the emitter, the relationships would be reestimated.[35]

---

[35]Anderson et al. (1974, pp. 103–04) point out that this method is already being used for water pollution monitoring in both Germany and Czechoslovakia.

What is sacrificed by this approach is the monitoring of circumstances such as the temporal deterioration of the equipment or deviations from normal operating practice (for example, accidents or shutting the equipment off).

Anderson and his coauthors (1977) also address the question of whether the responsibility for the initial burden of monitoring should fall on the discharger or the pollution control authority. Appealing to the income tax system as a viable example they suggest that a self-reporting system complemented by punitive fines for false reports could play a significant role. As to the constitutionality of such a system they point out that it has survived constitutional challenges in at least one (unidentified) state (p. 93, n. 1).

In the longer run several promising technologies (e.g., lasers) are on the drawing board for remote sensing.[36] The development of this capability would significantly increase the feasibility of a self-reporting system by providing a random, surprise audit capability for the control authority. No doubt the implementation of control systems for which such capabilities would be useful would hasten the development of such systems by providing an obvious market for them.

Monitoring, however, is only the first of the requirements for an effective enforcement strategy. The second is the legal authority to deal with noncompliance, including effective sanctions. The most complete treatment of the legal issues is found in Irwin and Lirhoff (1974), although some pertinent insights are contained in Anderson et al. (1977), deLucia (1974) and Jaffe et al. (1978).

There are several legal constraints which might potentially inhibit an effective enforcement policy. The first is federal preemption. If there is a federal preemption, then states may not venture into what becomes, with preemption, a national jurisdiction. On this issue Irwin and Lirhoff (1974) conclude:

> The states and their subdivisions are not preempted from adopting disincentives applicable to other than new motor vehicles, aircraft or the use of fuels or fuel additives (p. 75).

The general constitutional constraints on state action are derived from the Fourth, Fifth and Fourteenth Amendments to the U.S. Constitution. The general impression given by Irwin and Lirhoff's extensive review is that a carefully conceived economic disincentive program (such as a transferable discharge permit system) would survive constitutional tests.[37]

Effective enforcement also requires the availability of legal sanctions

---

[36]For an extensive discussion of the possibilities see Staff of Research and Education Association (1978, chap. 18, pp. 49–65).

[37]One of the legal issues involving a TDP system involves the constitutionality of particular permit designs. These issues are reserved for the next section which discusses permit design.

which can be invoked when emitters fail to comply with the terms of the permits. Noncompliance can occur either when the emitter delays in installing the pollution control procedures and equipment or when the conditions of the permit are exceeded during the normal operation of these procedures and equipment either intentionally or unintentionally.

The most complete early work on the economics of air enforcement of air pollution control policies can be found in Downing and Watson (1973). They conduct a theoretical and (limited) empirical analysis of the air pollution control enforcement system as it currently exists and as it might exist if alternative approaches were pursued. Unfortunately, they do not consider transferable discharge permit systems, but they do introduce a number of ideas which are relevant to such a system. [38]

One of these ideas concerns the type of sanction to be employed. Downing and Watson (1973, p. 26) point out that there are three basic types of sanctions which can be imposed: (1) cease and desist orders, (2) financial penalties of various types and (3) a shutdown order. The first of these typically is insufficient since it provides no penalty for past noncompliance and therefore weakens the compliance incentives in the future. The shutdown order suffers the opposite kind of problem; it is such a harsh response that emitters do not expect it to be invoked. It does not represent a credible threat. [39]

Most attention therefore has focused on using some sort of monetary penalty to induce compliance. These penalties can take several forms. They can be predetermined or determined after the fact and tailored to the circumstances of the permit violation. They can represent a charge per unit of emissions, more or less continuously imposed, or a lump sum compensation imposed in infrequent judicial proceedings.

Spence and Weitzman (1978)* focus on the predetermined charge-per-pound sanction. They recommend that the predetermined fee be set equal to the "marginal damages at the level of effluents prior to the imposition of regulation" (Spence and Weitzman 1978, p. 213). Thus the emitter is assessed a penalty which is crudely related to the damages that would be caused if the conditions of the permit are violated. The emitter knows precisely what the unit cost of noncompliance will be prior to its occurrence.

Quite a different approach has been developed in what is known as "the

---

[38]One of the other useful insights of this document (p. 6) is the demonstration that an effective control policy is not very sensitive to the accuracy of monitoring devices as long as the measurement error is known. They conclude that research and development money should be spent on extending the number of pollutants which can be monitored rather than improving the accuracies of existing systems.

[39]A good example of sanction overkill is the repeated failure of the EPA to stand firm in the face of repeated successful demands of the automobile manufacturers to delay the implementation dates for the new car emission standards. See Grad et al. (1975, p. 375).

*Chapter 13 in this collection—*Eds.*

Connecticut plan." This plan (described by Clark 1978 and Drayton 1978), which has currently been adopted by Connecticut, relates the non-compliance penalty imposed to the costs of pollution control avoided by the emitter who is not in compliance. This approach obviously is designed to remove any financial advantages to the emitter from noncompliance.[40]

The philosophical differences between these two approaches can be summarized as being an argument between optimal noncompliance and complete compliance. The Spence and Weitzman approach attempts to relate compliance costs to the damages incurred. When compliance costs are higher than the damages, noncompliance will be the expected (even desired) outcome. The Connecticut plan, on the other hand, attempts to insure that noncompliance will never be in the interest of the emitter.

The former approach has much to recommend it; it can be handled administratively with a minimum of cost. As Anderson et al. (1977, pp. 123–28) point out, however, it is not clear that the courts would allow the administrative imposition of a charge as being consistent with the due process requirement of the constitution. There are two issues involved. The first deals with the constitutionality of executive (as opposed to judicial) imposition of noncompliance fees while the second involves the procedural safeguards involved in various forms of assessment. These are important issues because they have a drastic effect on the rapidity with which the system can be enforced and the amount of resources that would have to be committed to enforcement.

A court determination that these fees were criminal penalties would prompt the need for procedural safeguards that present serious obstacles to an effective enforcement policy. Thus the feasibility of this approach rests with the court system and remains to be established. Unfortunately as long as the statutes require the "Connecticut plan approach" that test will not be forthcoming.

III. DESIGNING PERMIT ENTITLEMENTS

Taxonomy of Permit Entitlement Choices

There are three basic permit designs discussed in the literature. In this section these designs will be described and the differences among them clarified. In the next section the relationships of these various designs to the achievement of the two goals will be discussed. The succeeding sections will then treat other matters which impinge on the permit design.

The major difference among permit designs concerns whether the per-

[40]This is the basic approach that the 1977 amendments require. The amendments require the penalty to be set at "the economic value of noncompliance." See 91 Stat 715.

mits are differentiated or not. An undifferentiated discharge permit (UDP) is one which conveys the same entitlement to emit to every emitter. Transfers among emitters take place on a one for one basis. Total emissions are unaffected by the trade.

In contrast to this stands the various types of differentiated permits. The first, an ambient differentiated permit (ADP), allows each permit holder to discharge one standardized unit of emissions in the air for each permit held. [41] The standardization procedure is specific to the receptor location. A number of key receptor locations are identified and the air quality control region is divided into a number of source zones. A separate permit system is created for each of these key receptor sites. A source has to have sufficient permits in each of these markets to justify its emissions. The trades within each zone take place on a one for one basis. The standardization procedure defines the exchange ratio for interzonal trades. It is worth noting that this standardization procedure has to be accomplished only once when the system is established. This contrasts sharply with current offset policy procedure which has to be reaccomplished every time a trade is proposed. [42]

Yet it is true that a separate permit system for each key receptor site is more administratively cumbersome than a single permit market. Therefore some attempts have been made to define a single permit market which will have desirable allocative properties. One of these, the emission discharge permit (EDP), was mentioned by Rose-Ackerman (1977) and developed in more detail by Montgomery (1972). Its legal implications are explored in Jaffe et al. (1978). It allows a fixed number of permits to be allocated to each of several zones within an airshed. These permits are freely transferable on a one for one basis within zones, but no transfers are permitted between zones. This system represents an attempt to gain the advantages of a single market while maintaining the cost reduction possibilities to be derived from a spatially differentiated permit system. As we shall see in the next section it is possible only in a limited sense.

Properties of Alternative Permit Designs

As Baumol and Oates (1975) and Hamlen (1977) have shown formally, under competitive conditions the UDP system can fulfill the ECE criterion. The transferability of the permits insures that they will be reallocated

---

[41]The standardization procedure is described in some detail in Tietenberg (1974, pp. 288–89). It is a quantity standardization procedure. Anderson et al. (1979, pp. 7–11) suggest a price differentiation process. Mathematically they are equivalent, but administratively they have rather different properties.

[42]Since legal challenges can arise every time a new standardization procedure is proposed (which is with every proposed trade under the current system), this could be a significant advantage of the ADP system.

by the market until the marginal costs of control are equalized across emitters. This in turn guarantees that the responsibility for achieving the aggregate emission target will be allocated among emitters so as to insure that the target is met with a minimum commitment of pollution control resources. This conclusion is an important one because it implies that a UDP system is a cost-effective means of dealing with global pollutants. [43]

As Tietenberg (1973a) has demonstrated formally, a UDP system will not fulfill the ACE criterion. Furthermore the magnitude of the cost increase from using a UDP system to achieve an ACE criterion is apparently substantial (Tietenberg 1978a) and there is reason to believe that the cost advantage of differentiated permit systems over the undifferentiated systems will be even larger in the future than it is currently. This is due to the effect of each system on location incentives. With a UDP system the location of the emitter within the airshed does not affect its pollution control cost while for the differentiated permit systems it does. With differentiated permit systems emitters have the incentive to consider the air quality impacts of their location decision while with UDPs they do not. The long run cost-effective achievement of the ambient standards requires that both relocation and the adoption of pollution control equipment be considered as ways to reduce the costs of achieving the standards. While clearly no wholesale relocation of emitters should or will take place, for some emitters, particularly new entrants to the area, the choice of location may greatly affect both their costs and the costs of other emitters in the area. [44]

As Montgomery (1972) has demonstrated, both the ADP permit design and the EDP permit design can fulfill the ACE criterion at any point in time, but in order to do so the EDP system requires much more information on the part of the control authority and this particular design allows much less flexibility to the control authority in its capacity to influence the distribution of the costs. The ADP design allows the control authority to choose any initial allocation of permits and therefore to use this allocation as a means of choosing an agreeable distribution of the costs. Regardless of what initial allocation is chosen, because of the complete transferability of the permit across zones, the ultimate allocation will be cost-effective. This cost-effective allocation of permits will imply a particular distribution of emissions across the airshed. If the control authority were to allocate the permits in an EDP system so that the initial allocation was consistent with this distribution of emissions, then that EDP system would fulfill the ACE criterion as well.

However, two aspects of this conclusion are worth pointing out. First, for the initial allocation to fulfill the ACE criterion in the EDP system the

[43]The Rand Corporation is currently engaged in the examination of the feasibility of a TDP market for fluorocarbons.
[44]See, for example, the empirical work presented in Kohn (1974).

control authority would have to know the control costs for every emitter in the airshed. This information is not required for an ADP system to be cost-effective. Secondly, even if the control authority possessed that rather unrealistically large amount of information it would have only *one* distribution of permits which would be consistent with the ACE criterion (as opposed to the infinite number of ADP initial allocations which are consistent with the ACE criterion).[45] Even in the rather unlikely event that the control authority was able to discover this one initial EDP allocation it would have no flexibility in trying to distribute the costs fairly.

All of these differentiated systems are based upon source zones. The number of these zones and their size is one variable which can be manipulated by the control authority when the system is designed. Since the administrative complexity of the system rises with the number of zones, it is tempting to use very large zones (and hence only a few of them) to characterize a particular geographic area. This temptation should be resisted, however, since an increase in the size of the zone will introduce the possibility that "hot spots" or high pollution levels will occur within the zone, since one-for-one transfers within zones could result in a clustering of emitters. On the other hand, in the EDP system smaller zones reduce the transferability of permits, a prime source of cost reduction.

These types of permit designs should not be considered as either/or choices. It may well be that for small sources a UDP system is appropriate when these sources are relatively ubiquitous. In this case the additional complexity of spatial differentiation is not warranted. On the other hand, for large sources, where their location makes a big difference, one of the differentiated permit systems could be used.[46]

Incorporating the Temporal Dimension

Although it would be possible to define an aggregate emission standard with a time dimension the use of an aggregate emission standard as the policy target normally implies that the temporal pattern of emissions is not important and, therefore, is of no consequence in the design of the permit. This is not true when an ambient standard policy target is pursued using an ADP system. Both the emission intensity of contributing sources and meteorological conditions have a temporal component that can be integrated into the design of an ADP system. Several options exist for the

[45]It would be pure coincidence if this unique allocation coincided with the minimum disruption allocation discussed above. Therefore it is not likely that an EDP system could fulfill the ACE criterion while simultaneously guaranteeing no emitter would be worse off by leaving the current system. The ADP system can simultaneously achieve those goals.

[46]This multiple-tiered approach in which a UDP system for small emitters is combined with an ADP approach for large emitters was suggested to me by Barbara Ingle in a telephone conversation.

manner of integration, each incurring quite different public control costs. In general, the most complete control over the system requires the largest outlay of funds; therefore, some trade-off between completeness and minimizing costs appears in order.

For the integration of a changing meteorology into the control process a continuum of options is available, at one end of which would be a control policy based on current meteorological conditions. Here the standardization coefficients (which determine the exchange rate between emission in one zone and emission in another) would have to be changed at least daily. Although this policy would increase the degree of control over the process in the sense that air quality forecasts would be based on actual rather than annual average conditions, it would also incur significant and, at present, even prohibitive costs both for the public and private sectors. Public costs would be high because of the necessity for continuously updating the table and transmitting this information to the firm; private costs would be high because it would be very difficult to plan production schedules.

The opposite end of the control spectrum in terms of temporal complexity, a system of controls that does not vary over time, is much easier to implement, administer, and enforce. Such constant control is implemented by using annual average meteorological conditions in constructing the table of equivalences and choosing a level of control such that even under somewhat adverse conditions the desired concentration level would not appreciably exceed the standard. The problem with this approach is that it contains no special provisions for handling those rare, but devastating, occasions when thermal inversions prevent the normal dispersion and dilution of the pollutants. The damage that can occur on these occasions to humans, animals, materials, and vegetation is so severe as to warrant a special procedure for such days. One solution is to establish a two-tier kind of control in which the permits allow two different emission rates— one for days characterized by thermal inversions and one for all other days. The emission rates allowed during thermal inversions would be substantially less.

One other element in the temporal nature of the meteorological processes can be integrated into the design of an ADP: the regular periodic nature of wind patterns. The conditions that go into the derivation of the equivalency remain relatively constant within seasons but tend to change as the seasons change. Since accurate records are kept by the weather bureau for wind patterns, it is possible to derive a separate set of equivalences for each season. With this table of equivalences defined the administrator could either design specific permits to be sold in four seasonal markets, or sell an annual permit that would be adjusted for each season. As long as these permits were transferable among emitters, the allocation could change over the seasons without the need for different permits in each season.

One benefit of incorporating meteorological seasonality into the design is that it provides incentives to emitters to shift their emissions to time periods and locations where they will do the least harm. Consider a hypothetical firm, moving into a metropolitan area, that emits only during the winter months because of a large space-heating need then. If a seasonal ADP market is operative, the firm will have an incentive to locate where its costs per unit of emissions is the lowest. According to the design of the market, this location is where winter emissions do the least harm in terms of meeting the desired air quality standards. On the other hand, if a nonseasonal, annual ADP market is in existence, the firm will still locate where the ADPs are the cheapest, but this will no longer coincide with the location where *winter* emissions do the least damage. Rather it will coincide with the location where a steady emission rate all year would do the least damage. Because of the quite large differences in seasonal wind patterns, these locations will not generally coincide.

Another temporal pattern exists for emissions. Our society concentrates its economic activity into the daylight hours, resulting in higher concentrations during the daytime. The impact of this on the design of the ADP is that if the permits were made specific to an eight-hour period then presuambly the price of the permit during the 12:00 A.M.–8:00 A.M. period would be much lower than those for the two other shifts. This would provide an incentive for emitters to shift their emissions to this period, partly offsetting some existing incentives in the opposite direction, such as the higher wage rates required for the "graveyard" shift. If this incentive were successful in shifting some production into these hours, a beneficial result may be observed on automobile-produced pollutants as well. The portion of the rush-hour traffic representing the workers affected by this shift would be channeled into a less active period.

The 1977 legislation would appear to rule these refinements out. The Clean Air Act Amendments of 1977 state unambiguously:

> The degree of emission limitation required for control of any air pollutant under an applicable implementation plan under this title shall not be affected in any manner by . . . any other dispersion technique. . . . For the purposes of this section, the term "dispersion technique" includes any intermittent or supplemental control of air pollutants varying with atmospheric conditions.[47]

Unlike many other characteristics of an expanded TDP system this refinement, if implemented, would require new legislation.

---

[47]91 Stat 721.

Duration of the Permit

The TDP could be designed either to entitle the owner to discharge a specified emission rate into the air in perpetuity or for some finite time period. In theory, either option is acceptable, because their transferability would insure that they were reallocated in response to a changing environment. New firms would bid them away from firms who no longer had need for them, and the government would have to set up the auction market only once if the TDPs were perpetuities.

David et al. (1980), deLucia (1974) and Tietenberg (1974) argue for a limited term permit on the grounds that it provides more administrative flexibility in changing the allowed level of pollution. However, even with perpetual rights the control authority can increase or decrease the level of pollution by entering the market as a buyer of existing permits or a seller of new permits. The increased flexibility to the control authority offered by a limited duration permit is offset by an increase in the uncertainty of the emitter.[48]

The upshot of the argument is that it probably makes sense for the control authority to issue permits with different terms. Those emitters who feel the need for a long term permit and are willing to pay for that security can purchase those permits. Conversely those polluters who prefer lower cost permits and less security can purchase the shorter term permits. Staggered expiration dates would allow the control authority an annual option to increase or decrease the number of permits in the market.

Coverage of Pollutants

Administratively it would be desirable if all the different types of pollutants in a particular airshed could be handled within the context of a specific market. This would be possible with the aggregate emissions target as long as the target is defined in terms of equivalent emissions. Macintosh (1973, pp. 66–68) suggests this approach. He suggests choosing the weights for determining the equivalency on the basis of some historical contribution to pollution in the local area.

In general, however, a single market with temporally fixed weights is not consistent with the cost-effective achievement of the ambient air quality standards. The cost-effective weights will necessarily change over time since some pollutants will be cheaper to clean up than others and will be subject to different growth pressures due to the entry and exit of emitters. The differential costs of control faced by emitters would, with a single multipollutant market, cause the permits to be used for pollutants which

---

[48]Anderson et al. (1979) also suggest that governments lease the permits for limited durations rather than sell them.

were expensive to control, rather than for others. When some historical basis for choosing the weights is used, this will lead to a situation in which the ambient air quality standard for those pollutants which are less expensive to control will be met easily, but the standards for the other pollutants will be violated. To compensate for this the control authority would have to reduce the number of permits until the ambient air quality standards were met for all pollutants, but this would generally result in a very expensive, excessive control of the other pollutants. The problem with this approach is that the final amount of each pollutant in the air is determined by the emitters; the control authority will have lost the capability to control each pollutant. The retention of that capability, in the absence of detailed knowledge on control costs, requires a separate market for each pollutant. [49]

Administrative Feasibility of ADP and EDP Systems

The ADP system and some versions of the EDP system require the use of air diffusion modeling. The ADP system uses these models to define the exchange rate between the permissible emission rates allowed by a permit in any two zones. This is essential to insure that regardless of which emitters use the permits the air quality standards will be met. It is the mechanism which makes complete transferability compatible with the achievement of the air quality standards at minimum cost. The EDP system may use it to define an initial allocation of permits which is compatible with the air quality standards. If the state of the art in diffusion modeling is not equal to the task, then, by default, alternative "second best" strategies become attractive.

The adequacy of the state of the art of diffusion modeling for policy purposes cannot definitively be affirmed; reasonable people may disagree. Yet, on balance, for nonreactive and relatively nonreactive pollutants it does seem adequate, particularly in light of the tremendous reductions in cost which can be achieved with its use. For highly reactive pollutants such as volatile organic compounds this conclusion clearly is not warranted.

The most common perception of those who know nothing about diffusion modeling is that the task is hopeless because air flow is such a complicated dynamic process. To be sure it is a complicated process, but annual average air flow patterns show a striking regularity—and annual averages are all that are needed to implement the ADP and EDP systems! [50] The diffusion models which predict annual averages are, on the

---

[49] It is worth noting that the offset policy explicitly rules out interpollutant trades. See 44 FR 3284, §4.A.3. (16 January 1979).

[50] The method for relating annual averages to the standards is given in Tietenberg (1974, p. 285).

whole, quite accurate. [51] They are already authorized by the Environmental Protection Agency for use in developing state implementation plans and more important for our purposes they are authorized for defining permissible offset trades.

Administratively the ADP system requires only one type of information to implement it over and above that required to implement a UDP system—a matrix defining the amount of emissions a permit allows depending on where the emissions are injected into the airshed. The data requirements for deriving this matrix are available in urban areas and rural areas served by an airport. [52] Because of the stability of annual averages from year to year, the matrix need only be constructed once. The cost of the exercise, therefore, when compared to the saving to be achieved by conducting it, is trivial.

The ultimate test is, of course, whether the state of the art is sufficiently well developed to withstand effective legal challenge. Local areas would invoke these permit systems under their general police power to protect public health and welfare. The legal challenge, if it were to be mounted, would probably question whether an ADP or an EDP system violated the equal protection clause of the Constitution. The equal protection clause precludes discriminating among persons or organizations in the same classification. The question of interest is whether or not a differentiated permit system would violate this requirement. If the classification system is not arbitrary or capricious and is consistent with some broader social purpose it can survive. While no direct precedent of the ability of a differentiated permits system to meet this test exists, it does appear that air dispersion models are now sufficiently well developed to allow them to serve as the basis for an acceptable classification system within an airshed. Pierce and Gutfreund (1975), for example, conclude that air dispersion models are likely to be admissible provided they are used by trained professionals and are based upon the best available data and methodology. [53] Similarly, for the EDP system Jaffe et al. (1978, p. 95) conclude:

> The developing law in land-use controls does not yet provide all the answers to these problems, but it does suggest that careful implementation of an emission quota strategy can escape constitutional challenge except in extreme cases.

It is much less clear that the models used to forecast the long range transport of pollutants and those forecasting the concentrations of the highly reactive pollutants would be similarly admissible.

---

[51] A survey of the state of the art in air diffusion modeling is provided in Staff of Research and Education Association (1978, chap. 2). Some evidence on the accuracy of these models is also provided in Grad (1975, pp. 220–24).

[52] The data requirements are described in Grad (1975, pp. 217–19).

[53] Interestingly enough, the legal issues for spatially differentiated air pollution emission charges are more complicated. See Tietenberg (1978*b*).

*   *   *

IV. CONCLUSIONS

The use of a transferable discharge permit control system would appear to offer real potential for achieving our air quality goals at a minimum cost both in allocating the offsets in nonattainment regions and in allocating the PSD increments in attainment regions. The current reforms embodied in the bubble concept and the offset policy represent substantial moves in this direction, but they contain important restrictions on transferability. How much is accomplished by these reforms in practice will only be determined after we have sufficient experience with them to see how constraining these restrictions turn out to be.

Similarly, it would be a mistake to ignore the host of issues which must be resolved if the current system is to be replaced by a regularized TDP market. In this paper I have tried to survey those issues and what we know both theoretically and empirically about them. What has emerged is not only a menu of approaches which can be taken, but also a sense that the general approach can be usefully tailored to individual circumstances.

It has also been pointed out that the ultimate success of the 1977 reforms as well as any potential further reforms will depend on future events at present only dimly perceived. These include legal issues, the development of monitoring instrumentation and further empirical work toward defining a balanced compromise between administrative complexity and cost-effectiveness.

The theoretical and empirical case for transferable discharge permit systems is extensive and persuasive; it is also incomplete. Simulations are suggestive, not conclusive; many small details that could serve to undermine the central strengths of the proposal are omitted by assumption from these simulations. We should not let the enthusiasm built up by these initial positive results blind us to potential flaws as TDP systems move closer to becoming operational.

REFERENCES

Anderson, Frederick R. et al. 1977. *Environmental Improvement through Economic Incentives.* Baltimore, Md.: Johns Hopkins University Press.

Anderson, Robert J., Jr. et al. 1979. *An Analysis of Alternative Policies for Attaining and Maintaining a Short Term NO$_2$ Standard.* Report to the Council on Environmental Quality. Princeton, N.J.: MATHTECH, Inc.

Atkinson, Scott E. and Lewis, Donald H. 1974. "A Cost-Effective Analysis of

Alternative Air Quality Control Strategies." *Journal of Environmental Economics and Management* 1 (Nov.): 237–50.

Barnett, Andy H. and Yandle, Bruce, Jr. 1973. "Allocating Environmental Resources." *Public Finance* 28(1): 11–19.

Bath, C. R. 1978. "Alternative Cooperative Arrangement for Managing Transboundary Air Resources along the Border." *Natural Resources Journal* 18: 181–98.

Baumol, William J. and Oates, Wallace E. 1975. *The Theory of Environmental Policy.* Englewood Cliffs, N.J.: Prentice-Hall, Inc.

Bohm, Peter. 1972. "Estimating Demand for Public Goods: An Experiment." *European Economic Review* 3(2): 111–30.

Buchanan, James M. 1969. "External Diseconomies, Corrective Taxes and Market Structure." *American Economic Review* 59 (March): 174–77.

Clark, Edwin H., II. 1978. "Regulatory Strategies for Pollution Control: Comment." In *Approaches to Controlling Air Pollution,* ed. Ann Friedlander. Cambridge, Mass.: The MIT Press.

Clarke, E. H. 1971. "Multipart Pricing of Public Goods." *Public Choice* (Feb.): 17–33.

Cleveland, William S. and Graedel, T. E. 1979. "Photochemical Air Pollution in the Northeast United States." *Science* 204 (June): 1273–78.

Crocker, Thomas D. 1966. "The Structuring of Atmospheric Pollution Control Systems." In *The Economics of Air Pollution,* ed. Harold Wolozin. New York: W. W. Norton.

Dales, J. H. 1968a. "Land, Water, and Ownership." *Canadian Journal of Economics* 1 (Nov.): 797–804.

Dales, J. H. 1968b. *Pollution Property and Prices.* Toronto: University Press.

David, Martin H. et al., 1980. "Marketable Permits for the Control of Phosphorus Effluent into Lake Michigan." *Water Resources Research* 16 (Apr.): 263–70.

deLucia, R. J. 1974. *Evaluation of Marketable Effluent Permit Systems.* Office of Research and Development, U.S. Environmental Protection Agency. Washington, D.C.: Government Printing Office.

Downing, Paul B. and Watson, William D., Jr. 1973. *Enforcement Economics in Air Pollution Control.* U.S. Environmental Agency, EPA 600/5-73-014.

Drayton, William, Jr. 1978. "Regulatory Strategies for Pollution Control: Comment." In *Approaches to Controlling Air Pollution,* ed. Ann Friedlander. Cambridge, Mass.: The MIT Press.

Grad, Frank P. et al. 1975. *The Automobile and the Regulation of Its Impact on the Environment.* Norman, Okla.: University of Oklahoma Press.

Hamlen, William A. 1977. "The Quasi-Optimal Price of Undepletable Externalities." *The Bell Journal of Economics* 8 (Spring): 324–34.

Irwin, William A. and Lirhoff, Richard A. 1974. *Economic Disincentives for Pollution Control: Legal, Political and Administrative Dimensions.* U.S. Environmental Protection Agency, EPA 600/5-74-026 (July).

Jaffe, Martin et al. 1978. *Legal Issues of Emission Density Zoning.* U.S. Environmental Protection Agency, EPA 450/3-78-049 (September).

Kneese, Allen V. 1977. *The Economics of the Environment.* New York: Penguin Books.

Kneese, Allen V. and Bower, Blair T. 1968. *Managing Water Quality: Economics, Technology and Institutions.* Baltimore, Md.: Johns Hopkins University Press/Resources for the Future.

Kneese, Allen V. and Schultz, Charles L. 1975. *Pollution Prices and Public Policy.* Washington, D.C.: The Brookings Institution.

Kohn, Robert E. 1974. "Industrial Location and Air Pollution Abatement." *Journal of Regional Science* 14 (April): 55–63.

Mar, B. W. 1971. "A System of Waste Discharge Rights for the Management of Water Quality." *Water Resources Research* 7 (Oct.): 1079–86.

Macintosh, Douglas R. 1973. *The Economics of Airborne Emissions: The Case of an Air Rights Market.* New York: Praeger Publishers.

Mishan, Ezra J. 1967. *The Costs of Economic Growth.* New York: Frederick A. Praeger.

Montgomery, David W. 1972. "Markets in Licenses and Efficient Pollution Control Programs." *Journal of Economic Theory* 5 (Dec.): 395–418.

Nulty, Peter. 1979. "A Brave Experiment in Pollution Control." *Fortune* 99 (Feb. 12): 120–23.

Pierce, D. F. and Gutfreund, P. D. 1975. "Evidentiary Aspects of Air Dispersion Modeling and Air Quality Measurements in Environmental Litigation and Administrative Proceedings." *Federation of Insurance Council Quarterly* 25 (Spring): 341–53.

Randall, A. et al. 1974. "Bidding Games for Valuation of Aesthetic Environmental Improvements." *Journal of Environmental Economic Management* 1 (Aug.): 132–49.

Rassin, A. David and Roberts, John J. 1972. "Episode Control Criteria and Strategy for Carbon Monoxide." *Journal of the Air Pollution Control Association* 20 (April): 254–59.

Roberts, Marc and Spence, Michael. 1976. "Effluent Charges and Licenses Under Uncertainty." *Journal of Public Economics* 95 (April/May): 193–208.

Rose-Ackerman, Susan. 1977. "Market Models for Water Pollution Control: Their Strengths and Weaknesses." *Public Policy* 25 (Summer): 283–406.

Rose, Marshal. 1973. "Market Problems in the Distribution of Emission Rights." *Water Resources Research* 5 (Oct.): 1132–44.

Scherr, B. A. and Babb, E. M. 1975. "Pricing Public Goods: An Experiment with Two Proposed Pricing Systems." *Public Choice* 23 (Fall): 35–48.

Spence, A. Michael and Weitzman, Martin L. 1978. "Regulatory Strategies for Pollution Control." In *Approaches to Controlling Air Pollution,* ed. Ann Friedlander. Cambridge, Mass.: The MIT Press.

Staff of Research and Education Association. 1978. *Modern Pollution Control Technology: Vol. 1, Air Pollution Control.* New York: Research and Education Association.

Teller, Azriel. 1970. "Air Pollution Abatement: Economic Rationality and Reality." In *America's Changing Environment,* eds. Roger Revelle and Hans H. Landsberg. Boston, Mass.: Beacon Press.

Tietenberg, Thomas H. 1973a. "Controlling Pollution by Price and Standards Systems: A General Equilibrium Analysis." *Swedish Journal of Economics* 75 (June): 193–203.

———. 1973b. "Specific Taxes and the Control of Pollution: A General Equilibrium Analysis." *Quarterly Journal of Economics* 87 (Nov.): 503–22.

———. 1974. "Design of Property Rights for Air Pollution Control." *Public Policy* 22 (Summer): 275–92.

———. 1978a. "The Quasi-Optimal Price of Undepletable Externalities: Comment." *The Bell Journal of Economics* 9 (Spring): 287–91.

————. 1978*b*. "Spatially Differentiated Air Pollutant Emission Charges: An Economic and Legal Analysis." *Land Economics* 54 (Aug.): 265–77.

Yandle, Bruce. 1978. "The Emerging Market in Air Pollution Rights." *Regulation* (July/August): 21–29.

ADDITIONAL READINGS

The articles below are listed for those interested in more recent readings on the topic.—*Eds.*

Hahn, R. W. and Noll, R. 1982. "Designing a Market for Tradeable Emissions Permits." In *Reform of Environmental Regulations,* ed. W. Magat. Cambridge, M.A.: Ballinger. 119–46.

Hahn, R. W. 1989. "Economic Prescriptions for Environmental Problems: How the Patient Followed the Doctor's Orders." *Journal of Economic Perspectives* 3, no. 2: 95–114.

Montgomery, David W. 1972. "Markets in Licenses and Efficient Pollution Control Programs." *Journal of Economic Theory* 5: 395–418.

Oates, Wallace E. 1981. "Corrective Taxes and Auctions of Rights in the Control of Externalities: Some Further Thoughts." *Public Finance Quarterly* 9, no. 4: 471–77.

Roberts, Marc and Spence, Michael. 1976. "Effluent Charges and Licenses under Uncertainty." *Journal of Public Economics* 95.

Tietenberg, Thomas H. 1985. *Emissions Trading: An Exercise in Reforming Pollution Policy.* Washington, D.C.: Resources for the Future.

Yandle, Bruce. 1978. "The Emerging Market in Air Pollution Rights." *Regulation:* 21–29.

16

Economic Incentives in the Management of Hazardous Wastes

CLIFFORD S. RUSSELL*

Clifford S. Russell is Professor of Economics and Director
of the Vanderbilt Institute for Public Policy Studies at
Vanderbilt University.

The management of hazardous waste in the United States is currently the responsibility of local, state and federal agencies, with the U.S. Environmental Protection Agency (EPA) having overall responsibility for setting standards, coordinating activities, and approving plans under the terms of the Resource Conservation and Recovery Act (RCRA) of 1976 as subsequently amended.[1]

This paper will briefly explore the management system created under the authority of RCRA, suggest why society can expect that system to be inadequate, evaluate arguments for the type of system established and suggest alternative approaches, in particular, the use of economic incentives that in certain important situations will better promote compliance with the disposal goals of RCRA.

"Economic Incentives in the Management of Hazardous Wastes," by
Clifford S. Russell, from *Columbia Journal of Environmental Law*, v.
13 (1988), pp. 257–74. Reprinted by permission.

*I am grateful for the extensive comments Dick Stewart made on an earlier draft. See Appendix (p. 287) for a brief history of hazardous waste management regulations in the United States.

[1] 42 U.S.C. §6901 (1982 E Supp. III 1985). For a brief history of federal laws pertaining to hazardous waste management, see the Appendix (p. 287) at the end of this article.

Regulations promulgated by EPA define hazardous wastes as wastes displaying one or more of four properties: ignitability, corrosivity, reactivity and toxicity.[2] Thus, a household's waste lubricating oil and cadmium batteries, the service industry's spent solvents and pesticide residues, and manufacturing's pickling acids, plating wastes and drilling muds are all categorized as hazardous. But industry is by far the largest source of such wastes.[3] Table 1 shows the estimated amounts of hazardous wastes generated in 1983 by industrial sources, where the wastes are categorized by type. Table 2 shows the estimated distribution of the mean 1983 industrial generation across industries.

The major elements of the management system established by RCRA and corresponding regulations are:

- a manifest system[4] for tracking hazardous wastes that leave the premises of the generator, designed to discourage illegal disposal;
- design and performance standards for treatment, storage and disposal facilities that will handle hazardous wastes;
- post-closure, financial responsibility and liability insurance requirements for such facilities.[5]

The amendments of 1984 expanded the scope of the regulations by bringing into the system an estimated 600,000 additional generators—those which generate between 100 and 1,000 k.g. per month, a group that

[2]Identification and Listing of Hazardous Waste, 40 C.F.R. §261.20–4 (1987). In RCRA the definition of hazardous waste is broader, reading:
    The term "hazardous waste" means . . . waste . . . which because of its quantity, concentration, or physical, chemical, or infectious characteristics may—
    (A) cause, or significantly contribute to an increase in mortality or an increase in serious irreversible, or incapacitating reversible, illness; or
    (B) pose a substantial present or potential hazard to human health or the environment when improperly treated, stored, transported, or disposed of, or otherwise managed.
Resource Conservation and Recovery Act §1004(5), 42 U.S.C. §6903(5) (1982 & Supp. III 1985).
[3]11 Council on Envt'l Quality Ann. Rep. 216 (1981).
[4]A "manifest system" was contemplated in RCRA as a means "to assure that all such hazardous waste generated is designated for treatment, storage, or disposal in, and arrives at treatment, storage, or disposal facilities (other than facilities on the premises where the waste is generated) for which a permit has been issued as provided in this subchapter. . . ." 42 U.S.C. §6922 (a)(5) (1982 & Supp. III 1985). The word "manifest" specifically refers to the form used to identify the waste being transported, its origin and its destination. 42 U.S.C. §6903(12) (1982 & Supp. III 1985). Transporters are not supposed to accept wastes from generators without a proper manifest, nor are storage or disposal facility operators supposed to accept wastes from a transporter without that manifest. 42 U.S.C. §§6923(a), 6924(a) (1982 & Supp. III 1985).
[5]See, generally, F. Grad, Environmental Law 646–51 (3d ed. 1985) (overview of RCRA); see also 40 C.F.R. §§262, 263, 264 (1987).

TABLE 1.   **Estimated Generation of Industrial Hazaedous Waste in 1983, Ranked by Waste Quantity**
(in thousands of metric tons)

WASTE TYPE	ESTIMATED RANGE		MEAN QUANTITY	PERCENT OF TOTAL
	LOWER	UPPER		
Nonmetallic Inorganic Liquids	68,102	96,420	82,261	31
Nonmetallic Inorganic Sludge	23,285	32,837	28,061	11
Nonmetallic Inorganic Dusts	19,455	22,784	21,120	8
Metal-Containing Liquids	14,125	25,394	19,760	7
Miscellaneous Wastes	14,438	16,393	15,415	6
Metal-Containing Sludge	13,246	15,748	14,497	6
Waste Oils	9,835	18,664	14,249	5
Nonhalogenated Solvents	11,325	12,935	12,130	5
Halogenated Organic Solids	9,321	10,246	9,784	4
Metallic Dusts and Shavings	6,729	8,738	7,733	3
Cyanide and Metal Liquids	4,247	10,520	7,383	3
Contaminated Clay, Soil, and Sand	5,092	5,830	5,461	2
Nonhalogenated Organic Solids	4,078	5,078	4,578	2
Dye and Paint Sludge	4,035	4,438	4,236	2
Resins, Latex, Monomers	3,451	4,585	4,018	2
Oily Sludge	2,965	4,502	3,734	1
Halogenated Solvents	2,774	4,185	3,479	1
Other Organic Liquids	2,866	4,003	3,435	1
Nonhalogenated Organic Sludge	2,179	2,305	2,242	1
Explosives	508	933	720	<1
Halogenated Organic Sludge	583	848	715	<1
Cyanide and Metal Sludge	537	577	557	<1
Pesticides, Herbicides	19	33	26	<1
Polychlorinated Biphenols	1	1	1	<1
Total	223,196	307,997	265,595	

Source: Congressional Budget Office, 1985. *Hazardous Waste Management: Recent Changes and Policy Alternatives.* Washington, D.C., USGPO, p. 18.

had been exempted previously.[6,7]Further, the amendments indicated a change in regulatory approach, from what some saw as a bias toward (cheap) land disposal to a definite anti-land disposal approach. This shift was accomplished by requiring that a waste could only be disposed of in

[6]Bureau of National Affairs, *U.S. Environmental Laws* 175 (1986).
[7]42 U.S.C. §6921(d) (1982 & Supp. III 1985).

TABLE 2. **Estimated National Generation of Industrial Hazardous Wastes Ranked by Major Industry Group** (in thousands of metric tons)

MAJOR INDUSTRY	ESTIMATED QUANTITY IN 1983	PERCENT OF TOTAL
Chemicals and Allied Products	127,245	47.9
Primary Metals	47,704	18.0
Petroleum and Coal Products	31,358	11.8
Fabricated Metal Products	25,364	9.6
Rubber and Plastic Products	14,600	5.5
Miscellaneous Manufacturing	5,614	2.1
Nonelectrical Machinery	4,859	1.8
Transportation Equipment	2,977	1.1
Motor Freight Transportation	2,160	0.8
Electrical and Electronic Machinery	1,929	0.7
Wood Preserving	1,739	0.7
Drum Reconditioners	45	<0.1
Total	265,595	100.0

Source: Congressional Budget Office, 1985. *Hazardous Waste Management: Recent Changes and Policy Alternatives.* U.S. Government Printing Office: Washington, D.C. p. 20.

a land fill operation, however carefully designed, if the EPA Administrator certified by a certain future date that such disposal would satisfy some very restrictive safety requirements. In addition, the amendments imposed a set of tight deadlines by which the Administrator was to promulgate rules establishing performance standards governing these requirements. [8]

It is too early for a full evaluation of a management system that is still being put in place, but thoughtful commentators and experience with the act to date suggest a few observations:

- The manifest system apparently neither applies to much of the waste being generated nor produces information crucial to discovering dumping violations. [9]

[8]42 U.S.C. §§ 6924, 6925 (1982 & Supp. III 1985).

[9]The Congressional Budget Office estimates that 96 percent of industrial hazardous wastes are dealt with on the site of generation and thus are not now subject to manifesting. This is not to say they will always be stored where generated; in the long run, shipment to incinerators or other advanced disposal facilities will have to be undertaken. Congressional Budget Office, U.S. Congress, *Hazardous Waste Management: Recent Changes and Policy Alternatives* 26 (1985). The United States General Accounting Office could not find, in its study of four states, any illegal disposal cases identified through manifest exception reports (reports

- Advance disposal techniques that meet EPA performance criteria (such as 99.99 percent destruction of hazardous chemicals in incineration) and sites fitted with these technologies may not be available soon enough to prevent the required land disposal bans from effectively closing all legal disposal options for some generators in some places. [10]

Even if sites are found and equipment approved, problems with obtaining required liability insurance for operational and post-operational periods may well result in severe pinching of legal disposal capacity over the next five to ten years. [11]

All together, there is reason to be concerned that under the existing RCRA management system, pressures for illegal disposal of hazardous wastes are building.

Note that there is a contrast between what is (or will be, after regulations are finalized) illegal with respect to hazardous waste disposal and what is illegal in the context of more conventional pollutants. For the most part, hazardous waste disposal is limited by regulations directing where such wastes may be disposed and by what method the disposal may proceed, but not limiting the quantities disposed of. [12] Discharges of more conventional pollutants, such as biochemical oxygen demanding organic material in waste water streams or particulate matter in combustion of gases, are limited to quantities specified in discharge permits. While the permit terms may be based on hypothetical application of particular technologies to plant raw waste loads, the actual choice of source reduction strategies is up to each source. [13]

The implication of this observation is interesting when assessing the systems and policies that guide the management of various types of currently generated wastes. To wit, it is in hazardous waste management, if anywhere in environmental policy, that the infamous "command and control" approach is to be found. That is, hazardous waste disposal facilities are required to meet very specific technological requirements—they are told how to do what they do—while conventional pollutant

---

to EPA indicating that a transported waste may not have reached a designated storage or disposal facility). U.S. Gen. Accounting Office, RCED-85-2, *Illegal Disposal of Hazardous Waste: Difficult to Detect or Deter* iii–iv (1985).

[10]*See generally,* 6 *Inside EPA* (Inside Washington Publishers) No. 7, at 12–13 (Feb. 15, 1985).

[11]*Cf.* U.S. Gen. Accounting Office, RCED-88-2, *Hazardous Waste: Issues Surrounding Insurance Availability* 2–5 (1987) (noting general unavailability of insurance).

[12]Resource Conservation and Recovery Act § 3002, 42 U.S.C. § 6922 (1982 & Supp. III 1985); 40 C.F.R. § 262–4 (1987).

[13]For a discussion of strategy for the control of conventional air and water pollution, see Freeman, *"Air and Water Pollution Policy,"* in *Current Issues in U.S. Environmental Policy* 12 (1978).

dischargers are told what result (discharge reduction) to achieve, but not how to achieve it.

Critics of the command and control approach, such as Charles Schultze, argue that such an approach is too rigid and coercive.[14] The concerns voiced by these critics lead one to look for more flexible methods of managing hazardous wastes. In particular, perhaps hazardous waste management, or at least some part of it, might be done better using economic incentives[15] that promote desired behavior rather than using bureaucratic orders.

The usual reaction to such a suggestion, however, has been that while economic incentives might be excellent tools for encouraging the reduction of conventional pollution discharges, their use would be out of place where hazardous wastes are involved.[16] This view apparently rests on three related assumptions:

- That because economic incentives allow sources flexibility, those sources cannot be counted on to achieve any particular discharge limitation goals.[17]
- That the effects of hazardous wastes on the natural environment or on human health exhibit thresholds (i.e. concentrations below which no damages are observed, but above which very large damages occur).[18]
- That when regulations are imposed they will be obeyed, either because the sources are good citizens or because sufficient monitoring is done to induce compliance.

Thus, if there is a threshold for damages, society will want to make sure that the threshold is not crossed. Discharge reduction or treatment method regulations can be written to ensure that the threshold is not crossed. Perfect compliance with such orders is assumed. But with an incentive system, sources may produce any discharge level, so the threshold might be passed. Charges become too dangerous to try.

Now, the assumption that under a system of charges sources might respond in unpredictable ways must in turn rest on one or the other of two alternative prior assumptions:

[14]*See* C. Schultze, *The Public Use of Private Interest* 5–6 (1977).
[15]Economic incentives may take several forms in the context of pollution control, but a common suggestion is that each unit of pollutant discharged be subject to a charge payable to the environmental management agency. *See generally,* Bohm and Russell, *"Comparative Analysis of Alternative Policy Instruments,"* in *Handbook of Natural Resource and Energy Economics* (A. Kneese & J. Sweeney eds. 1985).
[16]*See, e.g.,* W. Baumol and W. Oates, *Economics, Environmental Policy, and the Quality of Life* 312–13 (1979).
[17]*Ibid.*
[18]R. Dorfman and N. Dorfman, *Economics of the Environment* 36 (2d ed. 1977).

i. That the agency setting the charge per unit of discharge does not know the marginal cost of discharge reduction for each source.[19]
ii. That a source might act against its own self-interest by discharging more (or, indeed, less) pollution than would be economically optimal for it.[20]

It would be an unusual economic argument, to say the least, that rested on a premise of sustained perversity of the sort envisioned by alternative assumption (ii). Even if individuals, and even individuals in their roles as owners and managers of firms, do occasionally engage in cutting off their noses to spite their faces, it is a fundamental part of the argument for the social superiority of a free market economy that such behavior cannot be engaged in indefinitely. It is more palatable to accept assumption (i)—that the agency may not have enough information about discharge reduction costs to set charges correctly. But the auxiliary assumption of perfect compliance presents its own problems. The most important of these by far is that the assumption is at odds with the evidence, which suggests widespread and serious violations of discharge permit terms.[21]

It is important to emphasize that this evidence applies almost entirely to stationary sources of conventional pollutants as described above—that is, to such high volume and easily located discharges as particulates from coal-burning power plants and biochemical oxygen demanding organics from municipal and industrial waste water treatment plants. The problem is not that environmental agencies lack the technical ability to monitor such sources or that they lack information about where these sources are located. Such evidence as is available strongly suggests that the problem is simply one of lack of resources to devote to monitoring and enforcement.[22] If this is a problem for conventional, stationary, point-source

---

[19]It can be shown that if the environmental agency knows that the marginal social damages from additional units of pollution rise sharply with increasing pollution levels but is uncertain about the level of marginal discharge reduction costs, then setting a discharge standard is a risk-averting strategy relative to trying to use a charge (assuming perfect compliance, of course). That is, with a standard, the maximum possible social loss from being wrong about the marginal cost curve is less than with a charge. *See* Roberts and Spence, *"Effluent Charges and Licenses Under Uncertainty,"* 5 *J. Pub. Econ.* 193–208 (1976).

[20]From each source's point of view, the economically optimal response to a charge per unit of a certain kind of pollution discharge is to reduce its discharge level until the charge equals the additional cost of reducing discharges by one further unit (the marginal cost of discharge reduction). For qualifications and complications, see generally Bohm and Russell, above, note 15.

[21]For a catalog of evidence of the extent of violations of pollution discharge permit terms based on government reports, see Russell, *"Monitoring and Enforcement,"* in *Environmental Regulation in the U.S.: Public Policies and Their Consequences* (P. Portney ed. 1988) Resources for the Future, Washington, D.C.).

[22]*Ibid.; see also* C. Russell, W. Harrington and W. Vaughan, *Enforcing Pollution Control Laws* 16–44 (1986).

pollution, it is likely to be an even greater one for the prototypical hazardous waste setting in which the volume of waste generated is small enough that the waste need not be "discharged" or treated where and when it is generated but can be containerized, concealed and "discharged" anywhere.[23] Examples include waste oil from service stations, sludge from batch-process chemical reactors, spent solvent from electronic part-cleaning operations and biologically contaminated solid wastes from hospitals.[24,25]

This paper concentrates on these small volume situations. For these wastes, monitoring is no longer a matter of measuring the output of an unconcealable smoke stack or even of an easily located if not so obvious river discharge pipe. It is in principle more a matter of poking bayonets into hay carts looking for escaping "prisoners"—searching everything that leaves a plant to discover the concealed drums and checking every existing tank or rubbish truck for waste composition. Exacerbating the problem are the unfortunate facts that many toxic wastes can cause damage in tiny concentrations, do not break down in the natural environment and can be transported through the environment by ground or surface water—whether because the chemical enters into solution or because it somehow becomes suspended in the water column. Thus, if the bayonets miss even one prisoner, society may suffer significant losses.

As a matter of fact, very little monitoring that would correspond to the bayonetting image seems to be occurring. Inspections of hazardous generators and transporters are reported to be largely of the administrative and safety varieties. The point is made by responsible state officials that to substantially increase the probability of detecting illegal transportation and dumping of hazardous wastes would be enormously expensive.[26] While it is almost by definition impossible to know the extent of the illegal dumping that is going on in the current circumstances, the few spectacular reported incidents seem unlikely to be the whole of the problem.[27,28]

One possible reaction to the situation just described is to look for an

---

[23]In reality, there are a number of different hazardous waste management situations. One is very similar to that of conventional pollutants: the hazardous substance is entrained in a water or gas stream. While there may be special technical problems of measurement and control, such cases do not present management problems that are conceptually different from those of conventional pollutants.

[24]Other possibilities exist. For example, the hazardous material may become an environmental pollutant in the very act of use, as with a pesticide or herbicide, or, in a different way, as with asbestos in brake linings.

[25]U.S. Gen. Accounting Office, above, note 9, at 32–36.

[26]*Ibid.*

[27]*Ibid.*

[28]The logical difficulty is that obtaining the information on which to base a sound estimate of the extent of illegal dumping activity would itself involve a substantial increase in some sort of monitoring activity and thus would only tend to show how much illegal dumping would go on if the monitoring effort were greater than it currently is.

alternative system for encouraging compliance with desired goals. If the goal is to discourage disposal in unapproved sites and encourage it at approved facilities, perhaps a promising alternative is a positive incentive system. Rather than fining a generator or transporter for instances of detected improper disposal, why not pay for proper disposal?[29] The idea would be to change the balance of cost considerations that currently may make an attempt at illegal disposal attractive: "If you do what we want with your specific toxic waste, we'll pay you for having done so." Then, if the amount of the payment is tuned correctly, the source should have an incentive not to try to conceal its waste—to sneak it out the back gate and have it dumped in some distant woods—but rather to work to collect the reward. Presto! The terribly difficult monitoring problems seem to be solved. Toxic wastes all end up in the right places, whether these be recycling centers, high temperature incinerators, or specifically designed landfills.

The essence of the contribution made by shifting from a stricture (or a tax) on an undesirable activity to a reward for desirable activity is the shift in burden of proof it appears to make possible. A prohibition on dumping toxic waste material by the side of the road, enforced with a fine or an administrative penalty for violation, is only as effective as the effort put into discovering violations. The regulated party must be caught to be fined. The easier concealment, the tougher the challenge. With a reward, the party must prove that it has done the correct thing. Of course, the size of the reward must be large enough to make an action, including the proving of it, worthwhile.

However, it turns out that the key element, the burden on the regulated party, is not dependent on the existence of a reward. With an important qualification to be explained in a moment, the shift of the burden is compatible in incentive terms with a tax as well. Thus, if generation is known, a tax can be charged for material generated but not turned in to the approved disposal site. The problem is, of course, that generation must be known. Knowing generation requires knowledge of the processes used by the firm or facility (to know where and what to look for) and access to

---

[29]Another alternative, the "waste reduction" movement, holds some appeal in the current situation because if there is less waste generated, there will be less to dispose of in whatever way. *E.g.,* Office of Technology Assessment, U.S. Congress, *Serious Reduction of Hazardous Waste* (1986). And one can hope that that translates into less waste disposed of illegally. The central questions, of course, are how and how much reduction will come about. If waste reduction always paid off in profit increases (or cost decreases to publicly owned facilities) and if it continued to pay off as percentage reduction approached 100, there would be no problem today. But if it requires public intervention to achieve large scale waste reductions, as in the imposition of a waste end tax, we are back to the problem in which evading the charge may be relatively easy because of the nature of the wastes. For a discussion of different possible waste end taxes structured to encourage changes in disposal method, to encourage waste reduction, or to produce a steady stream of revenue for cleanup of abandoned disposal sites, see Congressional Budget Office, above, note 9, at 64–83 (1985).

the facility so that unannounced random sampling of the particular residues may be undertaken. Thus, determining generation of raw waste loads is probably only slightly less difficult as a monitoring problem than measuring unapproved disposal directly. Thus we find ourselves talking monitoring again.

In short, the major advantage of the positive incentive is that it encourages the firm or facility to meet the conditions imposed by the agency, be these matters of disposal place, timing, or form, or all three, and reduces the agency's need to measure what is being accomplished inside the establishment itself. There are, however, potential disadvantages as well. The two most serious seem to be:

- That the payments will be a drain on the treasury, and therefore taxes will have to be increased somewhere in the system to offset the incentive payments.
- That it may be difficult to arrive at an incentive level that is high enough to encourage the desired actions but not so high as to encourage what might be called counterfeiting of wastes.

The latter problem in its starkest form would involve a firm manufacturing a compound simply in order to claim the reward for turning it into an approved disposal site. But less drastic possibilities exist, such as diluting a mix that contains the waste so that, without measurement, the agency may be duped into paying for an amount larger than that actually disposed of. These problems will be discussed again below, but for now let us consider the matter of revenue.

Because positive incentives, by themselves, mean a net increase in public spending, many, and I include myself in this group, have been drawn to deposit-refund (DR) systems.[30] Most people have had some experience with such an arrangement. For example, in most of the New England states today, when you buy a bottle or can of soda or beer you pay a deposit at the store. When that bottle is returned to a designated collecting facility, which will probably itself be a store but need not be the same one or even one in the same state, the person returning it collects the deposit as a refund, so the system creates no net call on the treasury. The idea of a deposit-refund for toxic wastes is often supported by appeals to experience of various U.S. states with bottle and aluminum can DRs, of Sweden and Norway with DRs for automobile hulks, and of West Germany with a lubricating oil tax and a waste oil rebate system.[31]

The bottle, can, and auto hulk systems seem quite to the point. When a purchase is made, a deposit is added to the price. When the object in

[30]Peter Bohm, *Deposit-Refund Systems* (1981) (for a discussion of DR systems).
[31]For a description of the latter two systems, see, generally, Organization for Economic Cooperation and Development, *Economic Instruments in Solid Waste Management* (1981).

question is returned to a designated place a refund is paid. The place of return for bottles may be any store or may be a specific, designated receiving store; for auto hulks, a dismantler and press.

The German lubricating oil system is somewhat different. In that system a tax is charged on all lubricating oil purchases. The proceeds go to a reserve fund from which aid is paid to firms that engage in collection and non-polluting disposal of waste oil from various uses. The amount of aid depends on the fate of the oil, whether it is burned or cleaned up for recycling. The aid is designed to make up the difference between the firms' costs and their proceeds. Table 3 provides a summary of existing (in 1984) and proposed "product charges," including the DR systems discussed above.

So far so good. The available commentaries have positive tones, though no data on the before and after situations are offered.[32] There are other advantages of the DR systems as well—for example, that they can provide decentralized incentives to achieve the desired end.[33] That is, the refund on a bottle goes to the person who returns it, regardless of who purchased it. Thus, some people may spend time collecting littered bottles—even, conceivably, collecting auto hulks—as an income supplement. Applied to toxic wastes, this might mean that, with some specialized equipment, scavenging firms could pick up discarded drums and turn them in, either determining their contents or letting the collection center do so. Of course, that feature would not help if the toxic has been dispersed, as some have been, by spraying along rural roads, for example.

A natural question at this point is: what are the real chances that positive incentives, perhaps in the form of DRs, could be successfully applied to hazardous wastes? First, stressing the positive reward aspect rather than the self-financing via deposit, consider some of the attributes of bottles and auto hulks that can be presumed to contribute to the success of these systems. Most important, auto hulks and bottles are easy to identify. No sophisticated or lengthy chemical tests are necessary. There is no danger that a bicycle frame could be fraudulently passed off as an auto hulk or that a soup can could be a beverage bottle. Second, the auto hulk, at least, is most unlikely to be fabricated just to get the refund. That is, there is little risk of finding that an auto hulk producing industry has been created and is producing auto hulks purely for refunds or that car thefts would occur to produce hulks rather than drivable cars. With bottles there may be a fine line between a deposit high enough to pull in returns but not so high as to encourage counterfeiting. A final attribute worth mentioning is that it takes effort to destroy and dissipate hulks and bottles. The original object can survive some rough handling and still exist to be returned.

---

[32]*Economic Instruments in Solid Waste Management,* above, note 31, at 42–45 and annexes.
[33]Bohm, above, note 30, at 6.

TABLE 3.   Product Charges and Deposit-Refund Systems in Europe[a]

CURRENT	PROPOSED
FRANCE	
Product charges are not currently used in France.	None
For a short period (1979–80) there was a charge on lubricants used to subsidize the re-refining industry. This was later replaced by a system of regulatory controls designed to provide regenerators with waste oil at low cost.	
GERMANY	
Waste oil charge. A levy is raised on all lubricants put on the market and the proceeds of this levy are used to provide financial assistance to waste oil collectors in order to facilitate recovery.	There are recurring proposals for a charge on beverage containers. The suggestion of its introduction has been used to encourage industry to operate container recovery and re-use systems.
This levy (and the subsidy scheme) is to be phased out gradually up to 1990. The scheme has helped to set up an established collection and recovery industry, and oil prices are now sufficiently high that the value of waste oil itself provides an incentive to recycling.	
NORWAY	
Charge on non-refundable beer and mineral water containers	Charge on heavy metal batteries
Deposit-refund system on automobile bodies. Deposit paid as part of import duty on new cars. (In 1979 this amounted to 1% of sales price). Refund paid at any of 100 collection points.[b]	Charge on chlorofluorocarbons
SWEDEN	
Charge on beverage containers. This was introduced in 1973, and intended to reduce the use of non-returnable beverage containers. This charge is to be discontinued and replaced by a more comprehensive deposit system.	The beverage container charge will be replaced by a deposit system to include cans as well as bottles. The scope of the deposit system will be larger than that of the container charges and deposits will be set at SKr. 0.25 rather than SKr. 0.10 which was the charge made.

TABLE 3. *(continued)*

CURRENT	PROPOSED
Charge on fertilizers. A charge, which adds about 10% to the price, is levied on fertilizers based on nitrogen and phosphorus content. The intention of the charge is to encourage reduced fertilizer use. This charge is to be gradually increased over the next few years to 25% of fertilizer prices, and further increases to 50% in 1990 are under discussion.	The funds generated will be used to finance a collection and recycling system intended to recycle 75% of aluminum cans. Manufacturers have also guaranteed to maintain systems for recycling bottles.
Vehicle scrapping deposits. Since 1976 a charge of Skr. 250 has been made on sales of new cars. When a car is delivered to an authorized scrap dealer, the final owner receivers SKr. 300 and all liability to car tax comes to an end.	Charge on heavy metal batteries. A charge of SKr. 0.5 will be paid on each battery imported. Charge revenue will be paid to a collection and recycling company who will in turn offer an incentive to consumers to return used batteries.

a Adapted from table 3.6(a) in Environmental Resources Ltd, "Cost Effectiveness: Experience and Trends" prepared for the government of the Netherlands, June, 1984.
b Organization for Economic Cooperation and Development (OECD) 1981. *Economic Instruments in Solid Waste Management.* Paris, OECD.

It seems possible that, in certain circumstances, refinements could be introduced into the DR system to reduce the chances that the agency will be defrauded. For example, the purchaser of a machine or compound could be given a piece of paper—a title or manifest—that would set out just how much of what was bought. This would have to be produced along with the actual item(s) to qualify for the refund. Then counterfeiting would involve both item and title. Chances of detection would be increased. Such a system would perhaps be justified for large purchases such as refrigerators (containing fluorocarbon compounds in their cooling systems). But it seems clear that at some point, as the cost of the item(s) purchased decreases, the system would cease to be justified. [34]

In thinking about applications of positive incentives in general and of DR systems in particular, it will be useful, if not essential to ensure the success of the system, to begin by cataloging the hazardous wastes that display the characteristics identified above. For example, refrigeration

---

[34] Applying the manifest idea further up the chain, say at the wholesale level, would probably imply that the ultimate purchaser/user would have to bring the item to a particular place so that it could be matched against the wholesaler's record. This might not be a problem for, say, containers.

units containing freons (toxic only indirectly via their effect on ozone) seem to fit these categories. They are easy to identify, unlikely and difficult to counterfeit and difficult to destroy. Mercury batteries may be hard to destroy, but identifiability could be a problem and counterfeiting would seem a definite threat. Liquid chemicals and sludges—for example, the halogenated hydrocarbon solvents and chemical reactor "bottoms"—are more problematic still. It will be difficult to determine whether a compound presented at a disposal site is actually that for which the refund is offered. It will be harder still to tell whether the compound has been diluted or contaminated with other substances. It may be that counterfeiting will prove worthwhile for some compounds or at least that some cheap dilutant may be profitably substituted for the compound of greatest interest. In short, to the extent that the identity of the waste being returned is in doubt, monitoring will be required and will be more demanding than that required for eyeballing a bottle or an auto hulk. This will be true whenever the system has to deal with containerized materials that may not be what the label asserts.

The above cautions are not meant to suggest that DR systems should be relegated to the scrap heap, if the expression will be pardoned. Rather, the lesson seems to be that, as in every policy issue, the closer we examine the facts, the more complicated the task becomes. Thus, when examining any particular disposal problem, it will be helpful to bear in mind the following questions:

1. What do we want to achieve:
   - Reduction in use, as we might for a pesticide;
   - Recycling and recovery of (nearly) all of what is used, as we might for a solvent, lubricant, or refrigerant;
   - Relatively safe disposal, as we might for a compound that breaks down but does not dissipate in use?
2. What are the "characteristics" of the use of the hazardous material:
   - Inevitably dissipative, as for herbicides and pesticides that have been introduced into the environment;
   - Naturally conservative, as for a refrigeration unit?
3. What are the characteristics of the source:
   - Fixed, as are classical point sources of air and water pollution;
   - Moveable, if not actually moving, as in small volumes of liquids generated in ways that make capture and containerization feasible?
4. How hard is it to identify the waste stream?
   - Will simple inspection suffice, as it does for car hulks and bottles?
   - Will it be necessary to perform simple tests, such as a weight check, or a straightforward qualitative analysis to see if the material is, for example, an acid?
   - Will a difficult quantitative analysis be required to determine composition and contamination levels?

5. What is the origin of the waste:
   - Contaminated residue of something toxic introduced into a process, such as chromium in the plating industry;
   - A product of the process itself such as dioxin;
   - A product that once used becomes a toxic discharge, such as a pesticide or herbicide?

The application of these questions to specific design problems may be illustrated with a few examples:

- If the hazardous material is inevitably dissipated in use, society cannot reasonably have as a goal its recovery. But the goal of decreased use is reasonable. As a general rule, orders to reduce use will involve enforcement difficulties. But a tax on the material itself has the decentralized effect of encouraging each user to decrease use. The size of the tax can be tailored to produce the desired decrease in use if the appropriate demand relation(s) is(are) known.
- If recycling is the desired goal, a reward for turning in the recyclable material sounds promising. As suggested above, the easier it is to be sure of the identity and suitability of what is returned, the easier it will be to administer the reward.
- If the waste is generated in large volumes in a production process (spent pickling acid from a steel mill, for example) the problem begins to look a lot like a standard point source problem. The sources' options for evading regulations and thus avoiding negative incentives will be limited. The definition of waste as hazardous need not determine the approach to achieving a goal of safe disposal.

Beyond this common sense sort of analysis, what can be said about the potential impact of economic incentives in toxic waste control? For example, what about static economic efficiency, a major concern of commentators exploring such incentives in other environmental pollution contexts? The answer to this apparently straightforward question is a complicated one: whether economic efficiency is even a well-defined concept, let alone a benchmark we can conceivably approach, depends on answers to the above questions. Thus, for example, if society wants to achieve 99 percent recycling of spent auto lubricating oil, regardless of source (service station or backyard), then, a deposit-refund system has promise, and the size of the refund need not vary with location of the source (though it might vary depending on the type of source—whether private person or repair shop). Therefore, trial and error could conceivably be used to establish the appropriate refund, which would be that refund resulting in exactly the desired percentage of recycling. The cost of achieving the goal would, at least in theory, be minimized by applying this deposit-refund level.

If, on the other hand, the goal were to keep groundwater contamination below some upper limit in a particular aquifer, source location would matter because the relation of discharge point to the flow of water in the aquifer would determine the contribution of any particular source to contamination at any "downstream" point. The relationships between specific source discharges and observed downstream contamination are difficult to determine for aquifers with more than one or two sources of contamination or with complicated flow patterns. Because source location matters to relative contribution to the problem, the size of the optimal refund per gallon for each source would in principle have to vary to achieve economic efficiency. But finding a set of location-specific deposit-refund levels would be very hard to do given the current state of knowledge of most aquifers, as just discussed. Further, trying to determine the efficient (cost-minimizing) deposit-refund levels by trial and error would be costly to undertake and probably doomed to failure in any case.[35]

CONCLUSION

The bottom line amounts to this. Economic incentives seem to have some unique promise in the hazardous waste field. This promise grows out of the chance to change the terms of the monitoring and enforcement problem of the environment watchdog, not out of the promotion of economic efficiency. It is potentially greatest where the wastes involved are segregated and generated in small volume, precisely those situations in which the threat of improper disposal is most serious.

Although positive reward systems are not required to reap these benefits, they are most likely to be the systems of choice. A taxation system requires knowledge of raw waste generation so that the difference between amounts generated and amounts properly disposed of may be taxed. Deposit-refund systems have the advantages of being self-financing and of promoting decentralized actions to correct improper disposal. But they are effective only where the waste is in the form of an appropriate deposit vehicle. It must be remembered that even a reward system leaves the agency with a monitoring problem. How difficult this problem is depends on how hard it is to ensure that what is submitted for the reward is what it purports to be.

[35]The extra expense of trial and error policymaking lies in the combination of lags in response by the sources to changes in the refund level and in the irreversible commitments of investment capital they make when they do respond. The first imply extra damages to society and the latter imply extra disposal costs—where "extra" is measured relative to a situation in which the correct refund level can be calculated in advance from knowledge of source costs and natural world (here aquifer) behavior. The effort would probabaly fail because even if the agency found a set of refunds that seemed to encourage the right level of waste recovery or destruction, it would be essentially impossible to determine merely by further trials whether this was the cost-minimizing set.

APPENDIX

Hazardous waste management began at the federal level with control of solid waste under the Solid Waste Disposal Act of 1965 . . . [That] law focused on garbage, particularly on restricting open burning, which was considered a fire hazard. In 1970, then-President Nixon signed an amended version of the solid waste law and renamed it the Resource Recovery Act. This law provided funds for collecting and recycling materials and required a comprehensive investigation of hazardous waste management practices in the United States. . . . In 1976, [Congress] passed the Resource Conservation and Recovery Act of 1976 (PL 94-580), which completely replaced the language of the Resource Recovery Act. The new law contained provisions on solid waste and resource recovery, including disposal of used oil and waste, and it closed most open dumps; it redefined solid waste to include hazardous waste and ordered EPA to require "cradle to grave" tracking of hazardous waste and controls on hazardous waste facilities. The Act required standards to be set for hazardous waste treatment, storage, and disposal facilities to provide for "the maintenance of operation of such facilities and requiring such additional qualifications as to ownership, continuity of operation, training of personnel, and financial responsibility as may be necessary or desirable." . . . The [Love Canal] event triggered the discovery of thousands of other dumpsites, alarming the public and mobilizing the Administration and the Congress . . . [EPA] issued the first two portions of the RCRA hazardous waste rules in 1980 . . . in an attempt to prevent creation of more toxic waste dumps. Also in 1980, Congress passed what is a logical complement to RCRA, the Comprehensive Environmental Response, Compensation, and Liability Act of 1980 (PL 96-510) (CERCLA), also known as the "superfund law", which assures financial responsibility for the long term maintenance of hazardous waste disposal facilities, and provides for the containment and cleanup of old, abandoned hazardous waste disposal sites that are leaking or endangering the public health.

Regulations governing the transport of hazardous wastes were developed jointly by EPA and the Department of Transportation (DOT). . . . Major provisions of the 1984 RCRA amendments (PL 98-616) call for banning land disposal of untreated hazardous waste within five and one-half years. . . . The new law also closed "loopholes" in previous hazardous waste rules. . . . Increasing the breadth of EPA's regulatory program, the amendments require the agency for the first time to regulate an estimated 600,000 generators of small quantities of hazardous substances and petroleum products.

—Bureau of National Affairs, *U.S. Environmental Laws* 173–76 (1986).

17

Weighing Alternatives for Toxic Waste Cleanup

PETER PASSELL

Peter Passell is a columnist on economic and financial issues for *The New York Times.*

A decade after Washington declared war on businesses that expose the public to hazardous wastes, environmental experts are questioning the unquestionable: Is it worth spending a staggering $300 billion to $700 billion to restore waste sites to pristine condition?

While acknowledging that no level of exposure to dangerous chemicals is desirable, the experts argue that the risks should be put in perspective.

Virtually all of the risk to human health, most analysts agree, could be eliminated for a tiny fraction of these sums. In a typical project, in Holden, Mo., $71,000 would be enough to isolate an abandoned factory containing residues of toxic chemicals, making it extremely unlikely that anyone would ever be harmed by the wastes. Another $3.6 million would clean up virtually all residues and bury remaining traces under a blanket of clay.

But state and Federal laws require a cleanup that would cost $13.6 million to $41.5 million. "The last couple turns of the screw cannot be justified on economic criteria," said Tom Grumbly, an environmentalist who is president of Clean Sites, a nonprofit organization in Virginia that advises communities on hazardous waste cleanups.

To be sure, the problem of weighing cleanup benefits against the costs is complicated by a lack of information about how dangerous individual chemicals are, and in what concentrations. Nonetheless, experts insist that what began as a crusade against polluters has become a diversion, siphoning money and technical expertise from more pressing environmental concerns.

More stringent scientific criteria should be used to identify waste sites needing immediate attention, they say. And once identified, the cleanup should be carefully aimed at saving lives rather than restoring land to preindustrial condition.

Analysts acknowledge that redirecting Government policy will be nearly impossible without a radical change in the way Americans think about the risks from hazardous waste. Some would try to temper the zeal for large-scale cleanups by asking the communities that benefit to share in the costs. "Everybody wants a Cadillac as long as someone else is paying," said Robert Hahn, an economist at the American Enterprise Institute, a research organization in Washington.

But there is little confidence in Washington that cost-sharing would make a big difference in public attitudes. And no enthusiasm can be detected for asking voters to pay to protect themselves against hazards they did not create.

A first priority, argues Frank Blake, a former general counsel to the Environmental Protection Agency who now works for the General Electric Company, is more credible information that lifts the "fog of fear" from public perceptions. Specifically, he says, people need ways to compare the risks of exposure with the dozens of other risks that are accepted as part of daily life.

The best-known Federal waste initiative is Superfund, which has spent $11 billion in a decade on emergency measures at 400 abandoned sites and full-scale cleanups at 60 others. But the cleanup mandated by the Superfund legislation has barely begun.

Some 1,200 other sites are already on the Environmental Protection Agency's national priority list, and the Congressional Office of Technology Assessment expects thousands to be added. What is more, a slew of Federal laws require the eventual cleanup of tens of thousands of other sites polluted by government and business. Mr. Grumbly of Clean Sites says the bill for waste containment could reach $20 billion a year by the end of the decade.

The country could afford such sums, everyone agrees, if the problem demanded it. Businesses and taxpayers have managed to cope with other anti-pollution regulations that now cost $115 billion annually, by the E.P.A.'s reckoning. But little hard evidence exists to support a crash effort.

People living or working near identified toxic waste sites naturally worry that the chemicals will find their way into air, food and water. But many analysts believe the immediate dangers have been exaggerated.

"When you look for deaths from hazardous wastes, you just don't find them," said Bill Ralston, an analyst at SRI International, a consulting company in California.

The Big Cleanup
The cost of cleaning up several categories of polluted sites.

CATEGORY	NUMBER OF SITES (ESTIMATED)	ESTIMATED COST ($ BILLIONS)
Superfund abandoned sites	4,000	$80–120
Federally owned sites	5,000–10,000	75–250
Corrective action on active private sites	2,000–5,000	12–100
Leaking underground storage (tanks)	350,000–400,000	32
State law mandated cleanups	6,000–12,000	3–120 +
Inactive uranium tailings	24	1.3
Abandoned mine lands	22,300	55

Sources: Robert Hahn, American Enterprise Institute, Office of Technology Assessment

The Environmental Protection Agency estimates that roughly 1,000 cancer cases annually can be linked to public exposure to hazardous waste. That hardly makes abadoned waste sites a non-problem in anyone's book. But the environmental agency does rank hazardous waste far behind cancer risks like exposure to chemicals in the workplace and depletion of the atmosheric ozone layer. Moreover, many of these greater risks could be reduced at a tiny fraction of the cost of the hazardous waste fix mandated by law.

The E.P.A. estimates, for example, that 5,000 to 20,000 lung cancer deaths are caused each year by indoor exposure to radon gas leaking from underground rock formations. Most of these deaths could be prevented by performing inexpensive tests on buildings in radon-prone areas and ventilating basements where concentrations of the radioactive gas are high.

What explains this apparent distortion in environmental priorities? Political economists have long noted that the focused convictions of the few almost always dominate the weakly held views of the many. And Christopher Daggett, the former head of the New Jersey Department of Environmental Protection, argues that nothing focuses convictions like a nearby waste site. "Try telling people that the leaking drums across the street aren't a hazard," he said.

There are the questions of legal and moral responsibility. Some environmental hazards, like auto emissions, are everyone's fault—and therefore no one's. Some, like cigarette smoking, are no one's fault but one's own. But Mr. Gough notes that the public blames "greedy and thoughtless" corporations for the waste peril and is thus little inclined to weigh the benefits of cleanups against the costs.

It is not surprising, then, that communities with waste sites in their midst want them restored to how they were before industrial use. Nor is it surprising that the elected officials and administrators who write the rules for cleanups take their cues from an angry and anxious public.

Consider the 11-acre Rose Chemicals Company site in Holden, Mo. Until 1986 the site was used to store and process PCB's, toxic liquid chemicals prized by makers of electrical transmission equipment for their insulating properties and their stability at high temperatures. The barrels of PCB's were removed long ago, but residues remain in the buildings, the soil and the bed of a small stream.

An E.P.A. analysis says that for $71,000, the site could be permanently isolated from the community. The town's drinking water would be safe because there is no groundwater to tap near the site. But under Superfund regulations, this cheap fix would be unacceptable because there is one chance in 10,000 that someone would develop cancer from eating the cattle that grazed near the property's perimeter—and roughly the same risk that a determined trespasser could get cancer from repeated contact with building surfaces.

Other possible fixes range from cleaning the stream bed and capping the site with 10 inches of clay ($3.7 million) to removing 14,000 tons of contaminated soil and building materials and incinerating it elsewhere ($41.5 million). The E.P.A.'s preferred $13.6 million option: remove all of the suspect soil and materials, then incinerate the most contaminated debris and bury the rest in a specially designed landfill.

Mr. Grumbly of Clean Sites notes that the E.P.A.'s compromise would allow virtually any future use of the site. And he points out that it would cost just a third as much as the maximum effort.

But others may wonder whether the more flexible use of the land is worth the extra $900,000 an acre beyond the cost of scrubbing the stream and capping the property. And still others may ask whether a bare-bones fix—one that reduces neighborhood cancer risks to, say, one-thousandth the chance of getting cancer from a lifetime of normal exposure to the sun—would not be adequate.

Mr. Hahn of the American Enterprise Institute and others inclined toward less ambitious goals suggest using financial incentives to discipline the decision-making process. If those at risk from hazardous waste sites wanted more than a basic cleanup, they could be asked to share in the cost.

Dan Dudek, an economist at the Environmental Defense Fund, a mainstream environmental group, offers a variation based on the carrot rather than the stick. Instead of taxing communities to cover the extra cleanup effort, he would give them part of the savings associated with a less-than-total repair. The town of Holden, for example, might be offered 20 percent of the $9.9 million difference as an incentive to save lives in other ways—say, by purchasing more efficient firefighting equipment, or by testing every home for radon and lead contamination.

That idea appeals to Francis Brillhart, the mayor of Holden, who said he fears that the "E.P.A. is going a bit overboard," slowing the cleanup by choosing a method that requires incineration. If Holden could keep some of the savings generated by a faster cleanup, he says, so much the better.

But few experts think many communities would be likely to take the financial bait. The approach presumes that people would compromise on safety in return for lower taxes or improved public services. Everyone trades safety for money or convenience or simply pleasure in daily life— new tires are safer than worn ones; turkey breast is safer to eat than a cheeseburger. But people rarely acknowledge they are making choices, even to themselves.

That is why most analysts worried about the open-ended commitment to scrubbing the country pin their hopes on educating the public about relative risks. "We need more and better information about health risks," said Katherine Probst, an analyst at Resources for the Future, a nonprofit environmental research group.

The way in which the information is presented may also matter: knowing hazardous waste can kill is less useful than knowing it kills one-twentieth as many Americans as radon.

Mr. Grumbly of Clean Sites argues that the credibility of government turns on the transparency of its decision-making. One key to a rational public debate, he insists, is to "let the public in on the game."

But Mr. Grumbly is an optimist. "People understand," he says, "when they are wasting other people's money."

IV

BENEFIT-COST ANALYSIS AND MEASUREMENT

All the articles in this collection have been concerned in one way or another with the problem of achieving a socially optimal use of environmental resources. Whether the goal is pursued through the policy most often recommended by economists of "putting a price on pollution," or by means of direct controls of the sort embodied in federal environmental legislation, progress depends on a reasonably correct evaluation of the social benefits and costs that are likely to flow from varying amounts of environmental protection.

The theoretical justification for diverting resources from normal market-directed uses was developed in the first two sections of this volume. In principle, such redirection could be justified if the beneficiaries could afford to more than compensate the losers and still gain from the change. Thus the evaluation of environmental protection measures, as well as other governmental programs, depends on their benefits and costs as indicated by the community's willingness to pay for them.

Market prices measure willingness to pay for the vast majority of goods and services produced in our economy. But the benefits derived from environmental improvements frequently are inappropriable, meaning that it frequently is impractical or even impossible to deny them to anyone who wants them. For this reason, they cannot be marketed and do not have prices. Thus evaluating environmental programs is much more complicated than simply estimating the goods and services that they produce and consume, looking up the prices, and adding up the totals.

The first selection in Section IV, Robert Dorfman's "An Introduction to Benefit-Cost Analysis," introduces the concepts and principles of benefit-cost analysis. It pays particular attention to the methods used in practice to establish monetary values for the beneficial effects of environmental projects. This paper leads into Kenneth J. Arrow's discussion of the problems of measuring the benefits of social, as distinct from private, investments in "Criteria for Social Investment." These problems, arising from the absence of market prices, include the needs to impute "shadow prices" to the various types of benefit resulting from the investments, and to identify the appropriate social discount rate to be applied to deferred benefits. As Arrow points out, there is no difficulty about the concept of a shadow price, but there is an "intensely practical problem of measuring it." In the case of environmental benefits, the difficulties have proven especially intractable. Let us consider some of the reasons. The economist has several sets of tools in her kit for approximating what people would be willing to pay for unmarketed goods and services. To begin with, benefits may flow, in part at least, from "intermediate" goods or services whose values can be measured by their contribution to the production of commodities that are marketed. A case in point is a decrease in the cost of producing some marketed commodity resulting from reducing the atmospheric concentration of sulfur dioxide or other corrosive gases. Another example is the increase in the value of the fish catch when the fish population in a river or stream increases as a result of antipollution measures.

An alternative approach is to calculate the amount by which some identifiable costs attributable to environmental conditions would be lowered if those conditions were improved. Examples might be reductions in the costs of maintaining and replacing industrial equipment or in medical expenditures.

A third approach is to deduce values for nonmarketed goods and services from information about what producers or consumers pay for similar benefits that are bought and sold. This method requires that close substitutes for the environmental services be traded in the markets.

But many of the kinds of benefit that accrue to society from environmental improvements elude all these market-based techniques. The kinds of benefit that affect consumption activities directly are more subject to these difficulties than benefits that affect production. An important component of the demand for environmental protection, for example, is the public's desire to enjoy greater natural amenities. "Nonparticipatory values" can also be very important. For example, the millions of people who each place a small value on the preservation of the California redwoods without expecting ever to see them may account for more value in the aggregate than the few million who actually visit them. It is difficult to find good substitutes in the marketplace for such amenities.

In "Comparison of Methods for Recreation Evaluation," Jack L. Knetsch and Robert K. Davis evaluate some of the ways that have been tried for measuring the value of improving or preserving amenities of one category, outdoor recreation facilities. The paper by Alan Randall, Berry Ives, and Clyde Eastman is a classic exposition of the method now (inappropriately) called "contingent valuation," which is used increasingly to evaluate outdoor recreation and aesthetic facilities.

Perhaps the most insistent problem encountered in benefit-cost evaluation is that of finding appropriate monetary values to place on preventing premature deaths, or preventing or alleviating illnesses. J. Steven Landefeld and Eugene P. Seskin summarize critically the results of a wide variety of efforts to establish such values in "The Economic Value of Life." The next paper, Thomas C. Schelling's "The Life You Save May Be Your Own," has become famous for its analysis of philosophic and conceptual issues raised by estimating the value of life, as well as the practical difficulties.

As Schelling's paper exemplifies, it is often hard to place values on the results of environmental improvements, because those results are likely to be uncertain. It is nearly as hard to deal with risk and uncertainty in the analysis of decision making as it is to confront them in actual practice. Richard Wilson and E. A. C. Crouch's paper, "Risk Assessment and Comparisons," is a brief introduction to some of the conceptual issues. Space limitations permitted them barely to allude to practical methods for analyzing decisions with uncertain outcomes (which includes practically all decisions), such as the methods developed in decision theory.

Most of the foregoing considerations are pulled together and deployed in the section's final essay, by Alan J. Krupnick and Paul R. Portney. "Controlling Urban Air Pollution: A Benefit-Cost Assessment" exemplifies about as well as any brief selection can the complexity, messiness, and imprecision of real life benefit-cost analyses which, nevertheless, are indispensable for appraising past or proposed environmental protection undertakings.

18

An Introduction to Benefit-Cost Analysis

ROBERT DORFMAN

•

Robert Dorfman is David A. Wells Professor of Political Economy Emeritus at Harvard University.

Chapter 5 presented the basic concepts of economic welfare theory and derived from them the conclusion that for practical purposes the best broad indicator of economic welfare is the augmented gross domestic product, AGDP. Accordingly, in reaching decisions about environmental protection policies, the effects of the alternative policies on AGDP are an important consideration.[1] Fortunately, for decision-making purposes it usually isn't necessary to estimate what the entire AGDP would be if the various alternatives (including the status quo) were adopted, but only to estimate the differential effects of the alternatives on the components of AGDP that are affected significantly.

The effect of any measure on AGDP is the resultant of its favorable effects, called benefits, and its unfavorable effects, called costs. Benefit-cost analysis (B-CA) is the task of compiling and evaluating these effects and calculating the overall change in AGDP that each alternative would produce. Though the idea is simple and commonsensical, complications, subtleties, and perplexities abound in the execution.

Most of the complications derive from the circumstance that governments deal largely in services and goods that business firms eschew because they cannot conveniently be sold, or anyway not at market-determined prices. No one can be charged for breathing air with a low concentration of sulfates, or for benefiting from many other environmen-

[1]Since any decision about environmental protection is a political decision, no simple criterion such as effects on AGDP can be completely determinative.

297

tal programs. As a result, no market prices for evaluating the contributions of most governmental programs to AGDP are available, which makes estimating the social values of government-provided goods and services a principal, and difficult, task of B-CA. We shall consider it in some detail below.

There are two principal broad approaches to B-CA. One, which we shall call the standard approach, is to enumerate all the ways in which a proposed environmental policy would impinge on AGDP, to estimate how great each of these effects would be and how much it would affect AGDP, and finally to aggregate all these benefits and costs into the total effect of the policy on economic welfare as measured by AGDP.

The other approach is called "contingent valuation," though "hypothetical valuation" would be a more accurate description. Its basis is simple: The effect of any project on AGDP depends on how highly people value its results and costs, and if you want to know how highly people value any project, just ask them. Accordingly, in a contingent valuation study, a properly randomized and stratified sample of the people affected is drawn, the project and its expected results are explained to each respondent, and each is asked how much she/he would be willing to pay to have the project instead of the status quo or some other basis of comparison. (Negative quotations are permitted; they indicate that the respondent prefers the status quo.) The total willingness to pay expressed by the sample, inflated to represent the whole population affected, is then an estimate of the social value of the proposed project. This approach avoids the nasty problems of assigning dollar values to nonmarket consequences, but instead asks a sample of the affected public to do the assigning. We shall have to consider below how far the general public can be relied on to perform such evaluations.

The next section will present a brief outline of the standard approach to benefit-cost analysis. Since it is not always possible to find acceptable monetary equivalents for the social values of the benefits and costs of environmental projects, complete benefit-cost evaluations are not always feasible. Accordingly, the following two sections are devoted to alternatives to B-CA that are often used when a B-CA cannot be completed.

The fourth section is an extensive discussion of nonmarket benefits—the ones for which monetary values are likely to be elusive—and includes three examples of methods frequently used to find monetary values in the absence of explicit price quotations. After that, the alternative approach to benefit-cost evaluation, the contingent valuation approach, will be considered. Finally, there is a brief discussion of two complications that tend to be neglected in B-CA, followed by an even briefer concluding section.

BENEFIT-COST ANALYSIS, THE STANDARD APPROACH

The standard approach to B-CA amounts to constructing a model of the undertaking to be evaluated, and tracing through the effects of the measure on the pertinent components of the AGDP, translating those effects into monetary terms, and aggregating the resulting benefits and costs into an estimate of the net effect of the measure.

Figure 1 depicts a benefit-cost model for a proposed limitation on some harmful discharge. The chain of effects by which a governmental initiative (shown in the topmost box) affects public health, ecological integrity, and other conditions of concern is outlined in the top five boxes. In carrying out a standard B-CA, each of the reactions in the sequence (indicated by the arrows connecting the boxes) has to be estimated. These estimating tasks are usually technical, and draw on a wide range of specialties: public administration, several branches of engineering, meteorology, hydrology, public health, and biology with emphasis on several fields of ecology, to mention just a few. Questions on, or even beyond, the frontier of knowledge are likely to arise. The resultant estimates are subject to considerable ranges of uncertainty.

The sixth box, the social valuation of the changes achieved by the governmental initiative, is the most difficult and disputatious of all. Sometimes the social value of a particular kind of benefit is estimated by a public opinion survey, as in the contingent valuation approach. Because of skepticism about the reliability of people's reports about their "willingness-to-pay" for environmental improvements (as well as other things), most analysts prefer to use concrete, cash-on-the-barrelhead indications of how highly people value the changes. A great deal of ingenuity has been devoted to inferring how much people would be willing to pay for environmental goods from decisions in related markets, and a voluminous literature has accumulated. But we must finish outlining the tasks of B-CA before tackling the problem of valuing nonmarket benefits.

The final step in a B-CA is to take costs into account. As indicated in the figure, costs are incurred at several stages in the chain of responses. First, the government will have expenses for administering, monitoring, and enforcing the program. Then the firms and/or individuals whose behavior is constrained will have costs for complying with regulations, and reporting their discharges and the measures they have taken. Finally, the public at large is likely to incur some costs, both monetary,[2] and nonmonetary in the form of reductions in convenience and amenities.

These costs must be estimated, just as the various benefits were, and the

---

[2] Care must be taken at this point not to include both the costs the program imposes on business firms and any price increases that firms use to pass the cost along to users of the product.

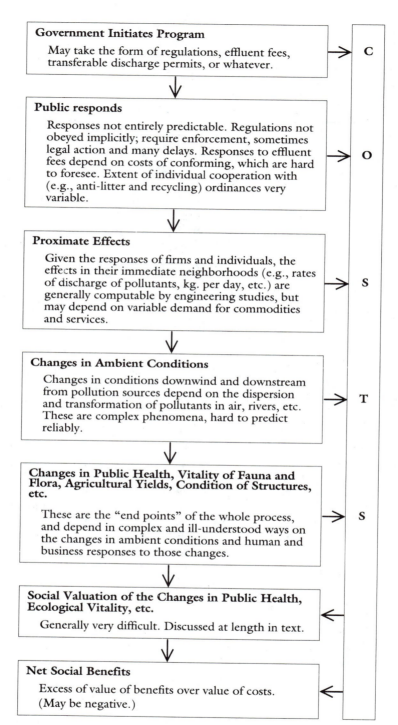

Government Initiates Program

May take the form of regulations, effluent fees, transferable discharge permits, or whatever. → C

Public responds

Responses not entirely predictable. Regulations not obeyed implicitly; require enforcement, sometimes legal action and many delays. Responses to effluent fees depend on costs of conforming, which are hard to foresee. Extent of individual cooperation with (e.g., anti-litter and recycling) ordinances very variable. → O

Proximate Effects

Given the responses of firms and individuals, the effects in their immediate neighborhoods (e.g., rates of discharge of pollutants, kg. per day, etc.) are generally computable by engineering studies, but may depend on variable demand for commodities and services. → S

Changes in Ambient Conditions

Changes in conditions downwind and downstream from pollution sources depend on the dispersion and transformation of pollutants in air, rivers, etc. These are complex phenomena, hard to predict reliably. → T

Changes in Public Health, Vitality of Fauna and Flora, Agricultural Yields, Condition of Structures, etc.

These are the "end points" of the whole process, and depend in complex and ill-understood ways on the changes in ambient conditions and human and business responses to those changes. → S

Social Valuation of the Changes in Public Health, Ecological Vitality, etc.

Generally very difficult. Discussed at length in text. ←

Net Social Benefits

Excess of value of benefits over value of costs. (May be negative.) ←

FIGURE 1. Chain of Effects of a Pollution Control Measure.

resultant total cost subtracted from the total value of the benefits to obtain the net social benefit.

In principle at least, the tasks just sketched have to be performed for each year during the expected duration of the program, since governmental programs, especially in the environmental field, generally extend over substantial periods of time, and it is not legitimate to assume that all years are the same. Usually, indeed, the costs will exceed the benefits during the early years when facilities are being constructed and installed, and the government, the firms affected, and the public are becoming acclimated to the program. Normally, it is only gradually, after a period of installation and running in, that benefits begin to exceed costs in individual years. Table 1 illustrates the kind of information that emerges for a typical year. It is, we hope, self-explanatory.

The next operation is to combine the net benefits for the individual years into an overall net benefit for the entire program. This phase raises

TABLE 1.   Benefits and Costs of Hypothetical Atmospheric Antipollution Measure, Year 19xx

(A) TYPE	(B) UNIT	(C) QUANTITY	(D) UNIT VALUE ($)	(E) TOTAL VALUE (C) × (D) ($)
Benefits				
Reduced days of illness	person-days	500,000	40	$20,000,000
Reduced medical expense	$	1,250,000	1	1,250,000
Reduced crop losses	bushels	. . .	. . .	. . .
Increased days of high visibility	days	. . .	. . .	. . .
etc.				
Total Benefits				$ . . .
Costs				
Municipal:				
Capital investment	$	. . .	1	. . .
Operating & maintenance	$	. . .	1	. . .
Business:				
Capital investment	$	. . .	1	. . .
Operating & maintenance	$	. . .	1	. . .
Monitoring and enforcement	$	. . .	1	. . .
etc.				
Total costs				$ . . .
Net benefits, 19xx				$ . . .

one of the most vehemently debated aspects of the whole procedure: the question of time-discounting. The question is: Can the social worth of a program or project be obtained by simply summing its net values year by year, or must consequences that emerge in future years be reduced by discounting, and if so, at what rate?

This is a moral question, an empirical one, and an economic one, all simultaneously. The moral aspect is beyond the scope of this essay. Empirically, it is clear that people do discount future events and experiences, i.e., that there is "a perspective diminution of the future."[3] Finally, economic reasoning indicates that future events should be discounted, and that the reasonable rate of discount is the marginal productivity of capital. For, consider any desirable event, say averting a premature death. If it costs $D at present to achieve the event (by cost-effective means), then achieving X repetitions at present would cost about DX if X is not unduly large. Now, if r is the marginal productivity of capital, DX could be invested to yield $(1+r)DX$ next year, which would be enough to achieve $(1+r)X$ repetitions of the desired event. So, if society now spends some resources on achieving the result this year and simultaneously saves some to be used next year, as is often the case, it will follow that one occurrence this year is deemed neither more nor less important socially than $(1+r)$ occurrences next year; i.e., that the social discount rate for that event, whatever it is, is r per annum, the same as the marginal productivity of capital. Conclusion: The current social values of future events, even matters of life and death, fall at the rate of r per annum, where r is the marginal productivity of capital. This is the justification for discounting delayed benefits and costs in B-CA. We find this argument persuasive; not everybody does.[4]

Once the yearly net benefits have been calculated, finding the overall net benefit is straightforward. Suppose, to be specific, that the project or program is intended to be effective over a period of 50 years and that a discount rate of $100r$ percent per year is appropriate. Suppose also that the yearly net benefits have been estimated and that they are, successively, B_1, B_2, ..., B_{50}. Then the overall net benefits are given by the formula:

$$\text{Net Benefit} = \frac{B_1}{(1+r)} + \frac{B_2}{(1+r)^2} + \cdots + \frac{B_{50}}{(1+r)^{50}}.$$

In practical execution there are many variations—both shortcuts and

[3]The phrase is from Eugen von Böhm-Bawerk *Kapital und Kapitalzins, Vol. 2* (1889). English translation, *Capital and Interest*, Vol. 2, translated by G. D. Huncke and H. F. Sennholz (1959).

[4]Notice that this argument does not contend with the philosophic aspect of the question, and that it does not deal with the empirical or psychological aspect very profoundly since the existence of a positive marginal productivity of capital remains unexplained.

complications—of the procedure just described, but all benefit-cost evaluations fit this same general format.

ALTERNATIVES TO BENEFIT-COST ANALYSIS

Frequently it is not possible to estimate how much people would be willing to pay for some of the kinds of benefits that a program yields, or to avoid some kinds of costs, without resorting to unacceptable assumptions. In such cases, the B-CA analysis cannot be completed; the best that can be done is to estimate the net value of the benefits for which monetary equivalents can be found and to note the magnitudes of the remaining consequences in the most meaningful units available. We shall describe below two alternatives to B-CA that do not make such severe demands for data in monetary form.

To appreciate how demanding B-CA's data requirements are, consider the task of estimating the monetary value of the health benefits of a regulation intended to reduce emissions of sulphur oxide into the atmosphere. Among the data required are:

1. The amount by which power plants and other emitting sources will actually reduce their discharges in response to the regulation,
2. Given the amount of reduction at the sources, the amount by which the concentrations of sulphur oxide and its chemical products will be reduced at various places in the city,
3. Given the amount of reduction at different places in the city, the numbers of people who are exposed to various concentrations for various lengths of time,
4. Given the numbers of people exposed to different concentrations and durations, the reductions in the number of days of illness (perhaps distinguished by severity) and in medical expenses,
5. Given the reduction in days of illness, the social value to attach to it.

All this for one item in the table. Particularly difficult are steps 2 (because the process of the diffusion of pollutants in the atmosphere is complicated and not well understood), 4 (because the health effects of exposure to airborne and waterborne pollutants, technically called the "dosage-response curves," are known only very roughly), and 5 (because of the difficulty of attaching monetary values to nonmonetary consequences).

Such perplexities abound in the evaluation of every environmental protection measure or program. These difficulties, along with some methods for contending with them, are explored further in the last few selections in the volume. It is clear that precision is not attainable and that large ranges of uncertainty should be attached to all estimates of benefits, costs, and their net difference. It would be good practice for evaluators to indicate the ranges of uncertainty that they believe appropriate, but they rarely do so.

The most difficult estimate of all is the social evaluation of changes in health, environmental amenities, and the quality of life generally. Many benefit-cost evaluations do not even attempt it. In those evaluations, the nonmonetary consequences are simply omitted from the tables, and the totals of benefits and costs shown include the economically measurable consequences only. There is discretion in declining to attempt the nearly impossible, but there is also danger since effects omitted from the benefit-cost calculation tend to be given insufficient weight in making decisions based upon the calculation. Perhaps because of this danger, the usual practice is to attempt to make the estimates called for by column (D), however tenuous they may be, and to include all the anticipated effects of the proposal in the calculation.

On the other hand, for many practical purposes it is not necessary to assign dollar values to such consequences as reducing mortality, increasing the clarity of waters used for recreation, or preserving scenic areas or endangered species. Two methods for getting around the problem will be sketched briefly. They are most effective when there are only one or two types of benefit that defy evaluation in monetary terms, as is frequently the case.

The simpler method is to find the lowest unit value of the problematic type of benefit (or cost) that would lead to a total of benefits greater than the total of costs. In terms of Table 1, the only major category that resists dollar valuation is reduction in days of illness. If it should turn out that a valuation as low as $10 a day would be enough to make total benefits exceed total costs, almost any public body would find the proposed measure worthwhile. Further, if the net benefits of the proposed measure exceeded those of any alternative whenever illness-days were valued at $10 or more, this measure would be preferred to any of the alternatives. The effect of this device is to avoid estimating the AGNP contribution of the measure, but, instead, to find a lower bound for it.

The second method to be discussed is to construct a "trade-off diagram" or table. One is illustrated in Figure 2. In this figure four alternative proposals for reducing atmospheric pollution in an area are compared. It doesn't matter what the proposals are, so we shall not specify them. The monetary benefits, those measurable in dollar terms, are plotted horizontally. Since they may be positive but more usually are negative (because atmospheric improvements cost money—i.e., resources or economic output), the zero point for economic benefits is in the middle of the diagram. The nonmonetary benefits, in this case days of illness avoided, are plotted vertically. Proposal O is the status quo; it neither costs anything nor reduces illness, but serves as a basis for comparison. Proposal I is clearly

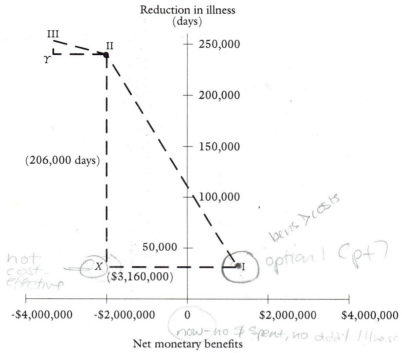

FIGURE 2. Trade-off Diagram between Monetary Costs and Reduction in Illness.

superior to Proposal O. It achieves some reduction in illness-days and, as a bonus, actually yields economic benefits in excess of its costs.

The other two proposals both have negative net economic benefits, that is, they entail economic sacrifices. But, to compensate, they lead to significant reductions in days of illness.

Such a diagram displays the major consequences of available alternatives without evaluating them in better-or-worse terms. Instead, it makes vivid the terms on which one kind of benefit or cost can be traded for the other; hence its name. The dashed lines in the diagram illustrate the comparisons among Proposals I, II, and III. In comparing Proposals I and II, the length of the line *II-X* (206,000 days) indicates the additional days of illness saved by adopting Proposal II rather than Proposal I, and the length of *I-X* ($3,160,000) represents the extra cost of Proposal II. Therefore the ratio of lengths, *I-X/II-X*, represents the cost per day of illness saved ($15.30) when the reduction is achieved by adopting Proposal II instead of Proposal I. The diagram doesn't tell whether reducing illness by this means is a bargain or exorbitant. That depends on the value that the community places on a day's reduction in illness, which is precisely the number that is virtually impossible to ascertain. Similarly, the diagram shows that Proposal III saves 12,500 more days of illness than Proposal II, at an additional cost of $1,333,000. Favoring Proposal III over Proposal II implies willingness to spend about $107 to avoid one day of illness in the community. Again the diagram says nothing about whether the additional cost is worthwhile.

Ultimately the choice among such proposals is a political decision dependent on social values, rather than one that is essentially dictated by a compelling measure of efficiency. A trade-off diagram can help make the decision an informed one.

These two devices (and there are others) show that even when definite prices cannot be assigned to nonmarket consequences, rational analysis and comparison of environmental programs is possible. A third device, called "cost-effectiveness analysis" (C-EA), is probably the most effective way to avoid resorting to unpersuasive estimates of the monetary value of nonmarket benefits.

COST-EFFECTIVENESS ANALYSIS

A reasonable question to ask of any proposed project is: Is it the cheapest way to do the job? The task of cost-effectiveness analysis is to answer that question. More formally stated, a project or a program is said to be "cost-effective" if it attains some specified physical or social goals and has net monetary benefits at least as great as those of any other project that attains them. The application of this test is cost-effectiveness analysis (C-EA). In terms of Figure 2, Project X is not cost-effective. It reduces illness the same amount as Project I but costs $3,160,000 more.[5] In general, a cost-effective project is necessarily on the production possibility frontier; if not, there would be some project that provided the same nonmonetary benefits and also either provided more benefits measurable in terms of money or cost less.

In operation, C-EA is much like B-CA. A reasonable range of alternatives that attain the designated goals have to be inspected, and for each the monetary values of benefits and costs have to be estimated, year by year, for all kinds of benefits and costs except the ones for which minimum levels have been prescribed. The advantage of C-EA is that it avoids the baffling need to find monetary equivalents for every type of benefit. But with that advantage comes a serious drawback, namely, the need to specify target levels for the critical kinds of benefit in advance. Usually there is not much justification for the target levels chosen, and, strictly speaking, the evaluation is valid for only those targets.

---

[5]That is just another way of saying that its nonmonetary benefits are that much lower.

In effect, one baffling task, specifying target levels, has been substituted for another, estimating monetary values of marginal changes in the levels of the benefits. It is often easier to specify reasonable target levels (e.g., the minimum level of dissolved oxygen in a stream) than to estimate the corresponding monetary value of the estimate (e.g., the social value of a marginal increase in dissolved oxygen). On the other hand, a C-EA generally provides less useful information than a B-CA unless there is some strong justification for the choice of the target level.

This limitation of C-EA can usually be relaxed by estimating the marginal social cost of changes in the prescribed targets. For example, in terms of Table 1, one might estimate the net monetary social cost of abating polluting emissions enough to reduce illness-days by 500,000 per year and also enough to reduce illness-days by 550,000. If the cost turns out to be very low, say $10 per illness-day, the more stringent abatement level would probably be judged worthwhile. Indeed, it might seem advisable to test the cost of still more stringent abatement goals until an abatement level is found at which a further reduction in illness-days is no longer an obvious bargain. In this way, practical judgments can be reached about abatement levels without expressing the social values of the results in terms of money. Of course, groping in this way to find a socially desirable critical level may prove very expensive.

VALUING NONMARKET BENEFITS

We have already noted that the most perplexing obstacle to the performance of B-CA is the need to find monetary values for nonmarket benefits, and that C-EA ameliorates this difficulty, though only partially. Most of the papers that follow in this section discuss different methods for valuing nonmarket benefits. In Chapter 20, Knetsch and Davis compare two methods, the "travel cost method" and contingent valuation. (Both will be discussed below.) Randall, Ives, and Eastman's paper on "bidding games," Chapter 21, is a classic and sober exposition of the contingent valuation approach. Chapter 22, by Landefeld and Seskin, summarizes the results of a large and varied set of efforts to place a value on measures that avert premature deaths; and Schelling's Chapter 23 discusses some of the philosophic issues that such efforts raise. In Chapter 25, Krupnick and Portney illustrate how the resulting estimates of values of nonmarket benefits are applied to appraising and evaluating a major, and very ambitious, environmental protection program. In Chapter 19, Arrow treats the specialized but important question of choosing the social rate of discount to be used in evaluating deferred benefits and costs. Finally, Wilson and Crouch (Chapter 24) discuss the problem of allowing for risk and uncertainty in benefit-cost evaluation, as well as elsewhere.

The ingenuity and effort devoted to establishing monetary equivalents to the social values of nonmarket benefits and costs are impressive. We now face the ungrateful task of appraising the results of that effort, of asking how valid and useful the estimates derived by the various approaches are. As a preliminary, we have to be clear about what the estimates are intended to represent and how they are supposed to be used.

The best starting point is probably the notion of consumers' sovereignty, which serves so well as a basis for estimating the social value of private goods. The skeleton of the argument on which consumers' sovereignty rests is: (1) the social value of any good lies in its contribution to the welfares of individual citizens, (2) every citizen (aside from children and a few other exceptions) is the authoritative judge of how much any good contributes to her welfare, (3) for private goods this judgment is conveyed by the amount the citizen is willing to pay for a marginal unit of the good or its services, (4) firms in competitive markets are responsive to the amounts citizens are willing to pay for private goods, and (5) the firms' responses result in an efficient allocation of production, i.e., one such that it is impossible to satisfy any citizen more without reducing some other citizen's level of welfare.

Now, if this doctrine does so well for private goods, shouldn't it be applied to public goods and the services of common property resources also? The major obstacle seems to be to discover the data conveyed in step 3, namely, how much citizens are willing to pay for a marginal unit of each public good or common property service. This is precisely the datum needed to fill the gap in a B-CA. We must therefore ask of each method for obtaining monetary expressions for the social value of public goods how well its results are likely to represent a public goods analog to the competitive market price of a private good.

REVEALED PREFERENCE METHODS

The revealed preference methods are a collection of more or less ad hoc devices connected only by the fact that they all infer consumers' valuations of environmental improvements from their choices in markets affected by them. We shall illustrate this diverse collection by discussing three frequently used methods.

"VALUE OF A LIFE." Since time immemorial, public health, as reflected in death rates and illness rates, has been a major concern of governments in general, and of environmental protection in particular. All such programs are expensive and, since budgets are always limited, judgments have to be made about which programs justify their draft on the budget. For this, as well as for other purposes, economists have searched assiduously for indications of the amounts that people are willing to pay to reduce the frequency of premature deaths and illnesses. The estimates cited by Landefeld and Seskin in Chapter 22 attest to how wide and thorough this search has been.

The "labor market" method for estimating this willingness-to-pay is very likely the prevalent one. It rests on Adam Smith's assertion that in equilibrium the net balance of pecuniary and nonpecuniary advantages and disadvantages must be the same in all occupations. On this ground, then, other things being equal, the more dangerous an occupation, the greater should be its wage. More exactly, we should be able to estimate how highly workers esteem their personal safety by analyzing how much they have to be paid to be willing to engage in occupations with different degrees of hazard. The principal information required for making such estimates are data on wage rates in several occupations with differing levels of occupational hazard, and the probabilities of fatal accidents (or illnesses) in those industries. As can be seen in Landefeld and Seskin's Table 2 (p 383), the results of those studies vary widely.[6] A more recent, and in some respects more sophisticated, estimate by Moore and Viscusi (1988) found that workers were paid an average of $3,400 per year (in terms of 1977 dollars) for an increase of 1/1,000 in the probability of suffering a fatal accident during a year.[7]

This very ingenious method for inferring the monetary value that people place on risks of death shares with the other methods to be discussed below a heavy reliance on some strong, and perhaps inapplicable, assumptions of economic theory. It assumes that the labor market is a perfect market, in which the workers are well informed about the risks and rewards of different occupations and in which any qualified worker can choose without impediment the job that suits her best. It assumes that there are no frictions, that when the wages or risks of an occupation change workers can quickly and at small expense enter or leave it. And it presumes that they have accurate psychological perceptions of very small risks, of the order of 1 in 10,000.

There is also a technical statistical problem, called the "specification problem," in this and the other methods to be discussed. Clearly riskiness is not the only characteristic that accounts for differences in wage rates

[6]Each study has its own technical peculiarities. Some relate to the risk of fatal injury only, others relate to all occupational accidents; only Thaler and Rosen include deaths from occupational diseases, and that inaccurately. In some studies, the workers covered are classified by industry, in others by occupation. The different studies also are based on samples from somewhat different working populations. The investigators should not be blamed for these discrepancies; they all had to rely on data gathered by government agencies and insurance companies for administrative rather than research purposes.
[7]This is equivalent to a "value of life" of $3.4 million. It should be noted that all these "value of life" estimates are really based on estimates of compensation paid for small increases in small probabilities of death.

among occupations. The effects of other factors, such as skill level, regularity of employment, pleasantness or unpleasantness of the tasks, section of the country, and many more have to be allowed for by statistical manipulation in order to isolate the effect of riskiness. The different studies accomplish this differently, depending on the data that were available and the judgment of the investigator.

The problem of specifying the relationships assumed in statistical analyses afflicts all empirical research in the social sciences. The best way to handle it is to try a number of plausible specifications in search of the one that fits the data best and to test whether the conclusions of the study are "robust," i.e., essentially the same under all plausible specifications. In each case the author has to choose the most satisfactory assumptions that her data permit, and the reader has to rely on her own judgment to decide whether to be satisfied with the specification chosen.[8]

VALUING IMPROVED URBAN ENVIRONMENT. A frequent problem in evaluating environmental programs is estimating how much the public is willing to pay for the improvement in environmental conditions that such programs achieve. One strategy is to analyze the results of the improvement into components such as reduced mortality rates, increased useful lives of structures exposed to the weather, reduced frequency of smog episodes, etc., and to estimate the public's willingness to pay for each of these. An alternative strategy, the one we shall explain here, is to cut through all those details and try to estimate directly the value that the public attaches to the improved environmental conditions.

The best indicator we have of the value the public places on environmental conditions is the observable relationship between them and property values. Table 2, which is taken from an important paper on methods for valuing nonmarket benefits,[9] gives some idea of what is involved in detecting and measuring the effect of environmental conditions on home values and rentals. There are two columns, each of which presents the coefficients of fourteen variables in a linear equation that describes the effects of those variables on the selling price of homes.[10] Thirteen of the variables are "nuisance variables," whose effects on the value of the house have to be eliminated in order for the effect of the environmental variables—the atmospheric concentration of nitrogen dioxide on the left and of suspended particulates on the right—to stand out clearly. In somewhat less elliptical notation, the equation for nitrogen dioxide concentration says:

[8]If you've ever wondered why so much of social science is controversial, here is a large part of the answer.
[9]See D. S. Brookshire, M. A. Thayer, W. D. Schulze, and R. C. d'Arge (1982).
[10]These equations are "least squares fits" to data recorded for the sales of 634 single-family dwellings in the Los Angeles metropolitan area in 1977–78.

TABLE 2.   Estimated Hedonic Price Equations. Dependent Variable = log (Home Sale Price).

INDEPENDENT VARIABLE	NO_2 EQUATION	TSP EQUATION
Housing Structure Variables		
Sale date	.018591	.018654
Age	−.018171	−.021411
Living area	.00017568	.00017507
Bathrooms	.15602	.15703
Pool	.058063	.058397
Fireplaces	.099577	.099927
Neighborhood Variables		
Log (Crime)	−.08381	−.10401
School quality	.0019826	.001771
Ethnic composition	.027031	.043472
Housing density	−.000066926	−.000067613
Public safety expenditures	.00026192	.00026143
Accessibility Variables		
Distance to beach	−.011586	−.011612
Distance to employment	−.28514	−.26232
Air Pollution Variables		
log (TSP)		−.22183
log (NO_2)	−.22407	
Constant	5.4566	4.0527
R_2	.89	.89
Degrees of Freedom	619	619

Source: Adapted from Brookshire, Thayer, Schulze, and d'Arge (1982), with the kind permission of the authors.

Log(selling price) = 5.4566 + .01859*(Sale date, months since Dec. 1976) − .01817*(Age in yrs. at date of sale) + . . . − .22407*log (average annual concentration of NO_2 at nearest air-monitoring station, parts per billion).

The only number of interest in all of this is the coefficient −.22407, which asserts that on the average the value of a house fell by 2/9 percent for every 1 percent increase in the concentration of NO_2 in its neighborhood. Nevertheless, the data on all fifteen variables had to be compiled, and the coefficients estimated.

Before moving on to interpretation, it should be noted that the two equations shown are virtually the same, coefficient by coefficient, with only one or two exceptions. This is no coincidence; it indicates that the two pollution variables, NO_2 concentration and TSP concentration, are indistinguishable statistically, a frequent condition technically called "colinearity." It occurs whenever two or more variables tend to vary together, as the two pollution variables do in this instance; the NO_2 concentration tends to be high in areas where the TSP concentration is high, and low where it is low. Thus the two equations measure the same thing: the effect of changes in air pollution in general on housing prices.

But, of course, these studies are not undertaken out of interest in the determination of housing prices, but rather as a step toward evaluating people's willingness to pay for decreases in atmospheric pollution. At first glance, it might appear that all that is necessary after determining how much an average family is willing to pay for a house with improved atmosphere is to multiply by the number of houses. Not so. In the first place, improving a city's atmosphere affects its shopping districts, workplaces, recreation areas, etc., in addition to its residential areas. Cleaning up the air in the central business district may increase housing values throughout the city, but it will not be reflected adequately in the values of houses exposed to different levels of pollutant concentrations. The sum of the induced increases in housing values therefore omits these components of the social value of a reduction in urban air pollution.

But there is a more subtle effect, which is likely to work in the other direction. Consider a city which, like most, has its more and its less polluted neighborhoods, and suppose that the houses in the less polluted neighborhoods are more expensive for that reason and perhaps others. If an environmental regulation were to reduce pollution chiefly in the most polluted residential areas, house prices would rise there, but house prices in the less polluted areas might fall because the price differential would be likely to fall if the pollution differential did. The total value of real estate in the city might remain the same, or even decline. What, then, could be said about the social value of the atmospheric improvement?[11]

The step from the induced change in house values to the monetary measure of the change in social welfare is thus difficult. But the formula for the induced change in house values is itself suspect. Just as with the formulas for reductions in the probability of fatal accidents, it depends on stringent and dubious economic assumptions. The house buyers must be well informed about the degree of atmospheric pollution near the houses they consider and its effects on health and maintenance costs. The costs and other obstacles in the way of changing houses when prices or environmental conditions change must be moderate. All householders must have the same utility or preference function. The equation describing consumers' preferences must be correctly specified and the variables in it must be measured accurately. There must be a single housing market, so that all prospective buyers can choose among all available houses, and it must

[11]For the answer, see Strotz (1968) or Lind (1973).

behave at least approximately like a competitive market. And so on. These strict requirements led K-G. Mäler (1977, p. 368) to conclude: "Together these difficulties show conclusively that there is no real possibility of estimating willingness to pay for environmental quality from property value studies."

THE TRAVEL COST METHOD. The preceding two examples both depended on using data about a market assumed to be in equilibrium to infer consumers' preferences concerning some nonmarket goods. The travel cost method is based on an entirely different principle. Its basic idea was discovered in 1844 by E. J. Dupuit, a French engineer who used it to estimate the economical amount to spend on proposed bridges. Dupuit reasoned that the maximum economical expenditure for any bridge is just the greatest amount that users of the bridge would be willing to pay to have it in place, and that this willingness-to-pay is equal graphically to the area under the demand curve for the use of the bridge up to the abscissa for the amount of traffic expected. Tersely stated, this area, usually called the "consumers' surplus," is the amount that consumers would be willing to pay for the use of the bridge (or other facility) in excess of the amount they are required to pay. [12] The social value of a public service or facility of any sort is just the consumers' surplus that it generates. To estimate it, all that is needed is to estimate the demand curve and compute the area under it.

Unfortunately there are no markets for nonmarket goods, so estimating the demand curve for one is not altogether straightforward or even possible in most cases. But it is possible if two conditions are satisfied: (1) though there is no explicit market, some inconvenience or expense is needed in order to acquire the good; and (2) the amount of inconvenience or expense is observable and different for different people. Use of national and state parks and other scenic sites and recreational areas satisfies these conditions, and the travel cost method is used frequently to estimate the social values of such facilities. [13]

The first step in estimating the social value of a scenic or recreation site by the travel cost method is to derive the demand curve for visits from data on the numbers of visits from several points of origin at different distances from the site. Though this task can be complicated in practice, it can be explained adequately in terms of the simple situation where several communities use a single, isolated recreation site.

Two special assumptions are needed. The first is the usual one that consumers behave reasonably rationally. By this assumption, each consumer will use the recreation site to the extent where an additional use

[12]Consumers' surplus is discussed in virtually all texts on price theory. The explanations in Dorfman (1978), pp. 134–36, and Mansfield (1988), pp. 99–101, are probably as good as any.
[13]For a practical example, see Knetsch and Davis's Chapter 20.

would cost her more than the experience would be worth. The second assumption is more special. It is that the residents of the cities and towns that use the site have similar enough tastes so that their responses to differences in the cost of using the site can be interpreted as responses along a common demand curve.

The key data for estimating the demand curve for the use of the site by use of these assumptions are the proportion of the population of each community served by the site that actually visits it. These are found from a sample survey of visitors to the site, in which the visitors are asked their point of origin, the frequency with which they visit the site, and any desired auxiliary information. At this point, a little notation will be helpful. Let V be the total number of visitors to the site in a season, v the number included in the sample, v_i the number of visitors from community i in the sample, and P_i the population of community i. Then $(V/v)v_i$ is an estimate of the total number of visitors from community i, and $(V/v)v_i/P_i$ is the proportion of the population of the community that uses the site. U_i will denote this proportion.

The rationality assumption implies that the people in community i who visit the site will be those, and only those, for whom the value of a visit is as great as the cost. It is thus reasonable to assume that the value of a visit to the least eager visitor from any community, i.e., marginal value of visits from that community, is approximately equal to the cost of those visits.

The cost of reaching and using the site from each community that uses it can be estimated by multiplying the number of hours of travel needed (round trip) by a unit-time cost equivalent to the opportunity cost per time-unit, and adding any entrance fee, other expenses at the site, plus mileage costs and out-of-pocket travel expenses. Call the cost TC_i. Then, for each community that uses the site, the coordinate pair (U_i, TC_i) shows the relation between the cost of using the site to people in that community and the proportion of the population who do so. These coordinate pairs form the desired demand curve. One is illustrated in Figure 3. By Dupuit's argument, the shaded area represents the consumers' surplus per 1,000 population for a community from which the average cost of a visit to Yosemite is $40.

Clearly many essential complications were ignored in this exposition, including the effects of competing recreation sites; the size, recreation facilities, and other characteristics of the site studied; complementary attractions and pleasures at and en route to the site; and the size of families, incomes, and other characteristics of the communities served. All these factors shift the demand curve. In principle, they can be allowed for by introducing nuisance variables, much as in the previous two examples, to the extent that the available data permit. In practice, estimating a demand curve from travel cost data is a delicate procedure, sensitive to specification error. Introducing complications like the foregoing would

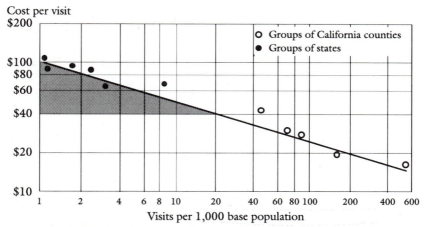

FIGURE 3. Estimated Demand Curve for Visits to Yosemite National Park, 1953. Notice the log-log scales. (Source: From Clawson and Knetsch (1966), p. 73.)

magnify the complexity of the calculations, vastly increase the requirements for data, and call for additional strong economic assumptions. In short, though the travel cost method is exceedingly ingenious, it is adapted to only very simple situations, and becomes awkward and unreliable when necessary complications are introduced.

CONTINGENT VALUATION METHODS

Since, as mentioned above, contingent valuation methods are controversial, an extensive literature is devoted to evaluating them. [14] The criticisms of contingent valuation all stem from the simple fact that a hypothetical question is not a real choice. In the literature, six types of bias that can afflict contingent valuation surveys have been distinguished and identified.

STRATEGIC BIAS occurs when a respondent chooses her answers in the hope of influencing the results of the survey. It is most likely to occur when respondents believe that the results of the survey will influence a pending political decision or have some other practical effect. Its presence can be detected by giving identical questionnaires to several samples, some of which are informed that a decision in which they are interested depended on the results. That kind of information had little or no effect on answers

[14]Probably the best general reviews are R. G. Cummings, D. S. Brookshire, and W. D. Schulze, et al. (1986), and R. C. Mitchell and R. T. Carson (1989).

to the questionnaire in several experimental tests. It appears that most people don't bother to falsify information in a survey for strategic reasons.

HYPOTHETICAL BIAS is the obverse of strategic bias. It occurs when respondents believe that their answers will not have any significant effects. It has several aspects, including: (1) people may be more generous when expressing willingness to make payments than in committing themselves actually to do so; (2) people may give offhand and ill-considered responses when they feel that nothing is at stake; and (3) especially when dealing with environmental questions, people may not visualize alternatives in a hypothetical situation as vividly as they do in a real one.

Experimental tests have revealed the reality of this source of bias. For example, when a sample of hunters was asked how much they would pay for a certain type of hunting license and a matched sample was allowed to bid in an auction of identical licenses, the hypothetical bids ranged from 30 percent to 65 percent higher than the actual ones. Similar experiments have disclosed the other types of hypothetical bias, but in some tests no biases were detected. Of course, hypothetical bias does not have to be detected on every occasion in order to be considered a dangerous pitfall when contingent valuation methods are used.

INFORMATION BIAS results when respondents are asked about unfamiliar options or contingencies. There is little point in asking ordinary people how much they would be willing to pay to have the concentration of sulphur oxides in their neighborhood reduced from 0.06 ppm to 0.03. On the other hand, it would strain the patience of most respondents if they were required to listen to an explanation of the meaning and effects of such a change. In consequence, the options that are of interest in a contingent valuation survey are likely to be vaguely delineated and even more vaguely understood. [15]

Quite apart from being ill-informed, respondents are confronted with an unfamiliar task when they are asked how much they would be willing to contribute to obtain or prevent a change in environmental conditions. Most people have never faced that choice in earnest, and it is hard for them to conjecture how they would respond if they should encounter it in earnest. Schelling, in Chapter 23, points out how difficult it is to induce a respondent to examine her own preferences for small changes in the risk of so serious an event as death.

VEHICLE BIAS refers to the effect of the mode of payment when respondents are asked "willingness-to-pay" types of questions. Respondents are likely to give different answers depending on whether they are told that the

---

[15]For examples, just think of the questions asked by Gallup or Roper or any other public opinion polling organization.

payment will be added to their tax bill, or deducted from their paychecks, or whatever. Thus the answers to such questions do not reflect the perceived value of the project in pure form, but rather an admixture of this value with the respondent's feelings about the mode of payment.

POPULATION SURVEY BIASES. In addition to those special problems of contingent valuation surveys, they are subject to the problems of public opinion surveys in general. For example: people like to make good impressions, even on interviewers.[16] Even when they are anonymous, they do not like to reveal themselves to be indifferent to nature, or stingy, or whatnot. People also prefer to be agreeable. Other things being almost equal, we'd rather say yes than no, and we are likely to at least shade our feelings if talking to a member of a minority or a female interviewer. The framing and form of a question is likely to affect the answer nearly as much as the substance. All in all, interpreting a human being's responses is inescapably tricky.

Finally, there is **NONRESPONSE BIAS.** In mail surveys, often fewer than half the questionnaires are returned with usable answers. There is always some nonresponse, and depending on its causes and extent, the sample of usable answers will depart more or less strongly from being truly representative of the population studied. There are correctives, such as following up on nonrespondents, but they are expensive and imperfect.

* * *

So much for weaknesses of contingent valuation surveys. They have two salient virtues. First, they are immune to most of the defects of the alternative approaches, which have been described. Second, they capture some aspects of social value that elude other methods, in particular the "non-use values." Non-use values, originally pointed out by Krutilla (Chapter 12) and Weisbrod (1964), are the values that people place on environmental or cultural amenities or conditions that they do not experience personally or expect to experience. If Niagara Falls were threatened with destruction, wouldn't you contribute at least a little to a fund that could save it for other people and future generations to see and marvel at?[17]

[16]Including the people who code the questionnaires in a mail survey.
[17]Before leaving contingent valuation, we should mention the continuing discussion of willingness-to-pay (WTP) vs. willingness-to-accept (WTA), since a contingent valuation survey has to choose between them in framing its questions. WTP is code for the most that the respondent would be willing to pay to obtain some environmental improvement. It corresponds to Hicks's concept of the compensating variation in income. WTA is the smallest recompense for which the respondent would be willing to forego an environmental improvement or consent to a degradation. According to consumption theory, WTP and WTA should be about equal when they are small in proportion to income; in contingent valuation surveys, WTA invariably turns out to be greater than WTP, and often four to six times as great.
  How to explain? Kahneman and Tversky (1979) confirmed long ago that marginal utility curves have a discontinuity at zero: a marginal increase in wealth has less psychological

The upshot of these considerations appears to be that while contingent valuation has some enthusiastic advocates, the bulk of the profession is holding its judgment in abeyance and evaluating projects and programs by the standard approach.

TWO UNWELCOME COMPLICATIONS

We have now reviewed the essentials of benefit-cost analysis as it is practiced, but have to mention two complications that ought to be incorporated in practice. One is allowance for considerations of equity and political feasibility in the distribution of the benefits and costs, and the other is allowance for the uncertainty and inaccuracy of estimates of benefits and costs and their components.

The need to consider equity and political feasibility arises from the circumstance that the benefits and costs of environmental improvement often accrue unequally to different segments of the population, with most of the benefits going to some segments of the population, and other segments getting most of the costs. There is no need to expand on the resentments, tensions, and feelings of injustice that are likely to result.

In 1936, when benefit-cost analysis was first formalized in the United States, the instruction from the Congress was that projects should be undertaken only "if the benefits to whomsoever they may accrue are in excess of the estimated costs." This standard, though nominally adhered to this very day, was early seen to be grossly inadequate, and has been circumvented habitually from the very outset. In politics, it won't do to be too candid about matters of distribution. Thus the admonitions of the Water Resources Council, which officially sets the standards for federal benefit-cost analyses, are widely disregarded. But we repeat them here because they are sound and because they have to be obeyed, albeit surreptitiously.

In brief, the Water Resources Council recommends that the benefit and costs in a benefit-cost analysis be disaggregated to show their incidence to meaningful population segments (not necessarily congressional districts, though that might be popular). It appears to be expedient, as well as conformable to prevalent standards of justice, to design projects to be

---

impact than a decrease of the same amount. I've noticed that myself. My personal explanation is that people form ego attachments to their possessions and entitlements, and psychologically regard and resist decreases as invasions of their established rights. On this ground, the observed discrepancies between WTP and WTA are not errors introduced by contingent valuation, but reflections of discontinuities in people's preferences that public decisions should take into account.

Pareto efficient insofar as possible, i.e., so that all interested population groups receive benefits that exceed their shares of the costs. It is, however, rarely possible to meet this standard completely.

The role of a benefit-cost analysis is then to exhibit the balance of benefits and costs for each significant population segment. As everyone recognizes, these, and not the grand totals, are the significant data for political decision making, and the political process will proceed more smoothly and effectively to the extent that these facts of life are recognized and agreed on by all concerned. This is good advice, but honored mostly deviously, to great social cost.

Our review of benefit-cost analysis has emphasized that the estimates on both sides of the account are uncertain and inaccurate. When estimates are checked by being reestimated by independent methods or by being audited in the light of history, it is entirely usual for discrepancies of 200 percent or 300 percent or even an order of magnitude to be disclosed. Such is the nature of the beast. Therefore, (1) the third "significant" figure should never be taken seriously and the second should generally be regarded with some skepticism, and (2) the report on a benefit-cost analysis should indicate the ranges of uncertainty of the principal estimates. Chapter 25, Krupnick and Portney's benefit-cost analysis of urban air pollution control programs, illustrates conscientious reporting of the uncertainties inherent in benefit-cost assessments. Another example of good reporting is A. M. Freeman's report (1982) on the costs and benefits of air and water pollution controls, from which Table 3 is taken. Notice that the range of uncertainty is an order of magnitude in several instances. Nevertheless, Freeman was able to reach interesting and significant conclusions. The important thing in writing a benefit-cost report is to avoid seeming to tell the reader more than you can possibly know, and in reading one, to recognize the inherent imprecision even when the writer doesn't remind you.

CONCLUSION

You must be aware by now that this essay is not intended to either puff or belittle benefit-cost analysis. Its goal is to portray it as it is, warts and all. There are plenty of warts. There are also plenty of strengths, enough to make benefit-cost analysis a necessary tool for any government that tries seriously to make effective use of the resources it requisitions from its community.

Benefit-cost analysis has an especially important role in the decision-making processes of a democratic community. To a large extent, the preparation of a benefit-cost analysis performs the staff work for all sides

TABLE 3. Air Pollution Control Benefits Being Enjoyed in 1978
(in billions of 1978 dollars)

CATEGORY	REALIZED BENEFITS	
	RANGE	MOST REASONABLE POINT ESTIMATE
1. *Health*		
Stationary Source		
Mortality	$2.8–27.8	$13.9
Morbidity	$0.3–12.4	$ 3.1
Total	$3.1–40.2	$17.0
Mobile Source	$0.0– 0.4	$ 0.0
Total Health	$3.1–40.6	$17.0
2. *Soiling and Cleaning*	$1.0– 6.0	$ 3.0
3. *Vegetation*		
Stationary Source	0	0
Mobile Source	$0.1– 0.4	$ 0.3
Total Vegetation	$0.1– 0.4	$ 0.3
4. *Materials*		
Stationary Source	$0.4– 1.1	$ 0.7
Mobile Source	$0.0– 0.3	$ 0.0
Total Materials	$0.4– 1.4	$ 0.7
5. *Property Values*		
Stationary Source	$0.9– 6.9	$ 2.3
Mobile	$0.0– 2.0	$ 0.0
Total Property Value	$0.9– 8.9	$ 2.3

Source: A. M. Freeman III (1982), p. 128.

of the inevitable debate about a proposal for environmental protection or improvement. It gathers the essential data and sets forth reasonable economic, demographic, and technical assumptions, which often serve as ground rules that help make the ensuing debate coherent and intelligible. At the very least, the analysis will rule out some of the blatant misrepresentations that frequently mar political discourse. Though the analysis may be contested, the very acts of contesting and defending it focus the attention of all disputants on relevant and comprehensible issues. In short, the sources and foci of people's disagreements are exposed.

A spectacular example of this function of benefit-cost analysis concerns a proposal to build a canal across central Florida to provide a shortcut barge route from Texas and Louisiana to the Atlantic Coast. The plan envisaged dredging and "improving" almost 50 nearly pristine miles of the Oklawaha River, probably the largest stretch of wild river through a tropical rain forest in the United States. A protracted brouhaha naturally resulted. Finally, a benefit-cost analysis was undertaken. It was carefully disaggregated according to interest groups that would be affected, such as shippers who might use the canal, hunters, fishers, environmentalists who prized the wilderness, lumbermen, and so forth. The analysis showed that scarcely any significant group would benefit greatly from the canal, and most would be net losers. After that finding, the proposal was dropped without further debate, although construction had already begun.[18]

Counterbalancing such pleasant triumphs is the notorious fact that benefit-cost analyses can be slanted, and often are. The technique is treacherously simple. If in favor of a proposal, overvalue the benefits and underestimate the costs; if opposed, do the opposite. The discussions in this paper of how benefit-cost analyses are conducted and how benefits and costs are valued should be helpful both in distorting analyses and in detecting the biases. It is the unfortunate case that truly neutral analyses are hard to find, though the cross-Florida canal episode may have been one. When dealing with benefit-cost analyses, one should never forget the maxim, *Caveat lector.*

The limitations of benefit-cost analysis have been emphasized sufficiently by now to dispel hope that it can provide decisive guidance. There is one fundamental limitation, however, that probably ought to be made more explicit. The goals and considerations that enter into any real public decision are varied, and subtle, and often left discreetly unstated. So there are almost invariably some goals and some restrictions that cannot be fitted into the benefit-cost format, but cannot be ignored either. In the end, therefore, benefit-cost analysis can be an important ingredient of the decision process, but the final decisions elude the benefit-cost accountants.

REFERENCES

Böhm-Bawerk, Eugen von (1889), *Kapital und Kapitalzins,* Band 2. Translated as *Capital and Interest,* Vol. 2, by G. D. Huncke and H. F. Sennholz (1959).

Brookshire, D. S., M. Thayer, R. Schulze, and R. d'Arge (1982), "Valuation of public goods," *Amer. Econ. Rev.,* 72:165–77.

Carter, Luther J. (1975), *The Florida Experience: Land and Water Policy in a Growth State.* Baltimore: Johns Hopkins University Press for Resources for the Future.

Clawson, Marion, and Jack L. Knetsch (1966), *Economics of Outdoor Recreation.* Baltimore: Johns Hopkins University Press for Resources for the Future.

---

[18]Luther Carter (1975, Chap. 9) is a vivid narrative of the first dozen years of this controversy. Unfortunately, it was written before the denouement sketched in the text was reached.

Cummings, R. G., D. S. Brookshire, W. D. Schulze, et al. (1986), *Valuing Environmental Goods: An Assessment of the Contingent Valuation Method.* Totowa, NJ: Rowman & Allenheld.

Dorfman, Robert (1978), *Prices and Markets,* 3rd edn. Englewood Cliffs, NJ: Prentice-Hall.

Dupuit, E. J. (1844), "De la mésure de l'utilité des travaux publics," *Annales des ponts et chausées, 2me ser.,* vol. VIII.

Freeman, A. Myrick III (1982), *Air and Water Pollution Control: A Benefit-Cost Assessment.* New York: John Wiley.

Kahneman, Daniel, and Amos Tversky (1979), "Prospect theory: An analysis of decision under risk," *Econometrica,* 47:263–91.

Lind, Robert C. (1973), "Spatial equilibrium, the theory of rents, and the measurement of benefits from public programs," *Quart. J. Econ.,* 87:188–207.

Mäler, Karl-Göran (1977), "A note on the use of property values in estimating marginal willingness to pay for environmental quality," *J. of Environmental Econ. and Mgt.,* 4:355–69.

Mansfield, Edwin (1988), *Microeconomics, Theory and Applications,* 6th edn. New York: W. W. Norton.

Mitchell, R. C., and R. T. Carson (1989), *Using Surveys to Value Public Goods.* Washington, DC: Resources for the Future.

Moore, M. J., and W. K. Viscusi (1988). "The quantity-adjusted value of life," *Economic Inquiry,* 26:369–88.

Strotz, Robert H. (1968), "The use of land rent changes to measure the welfare benefits of land improvement," in Joseph E. Haring, ed., *The New Economics of Regulated Industries.* Los Angeles: Occidental College.

Weisbrod, Burton A. (1964), "Collective consumption services of individual consumption goods," *Quart. J. Econ.,* 77:471–77.

19

Criteria for Social Investment

KENNETH J. ARROW

Kenneth J. Arrow is Joan Kenney Professor of Economics at Stanford University and a Nobel Laureate in Economics.

INTRODUCTION

The following paper was originally prepared as an exposition of the general principles of social investment. It is clear, however, that it has been most strongly influenced by the United States' experience in investment in water resources. The three major problems treated are the discounting of future benefits, the measurement of benefits, and the measurement of costs; the last category, however, is much less complex than the first two. As will be seen, the problems of discounting relate to the general economy rather than to the particular project at hand; thus the discussion is as applicable to water investments as to any other investments. The different problems in the measurement of benefits are important for different classes of water projects. For example, the divergence between social and private costs and the problem of appropriability is liable to be most acute for water purification. On the other hand, issues relating to economies of scale and the difficulties that they raise for the evaluation of benefits are likely to be much more important in irrigation projects. The importance of consumption (as opposed to production) benefits is less in water resource problems than in health or education, but it is illustrated by recreation. Thus, all the issues raised in this paper have applicability to the water resources area.

"Criteria for Social Investment," by Kenneth J. Arrow, from *Water Resources Research*, vol. 1, no. 1, pp. 1–8, 1965. Copyright by American Geophysical Union. Reprinted by permission.

323

REMARKS ON INVESTMENT IN GENERAL

Investment is the allocation of current resources, which have alternative productive uses, to an activity whose benefits will accrue over the future. The benefits take the form of production of goods and services.

The cost of an investment is the benefit that could have been derived by using the resources in some other activity.

An investment, then, is justified if the benefits anticipated are greater than the costs. This, of course, is an optimality condition for any productive activity. It takes particular forms for investment activities and, more specifically, for the special class of activities referred to as social investment.

The central problem in the evaluation of investments in general is commensuration over time. Benefits accrue at different times from each other and from the costs. To add up the benefits, we must establish rates of exchange between benefits at different times, weights to be assigned to the benefits before adding them together; the same procedure must be followed for costs.

One possibility, indeed, is simply to add benefits without regard to time period, i.e., to weight the benefits in all future time periods equally with each other and with the present. This practice is, however, unsatisfactory for two reasons: (1) time preference and (2) the opportunity cost of capital. (1) It can be taken as a datum that, from almost any point of view, present benefits are preferred to equal future benefits, especially if they are sufficiently removed in time. (2) The given investment must be compared with other investments also capable of yielding deferred benefits. If there exists an alternative investment capable of yielding a benefit of, say, 1.10 units of benefit a year hence for a present cost of 1 unit, then the given investment, to be justified, must be capable of yielding at least as much. This proposition is a straightforward statement of technical efficiency and is independent of any value judgments as to time preference. The most convenient way of expressing this demand is to define the present value of a future benefit as the current expenditure of resources which, if invested in the alternative manner, could yield the same benefit. If r is the rate of return on the alternative investment, then 1 unit of resources invested there could yield $1 + r$ units of benefit in 1 year, $(1 + r)^2$ units in 2 years (including reinvestment of the first year's return), and, in general, $(1 + r)^t$ units of benefits after t years. Hence, the present value of a benefit B_1 due in 1 year is

$$B_1 / (1 + r)$$

The present value of a benefit B_t to occur in the tth year ahead is

$$B_t/(1 + r)^t$$

Finally, the total present value of an investment which will yield benefits $B_1, \cdots, B_T$ in the first, $\cdots$, Tth year ahead, where T is the last year in which benefits are expected (the horizon), is

$$V = \sum_{t=1}^{T} B_t/(1 + r)^t.$$

The precise operational sense in which V is the present value of the given investment is now given. Suppose the amount V were invested today in the alternative investment. From the proceeds available at the end of 1 year, including recapture of some or all of the principal and additional income, withdraw benefits to the extent of B_1 and reinvest the remainder in the alternative investment. Repeat the process in each subsequent year, t. The result will be that precisely B_T can be withdrawn in year T with nothing else to reinvest. Thus, it can be seen that for an initial investment of the quantity of resources V in the alternative investment, it is possible to achieve the same benefits at each point of time as in the given investment. It follows that the given investment is justified (as against making an equal investment in the alternative) only if the cost C of the given investment is less than V. Thus, the condition that the present value of future benefits from a given investment, discounted by the rate of return on alternative investment opportunities, exceed cost is necessary for the efficient use of resources.

The efficiency or opportunity-cost interpretation of discounting is independent of any questions of time preference. In and of itself, however, it is only a partial solution to the determination of investment. If the present value of an investment, discounted at the rate of return of an alternative, falls short of the cost, it should certainly not be undertaken; if any investment at all is made, it should be in the alternative. On the other hand, if the alternative investment is in fact being made and if the present value of the given investment is greater than the cost, then the given investment should certainly be undertaken, at the expense of the alternative if necessary. But the opportunity-cost criterion is not an answer to the question of whether both the given and the alternative investment should be undertaken. The answer to this question depends on the aggregate volume of resources that will be devoted to investment as against current consumption and, hence, basically on the relative preferences for present and future benefits in relation to the rates of return on investments.

At an over-all optimum of the economy, therefore, the discount of future benefits according to opportunity costs must equal the discount according to time preference. However, in establishing criteria for a relatively limited body of investments, such as social investments will frequently be, the opportunity-cost criterion may be adequate if it can be presupposed that time preference has already been allowed to operate in the determination of the over-all volume of investment and, therefore, indirectly in the determination of the rate of return on alternative investments.

THE SPECIAL CATEGORY OF SOCIAL INVESTMENT

The majority of investments yield their benefits in the form of identifiable goods which can be marketed or withheld. These benefits are in a very natural way *appropriable,* in the sense that the organization producing them can without difficulty charge individual consumers for them, so that those who want and need the product can buy and others can refrain. The production of food and clothing provides, perhaps, the purest example of appropriable benefits. The future benefits from such an investment can be fairly measured by the output evaluated at the price at which it can all be sold less, of course, all current production costs (wages and materials).

But a wide and important class of investments yield benefits which, in their very act of production, inure to a wide class of individuals. They cannot be excluded from the benefit, and, hence, a price cannot be charged that will effectively discriminate between those who want the service and those who do not. Water purification provides a simple example; if it is decided to install equipment that will improve the purity of the water, all users will receive the benefits over the lifetime of the equipment, whether or not they would be willing to bear the cost in a free choice. (This choice is not only a matter of individual tastes for pure water; some of the uses of household water, such as gardening, have much lower purity requirements than others, so that some individual consumers may, in fact, derive very little additional benefit.) The price system is not operative, for it would require that each consumer be given the freedom to buy water at both the older and the newer levels of purity or, at the very least, be given his option between the two, with price differences reflecting cost differences. Water purification is really of the same order as the general run of collective services provided by government. In this context it is differentiated from the rest only in that there is an investment component; i.e., the benefits and costs do not accrue at the same point of time.

There are other instances in which pricing of benefits would be technically feasible, but for other reasons it is not regarded as performing an appropriate social function. Elementary, secondary, and, to a considerable extent, higher education have begun to belong to this category. The public schools could charge their pupils or their parents for the cost of education, but in the first place there may be a divergence of interest between the parents, who are capable of paying, and the children, who are receiving the benefit. This is part of a wider class of cases in which the beneficiaries are incapable of appreciating the benefit, either because of natural limitations of understanding (as in children or mental patients) or because the benefits would not really be understood until they have been experienced. The second reason, in the case of education, is that the benefits of education accrue not merely to the students but to the society of which they are a part.

In general, the line between appropriable and inappropriable cannot be drawn very sharply. There are very few acts, even of private consumption, which do not have some direct effect on the welfare of others. It is a matter partly of empirical and partly of value judgment as to when the external effects of benefits are sufficiently widespread to set aside the principle that the individual is the best judge of his own welfare.

There is another and very important reason rooted in the facts of technology for the treatment of wide-scale classes of benefits as inappropriable, even though it would be technically feasible to set prices, namely where there are increasing returns to the scale of operation. In that circumstance a collective agreement to undertake a productive enterprise and to share the costs in some way may benefit everyone, yet any ordinary pricing system would fail. For example, competition among electricity systems would certainly not ensure an optimal allocation of resources but instead would probably reduce the supply of electricity to small proportions. It is, to be sure, often possible, for example in irrigation to determine the benefits through a pricing calculation, but the supply must nevertheless be arranged through a monopoly; because of the dangers of monopoly in certain circumstances, the investment must actually be provided socially.

The remarks in this and the preceding section establish general outlines for the discussion to follow. In the next three sections we consider the following problems of evaluating the discounted benefits and costs: the measurement of benefits, the determination of the rate of discount, and the measurement of costs.

PROBLEMS IN THE MEASUREMENT OF BENEFITS

The benefits derived from social investment are by nature more difficult to measure than benefits from private investment. There is inevitably some failure in the extent to which the price system will be adequate. The price system, however, even in its ordinary form does have an important role in the estimation of benefits, and, in a more extended sense, there really is no benefit calculation possible that is not based on a set of at least hypothetical prices.

In our discussion we will distinguish between market prices and accounting or shadow prices. Market prices are prices actually charged for the benefits, leaving each consumer free to use or not use them and thereby incur or avoid the price. In investments lacking a social character, the highest market price that will lead consumers to buy the entire output is an adequate measure of the benefit per unit of product, and aggregate benefits are appropriately measured by the volume of aggregate sales less, of course, the aggregate of current operating costs. The process of allocation still involves the nontrivial element of forecasting and its inevitable counterpart, uncertainty. But it would take us too far afield to discuss these questions here. Instead, we wish to concentrate on those problems in the measurement of benefits that specifically differentiate private from social investments.

The basic reason for the difficulty of relying entirely on market prices is, as we have seen, the inappropriability of some benefits, the impossibility or, at any rate, difficulty of separating the creation of benefits for individuals from the act of production or, at least, the act of consumption by others. For these instances we must make a calculating equivalent of the price system, and to this equivalent the name of shadow or accounting price has variously been given.

Uses and Limits of Market Prices

For many purposes, and probably for more purposes than now realized, it would be feasible and useful to market the benefits from a project at a suitable price, as when the benefits are fully appropriable and the reason for the social character of the investment is rather the existence of economies of scale. In effect, the enterprise, though publicly operated, is still being required in the long run to pay its way. One may equivalently suppose that the investment is given a separate organization which borrows the initial capital and has to repay it with interest. The condition is that, with these costs added to current operating costs, the enterprise will still cover these total costs over the life of the investment.

The obvious advantage to such a system is the pressure for efficiency and responsibility, both in the initial act of investment and in the subsequent operation. At least after the event, there is a clear-cut determination of the profitability of the enterprise. This not only helps in supplying a record for the future but also imposes caution on the determination of the investment, since it is known that there will be such a check and in what terms the check will be. Finally, by making the benefits definable in fairly straightforward operational terms, it should improve the ability to forecast the profitability and, therefore, the desirability. To the extent that market prices are used in the distribution of benefits, we may speak of them as being recoverable. Recoverability is not an all-or-none proposition, of course. It is perfectly possible to charge a price to a direct beneficiary and still argue that the benefits that are covered are less than the total benefits because of indirect effects. For example, fees are frequently charged for public higher education, yet they are very far below the costs. This practice is presumably justified on the basis of indirect benefits.

Although the efficiencies gained by full, or even partial, recoverability of benefits are considerable, too much emphasis should not be placed on recoverability as a condition for investment. Well-known propositions of welfare economics tell us that the product of any investment should be offered at marginal cost up to the limits of its capacity. These marginal costs may or may not be adequate to recover the cost of the original investment. This point is of particular importance when there are widespread economies of scale, for then the marginal cost is almost sure to be below the long-run average cost, and recovery through market prices is necessarily less than total. It might be possible to achieve a greater recovery with a price above marginal cost (depending, of course, on the elasticity of demand), but then the investment, in effect, is not being used from the social point of view as well as it might be. By increasing the volume of recoverable benefits through higher prices, the aggregate benefits, recoverable and otherwise, have been reduced.

Divergence between Social and Private Benefits

A classic in economic theory is the case in which there is a direct beneficiary from whom the product can be withheld, but his act of consumption, or the act of production in order to achieve this consumption, yields benefits to other parties against whom no exclusion is possible. The water purification example cited above is an extreme case. To supply pure water to even one individual, it is necessary to supply it to everyone. Milder interactions are very common. Thus, treatment of infectious diseases is beneficial to the patient, but, in addition, the possibility of spread to other individuals is also reduced. Under these conditions it would be necessary to take into account the fact that the aggregate benefits are greater than the part that can be allocated privately with ease. Hence, to justify itself an investment in public health, including medical care of infectious diseases, need not expect to be fully made up from fees charged to patients.

Shadow Pricing

In the absence of market prices, it is necessary to impute value to the benefits. In the case of water purification, it might be asked what price, *if it could be charged,* would suffice to clear the market. More generally, the shadow price should be estimated for all beneficiaries, not only the primary ones. Computation replaces the market.

There is no difficulty with the concept of a shadow price; but there is the intensely practical problem of measuring it. Whereas market prices are operationally revealed in the market, shadow prices must be indirectly estimated, often by introspection on the part of a questionnaire-answering public or its governmental representatives. The difficulty of estimating a shadow price that represents the value of social, as opposed to private, benefits has led in practice to two opposite errors: (1) ignoring the additional benefits not representable by market prices and thus failing to make socially desirable investments, or (2) introducing nonprice and nonquantifiable justifications for projects which make difficult the rational weighing of alternatives (e.g., justifications such as the development of land as an end in itself or the provision of water as an absolute need).

Ultimately, a shadow price is a subjective valuation which must be made by individuals. In a democratic society it is perfectly proper that the valuation be made by the political process. It would, however, be a major improvement in the relevance of the discussion if the shadow price were the explicit subject.

Economies of Scale and Consumers' Surplus

Prices, shadow or market, strictly speaking are valuations only for small changes in quantities. Suppose, for example, it is contemplated to bring in a water supply for a desert region. The price in the absence of an aqueduct is prohibitive, but, once the aqueduct is installed, the marginal cost of water may be very small. If water is sold at its marginal cost, each individual will be better off than he would be if he had to pay the pre-aqueduct price; hence, there is a certain maximum lump sum payment (a fixed payment per year independent of the amount he consumes) which he could make and still be no worse off, according to his tastes and needs, than he was in the absence of the aqueduct. The benefits for the project are the aggregate of these hypothetical lump sum payments plus the benefits recovered from the marginal cost pricing less current operating costs. The measurement of these benefits for the purpose of making decisions about social investments is not necessarily tied to any attempt to recapture some or all of these benefits through taxes or a two-part price system.

The aggregate of lump sum payments is one of the many definitions of consumers' surplus. Although there are some conceptual difficulties that have been overlooked in this simplified exposition, the basic problem here, as with shadow prices, is one of measurement. All practical approximations are one form or another of the area under the demand curve as the price drops from its preproject to its postproject level. Considerations of this type are too well known to require further expansion here; it need only be remarked that they apply to shadow prices as well as to market prices. In each case the benefit per unit of output will lie between the preproject and postproject prices.

Production and Consumption Benefits

All benefits are, in the last analysis, benefits to individuals whom we may think of as consumers, but the relation may be direct or it may be indirect, through facilitating the production of goods desired by consumers. Most social investment activities yield benefits of both types. A highway increases the convenience of private automobile travel, a direct benefit to consumers; it also decreases the cost of trucking operations, which ultimately decreases the cost or increases the supply of consumers' goods.

Consumption benefits are those whose immediate beneficiaries are individuals in their capacities as consumers; production benefits are those whose immediate beneficiaries are economic units engaged in production for a market. The distinction has no significance from the point of view of determining the total benefits for evaluating a social investment project. The importance is rather that production benefits are far more easily measurable; in effect, the production unit imputes the market valuation of the final product back to the benefit yielded by the investment project. Thus, as a first approximation, the production benefit of a highway is the saving in cost on the volume of traffic originally carried. If the effect is large, the problem discussed in the preceding section arises, and some measure of surplus is needed; the cost saving will lead to a larger flow of traffic, and the benefit is measured instead by the cost saving on a volume of traffic intermediate between the original and final levels.

Hence, even if the production benefits are not recovered for one reason or another (in particular, under marginal-cost pricing of the benefits of a project with large economies of scale), they are fairly easily measurable. Any help in the measurement of benefits should be used. But there is a danger that the superior measurability of production benefits will lead to an underestimate or complete disregard of consumption benefits. A striking example has occurred in some discussions of the economics of public health. Benefits are measured in terms of the work-days saved, a production benefit, in complete disregard of the direct consumption benefit of prevention of or recovery from illness.

THE RATE OF DISCOUNT

The Measurement of the Opportunity Cost of Capital

As has been seen, it is a straightforward implication of efficiency that the sum of benefits, discounted at the rate of return on alternative investments, exceed the cost. But which of the many other investments in an economy is *the* alternative?

If the economy as a whole were operating at perfect efficiency before considering a possible new social investment, the rates of return on all alternative investments would be equal (at least at the margin, i.e. for small additions to investment). Further, each individual in such an economy has full access to the capital market, so that his division of current income between consumption and saving has been made on the assumption that the rate of return on saving equals the common rate of return on all investments. In such a situation there can be no doubt of the proper rate of discount; further, the mode of financing the social investment makes no difference either. If the resources are obtained by taxation, they are drawn either from consumption or from investment. A dollar drawn from private investment reduces aggregate output in future years by r dollars, where r is the rate of return; but an individual is, at the margin, indifferent between consuming and saving a dollar, which means that he is indifferent between consuming the dollar immediately and receiving a permanent income of r dollars per annum from it. Hence, the social investment must yield an equivalent income or be condemned as inefficient, whether the dollar invested came from investment or from consumption. Finally, if the money is borrowed, the rate of interest would have to be the same as the rate of return on private investment opportunities; otherwise no one would buy government bonds. Again the government would be confronted with the common rate of return on private investments as the appropriate discount rate to be applied to future benefits from a proposed social investment.

In free-enterprise societies it does not appear at first glance that the rate of return is, in fact, the same on all private investments. Savings-bank deposits, government bonds, and corporate bonds certainly yield different rates of return. Two interpretations of this observation have been offered: (1) Investments differ in riskiness, and the observed differences in rates of return are, in fact, compensations for these differences. It is argued in this interpretation that the risk-corrected rates of return are really the same. (2) There are imperfections and limitations of entry in capital and product markets so that rates of return really differ. Which of these explanations is in fact the better is an empirical question which cannot be dealt with here. The two hypotheses are not, of course, mutually exclusive, and both undoubtedly play a role in actuality—with what weights is certainly not easy to determine.

The two interpretations have very different implications for the determination of the opportunity cost of capital appropriate for discounting future benefits. (1) It remains true that the mode of financing has no influence on the appropriate discount rate, since each individual is presumed to have come into equilibrium as among consumption and the investments of varying degrees of riskiness. There is a uniquely defined pure rate of return, corresponding to riskless investments, which is usually taken to be the rate on government bonds of long maturity. Clearly, riskless social investment should be discounted at the pure rate of return under the hypothesis that observed rates of return differ because of riskiness. The implications for the rate of discount to be applied to benefits from risky social investments are less clear. According to one view, the rate to be applied is that which obtains in the private sector for investments of equal riskiness. Another position is that the government is necessarily in a better position to bear risks than any private investor. In fact, since it is involved with so many risky ventures, the law of large numbers ensures an aggregate certainty. It is therefore argued that the rate of discount should be the pure rate. The benefits to which the discount rate should be applied are uncertain; this uncertainty is what is meant by saying that the investment is risky. The single number that should represent a benefit is the expected value.

(2) If the view that differences in observed rates of return are due to market imperfections is correct, the mode of financing investments becomes important. The rate of discount is the rate on the alternative private investments available to the particular individuals from whom the money was drawn, whether taxpayers or bondbuyers. This alternative rate would be different for different individuals; thus the discount rate used by the government would have to be an approximate average.

It should be noted that the second view has implications not merely for the determination of the rate of discount but also for the preferred mode of financing. To the greatest extent possible, funds should be drawn from those individuals and fields of economic activity for which the rates of return on alternative investments are the lowest.

Social and Private Time Preference

As noted in a previous section, full optimality requires that the rate of time preference and the opportunity cost of capital be equal. This condition is, however, only of significance if the volume of social investment is not completely infinitesimal compared with private investment.

Suppose again that there is a perfect capital market; then for each individual, his rate of time preference equals the opportunity cost of capital. It has, however, been argued that even in this case it is not correct for society to be governed by the private rate. The government has an obligation to the future and, in particular, to unborn generations who are not represented in the current market. It should be made clear that this interest must be over and above the interest felt by individuals in the future welfare of their own heirs, born and unborn, for the latter is already reflected in individual time preference. This argument has been put in the form of divergence between private and social benefit. Each individual derives satisfaction from having wealth added to future generations. Each one can, by his own actions, add only infinitesimally to this wealth, but a collective agreement to do so will increase everyone's welfare.

If this argument is accepted, it does not necessarily lead to a special rate of discount for social investment. Indeed, the optimal solution would be to lower the required rate of return on all investment, private and social, for example by lending to private business at a lower rate than the market or by driving the rate of interest down through a budgetary surplus and debt retirement. More private investment would be undertaken, so that the marginal investment would have a lower rate of return and the opportunity cost of capital would be lower. Then, without changing the rule of discounting the benefits of social investment at the opportunity cost, the interest rate would be lowered.

If, however, it is accepted that there is an institutional limit on the extent to which the government can engage in direct or indirect financing of private investment, the social rate of time preference may remain below the common value of the opportunity cost of capital and the individual rate of time preference. It would clearly not be socially advantageous to withdraw resources from private investment to social investment with a lower rate of return, but it would be socially advantageous to withdraw some resources from consumption to social investment because of the divergence between social and individual time preference. Under these assumptions the appropriate rate of discount on future benefits from social investment will depend on the source of financing; it will be an average of the social rate of time preference and the opportunity cost of capital, with the weights depending on the extent to which resources are drawn from consumption or investment.

If we drop the assumption of a perfect capital market and admit differences in rates of return, we consider again the implications of the two alternative hypotheses of the preceding subsection. The first leads to much the same results as the case of a perfect market. The second hypothsis tends to reinforce the statements of the preceding paragraph.

The argument for a social rate of time preference distinct from individual rates is basically a matter of value judgment. Its validity and its importance, if valid, are both subject to considerable dispute.

THE MEASUREMENT OF COSTS

The analysis of costs offers fewer difficulties than those of benefits and the rate of discount. For most purposes even a large volume of investment will have little effect on the costs of the inputs, so that evaluation at market prices is usually satisfactory. There are, however, two major qualifications, arising out of the possible presence of unemployment and out of neighborhood or amenity effects of certain classes of costs.

Unemployed Resources

During a period of unemployment of labor or capital, the market price of an input will exceed its true social cost. Putting the idle worker or machine to work costs society nothing, but there is a wage or other price to be paid. The government, as guardian of the nation's economic welfare, should properly reckon only with true social costs. Hence, during a depression the cost of an investment should exclude costs of labor and plant that would otherwise be unemployed. Even in times of generally high employment there may be local areas of unemployment; the same rules should hold for projects in such an area. The allowance for unemployment applies only to the initial investment cost; in estimating future operating costs to be deducted from future benefits, it should normally be assumed that full employment will prevail.

Amenity Costs

Just as there can be a divergence between private and social benefits, so there can be divergence between private and social costs. A highway may conduce to an increase in air pollution; it may also, through its noise, be a source of disutility to the neighborhood. Since social investment projects are frequently large in magnitude, they frequently are so physically large that they impinge upon human sensibilities in a major fashion. Although it is hard to frame any general statement about amenity costs, they should in principle be assigned shadow costs which should be deducted from benefits or added to costs in making a benefit-cost calculation.

20

Comparisons of Methods for Recreation Evaluation

JACK L. KNETSCH

ROBERT K. DAVIS

•

Jack L. Knetsch is Professor of Economics at Simon Fraser
University in British Columbia. Robert K. Davis is Senior
Associate in the Institute of Behavioral Science at the
University of Colorado.

Evaluation of recreation benefits has made significant headway in the past
few years. It appears that concern is increasingly focusing on the hard core
of relevant issues concerning the economic benefits of recreation and how
we can go about making some useful estimates.

The underlying reasons for this sharpening of focus are largely pragmatic. The rapidly increasing demand for recreation, stemming from the
often-cited factors of increasing population, leisure, incomes, mobility,
and urbanization, calls for continuing adjustments in resource allocations.
This is the case with respect to our land and water resources in general; but
more specifically it bears on such matters as the establishment of national
recreation areas, setting aside or preserving areas for parks and open
spaces in and near expanding urban areas, and clearly on questions of
justification, location, and operation of water development projects.

Recreation services have only recently been recognized as products of
land and water resource use. As such, they offer problems that do not

occur when resolving the conflicting uses of most goods and services—for example, steel and lumber. Conflicting demands for commodities such as these are resolved largely in the market place of the private economy, where users bid against each other for the limited supplies.

Outdoor recreation, however, has developed largely as a non-market commodity. The reasons for this are quite elaborate, but in essence outdoor recreation for the most part is produced and distributed in the absence of a market mechanism, partly because we prefer it that way and have rejected various market outcomes, and partly because many kinds of outdoor recreation experience cannot be packaged and sold by private producers to private consumers. This absence of a market necessitates imputing values to the production of recreation services. Such economic benefits can be taken into account in decisions affecting our use of resources.

MISUNDERSTANDINGS OF RECREATION VALUES

Discussions of values of outdoor recreation have been beset by many misunderstandings. One of these stems from a lack of appreciation that the use of outdoor recreation facilities differs only in kind, but not in principle, from consumption patterns of other goods and services. Another is that the market process takes account of personal and varied consumer satisfactions.

It is, furthermore, the incremental values that are important in making decisions relative to resource allocations. The incremental values of recreation developments of various kinds are a manageable concept which can be used for comparisons, in spite of the very great aggregate value that some may want to attribute to recreation. Nothing is gained—and no doubt a great deal has been lost—by what amounts to ascribing the importance of a total supply of recreation to an added increment, rather than concentrating on the added costs and the added benefits.

A similar difficulty arises with respect to questions of water supply. That man is entirely dependent upon the existence of water is repeatedly emphasized. While true, the point does not matter. Decisions necessarily focus on increments and therefore on the added costs and the added benefits that stem from adding small amounts to the existing total.

Further, no goods or services are priceless in the sense of an infinite price. There is an individual and collective limit to how much we will give up to enjoy the services of any outdoor recreation facility or to preserve any scenic resource. The most relevant economic measure of recreation values, therefore, is willingness on the part of consumers to pay for outdoor recreation services. This set of values is comparable to economic values established for other commodities, for it is the willingness to give up on the part of consumers that establishes values throughout the economy.

Failure to understand these value characteristics results in two types of error. The first is the belief that the only values that are worth considering are those accounted for commercially. A second and related source of error is a belief that outdoor recreation experience is outside the framework of economics, that the relevant values have an aesthetic, deeply personal, and even mystical nature. We believe both of these to be incorrect. In particular, the notion that economic values do not account for aesthetic or personal values is fallacious and misleading. Economically, the use of resources for recreation is fully equivalent to other uses, and the values which are relevant do not necessarily need to be determined in the market place. This last condition does indicate that indirect means of supplying relevant measures of the values produced may be necessary. But this is an empirical problem, albeit one of some considerable dimension, and the primary concern of this paper.

The problem of using imputed values for value determination has been met with a considerable degree of success for some products of water resource development. Procedures have been developed to assess the value of the flood protection, irrigation, and power services produced by the projects, even though in many cases a market does not in fact exist or is inadequate for the actual benefit calculations. Without commenting on the adequacy of these methods, it is generally agreed that such measures are useful in evaluating the output of project services.

NATIONAL AND LOCAL BENEFITS

Discussions of these topics have often been further confused by failure to separate two types of economic consequences or benefit. This has led to improper recognition of relevant and legitimate economic interests, and to inferior planning and policy choices.

There are, first, what we may call primary benefits, or national benefits. Second, there are benefits we may refer to as local benefits, or impact benefits. Both sets of values resulting from investment in recreation have economic relevance, but they differ, and they bear differently on decision.

The primary recreation benefits, or values, are in general taken to be expressions of the consumers' willingness to pay for recreation services. These values may or may not register in the commerce of the region or in the commerce of the nation, but this does not make them less real. When appropriately measured, they are useful for guiding social choices at the national level. The other set of accounts is concerned with local expenditure of money for local services associated with recreation. While outdoor recreation is not marketed—in the sense that the services of parks, as such, are not sold to any great extent in any organized market—money does indeed become involved in the form of expenditures for travel, equipment, lodging, and so forth. The amount of money spent in connection with outdoor recreation and tourism is large and growing, making outdoor recreation expenditures of prime concern to localities and regions which may stand to benefit. Our concern is with measuring the more difficult of the two types of benefit just mentioned—national recreation benefits. While these are measured essentially by the consumers' willingness to pay, in some cases the benefits extend to the non-using general public.

ALTERNATIVE MEASUREMENT METHODS

There are obvious advantages to evaluating recreation benefits by market prices in the same manner as their most important resource competitors. However, as we have indicated, past applications have been hampered by disagreement on what are the meaningful values. In spite of growing recognition that recreation has an important economic value, economists and public administrators have been ill-prepared to include it in the social or public calculus in ways that lead to better allocations of resources.

The benefits of recreation from the social or community viewpoint are alleged to be many and varied. Some of the descriptions of public good externalities arising from recreation consumption are gross overstatements of the real values derived from the production of recreation services. But recreation benefits do in fact exist. Where externalities are real—as in cases of recreation in connection with visits to various historic areas or educational facilities, or where preservation of unique ecological units has cultural and scientific values—they should be recognized in assigning values to the development or preservation of the areas. However, it is our view that, by and large, recreation is a consumption good rather than a factor of production, and the benefits to be enjoyed are largely those accruing to the individual consumer participating. This is even more likely to be the case with recreation provided by water projects. The large bulk of primary recreation benefits can be viewed as the value of the output of the project to those who use them. This view stems from the concept that recreation resources produce an economic product. In this sense they are scarce and capable of yielding satisfaction for which people are willing to pay. Finally, some accounting can be made of this economic demand.

As the desirability of establishing values for recreation use of resources has become more apparent over the past few years, a number of methods for measuring or estimating them have been proposed and to some extent used. Some of the measures are clearly incorrect; others attempt to measure appropriate values, but fall short on empirical grounds [ref. 1, 2, 3].*

Gross Expenditures Method

The gross expenditures method attempts to measure the value of recreation to the recreationist in terms of the total amount spent on recreation by the user. These expenditures usually include travel expenses, equipment costs, and expenses incurred while in the recreation area. Estimates of gross recreation expenditures are very popular in some quarters; for one thing, they are likely to produce large figures. It is argued that persons making such expenditures must have received commensurate value or they would not have made them. The usual contention is that the value of a day's recreation is worth at least the amount of money spent by a person for the use of that recreation.

These values have some usefulness in indicating the amount of money that is spent on a particular type of outdoor recreation, but as justification for public expenditure on recreation, or for determining the worth or benefit of the recreation opportunity afforded, they are of little consequence.

The values we seek are those which show not some gross value, but the net increase in value over and above what would occur in the absence of a particular recreation opportunity. Gross expenditures do not indicate the value of the losses sustained if the particular recreation opportunity were to disappear, nor do they show the net gain in value from an increase in a particular recreation opportunity.

Market Value of Fish Method

A proposed method for estimating the recreation benefits afforded by fishing imputes to sport fishing a market value of the fish caught. The main objection to this procedure is the implied definition that the fish alone are the primary objective of the activity.

Cost Method

The cost method assumes that the value of outdoor recreation resource use is equal to the cost of generating it or, in some extreme applications, that it is a multiple of these costs. This has the effect of justifying any contemplated recreation project. However, the method offers no guide in the case of contemplated loss of recreation opportunities, and allows little or no discrimination between relative values of alternative additions.

---

*References in brackets are identified at the end of the chapter.

Market Value Method

Basic to the market value method measure is a schedule of charges judged to be the market value of the recreation services produced. These charges are multiplied by the actual or expected attendance figures to arrive at a recreation value for the services.

The method is on sound ground in its emphasis on the willingness of users to incur expenses to make choices. However, the market for outdoor recreation is not a commercial one, certainly not for much of the recreation provided publicly and only to a limited extent for private recreation. It is in part because private areas are not fully comparable with public areas that users are willing to pay the fees or charges. It seems, therefore, inappropriate to use charges paid on a private area to estimate the value of recreation on public areas. Also a single value figure or some range of values will be inappropriate for many recreation areas. Physical units of goods and services are not everywhere equally valuable, whether the commodity be sawtimber, grazing, or recreation. Location in the case of recreation affects value greatly. Moreover, differences of quality and attractiveness of recreation areas are not fully comparable or recognized by the unit values.

There are other methods, but few have received much attention. Where does this leave us? The only methods to which we give high marks are based on the concept of willingness to pay for services provided.

METHODS BASED ON WILLINGNESS TO PAY

We have alluded to two kinds of problems we face in measuring the benefits of outdoor recreation: the conceptual problems and the measurement problems.

Conceptually, we wish to measure the willingness to pay by consumers of outdoor recreation services as though these consumers were purchasing the services in an open market. The total willingness of consumers to pay for a given amount and quality of outdoor recreation (that is, the area under the demand curve) is the relevant measure we seek. Our conceptual problems are essentially that any measurement of effective demand in the current time period, or even an attempt to project effective demand in future time periods, must necessarily omit from the computation two kinds of demand which may or may not be important. These are option demand and demand generated by the opportunity effect.[1]

---

[1]These concepts are developed by Davidson, Adams, and Seneca in "The Social Value of Water Recreational Facilities Resulting from an Improvement in Water Quality: The Delaware Estuary," Allen V. Kneese and Stephen C. Smith, eds., *Water Research,* Johns Hopkins Press, Baltimore, 1966.

Option demand is that demand from individuals who are not now consumers or are not now consuming as much as they anticipate consuming, and who therefore would be willing to pay to perpetuate the availability of the commodities. Such a demand is not likely to be measured by observance or simulation of market phenomena. The opportunity effect derives from those unanticipated increases in demand caused by improving the opportunities to engage in a recreational activity and thereby acquainting consumers with new and different sets of opportunities to which they adapt through learning processes. To our knowledge no methods have been proposed which might be used to measure those two kinds of demand for a good.

Notwithstanding the undoubted reality of these kinds of demand, our presumption is that effective demand is likely to be the predominant component of the aggregate demand for outdoor recreation of the abundant and reproducible sorts we have in mind. We further presume that this quantity can be estimated in a useful way, although by fairly indirect means, for we have no market guide of the usual sort. Two methods—a direct interview, and an imputation of a demand curve from travel cost data—currently appear to offer reasonable means of obtaining meaningful estimates.

Interview Methods

The essence of the interview method of measuring recreation benefits is that through a properly constructed interview approach one can elicit from recreationists information concerning the maximum price they would pay in order to avoid being deprived of the use of a particular area for whatever use they may make of it. The argument for the existence of something to be measured rests on the conception that the recreationist is engaged in the utility maximizing process and has made a rational series of allocations of time and money in order to participate in the recreation being evaluated. Since the opportunity itself is available at zero or nominal price, the interview provides the means for discovering the price the person would pay if this opportunity were marketed, other things being equal.

The chief problem to be reckoned with in evaluating interview responses is the degree of reliability that can be attached to the information the respondent provides the interviewer. Particularly on questions dealing with matters of opinion, the responses are subject to many kinds of bias.

One such bias of particular interest to economists stems from the gaming strategy that a consumer of a public good may pursue on the theory that, if he understates his preference for the good, he will escape being charged as much as he is willing to pay without being deprived of the amount of the good he now desires. This may be a false issue, particularly when it comes to pursuing recreation on private lands or waters, because the consumer may be well aware that the owner could, through the exercise of his private property rights, exclude the user from the areas now occupied. An equally good case can be made that, on state and national park lands to which there is limited access, particularly when at the access points the authority of the state is represented by uniformed park patrolmen, recreationists would have no trouble visualizing the existence of the power to exclude them. This being the case, it is not unreasonable to expect the recreationist to be aware of some willingness to pay on his part in order to avoid being excluded from the area he now uses.

Counterbalancing the possibility that the recreationist may purposely understate his willingness to pay in order to escape charges is the possibility that he may wish to bid up his apparent benefits in order to make a case for preserving the area in its current use, a case equally appropriate on private or public lands and waters.

The problem, to continue the argument, is narrowed to one of phrasing the question in such a way that the recreationist is not asked to give his opinion on the propriety of charging for the use of recreation areas.

It has become something of a principle in survey methodology that the less hypothetical the question, the more stable and reliable the response. By this principle, the respondent ought to be a consumer of the product rather than a potential consumer, thus distinguishing the data collected as pertaining to effective demand rather than to option or potential demand. It may also be preferable to impose the conditions on the interview that it occur at a time when the respondent is engaged in the activity. This may contribute to the accuracy of the responses by reducing the requirement that he project from one situation to another. (Admittedly, it is desirable to experiment with the methodology on this question, as well as others, in order to determine its sensitivity to such variations.)

In sum then, we expect to discover the consumer's willingness to pay through a properly constructed interview, and further, we expect that this measure will be the same quantity as would be registered in an organized market for the commodity consumed by the respondent. In other words, we hold a deterministic view that something exists to be measured, and is a sufficiently real and stable phenomenon that the measurement is useful.

THE INTERVIEW PROCEDURE.   The willingness to pay of a sample of users of a forest recreation area in northern Maine was determined in interviews on the site [ref. 4]. The interviews included a bidding game in which respondents could react to increased costs of visiting the area. Bids were systematically raised or lowered until the user switched his reaction from inclusion to exclusion or vice versa. At the beginning of the interview rapport was established with the respondents largely through objective questions inquiring into their recreation activities on the area, on other areas, and the details of their trips. The bidding questions were interspersed with a series of propositions for which the respondent was to

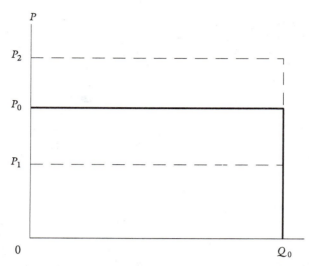

FIGURE 1. At prices in the range 0–P_0 the constant amount Q_0 will be demanded. Above P_0 demand will fall to zero. The individual may be in one of three states depending on the reigning price. Consider three individual cases with market price at P_0: The user paying P_1 is excluded; P_0 is associated with the marginal user; and P_2 is the willingness to pay of the third user who is included at the reigning price, P_0.

indicate his opinion in the form of a positive, negative, or neutral reaction. His reactions to increased expenses connected with the visit constituted the essence of the bidding game. Personal questions regarding income, education, and the like were confined to the end of the interview.

The sampling procedure amounted to cluster sampling, since the procedure followed was to locate areas of use such as campgrounds and to systematically sample from the available clusters of users. The interviews were conducted from June through November by visiting areas in the privately owned forests of northern Maine and in Baxter State Park.

The data from the interviews is pooled to include hunters, fishermen, and summer campers. This pooling is defended largely on the grounds that no structural differences between identifiable strata were detected in a multiple regression analysis of the responses.

The procedure imputes a discontinuous demand curve to the individual household which may be realistic under the time constraints faced particularly by vacation visitors and other non-repeating visitors. This rectangular demand curve (Figure 1) reflects a disposition either to come at the current level of use or not to come at all if costs rise above a limiting value. Its realism is supported by a number of respondents whose reaction to the excluding bid was precisely that they would not come at all. It seems reasonable to view the use of remote areas such as northern Maine as lumpy commodities which must be consumed in five- or six-day lumps or not at all. Deriving an aggregate demand function from the individual responses so characterized is simply a matter of taking the distribution function of willingness to pay cumulated on a less-than basis. This results in a continuous demand schedule which can be interpreted for the aggregate user population as a conventional demand schedule.

For the sample of 185 interviews, willingness-to-pay-per-household-day ranges from zero to \$16.66. Zero willingness to pay was encountered in only three interviews. At the other extreme, one or two respondents were unable to place an upper limit on their willingness to pay. The distribution of willingness to pay shows a marked skewness toward the high values. The modal willingness to pay occurs between \$1.00 and \$2.00 per day per household.

Sixty per cent of the variance of willingness to pay among the interviews is explained in a multiple regression equation with willingness-to-pay-per-household-visit a function of income of the household, years of experience by the household in visiting the area, and the length of the stay in the area. (See Equation 1). While the large negative intercept of this equation necessitated by its linear form causes some difficulties of interpretation, the exhibited relation between willingness to pay, and income, experience, and length of stay appears reasonable. The household income not only reflects an ability to pay, but a positive income elasticity of demand for outdoor recreation as found in other studies. It is also significant that an internal consistency was found in the responses to income-related questions.

$$R^2$$

$$W = -48.57 + 2.85\,Y + 2.88\,E + 4.76\,L \qquad .5925 \quad (1)$$
$$ (1.52) \quad (0.58) \quad (1.03)$$

$$W = .74\,L^{.76}\ E^{.20}\ Y^{.60} \qquad\qquad\qquad .3591^* \quad (2)$$
$$ (.13)\ (.07)\ (.17)$$

Standard errors of regression equations: (1) 39.7957; (2) 2.2007.
Standard errors of coefficients are shown in parentheses.
W = household willingness to pay for a visit.
E = years of acquaintance with the area visited.
Y = income of the household in thousands of dollars.
L = length of visit in days.
F = ratios of both equations are highly significant.
*Obtained from arithmetic values of residual and total variances. (R^2 of the logarithmic transformation is .4309.)

The significance of years of experience in returning to the area may be interpreted as the effect of an accumulated consumer capital consisting of knowledge of the area, acquisition of skills which enhance the enjoyment of the area, and in some cases use of permanent or mobile housing on the area.

The significance of length of stay in the regression equations is that it both measures the quantity of goods consumed and also reflects a quality dimension suggesting that longer stays probably reflect a greater degree of preference for the area.

Colinearity among explanatory variables was very low. The general economic consistency and rationality of the responses appear to be high. Respondents' comments indicated they were turning over in their minds the alternatives available in much the same way that a rational shopper considers the price and desirability of different cuts or kinds of meat. Both the success in finding acceptable and significant explanatory variables and a certain amount of internal consistency in the responses suggest that considerable weight can be attached to the interview method.

THE SIMULATED DEMAND SCHEDULE. While providing an adequate equation for predicting the willingness to pay of any user, the results of the interviews do not serve as direct estimates of willingness to pay of the user population, because the income, length of stay, and years' experience of the interviewed sample do not accurately represent the characteristics of the population of users. Fortunately, it was possible to obtain a reliable sample of the users by administering a questionnaire to systematically selected samples of users stopped at the traffic checking stations on the private forest lands. A logarithmic estimating equation, although not as well fitting, but free of a negative range, was used to compute the willingness to pay for each household in the sample. (See Equation 2.) The observations were then expanded by the sampling fraction to account for the total number of users during the recreation season.

The next step in the analysis consists of arraying the user population by willingness to pay, and building a cumulative distribution downward from the upper limit of the distribution. Table 1 shows the resulting demand and benefit schedule. The schedule accounts for the total of about 10,300 user households estimated to be the user population in a 450,000-acre area of the Maine woods near Moosehead Lake, known as the Pittston area.

The demand schedule is noticeably elastic from the upper limit of $60.00 to about $6.00, at which point total revenues are maximized. The interval from $60.00 to $6.00 accounts for the estimated willingness to pay of nearly half of the using households. Total benefits at $6.00 are $56,000. The price range below $6.00 accounts for the other half of the using households, but only for $15,000 in additional benefits. Benefits are estimated as the cumulative willingness to pay or the revenues available to a discriminating monopolist.

WILLINGNESS TO DRIVE VS. WILLINGNESS TO PAY. An alternative expression of the willingness of recreationists to incur additional costs

TABLE 1. Demand and Benefit Schedules for Pittston Area Based on Alternative Estimates of Willingness to Pay

PRICE	INTERVIEW RESULTS		WILLINGNESS TO DRIVE (INTERVIEW METHOD)		WILLINGNESS TO DRIVE (TRAVEL COST METHOD)	
	HOUSEHOLD VISITS	BENEFITS[1]	HOUSEHOLD VISITS	BENEFITS[1]	HOUSEHOLD VISITS	BENEFITS[1]
$70.00	0	0				
60.00	11.36	$ 747.77				
50.00	15.35	983.56				
40.00	44.31	2,281.46				
30.00	150.22	6,003.19	11.36	$ 384.79	165	$ 3,800
26.00	215.80	7,829.71				
22.00	391.07	12,027.89				
20.00	536.51	15,099.31	76.96	1,890.12	422	12,134
18.00	757.86	19,275.95				
16.00	1,069.01	24,607.81				
14.00	1,497.75	31,027.17	392.29	7,287.06		
12.00	1,866.41	35,802.70				
10.00	2,459.70	42,289.68	2,157.91	28,921.93	1,328	26,202
8.00	3,100.99	48,135.01				
6.00	4,171.89	55,794.64				
4.00	5,926.94	64,436.36	5,721.06	53,531.68	3,459	44,760
2.00	7,866.02	70,222.66				
0.00	10,333.22	71,460.94	10,339.45	63,689.99	10,333	69,450

[1]Benefits are computed as the integral of the demand schedule from price maximum to price indicated. Willingness to drive computations are based on an assumed charge of 5¢ per mile for the one-way mileage.

in order to continue using an area may be found in their willingness to drive additional distances. This measure was first proposed by Ullman and Volk [ref. 5] although in a different version than is used here. (See also [ref. 6].)

Willingness to drive additional distances was elicited from respondents by the same technique used to elicit willingness to pay. If there are biases involving strategies to avoid paying for these recreation areas, then certainly willingness to drive is to be preferred over willingness to pay as an expression of value. Analysis of the willingness to drive responses shows that a partly different set of variables must be used to explain the responses. Equation 3 shows willingness to drive extra miles to be a function of length of stay and miles driven to reach the area.

$$Wm = 41.85 + 20.56\,L + .15\,M \tag{3}$$
$$(3.03) \qquad (.04) \qquad (R^2 = .3928)$$

Wm = willingness to drive additional miles.
L = length of visit in days.
M = miles traveled to area.

The respondents thus expressed a willingness to exert an additional driving effort, just as they expressed a willingness to make an additional money outlay if this became a requisite to using the area. Moreover, there is a significant correspondence between willingness to pay and willingness to drive. The simple correlation coefficient between these two variables is .5. Because of the correlation with length of stay, the reduction in unexplained variance produced by adding either variable to the equation in which the other variable is the dependent one is not very high. However, willingness to pay was found to increase about 5¢ per mile as a function of willingness to drive additional miles. This result gives us a basis for transforming willingness to drive into willingness to pay.

We may now construct a demand schedule for the Pittston area on the basis of willingness to drive, and compute a willingness to pay at 5¢ per mile. The resulting demand and benefit schedules appear in Table 1. The estimated $64,000 of total benefits is very close to that developed from the willingness to pay interview. While one may quibble about the evaluation of a mile of extra driving and about the treatment of one way versus round-trip distance, the first approximation using the obvious value of 5¢ and one-way mileage as reported by the respondents produces a result so close to the first result that we need look no further for marginal adjustments. The initial result strongly suggests that mileage measures and expenditure measures have equal validity as a measure of benefits in this particular case at least.

There are some differences between the respective demand schedules worth noting. The much lower price intercept on the willingness to drive schedule reflects the effect of the time constraint in traveling as well as our possibly erroneous constant transformation of miles to dollars when an increasing cost per mile would be more reasonable. The travel schedule is also elastic over more of its range than the dollar schedule—also perhaps a result of the constant transformation employed.

This initial success with alternative derivations of the benefits schedule now leads us to examine an alternative method for estimating the willingness to drive schedule.

Travel-Cost Method of Estimating User-Demand Curve

The direct interview approach to the estimate of a true price-quantity relationship, or demand curve, for the recreation experience is one approach to the benefit calculations based on willingness to pay. An alternative approach has received some recognition and has been applied in a number of limited instances with at least a fair degree of success. This uses travel-cost data as a proxy for price in imputing a demand curve for recreation facilities. [Ref. 7, 8, 9, 10.] As with the direct interview approach, we believe that estimates derived from this approach are relevant and useful for measuring user benefits of outdoor recreation.

The travel-cost method imputes the price-quantity reactions of consumers by examining their actual current spending behavior with respect to travel cost. The method can be shown by using a simple, hypothetical example. Assume a free recreation or park area at varying distances from three centers of population given in Table 2.

The cost of visiting the area is of major concern and would include such items as transportation, lodging, and food cost above those incurred if the trip were not made. Each cost would vary with the distance from the park to the city involved. Consequently, the number of visits, or rather the rate of visits per unit total population of each city, would also vary.

The visits per unit of population, in this case per thousand population, may then be plotted against the cost per visit. A line drawn through the three points of such a plot would have the relationship given by the equation of $C = 5 - V$, or perhaps more conveniently $V = 5 - C$, where

TABLE 2.   Visits to a Hypothetical Recreation Area

CITY	POPULATION	COST OF VISIT	VISITS MADE	VISITS/ 1,000 POP.
A	1,000	$1.00	400	400
B	2,000	3.00	400	200
C	4,000	4.00	400	100

C is cost of a visit and V is the rate of visits in hundreds per thousand population. This information is taken directly from the tabulation of consumer behavior. The linear relationship assumed here is for convenience. Actual data may very well show, for example, that $1.00 change in cost might have only a slight effect on visit rate where the visit is already high in cost, and a large effect on low-cost visits.

The construction of a demand curve to the recreation area, relating number of visits to varying cost, involves a second step. Essentially, it derives the demand curve from the equation relating visit rates to cost, by relating visit rates of each zone to simulated increases in cost and multiplying by the relative populations in each zone. Thus we might first assume a price of $1.00, which is an added cost of $1.00 for visits to the area from each of the three different centers used in our hypothetical example. This would have the expected result of reducing the number of visitors coming from each of the centers. The expected reduction is estimated from the visit-cost relationship. The total visits suggested by these calculations for different prices or differing added cost are given as:

PRICE (ADDED COST)	QUANTITY (TOTAL VISITS)
$0.00	1,200
1.00	500
2.00	200
3.00	100
4.00	0

These results may then be taken as the demand curve relating price to visits to the recreation area. While this analysis takes visits as a simple function of cost, in principle there is no difficulty in extending the analysis to other factors important in recreation demand, such as alternative sites available, the inherent attractiveness of the area in question or at least its characteristics in this regard, and possibly even some measure of congestion.

A difficulty with this method of benefit approximation is a consistent bias in the imputed demand curve resulting from the basic assumption that the disutility of overcoming distance is a function only of money cost. Clearly this is not so. The disutility is most likely to be the sum of at least three factors: money cost, time cost, and the utility (plus or minus) of driving, or traveling. The total of these three factors is demonstrably negative, but we do not know enough about the significance of the last two components. In all likelihood their sum—that is, of the utility or disutility of driving and the time cost—imposes costs in addition to money. To the extent that this is true the benefit estimate will be conservatively biased, for, as has been indicated, it is assumed that the only thing causing differences in attendance rates for cities located at different distances to a recreation area will be the differences in money cost. The method then postulates that if money cost changes are affected, the changes in rates will be proportional. What this bias amounts to is, essentially, a failure to establish a complete transformation function relating the three components of overcoming distance to the total effect on visitation rates. The resulting conservative bias must be regarded as an understatement of the recreation benefits which the approach is designed to measure.

APPLICATION TO PITTSTON AREA. The travel-cost method was applied to the same area as that used to illustrate the interview method of recreation benefit estimation. The same data were utilized to allow at least a crude comparison of the methods. In all, 6,678 respondents who said the Pittston area was the main destination of their trip were used in the analysis.

Visit rates of visitors from groups of counties near the area and from some states at greater distances were plotted against distance. The results were fairly consistent considering the rough nature of the approximations used in estimating distance. A curve was drawn through the points, giving a relationship between visit rates and distance. The demand curve was then calculated, giving a price-quantity relationship based on added distance (or added toll cost) and total visits. It was assumed initially that travel cost would be 5¢ per mile, using one-way distance to conform with our earlier analysis of travel cost by the interview method.

The results at this point were not comparable to the interview method because of a difference in the number of users accounted for. It will be recalled that in the analysis we are now describing only those respondents were used who had specifically stated that the visit to the Pittston area was the main destination of the trip. In order to make this number comparable to the total number of users accounted for in the interview estimate, we counted at half weight the 1,327 respondents who said that Pittston was *not* the primary destination of the trip, and also included in this group the non-response questionnaires and others with incomplete information. In this way we accounted for the same number of users as in the interview estimate. This very crude approximation points out the problems of the multiple-destination visit, but perhaps adequately serves the present purpose.

On the basis of these approximations, the benefit estimates on an annual basis were $70,000, assuming 5¢ per mile one-way distance. While the assumptions made throughout this analysis are subject to refinement, the exercise does seem to illustrate that the procedure is feasible from a practical standpoint and does produce results that are economically meaningful.

COMPARISON BETWEEN TRAVEL-COST AND
INTERVIEW METHODS

Having demonstrated that fairly close results are obtained from both the interview and imputation methods of estimating recreation benefits on the basis of reactions to travel costs, and further that the interview method of directly estimating willingness to pay agrees closely with both estimates based on travel costs, we can now begin to assess the meaning of these results. In some ways the task would be easier if the results had not agreed so closely, for the three methodologies may imply different things about the users' reactions to increased costs. At least, it is not obvious without further probing as to why the agreement is so close.

The interview and imputation methods of estimating benefits on the basis of willingness to incur additional travel costs do not, for example neatly imply the same relationship between distance traveled and willingness to incur additional travel costs. The estimating equation derived from the interviews (Equation 3) suggests that the farther one has traveled, the greater additional distance he will travel. Yet the imputation procedure implies that the willingness to drive by populations in the respective zones does not vary consistently with distance. Furthermore, according to the interviews, responses to the monetary measure of willingness to pay do not attribute any variance in willingness to pay to the distance factor, nor is an indirect relationship obvious. It seems relevant to inquire into the implied effects of these factors to discover why the alternative procedures appear to imply substantially different determinants of willingness to pay.

The superficial agreement in results may be upheld by this kind of further probing, but there are also some methodological issues which should not be overlooked. The travel-cost methods are obviously sensitive to such matters as the weighting given to multiple-destination visits and to the transformation used to derive costs from mileage values. Both methods are sensitive to the usual problems of choosing an appropriately fitting equation for the derivation of the demand schedule. The interview method has a poorly understood sensitivity to the various methodologies that might be employed in its use. Moreover, even the minimal use of interviews in studies of recreation benefits makes the method far more costly than the imputation method based on travel costs.

There are, however, complementarities in the two basic methods which may prove highly useful. In the first place, the two methods may serve as checks on each other in applied situations. One is certainly in a better position from having two methods produce nearly identical answers than if he has to depend on only one. There are also interesting possibilities that interviews may be the best way of resolving the ambiguities in the travel-cost method concerning the treatment of multiple-destination cases and for finding the appropriate valuation for converting distance into dollars.

Much can be said for letting the recreationist tell us how to handle these problems.

In sum, we have examined three methods of measuring recreation benefits. All three measure recreationists' willingness to pay. This, we argue, is the appropriate measure of primary, or national, benefits. Furthermore, the measures are in rough agreement as to the benefits ascribable to an area of the Maine woods. This may be taken as evidence that we are on the right track. There are, however, some rough spots to be ironed out of each of the methods—an endeavor we believe to be worthy of major research effort if benefit-cost analysis is to contribute its full potential in planning decisions affecting recreation investments in land and water resources.

REFERENCES

1. Lerner, Lionel. "Quantitative Indices of Recreational Values," in *Water Resources and Economic Development of the West.* Report No. 11. Proceedings, Conference of Committee on the Economics of Water Resources Development of Western Agricultural Economics Research Council with Western Farm Economics Association. Reno: University of Nevada, 1962.
2. Merewitz, Leonard. "Recreational Benefits of Water-Resource Development." Unpublished paper of Harvard Water Program, 1965.
3. Crutchfield, James. "Valuation of Fishery Resources," *Land Economics,* Vol. 38, No. 2 (1962).
4. Davis, Robert K. "The Value of Outdoor Recreation: An Economic Study of the Maine Woods." Ph.D. thesis, Harvard University, 1963.
5. Ullman, Edward, and Volk, Donald. "An Operational Model for Predicting Reservoir Attendance and Benefits: Implications of a Location Approach to Water Recreation," *Proceedings Michigan Academy of Sciences,* 1961.
6. Meramec Basin Research Project. "Recreation," Chap. 5 in *The Meramec Basin,* Vol. 3, St. Louis: Washington University, December 1961.
7. Clawson, Marion. *Methods of Measuring the Demand for and Value of Outdoor Recreation.* RFF Reprint No. 10. Washington: Resources for the Future, Inc., 1959.
8. Knetsch, Jack L. "Outdoor Recreation Demands and Benefits," *Land Economics,* Vol. 39, No. 4 (1963).
9. ———. "Economics of Including Recreation as a Purpose of Water Resources Projects," *Journal of Farm Economics,* December 1964. Also RFF Reprint No. 50. Washington: Resources for the Future, Inc.
10. Brown, William G., Singh, Ajner, and Castle, Emery N. *An Economic Evaluation of the Oregon Salmon and Steelhead Sport Fishery.* Technical Bulletin 78. Corvallis: Oregon Experiment Station, 1964.

21

Bidding Games for Valuation of Aesthetic Environmental Improvements[1]

ALAN RANDALL

BERRY IVES

CLYDE EASTMAN

Alan Randall is Professor of Agricultural Economics at Ohio State University in Columbus; Berry Ives is with the Middle Rio Grande Council of Governments in Albuquerque; and Clyde Eastman is a Professor in the Department of Agricultural Economics and Agricultural Business at New Mexico State University in Las Cruces.

It has proved a difficult and often forbidding task to ascribe economic values to environmental improvements. Yet, rational and informed social decision making requires, among other things, a consideration of the economic costs and benefits of environmental improvements. The difficulties in economic evaluation are compounded in the case of environmental improvements of an aesthetic nature. This article discusses the prob-

"Bidding Games for Valuation of Aesthetic Improvements," by Alan Randall, Berry Ives, and Clyde Eastman, from *Journal of Environmental Economics and Management,* 1:132–49. Copyright © 1974 Academic Press, Inc. Reprinted by permission.

[1]Journal Article 506, New Mexico State University, Agricultural Experiment Station, Las Cruces. The authors are grateful for helpful comments from Ralph d'Arge and two anonymous reviewers.

lems inherent in the valuation of aesthetic environmental improvements and presents a case study in which bidding games were used as the valuation technique.

THE THEORY

Aesthetic damage to an outdoor environment, to the extent that it diminishes the utility of some individuals, is a discommodity and its abatement is a commodity. Abatement of this kind of external diseconomy is both a nonmarket good, since it is nonexclusive, and a public good in the sense of Davis and Whinston [6], since it is inexhaustible at least over a very substantial range. That is, additional consumers of this kind of aesthetic environmental improvement can be added without diminishing the visibility or scenic beauty available to each (at least, until crowding occurs). Additional users can be added at near zero marginal cost, over a substantial range.

Bradford [2] has presented a theoretical framework for the valuation of public goods. Traditional demand curves are inappropriate for the analysis of demand for public goods, since the situation is not one of individuals responding to a parametric price per unit by choosing an appropriate number of units. Rather, the individual directly arrives at the total value to himself of various given packages. In the case of a public good, the individual is unable to exercise any choice over the quantity provided him, except as a member of the collective which makes a collective choice. Further, the nature of a public good such as aesthetic environmental improvements is such that increases in the quantity provided are not purely quantitative increases, but are more in the nature of improvements in quality. Thus, the individual values alternative packages of a public good, which may differ in quantity and quality.

Bradford proposes the concept of an aggregate bid curve for public goods. Individual bid curves are simply indifference curves passing through a given initial state, with the numeraire good (which can be dollars) on the vertical axis and the public good on the horizontal axis.[2] The aggregate bid curve is the algebraic (or vertical, in diagrammatic analyses) summation of individual bids over the relevant population.

The aggregate bid curve is an aggregate benefit curve, as it measures

[2]If different packages of a public good represented continuous quantitative increases, the individual bid curve would be smooth and would exhibit decreasing marginal utility of increasing quantities of the public good. However, Bradford's concept of different packages differing in quantity and quality logically implies that individual bid curves need be neither smooth nor of continually decreasing slope. Bradford insists that, *a priori,* nothing can be said about the slope of the "demand," or marginal bid curve, for a public good of this nature.

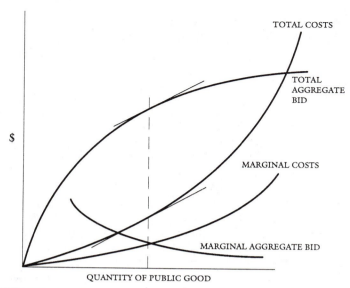

FIGURE 1. Collective optimization of the quantity of public good provided.

precisely what an accurate benefit-cost analysis of provision of a public good would measure as benefits. Using the approach of methodological collectivism, efficiency in the provision of a public good can be achieved by maximizing the excess of aggregate bid over total cost, or equating the first derivative of aggregate bid (i.e., marginal bid) with the marginal cost of provision.[3] Figure 1 shows the efficient level of provision of a public good.[4]

THE BIDDING GAME TECHNIQUE

It is possible to conceive of a number of techniques for estimating the aggregate bid curve for environmental improvements. Two general classes of techniques, direct costing techniques and revealed demand techniques, have been suggested in the literature and applied in empirical studies. Each of these has its difficulties, especially when adapted for valuation of aes-

---

[3]In the approach of methodological individualism, Pareto-efficiency is still not achieved since the price to the individual cannot equal the marginal cost to the individual (which is zero) and allow collection of sufficient funds to cover the total cost of provision.

[4]In Fig. 1, the aggregate and marginal bid curves are drawn as smooth curves consistent with diminishing marginal utility. As pointed out in footnote 2, this need not be even the typical case.

thetic environmental improvements. These techniques will be briefly discussed below. Then, a third type of technique, bidding games, will be proposed. Bidding game techniques are themselves not without difficulties, but we will argue that there may be applications for which they are the preferable or even the only feasible method for empirical studies. Methods of maximizing the reliability of bidding games will be discussed and an empirical study using bidding games will be presented.

DIRECT COSTING METHODS. Implicit in the concept of a "marginal value of damage avoided by abatement" curve, as proposed by Kneese and Bower [12], is the idea of estimating the benefits of abatement of environmental damage by directly estimating the costs attributable to that damage. Several workers have made progress along these lines. For example, Lave and Seskin [13] have had some success in relating the costs of impairment of human health to levels of air pollution. If all relevant costs of a particular incidence of environmental damage can be identified, evaluated and summed, a curve relating the value of damage avoided to levels of environmental improvements can be fitted. The first derivative of this curve is the "M.V.D.A." curve of Kneese and Bower [12].

These costing techniques are theoretically sound and may often be feasible in practice. However, difficulties may be introduced by the unavailability of information and the pricing and accounting problems inherent in this type of analysis. These techniques will have limited application in valuation of aesthetic environmental improvements, since the costs of aesthetic damages may seldom be directly reflected in the market.

REVEALED DEMAND TECHNIQUES. Revealed demand techniques have been widely used for estimation of the demand for outdoor recreation, often a nonmarket good.[5] A number of applications to valuation of the benefits of air pollution abatement have been made [1, 11, 14, 16, 18]. The principle is as follows. The benefits of provision of a nonmarket good are inferred from the revealed demand for some suitable proxy. In the case of air pollution abatement, the revealed demand for residential land is related by regression analysis to air pollution concentrations. In metropolitan areas, it is possible to obtain information on the concentration of specific air pollutants in different parts of the city. If all other variables relevant to the valuation of urban residential land can be identified[6] and measured, it ought to be possible to determine by regression analysis the extent to which air pollution concentrations affect observed land values.

---

[5]See [4].

[6]Some appropriate variables are size and value of structures on the land, distance from places where services and employment opportunities are concentrated, proportion of park land and open space in the neighborhood, density of population, proportion of various racial and ethnic minorities in the immediate vicinity, and the incidence of violent crimes.

In this way, a proxy measure of the benefits of air pollution abatement is obtained.

There are a number of difficulties with this type of analysis. Since the value ascribed to air pollution control is derived directly from the regression coefficient of the pollution concentration variable, accurate results require perfect and complete specification of the regression equation. In an interesting recent study, Wieand [17] claims that when such regression models are completely specified, the regression coefficient of the pollution concentration variable may not be significantly different from zero. Another difficulty, researchers in the field agree, lies in interpretation of the results. Are all of the benefits of air pollution abatement captured in residential land values? Most think not. For our purposes, the other side of that coin is of interest: Surely some benefits in addition to the aesthetic benefits are captured. Which additional benefits?

In the case study reported below, the geographical area affected by environmental damage includes urban areas, but also rural and agricultural areas, and substantial areas of Indian reservation and National Park, Monument, and Forest lands (which are typically not exchanged in the market). Thus, those revealed demand techniques currently available would seem to be inapplicable to the situation faced in our study.

BIDDING GAMES. In analysis of the demand for outdoor recreation, Davis [7] pioneered in the use of bidding games. During personal interviews, the enumerator follows on iterative questioning procedure to elicit responses which enable the fitting of a demand curve for the services offered by a recreation area. Respondents are asked to answer "yes" or "no" to the question: Would you continue to use this recreation area if the cost to you was to increase by X dollars? The amount is varied up or down in repetitive questions, and the highest positive response is recorded. Individual responses may then be aggregated to generate a demand curve for the recreation services provided by the area.

It seems reasonable that bidding games may be adapted to the estimation of the benefits from provision of an inexhaustible nonmarket good such as abatement of aesthetic environmental damage. Bidding games would seem to be the most direct method of estimating Bradford's aggregate benefit curve, which is derived from vertical summation of individual bid curves. The difficulties of interpretation which are inherent in the revealed demand techniques developed thus far do not occur when the bidding game technique is used. The data obtained with bidding games are not cost observations but individuals' perceptions of value. Thus, bidding games can be used in situations where direct costing techniques are ineffective for lack of data. These advantages of bidding games over revealed demand and direct costing techniques seem sufficient to justify attempts to adapt the bidding game technique for use in valuation of aesthetic environmental improvements.

Some General Considerations in the Design of Bidding Games

Bidding games are designed to elicit information on the hypothetical behavior of respondents when faced with hypothetical situations. In the case study presented below, the purpose of bidding games is to provide a measure of the benefits of aesthetic environmental improvements by measuring the willingness of a sample of respondents to pay for such improvements. The efficacy of bidding games used for this purpose depends on the reliability with which stated hypothetical behavior is converted to action, should the hypothetical situation posited in the game arise in actuality.

Willingness to pay is the behavioral dimension of an underlying attitude: concern for environmental quality.[7] Sociologists and public opinion researchers have built up a substantial body of literature which considers ways in which survey techniques of measuring attitudes and their behavioral component can be made as reliable as possible. Some desirable characteristics of such surveys have been identified [5, 9]. The hypothetical situation presented should be realistic and credible to respondents. Realism and credibility can be achieved by satisfying the following criteria for survey instrument design: Test items must have properties similar to those in the actual situation; situations posited must be concrete rather than symbolic; and test items should involve institutionalized or routinized behavior, where role expectations of respondents are well defined. Where the behavioral predisposition under study are affected by attitudes about a number of different things, the test instrument must be designed to focus upon those attitudes which are relevant. An example may be helpful. In the case study reported here, willingness to pay additional taxes to achieve aesthetic environmental improvement is affected by attitudes toward environmental quality, but also by attitudes toward the current tax burden and attitudes toward the idea of receptors of pollutants paying to obtain abatement of emissions. If the survey is carried out for the purpose of measuring the benefits of abatement, the test instrument must be designed to take cognizance of the various diverse attitudes which affect willingness to pay and to allow isolation of the relevant attitudinal dimensions.

Since abatement of aesthetic environmental damage is an inexhaustible, public good, bidding games intended to provide data for valuation of that good must be designed to avoid the effects of the freeloader problem, which encourages nonrevelation of preferences. One method would be to design games in which each respondent is told that all consumers of the good would pay for it on a similar basis, thus eliminating the possibility of freeloading.

With careful design of bidding games to ensure that the responses

---

[7]Three dimensions of attitudes are recognized [8]: (1) a cognitive dimension, (2) an affectual dimension, and (3) a behavioral dimension.

recorded are predictive of behavior, it should be possible to use the bidding technique to estimate the benefits of environmental improvements with reasonable accuracy.

AN EMPIRICAL APPLICATION:
ESTIMATION OF THE BENEFITS OF ABATEMENT OF AESTHETIC ENVIRONMENTAL DAMAGES ASSOCIATED WITH THE FOUR CORNERS STEAM ELECTRIC GENERATING PLANT

At New Mexico State University, research is under way to examine the socioeconomic impacts of development of the rapidly expanding coal strip-mining and steam electric generation industry in the Four Corners Region (southwestern United States), and to predict the impacts of alternative policies with respect to environmental management and economic development, as such policies would affect the industry. One facet of this research required estimation of the benefits of abatement of aesthetic environmental damage associated with the Four Corners power plant at Fruitland, NM, and the Navajo mine which provides its raw energy source—low energy, low sulfur, high ash, sub-bituminous coal.[8]

The mine–power plant complex causes several kinds of aesthetic environmental damage. Particulates, sulfur oxides and nitrous oxides are emitted into the air. The adverse effects of particulate pollutants on visibility is considered the most important aesthetic impact of the complex. The strip-mining process will create some aesthetic damage. Although the soil banks will be leveled, reclamation in the sense of reestablishing a viable plant and animal eco-system is uncertain. Transmission lines radiate from the plant in several directions, passing through the Navajo Reservation and bringing the paraphernalia of development to a landscape which is in some places very beautiful and otherwise untouched.[9]

---

[8]The following facts may provide some idea of the magnitude of this operation and its attendant environmental problems. In 1970, the power plant had a capacity of 2,080 MW. The mine provides coal at a rate of 8.5 millions tons annually. Over the 40 year projected life span of the mine, 31,000 acres will be stripped. In 1970, approximately 550 people were employed in the mine and power plant, total value of sales of electricity was $146 million, and 96,000 tons of particulates, 73,000 tons of sulfur oxides and 66,000 tons of nitrous oxides were emitted annually.

[9]To place this aesthetic environmental damage in perspective, it may be useful to point out that the Four Corners Interstate Air Quality Control Region includes the greatest concentration of National Parks and Monuments in the United States and a number of Indian reservations, the largest of which are the Navajo and Hopi reservations. The value of the region for tourism and recreation depends largely on its bizarre and unusual landscapes, the enjoyment of which requires excellent long distance visibility and depth and color perception. There exists a substantial minority of "traditional" Native Americans who have strong

It was decided to use bidding games to measure the benefits of abatement of the aesthetic environmental damage associated with the Four Corners power plant and the Navajo mine. [10] Considerable attention was devoted to the design and development of bidding games which provide a reliable estimator of these benefits.

Questionnaire Design

The bidding games were part of prepared schedules designed for use in a personal interview survey of samples of users of the Four Corners Interstate Air Quality Control Region environment (i.e., residents and recreational visitors to the region). In preparation for the bidding games, respondents were asked a series of questions about environmental matters, to focus their attention on that topic. Then, the subject of the coal-electricity complex in the Four Corners area was explicitly raised. The respondents were shown three sets of photographs depicting three levels of environmental damage around the Four Corners Power Plant, near Fruitland, NM.

Set A showed the plant circa 1969, prior to installation of some additional emissions control equipment, producing its historical maximum emissions of air pollutants. Another photograph depicted the spoil banks as they appear following strip-mining, but prior to leveling. A third photograph showed electricity transmission lines marring the landscape. Set A depicted the highest level of environmental damage, and accurately represented the actual situation in the early years of operation of the plant.

Set B showed an intermediate level of damage. One photograph showed the plant circa 1972, after additional controls had reduced particulate emissions (i.e., the type of emissions most destructive of visibility). Another showed the spoil banks leveled but not revegetated; a third showed the transmission lines placed less obtrusively (i.e., at some distance from major roads).

Set C was intended to depict a situation where the industries continued to operate, but with minimal environmental damage. One photograph showed the plant with visible emissions reduced to zero. [11] A second photograph showed a section of arid land in its natural state; it was

---

religious and cultural attachments to nature, and who resent the air pollution, strip-mining, and transmission lines; witness the prolonged litigation about location of the Tucson Gas and Electric Company transmission line from the San Juan power plant, which is under construction about 9 miles from the Four Corners plant.

[10] In that part of the overall study which deals with nonaesthetic environmental damage, direct costing techniques are used.

[11] This feat was accomplished by photographing the plant on a day when all units were shut down.

intended to depict a situation where the transmission lines were placed underground and the strip-mined land completely reclaimed.

The interviewers pointed out the salient features of each set of photographs to each respondent. For most of the respondents (with the exception of many recreationists), the situations were rooted in real experience: the residents of the region were familiar with the plant and mine, and their operation for the previous eight years. Most remembered situation A well, for that was exactly how it was only a few years earlier. Situation B was a fairly good approximation of the real situation at the time of the interviews. With the help of the photographs, situation C would be readily visualized.

Since the fitting of bid or benefit curves requires an expression of willingness to pay for abatement of aesthetic damages, it was necessary to design games based upon appropriate vehicles of payment. The vehicles for payment were chosen so as to maximize the realism and credibility of the hypothetical situation posited to respondents. As will be discussed below, it was necessary to design and use a series of bidding games, because no one vehicle of payment was appropriate for use with all of the subpopulations sampled. First, the general format applicable to all games is discussed. Then, the particular games used for particular subpopulations are discussed.

For each bidding game played, respondents were asked to consider situation A, the highest level of environmental damage, as the starting point. The bidding games were designed to elicit the highest amount of money which the respondent, an adult speaking for his or her household, was willing to pay in order to improve the aesthetic environment to situation C, and to situation B. Answers were elicited in terms of "yes" or "no" to questions expressed in the form "would you pay amount X . . . ?" A "yes" answer would lead the enumerator to raise the amount and repeat the question, maybe several times, until a "no" answer was obtained. A "no" answer would lead the enumerator to reduce the amount until a "yes" answer was obtained. The amount which elicited the highest "yes" answer was recorded as the amount the respondent was willing to pay.

It was emphasized that the respondent was to assume that the vehicle for payment used in a particular game was the only possible way in which environmental improvements could be obtained. This stipulation was designed to minimize the incidence of zero bids as protests against either the zero liability rule implicit in "willingness to pay" games or the particular method of payment used in a particular game. If a respondent indicated that he was willing to pay nothing at all, he was asked a series of questions to find out why. A respondent indicating that he did not consider his household to be harmed in any way by the environmental damage and, therefore, saw no reason to pay for environmental improvements was recorded as bidding zero. If a respondent indicated that his zero bid was in protest against the game, his answer was analyzed as a nonresponse to the bidding game, since he had refused to play the game by the stated rules. [12]

The selection of appropriate vehicles for payment provided a challenge. People are not accustomed to paying for abatement of air pollution and strip-mining damage. However, they are accustomed to paying for many other types of useful goods and services, many of which, such as parks and highway beautification, have aesthetic or "quality of life" components. So selection of realistic vehicles for payment was not impossible. However, the heterogeneous nature of the affected population meant that no single vehicle was suitable for data collection among all groups. In the Four Corners Region, the affected population can be divided into three broad groups: (1) the residents of Indian reservations, primarily Navajos, but also including members of several other tribes; (2) the residents of the nonreservation sections of the region, primarily Anglo-Americans, but with a sprinkling of Spanish-Americans, Native Americans living off the reservations, and other minorities; and (3) the tourists and recreationists who visit the area to enjoy its unique natural, historical and cultural attractions. Different versions of the questionnaire, using bidding games based on different vehicles for payment, were constructed for use with the three different subpopulations of the affected population.

The particular bidding games used are described below.

THE SALES TAX GAME.   Members of all three subpopulations are familiar with the practice of paying sales taxes. For most, this is a frequent occurrence. It is also understood by most that income collected in sales taxes is used to provide useful public services. It does not require much imagination to conceive of a public agency collecting a sales tax from residents of the affected region and using the income to finance environmental improvements.

The sales tax bidding game was used for both the resident samples. It was not used with the recreationist sample, since that group often purchased only a few items in the region, bringing most of their equipment and supplies with them. This would make a regional sales tax largely irrelevant for that group.

Respondents were asked to assume that a regional sales tax was collected on all purchases in the Four Corners Interstate Air Quality Control

---

[12]For the purpose of estimating the benefits of abatement, the treatment of "protest bids" as nonresponses is legitimate. By definition, a "protest bid" recognizes that positive benefits from abatement exist, but registers a protest against a particular method of financing abatement. We recognize that the elimination of "protest bids" from analyses aimed at estimating the benefits of abatement fails to remove all downward bias from the responses to particular games: some respondents may bid low (i.e., underestimate the benefits to themselves of abatement) in conscious or subconscious protest against the method of financing assumed in a game.

Region for the purpose of financing environmental improvements. [13] All revenue from the additional tax would be used for abatement of aesthetic environmental damage associated with the power plant and mine, and all citizens would be required to pay the tax. Recreational visitors to the region would contribute to environmental improvement through payment of additional users fees for facilities.

THE ELECTRICITY BILL GAME. The monthly electricity bill seemed to be a suitable vehicle for measurement of willingness to pay. It is the production of electricity which causes the environmental damage, and most people can readily comprehend that reduction of the damage may raise the cost of operating the industry and that passing these additional costs on to consumers of electricity is a not unlikely outcome. For the residents of those sections of the region outside the Indian reservations, payment of a monthly electricity bill is a routinized behavior. Therefore, a bidding game based upon the monthly electricity bill was played with the nonreservation resident sample.

This game was unsuitable for use with the other two samples. Many residents of Indian reservations do not have electricity available in their homes. Recreationists do not not pay monthly electricity bills while vacationing away from home.

The respondent was first asked the amount of his monthly household electricity bill. He was then asked to imagine that an additional charge was added to his electricity bill, and the electricity bills of everyone who uses electricity produced in the Four Corners Region, even people as far away as southern California. All of the additional money collected would be used to repair the aesthetic environmental damage caused as a result of electricity production and transmission in the Four Corners region.

THE USER FEES GAME. Measuring recreationists' willingness to pay for environmental improvements raised problems which prevented use of the electricity bill and sales tax games. For the recreationists, a satisfactory game would need to focus upon (1) the activities associated with vacationing, and (2) the collection of payments while they are in the region and using the regional environment. The payment of user fees for recreation services (i.e., campsite, utilities hook-up, boat launching), seemed to be a promising vehicle for a bidding game for the recreationists. If visitors were concerned about environmental quality in the places where they vacation, the payment of an additional sum along with their usual daily user fees would provide a suitable way to express that concern.

A sample of recreationists in the national parks, monuments and forests and state parks in the region played a bidding game based on user fees.

---

[13]The regional sales tax would be additional to current state and local sales taxes and would be charged on all commodities subject to existing state and local sales taxes.

Only recreationists who were not residents of the region were included. They were first asked the total sum of user fees they paid daily. They were then asked to suppose user fees in all the recreation areas in the Four Corners area were increased. All the additional money collected would be spent on environmental improvements. All recreators would pay and the year-round residents would pay, too, through additional regional sales taxes.

The Conduct of the Survey

The bidding games, as described above, were included in prepared schedules which also served as the instrument for collection of data for socioeconomic analysis of citizen environmental concern. Personal interviews were conducted by enumerators who were closely supervised and who had been carefully trained in formal sessions and in two separate field pre-tests of the questionnaire. Interviews were conducted during the summer of 1972.

Usable questionnaires were completed by 526 residents of nonreservation sections of the Four Corners Interstate Air Quality Control Region, 71 residents of Indian reservations and 150 recreators and tourists from outside the region who were using recreation sites within the region. The ratio of reservation residents to nonreservation residents sampled was proportional to their total numbers in the regional population; the size of the recreationist sample was chosen arbitrarily. Respondents from each subpopulation were selected by stratified random sampling. Stratification was based on concentration of air pollutants above the respondent's home or recreation site, as estimated by an atmospheric diffusion model developed as part of the larger research project. The population in higher pollution concentration zones was sampled more heavily.

Analysis and Results

For the *determination of three points on the aggregate bid curve,* corresponding to the situations, A, B, and C, the bidding game results were aggregated by methods appropriate to the stratified random sampling technique used, to provide estimates of the total bid for the relevant population. Two methods of aggregation were used, to generate two different aggregate bid curves.

1. The results of the sales tax game with area residents (reservation and nonreservation) were added to the results of user fee games played by recreators to estimate a total regional willingness to pay for three levels of environmental improvement.

2. The results of the electricity bill game were extrapolated over all consumers of power from the Four Corners plant to estimate consumer willingness to pay. This latter procedure involved the ethical premise that, since the production of electricity causes environmental damage, all citizens who consume Four Corners power ought to be willing to pay as much in additional electricity charges for environmental improvements as those who live in the region which suffers the damage. However appealing this ethical premise may be, our survey did not include people outside the region. Thus the consumer bid cannot be interpreted as an estimate of true "willingness to pay." It would be interesting to extend this research to include bidding games for these consumers of Four Corners electricity who do not live or recreate in the affected environment.

While both the regional and consumer aggregate bids are of interest, the authors believe that more faith may be placed in the regional bid since that bid was derived from samples of all segments of its relevant population.

Table 1 presents the estimated aggregate bids, standard errors, and 95% confidence limits at points A, B, and C. Regional and consumer bids are presented.

Using the estimated aggregate bids (Table 1), a *regional aggregate bid curve* and a *consumer aggregate bid curve* were fitted. To fit two-dimensional aggregate bid curves, it was necessary to select a single independent variable to serve as a proxy for the total package of aesthetic environmental improvements under consideration. Situations A, B, and C were de-

TABLE 1. Aggregate Bids for Abatement of Aesthetic Environmental Damage Associated with the Four Corners Power Plant, 1972

	SITUATION		
ITEM	A	B	C
Emissions (tons of particulates per year)	96,000	26,000	0
Level of abatement (tons of particulates per year)	0	70,000	96,000
Estimated regional aggregate bid ($ millions per year)	0	15.54	24.57
Standard error ($ millions per year)	—	1.24	1.52
95% Confidence limits ($ millions per year)	—	±2.43	±2.97
Estimated consumer aggregate bid ($ millions per year)	0	11.25	19.31
Standard error ($ millions per year)	—	0.68	0.98
95% Confidence limits ($ millions per year)	—	±1.33	±1.92

fined so that all three forms of aesthetic damage (air pollution, strip-mining, and transmission lines) were successively reduced together from their most obtrusive in situation A to virtual elimination in C. Of the three forms of damage, reduced visibility due to particulate air pollution was considered by respondents to be far and away the most serious. So, abatement of particulate air pollutant emissions (measured as the difference, in tons per year, between the level at A and the levels at B and C, respectively) was arbitrarily chosen to serve as a single independent variable for graphical analyses. [14]

The form of the curve requires some discussion. It has already been noted [footnote 2] that the usual restraints placed on the slope of demand curves are inappropriate for the first derivatives of aggregate bid curves for public goods, due to the impossibility of separating quantity and quality factors. Here we have a case in point. It seems reasonable that "consumers" of abatement of particulate emissions desire the attribute, visibility. Given the reasonable assumption that marginal utility of additional visibility is diminishing, one would expect the first derivative of the aggregate bid curve for visibility to be of negative slope.

Meteorologists have established that an inverse relationship exists between visibility and concentration of particulate pollutants. Visibility increases at an increasing rate as particulate pollution (measured in terms of weight) is abated [3, 10]. Therefore, the slope of the marginal aggregate bid curve for abatement of emissions (in tons per year) is *a priori* unpredictable, since the diminishing marginal utility of visibility and the increasing marginal visibility resulting from additional abatement influence that slope in opposite ways.

In terms of visibility, the aggregate bid curve form which provided the best fit of the three data points was

$$B = c \ln (v),$$

where B = aggregate bid in dollars, c = a constant, and v = visibility.

In terms of abatement of particulate air pollutants (measured in tons per year), the appropriate curve form was

[14]In the case study at hand, we recognize the inelegance introduced by this procedure. We do not believe it does serious violence to the truth, since most of the aesthetic environmental damage occurring is, in fact, due to particulate air pollutants. We emphasize, however, that this problem should not typically occur in the use of aggregate bid methodology and bidding game techniques. Rather, its occurrence here was a special case and is attributable to our need to value a package of different aesthetic environmental improvements within the following constraints: (1) a limited research budget, which confined us to one personal interview survey, and (2) the need to limit the length of each interview, to avoid exhausting the patience of respondents.

**TABLE 2. Tests of Hypotheses Concerning the Slopes of the
Aggregate Bid Curves**

	CONFIDENCE OF REJECTING H_0	
HYPOTHESIS	REGIONAL AGGREGATE BID CURVE (%)	CONSUMER AGGREGATE BID CURVE (%)
1. The aggregate bid curve is of linear positive slope[a]	99.9	99.9
2. The aggregate bid for situation B is one half of that for C[b]	99.9	94.5

[a]Rejection implies that the aggregate bid curve is of increasing positive slope.
[b]Rejection implies that the aggregate bid for B exceeds one half of that for C.

$$B(q) = c \ln \frac{k}{k - q},$$

where k = a parameter relating visibility to emissions, which is determined behaviorally, and q = tons of particulate emissions abated annually.

The aggregate bid curve fitted using this equation form passes through the origin, as logically it must, given that rational citizens would bid zero for zero abatement. The first derivative of the aggregate bid curve is of positive slope.[15] Statistical tests (Table 2) resulted in rejection of the hypotheses (1) that the aggregate bid curve was linear, or of decreasing positive slope, and (2) that the aggregate bid at point B was simply one-half of that at C. Regional and consumer aggregate bid curves are presented (Fig. 2).

[15]It must be emphasized that the curve form used provided the best fit, given the three data points available. It would have been desirable to have collected information adequate to generate more data points. The decision to collect data for only three points was made in recognition of limits to the patience of respondents. The multipurpose schedule was already quite lengthy, given the need to collect data relevant to the situation of the respondent, play the bidding games, and collect socioeconomic, sociological and attitudinal data.

It is recognized that, if more data points had been available, a different curve form may have been appropriate. The possibility of a sigmoid aggregate bid curve is logically appealing. Such a curve would have a segment of increasing slope, where the increasing marginal visibility from particulate abatement dominates the decreasing marginal utility of additional visibility then, as complete abatement is approached (i.e., somewhere to the right of our point B), the slope may become decreasing as the diminishing marginal utility of visibility becomes dominant. Such a curve form would be consistent with theoretical considerations and with the three data points available.

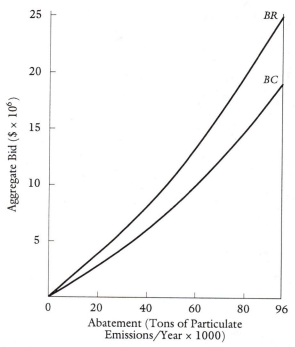

FIGURE 2. Estimated aggregate bid curves for abatement of aesthetic environmental damage, Four Corners power plant, 1972. BR, Regional aggregate bid; BC, Consumer aggregate bid.

The fitted aggregate bid curves were:

$$B_r = \$29{,}175{,}840 \; \ln \frac{168{,}890}{168{,}890 - q},$$ for the regional aggregate bid curve, and

$$B_c = \$15{,}396{,}700 \; \ln \frac{134{,}490}{134{,}490 - q},$$ for the consumer aggregate bid curve.

Marginal aggregate bid curves, or *price curves,* were generated by taking the first derivatives of the aggregate bid curves (Fig. 3). The derived price curves were:

$$P_r = \$\frac{29{,}175{,}840}{168{,}890 - q},$$ derived from the regional aggregate bid curve, and

$$P_c = \$\frac{15{,}396{,}700}{134{,}490 - q},$$ derived from the consumer aggregate bid curve.

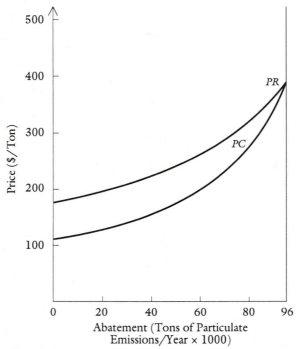

FIGURE 3.   Derived price curves for abatement of aesthetic
environmental damage, Four Corners power plant, 1972. PR, price
curve derived from regional aggregate bid; PC, price curve derived
from consumer aggregate bid.

These derived price curves are very useful for public policy analyses
with respect to optimal environmental management policies. In Fig. 4 a
hypothetical derived price curve is presented, along with a hypothetical
marginal cost of abatement curve. In this hypothetical example, the optimal level of abatement is S. A standard allowing maximum annual emissions of $(T-S)$ tons of particulates would be appropriate, and the penalty for violation of that standard should be set sufficiently high that the
polluter's expected penalty per ton of emissions in excess of the standard
would be at least P. An alternative institutional framework would call
for a fine or tax per ton of particulate emissions. The fine ought to be set
at least as high as P per ton. At the level P, the optimal level of abatement would be achieved. A fixed fine per ton of remaining emissions
would result in collection of the amount $XMTS$. However, since the
derived price curve is of positive slope, the sum of the fines collected
would be insufficient to compensate the receptors of the pollutants for
their loss in welfare. The necessary amount would be $XNTS$. This
amount could be collected, if full compensation were the accepted poli-

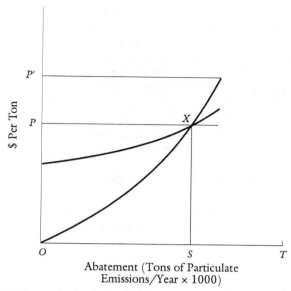

FIGURE 4. Optimal standards, penalties and per unit taxes on emissions, given hypothetical marginal cost and price curves.

cy,[16] by using a sliding scale of fines, ranging from P' for the first ton of emissions down to P for all emissions in excess of $T - S$.

If the marginal costs of abatement of aesthetic environmental damage associated with the Four Corners power plant were known,[17] the derived price curves presented in Fig. 3 could be used to perform policy analyses similar to those in the hypothetical example above.

The *relationship between willingness to pay and household income* is of

[16]Under a full compensation policy, a derived price curve generated from bidding games based on the concept of willingness to pay (which implicitly places the liability with the receptor of damages) would underestimate both the optimal level of abatement and the appropriate level of fines or taxes. Randall [15] and others have demonstrated that the demand for abatement of an external diseconomy is greater in the full liability situation than in the zero liability situation implicit in willingness to pay games.

[17]We are not yet in a position to present a complete benefit/cost analysis of the abatement of the aesthetic environmental damage associated with the Four Corners power plant and Navajo mine. Preliminary and tentative calculations indicate that, *if our attribution of most of the benefits reported here to abatement of particulate air pollutants is reasonable,* 99.7% abatement of particulate emissions (the current New Mexico standard for 1975) is economically justified on the basis of aesthetic considerations alone. Some additional abatement beyond the 1975 standard may be justified. The economic benefits from that abatement which has already taken place appear to far exceed the costs.

This conclusion is extremely tentative and subject to revision. It is presented in this footnote (at the request of an anonymous reviewer) to provide a "ball park" indication of the conclusions which may arise from our research.

interest. However, the concept of income elasticity of demand is inappropriate to the public good under study. The calculation of an income elasticity of quantity of abatement demanded would require consideration of the relationship between income and quantity of homogeneous units of abatement demanded at a constant price per unit. However, in this study there were no explicit unit prices for abatement; neither were there individual variations in quantity of abatement demanded, as the quantities were fixed as defined in situations A, B, and C. These conditions result from the inherent nonexclusive nature of abatement of aesthetic environmental damage: Everyone obtains the same quantity and there is no explicit price. This situation is the inverse of the market situation for private goods; dollar bids are the response to a quantity which is given.

Since there existed no market price at which to calculate the income elasticity of demand, an "income elasticity of bid" was estimated. The income elasticity of bid was defined as:

$$e_Y = \frac{dB}{dY}\frac{Y}{B} = b_1\frac{Y}{B},$$

where Y = household income, and B = the individual's total annual bid. A linear regression model was used to determine the statistic b_1. The mean value of Y and B were used, and the calculation was made at each level of abatement.

Calculated income elasticities of bid for the various subpopulations and bidding games are presented in Table 3. In all cases, income elasticity of bid was greater than zero, indicating that higher income households were willing to pay a greater amount than lower income households to achieve the same level of abatement of aesthetic damages. For the nonreservation residents, calculated income elasticity of bid ranged from 0.39 to 0.65, depending on the game and the level of abatement. Income elasticity of bid was significantly greater than zero at the 95% level of confidence. For the residents of Indian reservations and the recreational visitors to the region, lower positive income elasticities were recorded. These were not significantly greater than zero, at the 95% level of confidence. [18]

It was also found that willingness to pay an additional charge in the

[18]The estimates of income elasticity obtained with the nonreservation resident sample may be more reliable, for two reasons. First, the nonreservation sample was considerably larger than either of the other two samples. Second, the range of incomes encountered in the nonreservation resident sample more nearly approached that of society as a whole. The reservation resident sample was representative of its underlying population, in which incomes are concentrated at the extreme lower end of the national range. The visiting recreators had a mean household income about fifty per cent greater than the national average; very few recreators had incomes in the lower half of the national range.

TABLE 3.   Income Elasticity of Bid for Abatement of Aesthetic Damages Associated with the Four Corners Power Plant, 1972

SUBPOPULATION	GAME	LEVEL OF ABATEMENT	INCOME ELASTICITY OF BID	STANDARD ERROR	SIGNIFICANTLY GREATER THAN ZERO[a]?
Nonreservation residents	Sales tax	B	0.65	0.10	Yes
		C	0.65	0.08	Yes
Reservation residents	Sales tax	B	0.23	0.18	No
		C	0.24	0.18	No
Nonreservation residents	Electricity bill	B	0.54	0.09	Yes
		C	0.39	0.06	Yes
Recreators	Users fees	B	0.09	0.15	No
		C	0.16	0.11	No

[a]At the 95% level of confidence.

electricity bill for a particular level of abatement increased as the size of the electricity bill increased. Electric bill elasticity of bid, as defined as

$$e_b = \frac{dB}{d\text{Bill}} \frac{\text{Bill}}{B},$$

was calculated (for the nonreservation resident sample) to be 0.30 for situation B and 0.25 for situation C; at both points, it was significantly greater than zero at the 95% level of confidence. These estimates indicate that willingness to pay for a given level of environmental improvements increased as the size of the electricity bill increased, but at a lesser rate.

The Reliability of the Results

In the statistical sense, our estimates of the aggregate benefits from abatement of aesthetic environmental damage would seem to be of a high order of reliability. The 95% confidence limits of the aggregate bids are quite narrow, compared with the size of the estimated aggregate bids. Statistical estimates of the confidence which may be placed in these estimated aggregate bids are based upon the variance of the responses of the samples, and indicate the confidence with which sample results may be extrapolated to the whole population. These statistics, *per se,* are unable to give any indication of the reliability with which predispositions to behave, as measured by the bidding games, would be transmitted to actions should the hypothetical situation arise.

We argue, nevertheless, that our estimates of the benefits of abatement of aesthetic environmental damages associated with the Four Corners power plant are of a reasonable order of magnitude and, if anything, conservative. (1) We believe the design of the bidding games allows confidence in their efficacy. (2) The individual household bid for abatement, on average, is of the same order of magnitude as the estimates of the value of particulate pollution abatement obtained in revealed demand studies [1], when the latter are converted to a comparable basis. Mean individual household willingness to pay for abatement, measured by the sales tax game played with the nonreservation resident sample, was about $50 annually to achieve situation B and $85 annually to achieve situation C. (3) The estimated aggregate bids for abatement are relatively small given the scale of the operation at Four Corners, as indicated by its 1970 emissions rate and its total annual sales of $146 million [footnote 5]. (4) Theoretical analyses indicate that the demand for abatement of an externality will be lower under a zero liability rule than under intermediate or full liability rules [15]. The bidding games used were based on zero liability rules, and they should be expected to yield conservative estimates of the benefits of abatement.

It is recognized that three data points provide an inadequate basis on which to draw conclusions with respect to the shapes and slopes of the aggregate bid curves and their first derivatives. However, it is consistent with theoretical considerations and with the limited data available that the aggregate bid curves may have at least a segment with increasing slope.

It would seem that the income elasticity of bid and the electric bill elasticity of bid fall in the range from zero to −1. This result was consistent with our prior expectations.

CONCLUDING COMMENTS

In the case study reported, bidding games were used to estimate the benefits which would accrue from abatement of the aesthetic environmental damages associated with the Four Corners power plant and the Navajo mine. The problem situation was not amenable to the use of direct costing nor revealed demand techniques.

This study has revealed that substantial benefits may be gained from abatement of aesthetic environmental damage associated with the Four Corners power plant and Navajo mine. These potential benefits have not been revealed or realized in the market place. However, the process of political and institutional change has led to the imposition of increasingly rigorous control standards for particulate emissions from the plant, indicating a recognition, in some broad sense, that benefits may be gained from emissions controls. Our contribution has been to attempt a quantification of these benefits.

We believe that the use of bidding game techniques was successful in meeting the objective, valuation of these benefits. Bidding game techniques seem amenable to use as a research tool for valuation of a wide variety of nonmarket goods. It must be understood, however, that bidding games measure the hypothetical responses of individuals faced with hypothetical situations. Thus, considerable care must be exercised in the design of bidding games and the conduct of surveys for data collection, to ensure that the results obtained are as reliable as possible.

REFERENCES

1. R. J. Anderson, Jr. and T. D. Crocker, Air pollution and housing: Some findings, Paper No. 264, Institute for Research in the Behavioral, Economic and Management Sciences (January 1970).
2. D. F. Bradford, Benefit-cost analysis and demand curves for public goods, *Kyklos* 23, 775–791 (1970).

3. R. J. Charlson, N. C. Ahlquist, and H. Horvath, On the generality of correlation of atmospheric aerosol mass concentration and light scatter, *Atmos. Environ.* 2, 455–464 (1968).

4. M. Clawson and J. L. Knetsch, "Economics of Outdoor Recreation," Johns Hopkins Press, Baltimore (1966).

5. I. Crespi, What kinds of attitude measures are predictive of behavior? *Pub. Opin. Quart.* 35, 327–334 (1971).

6. O. A. Davis and A. B. Whinston, On the Distinction Between Public and Private Goods. *Amer. Econ. Rev.* 57, 360–373 (1967).

7. R. K. Davis, Recreation planning as a economic problem, *Natural Res. J.* 3, 239–249 (1963).

8. J. F. Engel, D. T. Kollat, and R. D. Blackwell, Attitude formation and structure, in "Consumer Behavior," Holt, Rinehart, and Winston, New York (1968).

9. H. Erskine, The polls: Pollution and its costs, *Pub. Opin. Quart.* 36, 120–135 (1972).

10. H. Ettinger and G. W. Roger, Particle size, visibility and mass concentration in a nonurban environment, Los Alamos Scientific Laboratory, LA-DC-12197 (1971).

11. J. A. Jaksch and H. H. Stoevener, Effects of air pollution on residential property values in Toledo, Oregon, Agricultural Experiment Station Special Report 304, Oregon State University, Corvallis (1970).

12. A. V. Kneese and B. Bower, "Managing Water Quality: Economics, Technology and Institutions," Johns Hopkins Press, Baltimore (1972).

13. L. Lave and E. Seskin, Air Pollution and Human Health, *Science* 169, 723–732 (1970).

14. H. O. Nourse, The effect of air pollution on house values, *Land Econ.* 43, 181–189 (1967).

15. A. Randall, On the theory of market solutions to externality problems, Agricultural Experiment Station Special Report 351, Oregon State University, Corvallis (1972).

16. R. G. Ridker and J. A. Henning, The determination of residential property values with special reference to air pollution (St. Louis, Missouri), *Rev. Econ. Stat.* 49, 246–257 (1967).

17. K. F. Wieand, Air pollution and property values: A study of the St. Louis area, *J. Reg. Sci.* 13 91–95 (1973).

18. R. O. Zerbe, Jr., The economics of air pollution: A cost-benefit approach, Ontario Dept. of Public Health, Toronto (1969).

22

The Economic Value of Life: Linking Theory to Practice*

J. STEVEN LANDEFELD

EUGENE P. SESKIN**

J. Steven Landefeld is Director of the Business Issues
Analysis Division and Eugene P. Seskin is Chief of the
Income Branch of the Bureau of Economic Analysis, both in
the U.S. Department of Commerce.

INTRODUCTION

Among the most hotly debated areas of public policy are those involving risks to human health and safety. For example, in the occupational area, industry representatives claim that exposure standards are so stringent they would fail any reasonable cost-benefit analysis. At the same time, public interest groups claim that cost-benefit analysis is a biased tool used to further the ends of industry to the detriment of workers' health and safety [1].

"The Economic Value of Life: Linking Theory to Practice," by J.
Steven Landefeld and Eugene P. Seskin, from *American Journal of
Public Health,* vol. 72 (1982), pp. 555–61. Reprinted by permission.

*The selection reprinted here omits the final section of the article on "Adjusted Willingness-to-Pay/Human Capital Estimates."—*Eds.*

**We would like to thank Thomas A. Hodgson for providing unpublished data and helpful comments, and Paul R. Portney, Nancy M. Gordon, and two anonymous referees for their useful suggestions. Any remaining errors are, of course, our responsibility; the opinions and conclusions expressed are not necessarily those of the Bureau of Economic Analysis or the US Department of Commerce.

Central to this debate is the valuation of human life. Although some claim the value of human life cannot be expressed in monetary terms, the competing demands on scarce public funds require that some value be placed on programs that save lives. Refusal to place an explicit value on life merely forces implicit valuations that are made as part of decisions to fund or not to fund public projects as well as decisions to take other regulatory actions.

Most economists writing on these issues agree that the conceptually correct method to value risks to human life in cost-benefit analyses should be based on individuals' willingness to pay (or on individuals' willingness to accept compensation) for small changes in their probability of surviv-al.[1] Despite this agreement, however, controversy continues on the appropriate technique for actually producing estimates for valuing risks to life. This paper reviews the major issues in this area.*

METHODS USED TO VALUE LIFE

Human Capital (HK)

The human capital (HK) approach to valuing life has a long history dating back to the works of Petty [2] and Farr [3]. Later studies by Fein [4], Mushkin and Collings [5], Weisbrod [6], and Klarman [7] polished and improved the theoretical and practical underpinnings of the approach. Finally, Rice [8] in her pathbreaking article, "Estimating the Costs of Illness," effectively codified the empirical application of the technique.

In the standard HK approach, it is assumed that the value to society of an individual's life is measured by future production potential, usually calculated as the present discounted value of expected labor earnings.[2]

---

[1]It should be recognized that the amount people are *willing* to pay depends, in part, on their *ability* to pay. Hence, any estimate of willingness to pay is dependent upon a given distribution of income. Similarly, because of these wealth effects, the amount people are willing to pay may differ from the amount they would be willing to accept as compensation.

*In the final section of this article, which is not reprinted here, the authors present a reformulated version of the human capital approach using a willingness-to-pay criterion, which they conclude produces "the only clear, consistent and objective values for use in cost-benefit analysis of policies affecting risks to life."—*Eds.*

[2]Rice [8] and Cooper and Rice [9] estimated the cost of illness in the United States by using the HK approach to calculate *indirect* costs in the form of forgone earnings due to sickness and death and then adding the *direct* costs based on medical expenditures for prevention, diagnosis, and treatment. Subsequently, Hartunian, Smart, and Thompson [10] suggested that in assessing prevention programs, incidence-based estimates of the costs of illness should replace the prevalence-based estimates used by Rice. More recently, it has been suggested that especially for fatal illnesses, an adjustment should be made to account for the fact that

Some analysts, like Weisbrod, have employed expected earnings net of consumption, based on the notion that when an individual dies not only is productive contribution lost, but also claims on future consumption [12]. Therefore, as would be the case for physical capital, the net loss to society is the difference between earnings and maintenance (consumption) expenditures. Whether the gross HK approach or the net approach— adjusting for consumption—is employed, each is implicitly based upon the maximization of society's present and future production.[3]

Since standard HK estimates are constructed from society's perspective, labor earnings are evaluated before taxes as representing the actual component of GNP, rather than after-tax earnings which represent the relevant magnitude to the individual. In addition, nonlabor income is excluded since individual capital holdings (and associated earnings) are not materially affected by an individual's continued existence. Thus, standard HK estimates incorporate a zero value for persons without labor income such as retired individuals with only investment or pension income.

By its emphasis on economic product, the HK approach also ignores other dimensions of illness and death as well as nonmarket activities that may be more important to an individual than economic loss. These include pain and suffering, aversion to risk, and loss of leisure which, itself, has value for the individual and perhaps for others as well. Furthermore, the only adjustment for nonmarket activities in HK estimates is the imputation of a value for housekeeping activities. These calculations are usually based on available information from time-use studies combined with data on wages of market substitutes for the relevant household activities.

An important issue that must be resolved to implement the HK approach involves the choice of an appropriate "social" discount rate to convert future earnings into present values. The problem amounts to determining what society forgoes when it invests in life-saving programs. The choice is made difficult because of the effects of taxes and risk aversion which cause the rate of return to society's investments to diverge from the rates of return to private investments. For example, taxation means that the before-tax rate of return to private investment (the marginal produc-

---

while those who die may suffer substantial costs before their deaths, society also avoids significant health-related expenses because of their deaths; on this point, see National Academy of Sciences [11].

[3]It should be noted, however, that if society merely wanted to maximize the gross national product (GNP)—a common measure of welfare—a less expensive course of action than investments in life saving might be to eliminate immigration barriers or discourage birth control. Furthermore, health measures that reduce death rates and add to GNP by increasing the labor force, could actually lower per capita GNP, if the number of workers increases relative to the amount of output they produce (see Mushkin and Landefeld [13]).

TABLE 1. Present Value of Future Earnings of Males by
Selected Age Groups, 1977[a]
(1977 dollars)

| AGE GROUP | REAL DISCOUNT RATE | | |
(YEARS)	2.5 PER CENT	6 PER CENT	10 PER CENT
1 to 4	405,802	109,364	31,918
20 to 24	515,741	285,165	170,707
40 to 44	333,533	242,600	180,352
65 to 69	25,331	21,801	18,825

[a]Dollar figures based on the present value of both expected lifetime earnings
and housekeeping services at 1977 price levels and an annual increase in labor
productivity of 1 per cent.
Source: Dolan, Hodgson, and Wun [14].

tivity of capital) will exceed the after-tax rate of return to the individual
investor (the individual's rate of time preference).[4] Risk aversion works in
the same direction, resulting in a risk premium that causes a divergence
between the marginal productivity of capital and the individual's rate of
time preference.

Although the problem of choosing a discount rate arises in the evaluation of investments in most public programs, the long life of "investments" in HK—with life expectancies on the order of 70 years—greatly
magnifies the difficulty. For example, Table 1 shows a range of HK values
when different *real* discount rates are used to estimate the present value of
forgone earnings of males by selected age groups.[5] As can be seen, the
values are significantly larger for low discount rates than for high ones,
especially in the case of children whose stream of earnings will be the
longest. At a real rate of 10 per cent—the rate of discount recommended
by the Office of Management and Budget (OMB)[6]—the value for males
aged 1 to 4 is less than one-tenth the value corresponding to a real rate of
2.5 per cent. At the same time, choice of a discount rate can affect the
relative valuations placed on persons in specific age groups. For example,
according to the Table, at a real rate of 6 per cent, males aged 20 to 24 are
valued higher than males aged 40 to 44; whereas, at a real rate of 10 per

[4]A standard result of capital theory is that, given perfect capital markets, rational individuals
will adjust their savings and consumption patterns so that their private rate of time preference will equal their private rate of return on investments.
[5]Since estimates of future earnings are generally made in constant or base-year dollars,
whatever nominal discount rate is chosen must be converted to a real rate by an adjustment
for inflation.
[6]According to OMB, "the prescribed rate of 10 per cent represents an estimate of the average
rate of return on private investment before taxes and after inflation" [15].

cent, the reverse is true. The usual response to these types of occurrences is for researchers (see, for example, Berk, *et al* [16]) to present sensitivity analyses using several discount rates. However, while this is notable on paper, public officials often complain, with some justification, that this practice is not only confusing, but it can also lead to abuses in which persons choose a rate that is most favorable to the outcome they desire.

Despite the conceptual problems associated with the HK approach, the technique is widely used. [7] To some extent this is an artifact of the relative ease in computation since necessary data are not difficult to obtain. More to the point, the standard HK approach has the virtue of providing numerical estimates that are indisputable measures of what they say they are: objective numbers based on life expectancy, labor force participation, and projected earnings.

Willingness to Pay (WTP)

Because of the inadequacies associated with the HK approach, economists have searched for alternatives. Mishan, in one of the early theoretical discussions on valuing human life, suggested that the only logically consistent basis for the valuation of loss of life in safety decisions should be the same criterion used by welfare economists in other areas of cost-benefit analysis, namely, the "potential Pareto improvement principle" [18].

A "potential Pareto improvement" is said to exist when individuals who gain from a social change are able to compensate those who stand to lose from the change and still leave a net gain. Thus, Mishan concluded that the relevant question is, What are individuals "willing to pay" (or accept as compensation) for a change that will affect loss of life? He went on to note that in most public safety decisions, the issue is not the value of an *identified* individual's life; rather, it is the value of a reduction in the probability of death for a given population. In other words, it is the aggregate value a population at risk places on programs that save "statistical" lives or the sum of the amounts individuals are willing to pay *ex-ante* to "buy" small reductions in the probability of death in the population. [8]

An example may help to illustrate this point. Suppose each person in a population of 100,000 is willing to pay $25 for a program that is expected to reduce the overall probability of death from 0.0009 to 0.0008. Since this is equivalent to a reduction in the death rate from 90 per 100,000 to 80 per 100,000, the implied value per each of the 10 "statistical" lives saved is $250,000.

---

[7] A recent survey of the literature by the Public Services Laboratory [17] turned up well over 230 separate cost-effectiveness/cost-benefit analyses of illness, many of which used the standard HK approach.

[8] See Mishan [18], pp 159–163.

In theory, WTP represents a comprehensive measure of the private valuation individuals place on small reductions in the risk (probability) of death. [9] Conceptually, everything that contributes to an individual's well-being would be captured in the measure, including nonlabor income, the value of leisure, aversion to risk, the value of avoiding pain and suffering, and the value he places on lives of others whose risk of death may be affected. In addition, the WTP would incorporate an implicit rate of time preference reflecting the weight given to future benefits of living.

A number of researchers have acted upon Mishan's counsel and attempted to calculate WTP estimates of the value of a statistical life. These attempts have proceeded along two general directions: 1) analyses of direct survey responses by individuals, and 2) statistical estimation of individuals' revealed preferences. Each approach has problems associated with it.

Acton [19], Jones-Lee [20], and Landefeld [21] have carried out survey estimates of individuals' WTP for reductions in risk of death (see Table 2). Acton, for example, asked individuals open-ended questions about their WTP for a coronary care unit that would reduce risk of death from heart attack by 0.002. He found that the average person in his sample was willing to pay approximately $76 for the unit. Stated differently, the aggregate WTP, in a community of N such individuals, would be $76N/0.002N = $38,000 per statistical life saved. [10] Jones-Lee [20], in a survey concerning safety and airline travel, employed similar methods and found a value per statistical life of $8.4 million. Landefeld [21], in another survey on the WTP for reducing cancer mortality, calculated a value of $1.2 million per statistical life saved.

Although the survey method is an improvement over the HK approach in that it is based on WTP, there remain serious problems in its application. For example, what individuals say they will do may vary considerably from what they will actually do when confronted with a true market test, often because of a lack of information. Furthermore, strategic behavior may occur: If, on the one hand, respondents believe that they will be assessed amounts equal to their WTP, they may deliberately understate their WTP, especially if provision of a "public good" is in question such as a program that decreases cancer incidence for all. If, on the other hand, respondents do not believe that they will be assessed according to their WTP, they may overstate their WTP in an attempt to promote the provision of a public good. [11] Finally, social psychologists, reviewing these

---

[9]Note, that people are not asked to place a value on saving (with probability one) their own, relatives', or any other identified lives. For many, the only limit to these latter valuations would be one's ability to pay.

[10]As noted in Table 2, all figures are in 1977 dollars.

[11]This difficulty in ascertaining the true WTP for publicly provided programs is known as the "free-rider problem."

TABLE 2. Willingness-to-Pay Estimates of the Value of Life[a]

METHOD	VALUE PER STATISTICAL LIFE (THOUSANDS OF 1977 DOLLARS)[b]
Survey Approach	
Acton [19]	38
Jones-Lee [20]	8,440
Landefeld [21]	1,200
Revealed Preference	
Labor Market	
Dillingham [22]	277
Thaler and Rosen [23]	364
Viscusi [24]	1,650
Smith [25]	2,045
Olson [26]	5,935
Consumption Activity	
Dardis [27]	101[c]
Ghosh, Lees, and Seal [28]	260
Blomquist [29]	342
Portney [30]	355

[a]Where a study included a "central" or "most reasonable" estimate, that is shown; where only a range was given, the lowest value is presented.
[b]Values were converted to 1977 dollars using the US Bureau of Labor Statistics Consumer Price Index (CPI).
[c]It is unclear from the Dardis study [27] what year's dollars apply, although the estimate presented here appears to be based on an average value for the period 1974–1979.

survey attempts to value life, have raised questions about the ability of individuals to respond rationally and consistently to the abstract and complex questions involving hypothetical risk [31, 32]. For example, an increase of 0.00002 in workers' risk of death would represent an increase in the overall work fatality rate of approximately 25 per cent. Yet, one would expect quite different WTP responses depending on which of these two ways were used to characterize this change in risk in a survey questionnaire.

The studies undertaken to determine WTP on the basis of revealed preferences have been based on observations of compensation necessary to induce individuals to voluntarily assume risk. Two categories of such studies are discussed below, one based on compensating differentials in the labor market, the other based on compensating differentials pertaining to consumption activities of more general populations. One might expect estimates based on labor market studies to differ from those based on consumption activities since the former relate more directly to WTP for a reduction in an individual's own risk, while the latter relate to WTP for a reduction in risk to others as well.

As can be seen from Table 2, the labor market studies that have examined the extra compensation necessary to induce workers to take risky jobs—Dillingham [22], Thaler and Rosen [23], Viscusi [24], Smith [25], and Olson [26]—have yielded dollar estimates for the value of a statistical life ranging from $277,000 to more than $5.9 million. One explanation for this wide range is that important characteristics of the worker remain unmeasured in the statistical analyses. This can bias the resulting estimates and cause instabilities in the calculated values for a statistical life. However, even when Brown [33] used a data set that included a number of worker characteristics omitted from, or poorly measured in, other studies (for example, educational attainment, marital status, and health problems), he was unable to narrow the range in the estimated values of life. [12]

The large range in the estimates of the value per statistical life from the labor market studies can be attributed to at least five general problems:

- First, wage premiums may not accurately reflect worker risk preferences if workers have incomplete information regarding the risks to which they are exposed. For example, young and inexperienced workers (who actually have the highest accident rates) will underestimate the risks to which they are exposed if they use information on risks to all workers.
- Second, wage premiums may not be accurate measures of worker preferences if there are significant imperfections in the labor markets. This may be the case if new, inexperienced workers have relatively little bargaining power to demand appropriate premiums for risk.
- A third but related problem is sample self-selection. That is, either because of low incomes, lack of economic opportunities, or specific individual preferences, those who work in risky jobs will exhibit less risk aversion than the population as a whole. [13] Thus, WTP valuations based on risk premiums paid to such persons will understate the correct values applicable to the general population.
- Fourth, statistical problems arise in attempting to separate risk of death from risk of injury since compensating wage differentials will

[12]As noted in Table 2, where the study included a "central" or "most reasonable" estimate, this is shown; where only a range was given, the lowest value is presented. In many cases the range of estimates was quite large and the size and significance of the risk coefficients (from which the estimates are derived) were extremely sensitive to the other variables included in the estimating equation and to the particular functional form used.

[13]Some may even be risk takers in that they will accept a wage premium that is less than the amount necessary to compensate them for the expected value of their loss associated with the increased risk.

be accounting for both types of risks in most hazardous jobs. Smith [25], for example, was unable to separate statistically the independent effects of risk associated with worker injuries from risk associated with worker fatalities.[14] Thus, the estimated wage premium associated with increased risk of death in certain jobs may have included a premium associated with increased risk of injury.

- Finally, data constraints may bias the statistical estimates. For example, estimates generated in the Viscusi [24], Smith [25], and Olson [26], studies are biased upward because aggregate industry data were used instead of individual (micro) data. Consider Viscusi's procedure. He assigned a risk premium based on industry data to blue-collar workers in his sample who perceived themselves to be in relatively risky jobs. Those workers who did not perceive their jobs as risky were assigned a zero risk premium. Yet, one would expect individual risk premiums for all blue-collar workers to be significantly higher than industry risk premiums since the latter are based on data that include office workers and others in less risky environments. Hence, by using the lower industry values instead of the actual compensation necessary to attract workers to the more risky jobs, Viscusi may have overestimated the implied value per statistical life.[15]

In order to circumvent some of the problems noted above, especially the issue of sample self-selection, researchers have employed more general populations in estimating compensating differentials associated with various consumptions activities: Dardis [27] examined purchases of smoke detectors; Ghosh, Lees, and Seal [28] analyzed time, fuel, and risk trade-offs in highway driving; Blomquist [29] investigated seat belt use; and Portney [30] examined housing values and environmental risk. Although the estimates from the consumption activity studies span a narrower range than those from the labor market studies (see Table 2), many of the same data and statistical problems remain. For example, quantitative information on the risk-reducing potential of the various activities is scarce. Thus, in a study such as the one by Dardis [27], risk estimates associated with the

---

[14]When Smith included a control variable for injury rates, its estimated coefficient was not statistically significant. He attributed this finding to the fact that expected uncompensated losses from injuries are very small compared to those associated with death. However, given the higher likelihood of injury than death and the incomplete coverage of workers under compensation laws, it is more likely that the lack of significance was an artifact of multicollinearity problems associated with the correlation between fatality rates and injury rates.

[15]Suppose Viscusi's sample consisted of workers in an industry, $9/10$ of whom were in zero-risk jobs, $1/10$ of whom were in jobs associated with an increased risk of death of 0.001. Use of the corresponding industry risk factor of 0.0001 (($9/10 \times 0$) + ($1/10 \times 0.001$)) will substantially overstate the compensation per unit of risk; each statistical life will be valued at \$1,650,000 (\$165/0.0001) when the actual value should be \$165,000 (\$165/0.001).

use of smoke detectors were based only on a personal communication with an employee of the National Bureau of Standards. The statistical difficulty of separating risk premiums from other confounding factors is also severe in these studies. For example, in estimating the risk premium associated with air pollution exposure, Portney [30] attributed the entire premium paid for improved air quality to reduced risk of death. However, as Portney recognized, [16] this overstates the "true" risk premium associated with death since "clean air also means lower cleaning, painting, and repair bills as well as enhanced aesthetic appeal."

Thus, in view of the large range of values associated with the survey results and despite some promising findings from the revealed-preference studies based on compensating differentials, practical application of these approaches is difficult. [17]

*   *   *

REFERENCES

1. Green M, Waitzman N: Business War on the Law: An Analysis of the Benefits of Federal Health and Safety Regulation. Washington, DC: Corporate Accountability Research Group, 1979.
2. Petty W: Political Arithmetick, or a Discourse Concerning the Extent and Value of Lands, People, Buildings, Etc. London: Robert Caluel, 1699.
3. Farr W: Contribution to 39th Annual Report of the Registrar General of Births, Marriages, and Deaths for England and Wales, 1876.
4. Fein R: Economics of Mental Illness. New York: Basic Books, 1958.
5. Mushkin SJ, Collings Fd'A: Economic costs of disease and injury. Public Health Rep 1959; 74:795–809.
6. Weisbrod BA: Costs and benefits of medical research: a case study of poliomyelitis. J Political Economy 1971; 79:527–544.
7. Klarman HE: Syphilis control programs. In: Dorfman R (ed): Measuring the Benefits of Government Investments. Washington, DC: Brookings Institution, 1965; pp 364–414.
8. Rice DP: Estimating the costs of illness. Am J Public Health 1967; 57:424–440.
9. Cooper B, Rice DP: The economic cost of illness revisited. Soc Sec Bull 1976; 39:21–36.

---

[16]See [30], p. 77.

[17]Blomquist [14] has defended the large variance in WTP estimates by noting that a pattern exists in which life values based on WTP decrease as the magnitude of the risk increases. In fact, he even suggests that this inverse ordering be used in public policy analysis. However, this ordering may be merely an artifact of the inability of individuals to perceive extremely low risks of death (or differences between low risks of death). That is, even if individuals are willing to pay more for larger risk reductions, they may be unwilling to pay proportionately more because they may be unable to distinguish between very small changes in risk of death. Hence, if life values were derived by dividing observations on relatively constant WTP by increasing levels of (relatively small) risk reductions, this pattern would always emerge.

10. Hartunian NS, Smart CN, Thompson MS: The incidence and economic costs of cancer, motor vehicle injuries, coronary heart disease, and stroke: a comparative analysis. Am J Public Health 1980; 70:1249–1260.
11. National Academy of Sciences, Institute of Medicine: Costs of Environment-Related Health Effects: A Plan for Continuing Study. Washington, DC: National Academy Press, 1980.
12. Weisbrod BA: Economics of Public Health. Philadelphia: University of Pennsylvania Press, 1961.
13. Mushkin SJ, Landefeld JS: Economic Benefits of Improvements in Mortality Experience. Washington, DC: Public Services Laboratory, Georgetown University, 1977.
14. Dolan TJ, Hodgson TA, Wun WM: Present values of expected lifetime earnings and housekeeping services, 1977. Hyattsville, MD: National Center for Health Statistics, Division of Analysis, February 1980.
15. Schultz GP: Discounting in the evaluation of projects with time delayed benefits. Circular no. A-94. Washington, DC: Office of Management and Budget, 1972; p. 3.
16. Berk A, Paringer L, Mushkin SJ: The economic cost of illness, fiscal 1975. Med Care 1978; 16:785–790.
17. Hu T-w, Sandifer F: Synthesis of Cost of Illness Methodology. Washington, DC: Public Services Laboratory, Georgetown University, 1981.
18. Mishan EJ: Cost-Benefit Analysis. New York: Praeger, 1971.
19. Acton JP: Evaluating Public Programs to Save Lives: The Case of Heart Attacks, Santa Monica, CA: Rand Corporation, R-950-RC, January 1973.
20. Jones-Lee MW: The Value of Life: An Economic Analysis. Chicago: University of Chicago Press, 1976.
21. Landefeld JS: Control of New Materials with Carcinogenic Potential: An Economic Analysis. College Park, MD: University of Maryland, Unpublished PhD dissertation, 1979.
22. Dillingham AE: The Injury Risk Structure of Occupations and Wages. Ithaca, NY: Cornell University, Unpublished PhD dissertation, 1979.
23. Thaler R, Rosen S: The value of saving a life: evidence from the labor market. In: Terleckyj NE (ed): Household Production and Consumption. New York: Columbia University Press for NBER, 1975; pp 265–298.
24. Viscusi WK: Labor market valuations in life and limb: empirical evidence and policy implications. Public Policy 1978; 26:359–386.
25. Smith RS: The Occupational Safety and Health Act. Washington, DC: American Enterprise Institute, 1976; Appendix B: pp 89–95.
26. Olson CA: An analysis of wage differentials received by workers on dangerous jobs. J Human Resources 1981; 16:167–185.
27. Dardis R: The value of life: new evidence from the marketplace. Am Econ Rev 1980; 50:1077–1082.
28. Ghosh D, Lees D, Seal W: Optimal motorway speed and some valuations of time and life. Manchester School Econ Soc Stud 1975; 43:134–143.
29. Blomquist G: Value life of saving: implications of consumption activity. J Political Economy 1979; 87:540–558.
30. Portney PR: Housing prices, health effects, and valuing reductions in risk of death. J Environ Econ Management 1981; 8:72–78.

Start at conclusion

bad

Applied economic model to life + death!

what would you be willing to pay to extend your life?

A cheerful article!

perception of risk

23

The Life You Save May Be Your Own*

THOMAS C. SCHELLING

Thomas C. Schelling is Professor of Economics at the University of Maryland.

This is a treacherous topic, and I must choose a nondescriptive title to avoid initial misunderstanding. It is not the worth of human life that I shall discuss, but of "life-saving," of preventing death. And it is not a particular death, but a statistical death. What is it worth to reduce the probability of death—the statistical frequency of death—within some identifiable group of people none of whom expects to die except eventually?

Worth to whom? Eventually I shall propose that it is to the people who may die, or who may lose somebody who matters to them. But the subject is surrounded by so much mystery, sentiment, moral consideration, husbandry, and paternalism that some of the fringe issues need to be discussed first, if only to identify what the subject is not. Some of these issues are exciting, more exciting than the economics of life expectancy. They involve the special qualities that make an individual's life unique and his death an awesome event, that make hangmen's wages a special market phenomenon and murder the only crime worth solving in a detective story.

The first part of this essay examines society's interest in life and death; the second surveys the economic impact of untimely death, viewing death

"The Life You Save May Be Your Own," by T. C. Schelling, from *Problems in Public Expenditure Analysis,* S. B. Chase, Jr., ed. Reprinted by permission.

*Reprinted with some deletions and minor editorial changes with the permission of the author—*Eds.

more as a loss of livelihood than as a loss of life, seeing how the losses and any possible gains are distributed among taxpayers, insured policyholders, and others who have no personal connection with the deceased. The third part deals with the consumer's interest in reduced mortality and how that interest can be identified, expressed, or allowed for in government programs that, at some cost, can raise life expectancy. It is here that we recognize that life as well as livelihood is at stake. So is anxiety, and the life at risk concerns the consumer personally.

SOCIAL INTEREST IN LIFE AND DEATH

"Pain, fear, and suffering," we are told, "are considered of great importance in a society that values human life and human welfare."[1] They are important, too, to ordinary people who do not like pain and suffering. We have been told that the value of a human life ought to be considered, at least partially, without regard to whether the person who might die is a producer or not, that this value should result from a collective decision concerning the "expense that the nation is willing—as a moral judgment— to undertake, to save one of its members."[2] Why a moral judgment? Why not a practical judgment—a consumer choice—by the members of society about what it is worth to reduce the risk of death? Is death so awesome, so frightening, and so remote, that in discussing its economics we must always suppose it is someone *else* who dies?

What is moral about wanting to live? People who do not care at all for each other, or for society, or for the value of human life, will take care to avoid pain and death. Why should it require a moral judgment for me to hire a policeman to protect my life, along with the lives of my neighbors who pay their share of his salary, but a purely economic judgment to hire him to protect my shop window, my payroll, or my automobile?

"For a variety of reasons it is beyond the competence of the economist to assign objective values to the losses suffered under [pain, fear, and suffering]."[3] The same is true of cola and Novocain, one of which puts holes in children's teeth and the other takes the pain out of repairing them. If they were not for sale it would be beyond the competence of economists to put an objective value on them, at least until they took the trouble to ask people. Death is indeed different from most consumer events, and its avoidance different from most commodities. There is no sense in being insensitive about something that entails grief, anxiety, frustration, and

[1]D. J. Reynolds, "The Cost of Road Accidents," *J. Royal Stat. Soc.,* 119 (1956), 393–408.
[2]Selma J. Mushkin, "Health as an Investment," *J. of Polit. Econ.,* Supplement, October 1962, 156. She cites an unpublished paper by E. E. Pyatt and P. P. Rogers.
[3]Reynolds, *op. cit.*

mystery, as well as economic privation. But people have been dying for as long as they have been living; and where life and death are concerned we are all consumers. We nearly all want our lives extended and are probably willing to pay for it. It is worth while to remind ourselves that the people whose lives may be saved should have something to say about the value of the enterprise and that we who study the subject, however detached, are not immortal ourselves.

Individual Death and Statistical Death

There is a distinction between individual life and a statistical life. Let a six-year-old girl with brown hair need thousands of dollars for an operation that will prolong her life until Christmas, and the post office will be swamped with nickels and dimes to save her. But let it be reported that without a sales tax the hospital facilities of Massachusetts will deteriorate and casue a barely perceptible increase in preventable deaths—not many will drop a tear or reach for their checkbooks. John Donne was partly right: the bell tolls for thee, usually, if thou didst send to know for whom it tolls, but most of us get used to the noise and go on about our business.

I am not going to talk about the worth of saving an identified individual's life. Amelia Earhart lost in the Pacific, a score of Illinois coal miners in a collapsed shaft, an astronaut on the tip of a rocket, or the little boy with pneumonia awaiting serum sent by dogsled—even the heretofore anonymous victims of a Yugoslavian earthquake—are part of ourselves, not a priceless part but a private part that we value in a different way, not just quantitatively but qualitatively, from the way we measure the incidence of death among a mass of unknown human beings, whether that population includes ourselves or not. If we know the people, we care. Half the entertainment industry and most great literature is built on this principle. But our concern in this essay will be statistical lives.

We must recognize, too, that the success of organized society depends on traditions, attitudes, beliefs, and rules that may appear extravagant or sentimental to a confirmed materialist (if there is one). The sinking of the *Titanic* illustrates the point. There were enough lifeboats for first class; steerage was expected to go down with the ship. We do not tolerate that any more. Those who want to risk their lives at sea and cannot afford a safe ship should perhaps not be denied the opportunity to entrust themselves to a cheaper ship without lifeboats; but if some people cannot afford the price of passage with lifeboats, and some people can, they should not travel on the same ship.

The death of an individual is a unique event. Even an atheist can wish he had been nicer to someone who recently died, as though the "someone" exists, which the atheist believes he does not. If death truly is final, it is only so for the person who dies. Whatever the source of the mystery, most of us have very special feelings about suicide and euthanasia, birth control and abortion, bloodsports and capital punishment, and there is no way to deny these feelings in the interest of "rationality" without denying most of what makes us human. We go to great lengths to recover dead bodies. We give a firing squad one blank cartridge so that every member can pretend he did not take a life.

Responsibility for death introduces special problems. A man can be sent on a mission or on repeated missions with small probability of survival, but sending a man to certain death is different. The "chance" makes the difference, apparently because people can hope—the people who go and the people who send them. Guilt is involved; one of the reasons for having a book of rules about when to run the risk and when not to—when to land the disabled aircraft and when to abandon it and take to parachute—is to relieve the man who gives the orders, the man in the control tower, of personal guilt for the instruction he gives. Safety regulations must be partly oriented toward guilt and responsibility. A window washer may smoke on the job until he gets lung cancer and it is no concern of his employer; but his safety belt must be in good condition.

To evaluate an individual death requires attention to special feelings. Most of these feelings, though, involve some connection between the person who dies and the person who has the feelings; a marginal change in mortality statistics is unlikely to evoke these sentiments. Programs that affect death statistically—whether they are safety regulations, programs for health and safety, or systems that ration risk among classes of people—need not evoke these personal, mysterious, superstitious, emotional, or religious qualities of life and death. These programs can probably be evaluated somewhat as we evaluate the commodities we spend our money on.

What is the alternative to death? It depends. For the paralytic it is a life of paralysis; for someone who escapes a highway accident it is the same life as before, unless the near miss changes his behavior. The type of risk that might be reduced is likely to be correlated with age, sex, income level, number of dependents, and life expectancy. Any program that reduces the risk of death will be discriminatory. Infant mortality affects infants and those who have them; motor accidents affect people who use the roads; starvation kills the poor; and a regulation that surrounds swimming pools with fences will affect different age groups according to the height of the fence. Even lightning is not random in its choice of victims—golfers are more at risk than coal miners—and any analysis that initially ignores the specific group affected has to be adaptable to the specific deaths that would be averted by a given program.

Where does the problem arise? It arises in disease, road accidents, industrial safety, flood control, the armed services, safety regulations, personal protection, and all the things that people do that affect their life expectancy. In the marketplace it arises in the choice of hazardous occupations, in home safety, in residential location and in risky everyday enterprises like diving and swimming. It is often hard to discern, though, or to separate, the things that people do to save their own lives or that governments do to save the lives of citizens, because mortality is so closely correlated with other things that concern people. We eat for satisfaction and avoid starvation, heat our homes to feel warm and avoid pneumonia; we buy fire and police protection to save economic loss, pain, embarrassment, and disorder, and in the process reduce the risks of death. When we ride in an airplane, death is about the only serious risk that we consider; but if we compare an advanced country with a backward one the difference in safety to life is correlated with so many comforts, amenities, and technological advances that it is hard to sort out life-saving and life-risking components. The impact on life expectancy of, say, the electric light, is so cumulative and indirect that it would hardly be worth sorting out if we could sort it out. The universal employment of snow-blowers would spare us all those heart attacks that the newspapers so faithfully report after a blizzard; but what number of us would eventually die younger for lack of exercise is not so readily estimated.

Who loses if a death occurs (or has to be anticipated)? First, the person who dies. Exactly what he loses we do not know. But, before it happens, people do not want to die and will go to some expense to avoid death. Beyond the privation that death causes the person who dies, there is the fear of death. The anxieties are visible and are real. Few who are sentenced to die, or have received the announcement that their deaths are inevitable, seem to consider those last few days or months the best of their lives. We may be in the grip of an instinct that has value for the race and not for the individual; but if we ask who is willing to make an economic sacrifice to prevent a death, in most societies there is at least one unequivocal answer: the person who is to die. By all the standards that economists take seriously, the prospective victim loses.

Second, death is an event—and the prevention of death a consumer good—that in our society inextricably involves the welfare of people close to the person who dies. Death is bereavement and disturbance of integral small societies—families—where people play roles that are often unique and always difficult for others to fill.

Finally, there is "society"—other people. They can lose money or save, as a result of a death with which they have no personal connection. In a few dramatic cases—the inventor of a wonder drug, a poet, statesman, or a particularly predatory criminal—the impact of a death may be out of all proportion to the victim's personal economics—to his earnings, expenditures, taxes, and contributions, and to his exploitation of public programs and facilities. The rest of us, though, are known to the economy mainly by the money we earn and spend and the money that is spent upon us; and an accounting approach will uncover most of the impact.

Death is a comparatively private event. Society may be concerned but is not much affected. There is a social interest in schools and delinquency, discrimination and unrest, infection and pollution, noise and beauty, obscenity and corruption, justice and fair practices, and in the examples men set; but death is usually a very local event. The victim and his family have an intense interest; society may want to take that interest seriously, but it is hard to see that society has a further interest of its own unless, as in military service or public orphanages, there is an acknowledged public responsibility. Society's interest, moreover, may be more in whether reasonable efforts are made to conserve life than in whether those efforts succeed. A missing man has to be searched for, but whether or not he is found is usually of interest—intense interest, to be sure—to only a very few.

But the taxes we pay and the school lunches we eat have their impersonal ramifications and can motivate someone else to take an economic interest in our longevity. The accounting for those ramifications is the subject of the next section.

ECONOMIC INTEREST IN LOST LIVELIHOODS

When we consider the costs of a death to society—the costs that might be decreased by a program that reduces deaths—it is as important to discover where the costs fall as to aggregate them. There is a convention that nations are the bases of aggregation, but costs can be local, regional, or national; they can fall on particular sectors of the economy, particular levels of income, particular groups of taxpayers or welfare recipients.

Especially if there is an opportunity to prevent the death—to reduce the incidence of death within some part of the population—there is as much interest in who would have borne the cost of the death as in what the total cost would be. First, interested parties may have to be identified, to persuade them that they should bear the cost of reducing some mortality rate. Moral judgments are fine, but in the end it may be airline passengers who want more air safety, parents who want children better protected, Oklahomans who want better tornado warning; it is worthwhile to identify the people who might care enough to do something about it. Second, if the losses are to be compensated, their location and size must be known. Third, if a sense of justice or social contract requires that the beneficiaries pay for the benefits, we want to know who benefits.

Someone may care about the effect of a death on the gross national product, though I doubt that anyone cares much. Still, the GNP is so often taken as the thing we care about that at least passing attention should be devoted to the aggregate effect of death and its postponement on the economy.

Population Economics

At the GNP level, death is mainly a matter of population economics. Population has both a territorial and a national significance. The GNP was raised when Hawaii and Alaska were assimilated; it could be raised more by bringing Canada into the United States. This is a purely "national" consideration, having to do with the virtues of being a big country.

There may be scale effects in efficiency or in the provision of public services, but it is hard to tell whether the United States is richer as it becomes more dense and more congested. Military considerations aside, it is not obvious that in a country like this the number of people makes much difference.

If it did, we would probably have a conscious policy of migration. We might also have a conscious policy of family incentives, subsidizing children or taxing their parents or designing social security programs to give incentives for larger or smaller families. It is hard to escape the conclusion that if people are what we want, programs to reduce mortality are a sluggish way to get them.

A question that has received some attention in efforts to put a wealth value on human life is how to calculate the worth of a child. There has been, it is often observed, some investment in the child, and, with accrued interest, this investment is lost if the child dies. Alternatively, the child will produce income in the future; and though the investment is sunk, the future income is lost if the child dies and his discounted net contribution, positive or negative, goes with him. This is complicated: if he lives he will produce and he will procreate; if he dies he may leave dependents of his own.

I doubt whether this kind of population economics is worth all the arithmetic. At best, it is the way a family will deal with the loss of a cow, not the loss of a collie. Though children are not pets, in the United States they are more like pets than like livestock, and it is doubtful whether the interests of any consumers are represented in a calculation that treats a child like an unfinished building or some expensive goods in process. At best, this would be relevant to a kind of replacement cost; but it tells little about the cost of replacement—whether a newborn baby is as good as a teenager if you cannot have the particular teenager whose death caused grief and loneliness.

No. Population economics is important, but if lifesaving deserves our attention, it is in some other context.

Assessing the Costs of a Death

If a lonely, self-sufficient hermit dies—a man who pays no taxes, supports no church, is too old for military service, and leaves no dependents, owning nothing but a burial plot and a prepaid funeral—there are no costs or benefits. Whatever he would have paid to make his life safer and to increase his life expectancy, he is dead now and no one knows the difference.

If a Harvard professor dies—a taxpaying man with a family, who contributes to the United Fund and owns twice his salary in life insurance, is eligible for social security, and has children who may go to college—the accounting of gains and losses is complicated.

The largest losses will fall on his family, and we should distinguish at once between his life and his livelihood. His family will miss him, and it will miss his earnings. We do not know which of the two in the end it will miss most, and if he died recently this is a disagreeable time to inquire. Let us for the moment leave aside the grief, the loneliness, the loss of direction or authority in the family, the emotional privation, and all the things the man represented except his income. The reason for leaving them out at this point is not that they are unmeasurable, or none of our business, but that they are nontransferable and nonmarketable, and there is no "accounting" way to estimate them. For the moment look at the material losses, and get the pure accounting out of the way.

How much of the loss of livelihood falls on the family depends on institutional and market arrangements. In an extremely communal society or an extremely individualist one, there may be a rule or tradition for sharing the loss: orphans may be supported by contributions, rotated among the neighbors, taken in by next of kin, absorbed in a communal orphanage, or otherwise supported at the expense of society at large or of a select responsible group. Alternatively, life insurance may accomplish somewhat the same thing. Whether a "protective benevolent society" is a genuinely fraternal institution or a modern insurance company with a quaint name, the effect is to share the costs.

It is somewhat arbitrary to say that the cost "really" falls on the family and the rest is redistribution, or the cost "really" falls on the committed members of the community, or on the policyholders whose premium payments will reflect the death. The family, the community, and the insurance market are all social institutions characterized by a system of enforceable or honored obligations; and it is a matter of social choice whether, in addition to identifying the child with its father, its consumption is identified with his earnings. The important question is who pays the costs or suffers the losses, not which losses are original and which are transferred.

* * *

CONSUMER INTEREST IN REDUCED RISK

The avoidance of a particular death—the death of a named individual—cannot be treated straightforwardly as a consumer choice. It involves anxiety and sentiment, guilt and awe, responsibility and religion. If the individuals are identified, there are many of us who cannot even answer whether one should die that two may live. And when half of the children in a hospital ward are to get the serum that may save their lives, half a placebo to help test the serum, the doctor who divides them at random and keeps their identities secret is not exclusively interested in experimental design. He does not want personally to select them or to know who has been selected. But most of this awesomeness disappears when we deal with statistical deaths, with small increments in a mortality rate in a large population.

Suppose a program to save lives has been identified and we want to know its worth. Suppose the population whose vulnerability is to be reduced is a large one, and approximately identifiable. The dimensions of the risk to be reduced are fairly well known, as is the reduction to be achieved. Suppose also that this risk is small to begin with, not a source of anxiety or guilt.

Surely it is sensible to ask the question, What is it worth to the people who stand to benefit from it? If a scheme can be devised for collecting the cost from them, perhaps in a manner reflecting their relative gains if their benefits are dissimilar, it surely should be their privilege to have the program if they are collectively willing to bear the cost. If they are not willing, perhaps it would be a mistake to ask anybody else to bear the cost for them; they, the beneficiaries, prefer to have the money or some alternative benefits that the money could buy. There are reasons why this argument has to be qualified, but there is no obvious reason why a program that reduces mortality cannot be handled by letting the beneficiaries decide whether it is worth the cost, if the cost falls on them.

There are two main ways of finding out whether some economic benefits are worth the costs. One is to use the price system as a test of what something is worth to the people who have to pay for it. It is possible to see what people are willing to pay for the privilege of sitting at tables rather than counters in a restaurant, what they are willing to pay to use library books or to save an elm tree in the front yard. Sometimes the market is poor; sometimes analysis is confused by joint products; sometimes consumer behavior is subject to inertia and the information is needed before the market adjusts. But at least we can try to observe what people will pay for something.

Another way of discovering what the benefits are worth is to ask people. This can be done by election, interview, or questionnaire; the more common way is to let people volunteer the information, through lobby organizations, letters to congressmen or to the newspapers, and rallies. There may be something a little like a price system here if people are allowed to show the trouble they will go to, or the expense they will incur, to lobby for or against something. Like the price system, these methods may be ambiguous.

It is sometimes argued that asking people is a poor way to find out, *prob* because they have no incentive to tell the truth. That is an important point, but hardly decisive. It is also argued, and validly, that people are poor at answering hypothetical questions, especially about important events— that the mood and motive of actual choice are hard to simulate. While this argument casts suspicion on what one finds out by asking questions, it casts suspicion, too, on those market decisions that involve remote and improbable events. Unexpected death has a hypothetical quality whether it is merely being talked about or money is being spent to prevent it. Asked whether he would buy trip insurance if it were available at the airport (or would decline to fly in an aircraft that had a statistically higher accident rate than another if it would save an overnight stop), a person may not give verbally the same answer as his actions in the airport would reveal. He still might not feel that his actual decision was authoritative evidence of his values or that, had mood and circumstance been different—even had the amount of time for consultation and decision been different—his action might not have been different too. This problem of coping, as a consumer, with increments in the risk of unexpected death is very much the problem of coping with hypothetical questions, whether in response to survey research or to the man who sells lightning-rod attachments for the TV antenna. If consumers regularly retained professional consultants in coping with such decisions, there might be a good source of information.

In any case, relying exclusively on market valuations and denying the value of direct inquiry in the determination of government programs (or even the programs of nongovernmental organizations) would depend on there being, for every potential government service, a close substitute available in the market at a comparable price. It would be hard to deduce from first principles that there is bound to be. Voting behavior is probably to be classed somewhere between a purchase and a questionnaire: an individual's vote is indecisive, while the election as a whole is conclusive.

Small Probability of Large Events

A difficulty about death, especially a minor risk of death, is that people have to deal with a minute probability of an awesome event, and may be poor at finding a way—by intellect, imagination, or analogy—to explore what the saving is worth to them. This is true whether they are confronted by a questionnaire or a market decision, a survey researcher or a salesman. It may even matter whether the figures are presented to them in percentage terms or as odds, whether charts are drawn on arithmetic or logarithmic scales, and whether people are familiar with the simple arithmetic of probability.

The smallness of the probability is itself a hard thing to come to grips with, especially when the increment in question is even smaller than the original risk. At the same time, the death itself is a large event, and until a person has some way of comparing death with other losses it is difficult or impossible to do anything with it probabilistically, even if one is quite willing to manipulate probabilities.

What it would be like to grow old without a companion, to rear a family without a mother or father in the house, to endure bereavement, is something that most of us have no direct knowledge of; and those who have some knowledge may not yet know the full effects over time. Many of us think about it only when we make a will or buy life insurance, suffer a medical false alarm or witness the bereavement of a friend or neighbor.

As consumers we can investigate the subject. It may be no harder to cope with than choosing a career, nor more painful than some of the medical decisions we actually have to take. But most of us have not investigated; the cost of doing so is high, and there is not much fun in it. In a program of interrogation, even a sales effort, some people will just not cooperate. Others, if they have to make a decision, would rather make a hasty one that may be wrong than a more painful or embarrassing decision that is more nearly right. Some of the reluctance may be unconscious, with a resulting bias that is hard to identify.

Furthermore, this is, more than most decisions, a family one, not an individual one. Nearly every death involves at least two major participants, typically the immediate family. It is not even clear who it is that has the greater stake in a person's not dying—himself, his spouse if he has one, his children if he is a parent, or his parents if he is a child—and the subject is undoubtedly a delicate one for the members of the consuming unit to discuss with each other. Whatever the motives of a respondent when being interviewed alone about a safety program or a hazardous occupation, his motives are surely complex when he talks to his wife about how much he would miss her or she would miss him, the likelihood of a happy remarriage, or which of them would suffer more if one of the children died.

Death Versus Anxiety

The problem is even harder if the risk to be attenuated is large enough, or vivid enough, to cause anxiety. In fact, the pain associated with the awareness of risk—with the prospect of death—is probably often commensurate with the costs of death itself. A person who sooner or later must undergo an operation that carries a moderate risk of being fatal will apparently sometimes choose to have the operation now, raising the stakes against himself in the gamble, in order to avoid the suspense. Wives of men in hazardous duty suffer; and most of us have sat beside someone on an airplane who suffered more with anxiety than if he had been drilled by a dentist without Novocain and who would have paid a fairly handsome price for the Novocain. Let me conjecture that if one among forty men had been mistakenly injected with a substance that would kill him at the end of five years, and the forty were known to the doctor who did not know which among them had the fatal injection, and if the men did not know it yet, the doctor would do more harm by telling them what he had done than he had already done with the injection.

This anxiety is separate from the impact of death itself. It applies equally to those who do not die and to those who do, to people who exaggerate the risk of death as much as if their estimates were true. It counts, and is part of the consumer interest in reducing the risk of death. It is not, or usually not, any kind of double counting to bring it into the calculation. But it is—except where knowledge of risk permits people to make better economic decisions, or exaggerations of risk lead them to hedge excessively and uneconomically—almost entirely psychic or social. Relief from anxiety is a strange kind of consumer good. What the consumer buys is a state of mind, a picture in his imagination, a sensation. And he must decide to do so by using the same brain that is itself the source of his discomfort or pleasure. However much "rationality" we impute to our consumer, we must never forget that the one thing he cannot control is his own imagination. (He can try though; this accounts for the business in tranquilizers, and for the readiness of airlines to serve their passengers alcoholic beverages.)

Consumer Choices and Policy Decisions

These, then, are some of the reasons why it is hard for our consumer to tell us intelligently what it is worth to him to reduce the risk of death, why it may even be hard to get him to make a proper try. These are also reasons why the consumer may be poor at making ordinary choices about death in the marketplace. He may not do much better in buying life insurance or seat belts, using or avoiding airplanes, flying separately or together with his wife when they leave the children behind, selecting cigarettes with or without filters, driving under the influence of liquor when he could have taken a taxi, or installing a fire detector over the basement furnace. Some of his marketplace decisions may be more casual (perhaps out of evading his responsibilities, not meeting them), but they may be no better evidence than the answers he would give to questions he might be asked.

Many parents try not to fly on the same plane (although they usually drive home together on New Year's Eve). I took for granted that this was sensible, though extravagant—a matter of the nuisance one would incur to reduce the risk of leaving the children without any parent at all, until Richard Zeckhauser suggested I think it over. Should a person double the risk of losing one parent to eliminate the risk of losing two? I decided then that the answer was hard to be sure of and probably sensitive to the number of children and their ages, even if only the welfare of the children is taken into account, and more so of course when the parents' welfare is considered as well. (Evidently, happily married childless couples should travel on the same plane, not just for company but to eliminate the risk of bereavement.) The point is not that I am right, now, where I was wrong before, but that I hadn't thought it through. Also that, now that I come to think about it, I'm not sure; and I still do not intend to discuss it with my wife, especially [this was written when they were still young] in the presence of the children.

Consumers apparently do often evade these questions when they have a chance. In matters of life and death doctors are not merely operations analysts who formulate the choice for the executive; they are professional decisionmakers, who not only diagnose but decide for the consumer, because they decide with less pain, less regret, cooler nerves, and a mind less flooded with alternating hopes and fears.

Still, in dealing with death-reducing programs, these are the kinds of decisions that somebody has to make. We can do it democratically, by letting the consumers decide for themselves through any of the market-place or direct-inquiry techniques that we can think of. Or we can do it vicariously, paternalistically, perhaps professionally, by making some of these highly introspective and imaginative decisions for them, briefing ourselves on the facts as best we can, or perhaps hiring out the decisions to people who have professional knowledge about the consequences of death in the family.

If then it turns out that the safety device or health program is a public good and not everybody wants it at the price, or that the tax system will not distribute the costs where the benefits fall, so that we are collectively deciding on a program in which some of us have a strong interest, some a weak interest, and some a negative interest, that makes it rather like any other budgetary decision that the government takes. We need not get all wound up about the "pricelessness" of human life nor think it strange that the rich will pay more for longevity than the poor, or that the rich prefer programs that help the rich and the poor those that help the poor. There may be good reasons why the poor should not be allowed to fly in second-class aircraft that are more dangerous, or people in a hurry should not be allowed to pay a bonus to the pilot who will waive the safety regulations; but these reasons ought to be explicitly adduced as qualifications to a principle that makes economic sense, rather than as "first principles" that transcend economics.

Some Quantitative Determinants

What results should we anticipate if we engage in the kind of inquiry I have described, or if we survey the market evidence of what people will pay to avoid their own deaths or the deaths of the people who matter to them? Is there any a priori line of reasoning that will help us to establish an order of magnitude, an upper or lower limit, a bench mark, or some ideal accounting magnitude that ought to represent the worth, to a reflective and arithmetically sophisticated consumer, of a reduction in some mortality rate? Is there some good indicator—life insurance, lightning rods, hazardous-duty pay—that will give us some basis for estimate? Is there some scale factor, like a person's income, to which the ideal figure should be proportionate or of which it should be some function? And to what extent should our estimate be expected to depend on social and economic institutions?

At the outset, we can conjecture that any estimate based on market evidence will at best let us know to within a factor of 2 or 3 (perhaps only 5 or 10) what the reflective individual would decide after thoughtful, intensive inquiry and good professional advice. This conjecture is based on the observation that most of the market decisions people make relate to contingencies for which the probabilities themselves are ill-known to the consumer, sometimes barely available to the person who seeks statistics, invariably applicable in only rough degree, and mixed with joint products that make the evidence ambiguous. What will somebody pay for a babysitter who, in case of fire, will probably save the children or some of them? With a little research one can find out the likelihood of fire or other catastrophe during the time that one is away from home, the likelihood that they would be saved if a babysitter were on guard, and the likelihood that they would save themselves or otherwise be saved if no one were home; an upper or lower limit of "worth" may be manifested in the price that one pays or refuses to pay to a babysitter. It would take a good deal more research to relate this to the age and type of furnace, the shape and composition of the house and the location of roofs and windows, the performance of babysitters of different ages and sexes, the ages and personalities of the children, the season of the year, the quality of the fire department, the alertness of neighbors, and the hour of day or night that one is going to be away. In addition, babysitters perform other services: they help get the children to bed, soothe the child that awakes from a bad dream, telephone the parents or doctor in case of sickness, let the dog out, guard against burglary, and sometimes even clean the dishes. (Readers with small children will appreciate that this is a theoretical paper.) What the family pays for the babysitter will depend, furthermore, on whether it is the husband or the wife who decides, on what the local custom is, and on what vivid experience some acquaintance had in recent years. The evidential value of this "market test" will barely give us an estimate to within a factor of 2 or 3.

WORTH AS A FUNCTION OF INCOME.   Is there some expectable or rational relation between what a person earns and what he would spend, or willingly be taxed, to increase the likelihood of his own survival or the survival of one of his family? Specifically, is there any close *accounting* connection between what he might spend and what he can hope to earn in the future, or what he owns?

So many examinations of the worth of saving a life are concerned with the fraction of a person's income that he in some way contributes, that there may be a presumption that the outside limit of the worth of saving his life is the entirety of his expected future earnings. It does seem that if we ask ourselves the worth of saving somebody else's life, and he is somebody who personally makes no difference to us, his net contribution to total production may be the outside limit to what we can interest ourselves in. But when we ask the question, What is it worth to him to increase the likelihood of his own survival (or to us, our own survival)?, it is hard to see that his (our) future lifetime earnings provide either an upper or lower limit.

There is no reason to suppose that what a person would pay to eliminate some specific probability, *P,* of his own death is more than, less than, or equal to, *P* times his discounted expected earnings. In fact there is no reason to suppose that a person's future earnings, discounted in any pertinent fashion, bear any particular relation to what he would pay to reduce some likelihood of his own death.

I am not saying that a person's expected lifetime income is irrelevant to an estimate of what he would pay to reduce fatalities in his age group. But discounted lifetime earnings are relevant only in the way that they are relevant to ordinary decisions about consumption, saving, quitting a job, or buying a house. They are part of the income and wealth data that go into the decisions. Their connection is a functional one, not an accounting one. What a person would pay to avoid death, to avoid pain, or to modernize his kitchen is a function of present and expected income but need bear no particular adding-up relation.

People get hung up sometimes on the apparent anomaly that if a person would yield 2 percent of his lifetime income to eliminate a 1-percent risk of death, he'd have to give up twice his entire lifetime income to save his own life—which he cannot do if his creditors are on their toes. But he doesn't have to. I'd pay my dentist an hour's income to avoid a minute's intense pain—even to prevent somebody else's pain—without having to know what I'd do if confronted with a lifetime of intense pain. This is why the "worth of saving a life" is but a mathematical construct when applied to an individual's decision on the reduction of small risks; it has literal meaning only if we mean that a hundred people would give up the equivalent of two lifetime incomes to save one (unidentified) life among them. [After this was written I found a helpful analogy. In counting work force the FTE (full-time equivalent) has come into common use; so we might refer to this lifesaving construct as FLE, or "full-life equivalent." If we'd each pay, on average, 2 percent of a year's income to eliminate a 1-percent risk of dying—that is, we'd pay together two annual incomes to eliminate one expected death among the hundred of us—we can say that the worth of one FLE is two years' income.]

Let me guess. If we ask people what it is worth to them to reduce by a certain number of percentage points over some period the likelihood that they will die, they will find it worth more than that percentage of the discounted value of their expected lifetime income. Arithmetically, if we tell a man that the likelihood of his accidental death over the next three years is 9 percent and we can reduce this to 6 percent by some measure we propose, and ask him what it is worth to reduce the probability of his death by 3 percent over this period (with no change in his mortality table after that period), my conjecture is it is worth to him a permanent reduction of perhaps 5 percent, possibly 10 percent, in his income.

This is conjecture. It is based on conversational inquiry among a score of respondents, and relates to fathers in professional income classes. The reader can add himself to my sample by examining what his own answer to the query would be.

DEATH ITSELF VERSUS ANXIETY. In conducting this inquiry it is important to make the distinction mentioned earlier between death itself and anxiety about death. If asked, for example, what it is worth to eliminate the fatality of certain childhood diseases (or—taking for illustration a problem that is commensurate with the problem of death—to reduce the danger of congenital defects and foetal injuries that cause infant deformities), one may discover that he is as preoccupied with the anxiety that goes with the risk as with the low-probability event itself.

A special difficulty of evaluating the anxiety and the event together is that they probably do not occur in fixed proportions. That is, their quantitative connection with the reduction of risk may be quite dissimilar. To be specific, there are good reasons for considering the worth of risk-reduction to be proportionate to the absolute reduction of risk, for considering a reduction from 10 percent to 9 percent about equivalent to a reduction from 5 percent to 4 percent. There is no reason for the anxiety to follow any such rational rule. Even a cool-headed consumer who rationally examines his own or his family's anxiety will probably have to recognize that anxiety and obsession are psychological phenomena that cannot be brought under any such rational control. If they could be, through an act of judgment, an act of self-hypnosis, a ban on disquieting conversations, or the avoidance of factual and fictional stimuli, through surgery or through drugs, the anxiety could perhaps be wholly disposed of. A family that lives with a "high" low probability of death in the family, high enough to cause anxiety but low enough to make it unlikely, may benefit as much from relief as from longevity if the risk can be eliminated.

The anxiety may depend on the absolute level of risk and the frequency and vividness of stimuli. There may be thresholds below which the risk is ignored and above which it is a preoccupation. It may depend on whether the risk is routine and continuous or concentrated in episodes. It undoubtedly depends on what people believe about risks and has no direct connection with what the risks truly are. The existence of one source of risk may affect the psychological reaction to another source of risk. Furthermore, the anxiety will be related to the duration of suspense and can even be inversely correlated with the risk of death itself.

In other words, decision theory, probability theory, and a rational calculus of risks and values will be pertinent—not compelling, but surely pertinent—to the avoidance of the event of death, but may have little or no relevance to choices involving fear, anxiety, and relief. People may, however, by engaging in enough sophisticated analysis of risk, change their sensitivity to the perception of risk, possibly but not surely bringing the discomfort into a more nearly proportionate relation to the risk itself.

There is a special reason why it is hard to separate the anxiety from the event itself. A person is unlikely to have pure or raw preferences involving small risks of serious events. He does not know what the elimination of a 0.0002 chance of death is worth to himself unless he can find some way of comparing it with the other terms of his choice, of making it commensurate with the other things that money can buy. One can hardly have a feeling about a 0.0002 chance of death quite the way one has a feeling about pain in the dental chair, the loss of an hour in a traffic jam, or even the loss of a favorite tree in the yard. It takes a little arithmetic even to remember that 1 chance in 5,000 is $\frac{1}{50}$th of 1 chance in 100, 20 times 1 chance in 100,000. A person may have to explore until he finds a magnitude of risk about which he has, or can imagine his way into, a feeling of the kind we associate with preferences and tastes. The risk may have to be brought above some threshold where the size has some feel or familiarity, where the intensity of his feeling is too strong to escape his efforts to respond to it. If he can find a favorite level of risk, a familiar bench mark, a degree of risk that he can in some way perceive directly rather than through pencil and paper, there may then be a possibility of scaling the risk and its worth to find a proper or rational valuation of smaller or larger risks. The anxiety associated with the risk, though, may be quite unamenable to any such scaling.

* * *

DISCRIMINATING FOR WEALTH. A special matter of policy is bound to arise here. If a government is to initiate programs that may save the lives of the poor or the rich, is it worth more to save the rich than to save the poor? The answer is evidently yes if the question means, is it worth more to the rich to reduce the risk to their own lives than it is to the poor to reduce the risk to their own lives. Just as the rich will pay more to avoid wasting an hour in traffic or five hours on a train, it is worth more to them to reduce the risk of their own death or the death of somebody they care about. It is worth more because they are richer than the poor. A hospital that can save either of two lives, but not both, has no reason to save the richer of the two on these grounds; but an expensive athletic club can afford better safety equipment than a cheap gymnasium; the rich can afford safer stoves in their homes than the poor; and a rich country can spend more to save lives than a poor one.

OTHER MEMBERS OF THE FAMILY. Most of this discussion has been focused on the person who supports the family. To deal comprehensively with the subject, the problem should be calculated from the point of view of the husband, putting wife or child at risk, and from the point of view of the wife, putting at risk her own life or her husband's or that of one of the children. (To get a proper feel for the subject, the children might be given a chance to express their views; their immaturity should not offhand make what they say irrelevant.)

There is a qualification about families and children: the values placed on lives by members of the family, as well as the costs to society involved in somebody's death, are not additive within the family. If death takes a mother, a father, and two children, each from a different family, the consequences are different from the death of a family of four in a single accident. This is true both of the costs to society, because of the differential impact of dependents' care, and of the personal valuations within the family.

If a family of four *must* fly, and has a choice among four aircraft, of which it is known that one is defective but not known which one, it should be possible to persuade them to fly together. The prospects for each individual's survival are the same, no matter how they divide themselves among the aircraft, but the prospects for bereavement are nearly eliminated through the "correlation" of their prospects. "Society's" interest, in support of the family's interest, should be to see that they are permitted and encouraged to take the same plane together. (Society's economic interest will coincide.)

CONCLUSION

We have looked now at several ways to approach the worth of saving a statistical life. We have had to distinguish between the life and the livelihood that goes with it. We have had to distinguish between the loss of that livelihood to the consuming unit—the family—and the loss of the share that went to other members of the economy—the taxpayers, insurance policyholders, and kin. We have considered some of the ways that reduction of the risk of death differs from other commodities and services that consumers buy.

To recapitulate: (1) Death is an awesome and indivisible event that goes but one to a customer in a single large size. (2) For many people it is a low-probability event except on special occasions when the momentary likelihood becomes serious. (3) Its effect on a family is something that many consumers have little direct acquaintance with. (4) In an already advanced economy many of the ways of reducing the risk of death are necessarily public programs, budgetary or regulatory. (5) Reduction of risk is often a by-product of other programs that lead to health, comfort, or the security of property, though there are some identifiable programs of which the saving of lives is the main result. (6) Death is an insurable event. (7) Death is more of a family event than most other casualties that one might like to avert; its analysis requires more than perfunctory recognition that the family is the consuming unit, the income-sharing unit, and the welfare-sharing unit.

Still, though these characteristics are important, they do not necessarily make the avoidance of death a wholly different kind of objective from others to promote the general welfare. Although it is important to be aware of how the avoidance of death differs from other programs, it is equally important to keep in mind in what respects it is similar. Society may indeed sometimes express its profoundest moral values in the way it deals with life and death, but in a good many programs to reduce fatalities society merely expresses the amount of trouble people will go to, or the money people will spend, to reduce the risks they run. There is enough mystery already about death, not to exaggerate the mystery.

A good part of society's interest in the livelihood that may be lost is no different from its interest in saving a man's barn or his drugstore. What are the costs of a fire that burns property? Everything that was said about taxes, saving, insurance, and contributions to the United Fund is equally pertinent to this case. The fact that an appraiser can value the barn more readily than a vocational analyst can appraise a person's livelihood simplifies the problem in only one dimension: estimating the value of the barn is only a point of departure for tracing out society's interest. As a statistical aggregate, the national wealth goes down if I lose my home and furniture, but who cares except me? If my bank cares, or my insurance company, or the taxing authorities of my town, we are on the track of some interests that matter. But society has no direct interest in the national wealth. It is not owned collectively, not in the United States.

What makes the barn or shop easier to evaluate than a life (not livelihood) is that it is less difficult to guess what it is worth to the person who owns it. Its replacement cost sets an upper limit. Even that, though, does not directly tell the worth of a small increment in a small probability of material destruction; it is the insurability of the structure, with a policy that pays off in the same currency with which one buys replacement, that makes it possible to estimate the worth to someone of an incremental change in the risk of fire, collision, or windstorm.

The difficult part of the problem is not evaluating the worth of a person's livelihood to the different people who have an interest in it, but the worth of his life to himself or to whoever will pay to prolong it. This is what is not insurable in terms that permit replacement. This is the consumer interest in a unique and irreplaceable good. His livelihood he can usually insure, not exactly but approximately, sharing the loss and making it a matter of diffuse economic interest; it is valuing his life that poses the problem.

And the difficulty is not just that, as with so many government budgetary and regulatory programs, the government has to weigh the divergent interests of various beneficiaries and taxpayers. Nor is it that, as with so many government programs, the government has to investigate how much the program is worth to people. The main problem is that people have difficulty knowing what it is worth to themselves, cannot easily answer questions about it, and may object to being asked. Market evidence is unlikely to reveal much.

Dealing with small changes in small risks makes the evaluation more casual and takes the pricelessness and the pretentiousness out of a potentially awesome choice. The question is whether the consumer, at this more casual level of straightforward risk reduction, has any sovereign tastes (or thinks he has) and can be induced to place his bets as calmly as he would fasten a seat belt or buy a lock for his door. If it appears upon inquiry—an inquiry that the consumer participates in—that he has been casually deceiving himself that his decisions are the right ones, it is necessary to decide whether that is his privilege and he wants it respected, or he should be goaded into an agonizing reappraisal or the reappraisal should be made for him. Scaring people is usually bad, and the airlines can hardly be expected to cooperate—or their passengers either—in a survey that quickens a passenger's appreciation of danger at the moment he settles into his seat.

In the end there may be a philosophical question whether government should try to adapt itself to what consumer tastes would be if the consumers could be induced to have those tastes and to articulate them. There may be a strong temptation to do the consumer's thinking for him and to come out with a different answer. Should one try to be guided by what the consumer would choose, when in fact the consumer may refuse to make the choice at all? If a doctor is asked to make a grave medical decision that a patient, or a patient's spouse, declines to make for himself, is the doctor supposed to guess what the patient, or the patient's spouse, would have decided if he'd had to decide for himself? Or is the doctor to decide as he thinks he would himself decide if he were in the patient's position? Or is he to make a welfare decision for the whole family or some other small society? Should the doctor ask the patient which among these criteria he wants the doctor to use, or does that merely upset the patient and lead to the doctor's having to decide how to decide on the criterion?

The gravity of decisions about lifesaving can be dispelled by letting the consumer (taxpayer, lobbyist, questionnaire respondent) express himself on the comparatively unexciting subject of small increments in small risks, acting as though he has preferences even if in fact he does not. People do it for life insurance; they could do it for lifesaving. The fact that they may not do it well, or may not quite know what they are doing as they make the decision, may not bother them and need not disfranchise them in the exercise of consumer-taxpayer sovereignty.

As an economist I have to keep reminding myself that consumer sovereignty is not just a metaphor and is not justified solely by reference to the unseen hand. It derives with even greater authority from another principle of about the same vintage: "no taxation without representation." Welfare economics establishes the convenience of consumer sovereignty and its compatibility with economic efficiency; the sovereignty itself is typically established by arms, martyrdom, boycott, or some principles held to be self-evident. And it includes the inalienable right of the consumer to make his own mistakes. . . .

24

Risk Assessment and Comparisons: An Introduction*

RICHARD WILSON

E. A. C. CROUCH
.

Richard Wilson is Mallinckrodt Professor of Physics, and E. A. C. Crouch is an Associate in the Department of Physics, both at Harvard University.

Every day we take some risks and avoid others. It starts as soon as we wake up. One of us lives in an old house that had old wiring. Each time he turned on the light, there was a small risk of electrocution. Every year about 200 people are electrocuted in the United States in accidents involving home wiring or appliances, representing a risk of death of about 10^{-6} per year, or 7×10^{-5} per lifetime. To reduce this risk, he got the wiring replaced. When we walk downstairs, we recall that 7000 people die each year in falls in U.S. homes. But most are over 65, so we pay little attention to this risk since both of us are younger than that.

How should we go to work? Walking is probably safer than using a bicycle, but would take five times as long and provide less healthful exercise. A car or, better, public transport would be both safer and faster. Expediency wins out, and the car comes out of the garage. Fortunately,

*The article as reprinted here contains a number of editorial revisions.

many daily risk,

the choice nowadays is not between horse or canoe—both of which are much more dangerous. The day has just begun, and already we are aware of several risks, and have made decisions about them.

Most of us act semi-automatically to minimize our risks. We also expect society to minimize the risks suffered by its members, subject to overriding moral, economic, or other constraints. In some cases these constraints will dominate, in others there will be trade-offs between the values assigned to risks and the constraints. Risk assessments, except in the simplest of circumstances, are not designed for making judgments, but to illuminate them (1). To effectively illuminate, and then to minimize, risks requires knowing what they are and how big they are. This knowledge usually is gained through experience, and the essence of risk assessment is the application of this knowledge of past mistakes (and deliberate actions) in an attempt to prevent new mistakes in new situations.

The results of risk assessments will necessarily be in the form of an estimate of probabilities for various events, usually injurious. The goal in performing a risk assessment is to obtain such estimates, although we consider the major value in performing a risk assessment is the exercise itself, in which (ideally) all aspects of some action are explored. The results, goals, and values of performing the risk assessment must be sharply contrasted with the cultural values assigned to the results. Such cultural values will presumably be factors influencing societal decisions and may differ even for risk estimates that are identical in probability.

RISK AND UNCERTAINTY

read

The concept of risk and the notion of uncertainty are closely related. We may say that the lifetime risk of cancer is 25%, meaning that approximately 25% of all people develop cancer in their lifetimes. Once an individual develops cancer, we can no longer talk about the risk of cancer, for it is a certainty. Similarly if a man lies dying after a car accident, the risk of his dying of cancer drops to near zero. Thus estimates of risks, insofar as they are expressions of uncertainty, will change as knowledge improves.

Different uncertainties appear in risk estimation in different ways (2). There is clearly a risk that an individual will be killed by a car if that person walks blindfolded across a crowded street. One part of this risk is stochastic; it depends on whether the individual steps off the curb at the precise moment that a car arrives. Another part of the risk might be systematic; it will depend on the nature of the fenders and other features of the car. Similarly, if two people are both heavy cigarette smokers, one may die of cancer and the other not; we cannot tell in advance. However there is a systematic difference in this respect between being, for instance, a heavy smoker and a gluttonous eater of peanut butter, which contains aflatoxin. Although aflatoxin is known to cause cancer (quite likely even in humans), the risk of cancer from eating peanut butter is much lower than that from smoking cigarettes. Exactly how much lower is uncertain, but it is possible to make estimates of how much lower and also to make estimates of how uncertain we are about the difference.

Some estimates of uncertainties are subjective, with differences of opinion arising because there is a disagreement among those assessing the risks. Suppose one wishes to assess the risk (to humans) of some new chemical being introduced into the environment, or of a new technology. Without any further information, all we can say about any measure of the risk is that it lies between zero and unity. Extreme opinions might be voiced; one person might say that we should initially assume a risk of unity, because we do not know that the chemical or technology is safe; another might take the opposite extreme, and argue that we should initially assume that there is zero risk, because nothing has been proven dangerous. Here and elsewhere, we argue that it is the task of the risk assessor to use whatever information is available to obtain a number between zero and one for a risk estimate, with as much precision as possible, together with an estimate of the imprecision. In this context, the statement "I do not know" can be viewed only as procrastination and not responsive to the request for a risk estimate (although this should not be read as condemning procrastination in all circumstances).

The second extreme mentioned, the assumption of zero risk, can arise because people and government agencies have a propensity to ignore anything that is not a proven hazard. We argue that this attitude is inconsistent if the objective is to improve the public health, may also lead to economic inefficiencies, and often leads to unnecessary contention between experts who disagree strongly. Fortunately, if risk assessors have been diligent in searching out hazards to assess, few hazards posing large risks will be missed in this way, so that there may be minor direct danger to human health from a continuation of the attitude.

RISK ESTIMATION BASED ON HISTORICAL DATA

The way in which risks are perceived is strongly correlated with the way in which they are calculated. Risks based on historical data are particularly easy to understand and are often perceived reliably. It is therefore easy to understand some characteristics of risk estimation by considering a risk calculated from historical data. There are plenty of data on automobile accidents (although never enough to make risk assessors happy). One thing that these data can tell us is the frequency of such accidents in the past and their trend through time. To make predictions, however, we must use a model. The simplest model is that there will be as many accidents next year as last, to within a statistical error of the square root of the number (assuming the Poisson model for the frequency distribution of accidents). A slightly more complicated, but perhaps more accurate, model might be to fit a mathematical function to numbers from previous years and to argue that next year's accidents will follow the trend given by this function. A possibly better and possibly more accurate model still might use all available information that might influence accident trends. For example, an oil embargo with a concomitant rise in oil price and reduction in automobile travel would be likely to reduce the risk of accident. In any event, it becomes clear that it is impossible to calculate any risk without a model of some sort, even the simple one that tomorrow will be like today.

RISK OF NEW TECHNOLOGIES

We can use the historical approach to estimating risks only when the hazard (for example, technology, chemical, or simply some action) has been present for some time and the risk is large enough to be measured directly (although when it is not large enough to be measured, an upper limit may be calculated, if one assumes some sort of model). If there is no historical database for the hazard (a new power plant or industrial facility, for instance), one approach is to consider it in separate parts, calculating the risks from each part and adding them together to estimate a risk for the whole. The sum of the risks is a good approximation to the total risk when the individual risks are small and not very numerous. For example, all possible chains of events from an initiator to a final accident are followed in an "event tree," with the probabilities of each event in the tree being estimated from historical data in different situations.

A particularly well-known example is the calculation of the probability of a severe accident at a nuclear power plant (3). That this procedure has at least a partial valididty is due to the fact that the design of nuclear power plants proceeded in approximately this factorable way: attempts were made to imagine all major accident possibilities, "maximum credible accidents" or "design basis accidents," and then to add an independent device for each type of accident to prevent it from having severe consequences. To the extent that the added safety device is independent of the component whose failure could cause an accident, the risk of a serious failure from that source is reduced to the component failure's probability multiplied by the probability that the back-up device would fail. The overall probability of a serious accident is then approximately the sum of the probabilities of the individual failure sequences.

RISKS BY ANALOGY: CARCINOGENIC RISKS

Some carcinogenic risks may be estimated from historical data. But this is complicated by the time delay between the insult and the final cancer, one reason why causality is hard to prove if the risk is small. This is one of the knottiest problems in epidemiology.

Although some of the largest cancer risks have been identified through the use of epidemiology (4), preventive public health suggests that we endeavor to estimate risks even where no historical data exist and the risk is small. This is often done by analogy with the cancer risks to animals, usually rodents, which are deliberately exposed to quantities of a pollutant large enough so that an effect is observed. To use these data to estimate the risk at low doses in people involves (to oversimplify matters) two difficult steps: the comparison of carcinogenic potency in animals and people (5–7), and the extrapolation from a high dose to a low dose. Both steps require a certain amount of controversial theory. Indeed, there are those who regard the uncertainty as so great that they prefer not to provide numerical estimates of risk (8, 9), although they may order materials according to carcinogenic potency. The difference between this and providing a numerical estimate is important, but is one of presentation rather than substance.

If there are no animal data, or if in an animal experiment there is no statistically significant effect, it does not necessarily mean that there is no risk. If the experimenters have been diligent, the risk is probably small, although never zero, even though that may be the best estimate. Various attempts are made to use data even less direct than the animal bioassays to estimate risks in such cases. These include simple analogies based on chemical similarity (10), and comparison with outcomes other than cancer—for example, mutagenesis (11) and acute toxicity (12, 13). Not surprisingly, these more indirect procedures arouse even more controversy than the animal bioassays.

There have been few attempts to perform risk assessments for biological end points other than cancer. However, it is known that the pollutants in cigarette smoke cause at least as many deaths through heart problems as by cancer (14), and we should not be surprised if other carcinogens were to produce chronic effects other than cancer. For now, the cancer risk assessment has to act as surrogate for these other risks also.

RISK VALUE VERSUS CERTAINTY OF INFORMATION

After risks of a number of situations have been assessed, we often want to order them in order to decide which should command our attention. It is not always the order of increasing risk that is used for such purposes. There have been proposals to order potential carcinogens on other factors (8, 15), such as the certainty of the implicating information. By "certainty" we here mean the width of the range within which the actual risk is believed to be almost sure to lie.

Vinyl chloride gas has been found to cause angiosarcomas both in people and in rats. Since an angiosarcoma is a rare tumor, the risk ratio (the ratio of the observed number of cancers in those exposed to the number expected in a similarly sized group of unexposed individuals) is of order 100 or more in some cases. If an angiosarcoma is seen in a vinyl chloride worker, the attribution to vinyl chloride exposure is almost certain. On the other hand, the number of persons who have been heavily exposed to vinyl chloride is small, so that only about 125 angiosarcomas have been seen among vinyl chloride workers worldwide in the last 20 years. Now that exposures in the workplace have been greatly reduced, no angiosarcomas attributable to recent occupational exposure have been seen. We do not know the dose-response relation, but it is generally believed that the response falls at least linearly as the exposure is reduced, so that no more than one cancer is expected in several years.

We can compare this with the possible cancer incidence that was predicted by the Food and Drug Administration (FDA) in 1977 from use of saccharin (16). This was based on experiments with rats, leading to an additional uncertainty. More people ate saccharin than were exposed to vinyl chloride, and nearly 500 cancers per year were estimated for the United States alone. We therefore have two situations. Each individual's risk of exposure to vinyl chloride is low, but there is considerable certainty of developing cancer if one is exposed. For saccharin the risk is higher, but there is more uncertainty about the value of the risk. Some persons, in some situations, may demand that more attention be given to the risk from vinyl chloride than to the risk from saccharin; for other persons or situations the reverse may be the case.

COMPARISONS OF RISKS

The purpose of risk assessment is to be useful in making decisions about the hazards causing risks, and so it is important to gain some perspective about the meaning of the magnitude of the risk. Comparisons can be useful. We are not born with an instinctive feeling for what a risk of one in a million per lifetime means, although we do learn that some risks are small and others large. It is particularly helpful to compare risks that are calculated in a similar way. For example, the risk of a fatal accident when traveling by automobile can be compared to that of traveling by horse with the use of historical data.

Another common procedure is to compare exposures only. Table 1 shows a list of radiation exposures in typical situations (17). The dose-response relation for radiations with similar energy deposition per unit track length will be similar, although there may be some correction required for dose-rate effects, so that ordering by exposure should be similar to ordering by risk. In estimating the number of lethal cancers on a linear hypothesis, we have here assumed that approximately one cancer is induced per 8000 man-rems of exposure at low doses, in itself uncertain by 30 percent or more.

An example of comparison of risks that are similarly calculated is the comparison of risks of various chlorinated hydrocarbons in drinking water. The risks to humans are estimated from carcinogen bioassays in rodents (rats and mice). Chloroform produces cancer in animals 20 times as readily as does trichloroethylene, an industrial solvent that is occasionally found in well waters as a result of accidental pollution. Since they are similar materials, we might expect that the dose-response relationships would have the same shape. Although neither is known to cause cancer in people, we might expect that chloroform would do so about 20 times as readily.

Table 2 shows a variety of risks calculated in various ways and our

TABLE 1. Comparison of Several Common Radiation Risks

ACTION	DOSE (MREM/YEAR)	CANCERS IF ALL U.S. POPULATION EXPOSED (ASSUMING LINEARITY)
Medical X-rays	40	1100
Radon gas (1.5 pCi/liter, equivalent dose)*	500	13,500
Potassium in own body	30	1000
Cosmic radiation at sea level	40	1100
Cosmic radiation at Denver	65	1800
Dose to average resident near Chernobyl first year	5000	Not relevant
One transcontinental round trip by air	5	135
Average within 20 miles of nuclear plant	0.02	< 1

*The radon exposure is to the lungs and cannot be directly compared to whole body external exposure. The comparison here is on the basis of the same magnitude of risk. The uncertainty of the radon number is at least a factor of 3.

TABLE 2. Some Common Risks
(mean values with uncertainty)

TYPE OF RISK	FATALITIES PER YR. (PER 100,000 EXPOSED)	UNCERTAINTY
Motor vehicle accidents:		
All	24	10%
Pedestrians	4	10%
Home accidents	11	5%
Electrocution (accidental)	0.5	5%
Air pollution, Eastern U.S.	20	(1)
Cigarette smoking (1 pack per day)	360	factor of 3
Background radiation (sea level, except radon)	2	factor of 3
All cancers	280	10%
Peanut butter (4 tablspoons per day)	0.8	
Drinking water w. EPA chloroform limit	0.06	factor of 10
Alcohol, light drinker	2	factor of 10
Police killed in line of duty:		
Total	22	20%
Killed by felons	13	10%
Frequent flying professor	5	50%
Mountaineering	60	50%

Source: Estimates by authors.
(1) = factor of 20, downward only.

estimate of the uncertainty. They are deliberately jumbled to provoke thought by juxtaposition. [Risk estimates quoted by the Environmental Protection Agency (EPA) for carcinogens tend to be greater than those shown in Table 2 by a factor approximately equal to the uncertainty factor—this is not accidental (5, 18).]

CONTRASTING RISKS

Objections have been raised to risk comparisons on the ground that they are misleading. This would be true if all risks of the same numerical magnitude were treated in the same way. But they are not. In some cases it is useful to contrast risks to indicate the different ways in which they are treated in society. In Table 3 we give an example by comparing and contrasting the carcinogenic effects of aflatoxin B1 and dioxin, both among the most carcinogenic chemicals known. The difference in treat-

TABLE 3. **Comparison of Two Very Toxic Chemicals, Aflatoxin B1** *(22)* **and Dioxin** *(23);* **CDC, Centers for Disease Control**

MEASURE	AFLATOXIN B1	DIOXIN
Acute toxicity	High	High
Carcinogenic potency to people [(kg · day)/mg]	∽500	Unknown
Carcinogenic potency to rats [(kg · day)/mg]	∽5000	∽5000
Mutagenic	Yes	No
Certainty of information on human carcinogenicity	High	Low
Activity (initiator or promoter)	Initiator	Promoter (?)
Possibility of threshold dose response	Low	High
Source	Natural	Artificial
Common knowledge	Little known	Agent Orange
FDA action level in peanuts (ppb)	20	
CDC level of concern in soil (ppb)	—	—

ment of these two materials is perhaps a reflection of different values assigned to various aspects of the problems caused by their presence.

Aflatoxin and dioxin have similar toxicities and carcinogenic potency (perhaps within a factor of 10, although both measures for both chemicals vary substantially with species tested). The certainty of information for aflatoxin is great. There is less information about carcinogenicity of dioxin. Dioxin may be a promoter and pose a minuscule risk at low doses, whereas aflatoxin is almost certainly an initiator also. Nonetheless such regulatory standards as there are appear to be more stringent for dioxin, possibly because dioxin is an artificial chemical and possibly because it was a trace component of a chemical mixture (Agent Orange) that was used in warfare.

The small risk of a large accident in a nuclear power plant can also be contrasted with the more numerous small accidents or events that occur every day in the mining, transport, and burning of coal. One feature that is brought out clearly here is that we do not always compare the risk averaged over time, but worry more about risks that are sharply peaked in time.

EXPRESSION OF RISKS

Just as a comparison of risks is an aid in understanding them, so is a careful selection of the methods of expression. It is hard to comprehend the statistical (stochastic) nature of risk. There are ways to mitigate this difficulty in comprehension. We are almost all used to one such statistical concept—the expectation of life. When we talk about the expectation of life being 79 years (for a nonsmoking male in the United States) we all know that some die young and that many live to be over 90. Thus the expression of a risk as the reduction of life expectancy caused by the risky action conveys some of the statistical concept essential to its understanding. One particular calculation of this type can be used as an anchor for many people, because it is easy to remember. The reduction of life expectancy by smoking cigarettes can be calculated from the increase per cigarette smoked in the risk of contracting cancer, one in 2 million, multiplied by the difference of the average life-span of a nonsmoker and a lung cancer victim. This turns out to be 5 minutes, or the time it takes to smoke the one cigarette.

It is important to realize that risks appear to be very different when expressed in different ways (19). One example of this can be seen if we consider the cancer risk to those persons exposed to radionuclides after the Chernobyl disaster. According to the Soviets (20), the 24,000 persons between 3 and 15 kilometers from the plant, but excluding the town of Pripyat, received and are expected to receive 1.05 million man-rems total integrated dose, or about 44 rems average. Even if we assume a linear dose-response relation, with 8000 man-rems per cancer, the risk may be expressed in different ways. Dividing 1.05 million man-rems by 8000 gives 131 cancers expected in the lifetimes of that population. This is larger than, and for some people more alarming than, the 31 people within the power plant itself who died within 60 days of acute radiation sickness combined with burns. Dividing the 131 again by the approximately 5000 cancer deaths expected in the same population from other causes, the accident caused "only" a 2.6% increase in cancer. This seems small compared to the 30% of cancers attributable to cigarette smoking. The difference is even more striking if we consider the 75 million people in Byelorussia and the Ukraine who received, and will receive, 29 million man-rems over their lifetimes. On the linear dose-response relation this leads to 3500 "extra cancers," surely a large number for one accident. But dividing by the 15 million cancers expected in this population leads to an "insignificant" increase of 0.0237%. Of course, none of the methods of expressing the risk can be considered "right" in an absolute sense. Indeed, it is our belief that a full understanding of the risk involves expressing it in as many different ways as possible.

REFERENCES AND NOTES

1. L. B. Lave, *Science* 236, 291 (1987).
2. R. Wilson, E. A. C. Crouch, L. Zeise, in *Risk Quantitation and Regulatory Policy* (Cold Spring Harbor Laboratory Press, Cold Spring Harbor, NY, 1985), Banbury Report, vol. 19, pp. 133–147.
3. N. C. Rasmussen *et al.,* "Reactor safety study—an assessment of accident risks in U.S. commercial nuclear power plants" (WASH 1400, NUREG 75/014, U.S. Nuclear Regulatory Commission, Washington, DC, 1975). See also D. Okrent, *Science* 236, 296 (1987).
4. R. Doll and R. Peto, *J. Natl. Cancer Inst.* 66, 1191 (1984).
5. E. L. Anderson *et al., Risk Anal.* 3, 277 (1983).
6. E. A. C. Crouch and R. Wilson, *J. Taxicol. Environ. Health* 5, 1095 (1979).
7. E. J. Calabrese, *Principles of Animal Extrapolation* (Wiley, New York, 1983).
8. R. Peto, in *Assessment of Risk from Low-Level Exposure to Radiation and Chemicals,* A. D. Woodhead, C. J. Shellabarger, V. Pond, A. Hollaender, Eds. (Plenum, New York, 1985), pp. 3–16.
9. B. N. Ames, R. Magaw, L. S. Gold, *Science* 236, 271 (1987).
10. "Control of trihalomethanes in drinking water," proposed rule, *Fed. Regist.* 43, 5756 (1968). See also the advanced notice [*ibid.* 41, 28991 (1976)] and the final rule [*ibid.* 44, 68624 (1979)].
11. M. Meselson and K. Russell, in *Origins of Human Cancer,* H. H. Hiatt, J. D. Watson, J. A. Winsten, Eds. (Cold Spring Harbor Laboratory, Cold Spring Harbor, NY, 1977), p. 1473.
12. S. Parodi, M. Taningher, P. Boero, L. Santi, *Mutat. Res.* 93, 1 (1982).
13. L. Zeise, R. Wilson, E. A. C. Crouch, *Risk Anal.* 4, 187 (1984).
14. *Smoking and Health, a Report of the Surgeon General* (PHS79-50066, Public Health Service, Washington, DC, 1979).
15. R. A. Squire, *Science* 214, 877 (1981).
16. "Saccharin and its salts," proposed rule and hearing, *Fed. Regist.* 42, 19996 (1977).
17. R. Wilson and W. J. Jones, *Energy, Ecology and the Environment* (Academic Press, New York, 1974), table 9-6. Other entries may be readily calculated from data in the reports of the United Nations scientific committee on the effects of atomic radiation ["Sources and effects of ionizing radiation" (United Nations, New York, 1977)] and the report of the Committee on the Biological Effects of Ionizing Radiations ["The effects on populations of exposure to low levels of ionizing radiations" (National Academy Press, Washington, DC, 1980)].
18. M. Russell and M. Gruber, *Science* 236, 286 (1987).
19. A. Tversky and D. Kahneman, *ibid.* 211, 453 (1981). See also P. Slovic, *ibid.* 236, 280 (1987).
20. U.S.S.R. State Committee for the Utilization of Atomic Energy, "The accident at the Chernobyl Nuclear Power Plant and its consequences," working document for the Post Accident Review Meeting, 25–29 August 1986, International Atomic Energy Agency, Vienna.
21. B. L. Cohen, *Health Phys.* 38, 33 (1980).

22. H. R. Roberts, "The regulatory outlook for nut products," paper presented at the Annual Convention of the Peanut Butter Manufacturers and Nut Salters Association, West Palm Beach, FL, November 1977.

23. R. D. Kimbrough, H. Falk, P. Stehr, G. Fries, *J. Taxicol. Environ. Health* 14, 47 (1984).

24. Our work on risk assessment has been supported by donations from Clairol, Inc., the Dow Chemical Company, the Cabot Corporation, the General Electric Foundation, and the Monsanto Corporation.

25

Controlling Urban Air Pollution: A Benefit-Cost Assessment

— Hal's lecture closely follows this article.

ALAN J. KRUPNICK

PAUL R. PORTNEY

Alan J. Krupnick is a Senior Fellow and Paul R. Portney is
Vice President and Senior Fellow at Resources for the
Future.

Environmental regulation is important to our health and well-being and
also is quite expensive. For these reasons, we must look carefully at our
environmental laws and regulations to see what they will accomplish and
at what cost.

Recently, three major changes were made to the Clean Air Act. First,
over the next decade electric power plants must make sharp reductions in
emissions of sulfur dioxide (SO_2)—10 million tons per year measured
from their 1980 level. Second, most major sources of what are called
hazardous air pollutants (less ubiquitous but still potentially harmful
substances such as benzene, acrylonitrile, beryllium, and coke oven emissions) must install state-of-the-art emissions control equipment and, eventually, further reduce any residual emissions that pose unacceptably high
health risks. Third, a number of new measures have been enacted to
improve air quality in areas where the national ambient air quality standards (NAAQS) are currently being violated (1).

"Controlling Urban Air Pollution: A Benefit-Cost Assessment," by
Alan J. Krupnick and Paul R. Portney, from *Science*, v. 252, pp.
522–528. Copyright 1991 by the AAAS. Reprinted by permission.

This third problem, referred to as nonattainment with regard to the NAAQS, is by far the most difficult to solve. According to the Environmental Protection Agency (EPA), in 1989 more than 66 million people in the United States lived in counties where the ozone (O_3) standard was being exceeded at one or more monitors (2). Another 27.4 million lived in areas violating the particulate standard; the corresponding totals were 0.1 million for SO_2, 33.6 million for carbon monoxide (CO), and 8.5 million for nitrogen dioxide (NO_2).

Most of the inexpensive pollution control measures were implemented during the last 20 years, so that future reductions in emissions are likely to be more expensive than earlier ones. Furthermore, the measures required to address remaining pollution problems will fall increasingly on individuals rather than on the large industrial facilities (called point sources) and motor vehicles, which shouldered most of the burden of the initial clean-up (3). More and more, air pollution problems are those associated with wood stoves, small dry-cleaning and degreasing operations, painting shops, bakeries, and other decentralized sources. As a result, the burden of pollution control will become more obvious to the public.

To help focus debate about the best use of society's resources, it is important to have estimates of the benefits and costs of further improvements in air quality. In this article we develop such estimates, focusing primarily on reductions in ground-level O_3, resulting from the control of volatile organic compounds (VOCs) (4); we also consider particulate control. We evaluate proposed efforts both at the national level and in the Los Angeles metropolitan area, where violations of air quality standards are most frequent and severe. In both cases, we first present point estimates of benefits and costs and then discuss uncertainties in a subsequent section. A brief background precedes this analysis.

BENEFIT-COST ANALYSIS AND AIR QUALITY REGULATION

BENEFIT-COST ANALYSIS. Benefit-cost analysis is used by economists to identify, quantify, and weigh the advantages and disadvantages of public policies designed to increase society's overall well-being. Originally used by the Army Corps of Engineers to judge alternative water resources projects, it has become an integral part of policy analysis at all levels and types of government.

The quantification of benefits and costs rests on the idea that an action has value if someone is willing to pay for it, and each individual is held to be the best judge of how a policy affects him or her (5). If an individual assigns a high or a low value to something about which others feel different, that value must be counted nonetheless. The overall value to society of a proposed policy change is measured by the sum of the individual valuations. Thus benefit-cost analysis is unabashedly anthropocentric in that things have value only to the extent that they provide well-being to individuals. These effects, however, are not restricted to purely financial gains and losses. They include the reduced well-being that results from aesthetic degradation, for instance, as well as that felt directly in the pocketbook.

The values that individuals would be willing to pay for the favorable impacts of a policy—whether increased agricultural output or intangibles such as cleaner air, purer drinking water, or the removal of hazardous wastes—often are difficult to ascertain, but economists have devoted much effort to developing valuation techniques consistent with the underlying principles of welfare economics (6). Costs are generally measured as the expenditures that private firms, governments, and individuals must make to comply with regulations. Analogous to benefits, however, costs should be measured by the amount of money required to compensate individuals for the unfavorable effects associated with a regulatory or other public policy; these might take the form of higher prices, reduced incomes, the inconvenience of forced car-pooling, or some other welfare-reducing policy (7). Because models frequently are not available to estimate costs in the preferred way, pollution control expenditures are usually used as proxies. That is the approach taken here.

In principle, benefit-cost analysis includes the value of even the most intangible effects of a policy. For instance, if everyone in the United States was willing to pay $4 per year to preserve a particular wetland area, annual benefits of its preservation would be about $1 billion; similarly, to the extent that people are willing to pay something for the preservation of species diversity, whether for commercial or philosophic reasons, it is counted as an economic benefit. It is important to understand that benefit-cost analysis is not restricted to goods bought and sold in private exchange.

Finally, although benefit-cost analysis is a technique for identifying efficient policies, economic efficiency is surely not the only basis on which policy decisions should be made. Distributional considerations, legal mandates, and ethical concerns are also of great importance, and benefit-cost analysis is generally (but not always) silent about such matters (8).

AIR QUALITY REGULATION.   When the Clean Air Act was amended in 1970, Congress directed the administrator of the new EPA to establish air quality standards to protect human health with an adequate margin of safety. The first standard for O_3 was set in 1971, the basis of which was total photochemical oxidants: The daily high 1-hour reading was not to exceed 0.08 part per million (ppm) on more than 1 day per year. In 1979 the EPA changed the basis of the standard from total photochemical oxidants to O_3 and relaxed the 1-hour standard to 0.12 ppm, not to be exceeded on more than 3 days in any 3-year period (in other words, the reading on the fourth highest day, called the design value, determines attainment status). The O_3 standard remains the same today.

In spite of the 1979 relaxation, violations of the O_3 standard are frequent. For instance, 101 metropolitan areas failed to meet the standard in 1988; 31 of these had more than 10 violations, and 3 (Los Angeles, Fresno, and Bakersfield) had more than 20 violations. Of these 3, Los Angeles led the pack with 148 days on which the 0.12-ppm 1-hour standard was exceeded at least one monitor.

In many of the areas where violations occur, the standard is exceeded only slightly. This is not the case everywhere, however. In Houston, New York, Chicago, and Philadelphia, for instance, the 1988 design values were about 0.22 ppm. In Los Angeles, the design value for 1988 was 0.33 ppm, nearly three times the level of the standard (9). Thus both the frequency and the severity of violations of the O_3 standard are fueling concern about the nonattainment problem nationwide.

REDUCING OZONE IN URBAN AREAS

To evaluate the benefits and costs of reducing ambient O_3 concentrations, one must first estimate the VOC reductions expected in various areas and the O_3 improvements that they imply. Then the costs of the measures to be used to obtain the VOC reductions and the benefits associated with the O_3 improvements can be estimated.

In 1989, the Office of Technology Assessment (OTA) released a major study of air quality problems in the United States (10). The study estimated the changes in emissions of VOCs and, subsequently, reductions in O_3 design values that would result in the years 1994 and 2004 from the application of all currently available VOC control technologies in nonattainment areas and some added control in clean areas (11). No transportation control plans or additional controls on nitrogen oxides (NO_x) were included.

As estimated by OTA, by the year 2004 these control measures would reduce total annual emissions of VOCs in nonattainment areas from about 11 million to about 7 million tons, representing a 35% reduction. Depending on the particular urban area in question, the annual VOC reduction would vary from 20 to 50% (12). Our benefit and cost estimates pertain to this predicted change in air quality.

These estimated emission reductions for each city were then passed through a set of EPA trajectory models, which predict peak ambient concentrations of O_3 (13). The VOC controls that OTA considered were projected to bring 31 of the 94 areas in mild violation of the O_3 standard in 1985 into attainment by 2004. Areas such as Los Angeles, however, with design values greater than 0.15 ppm, were predicted to remain in violation even after controls were implemented, although the predicted design values would be reduced somewhat. For instance, in Los Angeles the highest single reading was predicted to fall by about 20% by the year 2004 as a result of the controls that OTA considered.

COSTS. According to OTA, the annualized cost associated with this ambitious set of measures would be $6.6 billion to $10.0 billion in the nonattainment areas alone, or about $1800 to $2700 per ton of VOC reduced there. Adding in the costs that would be borne in attainment areas raises the estimated total to $8.8 billion to $12.8 billion per year.

Of all the control measures examined, reducing the volatility of gasoline accounts for the greatest emission reduction (about 14%) and would also be the most cost-effective control technology because this reduction would cost between $120 and $740 annually per ton of VOC reduced. At the other extreme, OTA found that using methanol (an 85% blend) to power fleet vehicles would be an expensive measure: $8,700 to $51,000 per ton of VOC reduction (14). A ranking of individual approaches by cost effectiveness shows that marginal costs increase sharply for obtaining any more than a 30% reduction in VOC emissions (Fig. 1).

BENEFITS. How does one ascertain the amount individuals would be willing to pay for the air quality improvements that OTA projects? We concentrate on acute health benefits because protecting health is the primary justification for setting air quality standards under the Clean Air Act

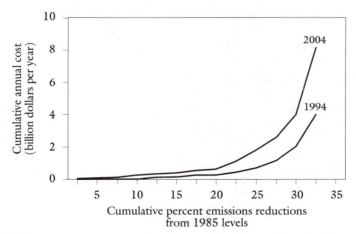

FIGURE 1. Escalation of cumulative annual cost above a 30% reduction in VOC emissions, using control methods analyzed by OTA (nonattainment cities only). [Adapted from *(10)*]

and because only acute health effects have been linked convincingly to O_3 concentrations. Other benefits could accrue in the form of reduced damage to exposed materials, crops, and other vegetation and possibly reductions in the prevalence of chronic illness.

To determine the acute health benefits associated with the estimated 35% reduction in VOC emissions in nonattainment areas, we used a county-level model developed for this purpose (15). The model predicts reduced baseline O_3 concentrations in each area for each day of the O_3 season on the basis of the percentage reductions in O_3 design values obtained from OTA (16). These air quality changes must then be mapped into improved human health. To do so, we combined area-specific data on air quality improvements and population with dose-response functions based on epidemiologic and clinical (controlled laboratory) studies relating ambient O_3 concentrations to various symptoms and other adverse human health effects (17). Thus, for example, the predicted air quality improvement in a particular urban area in the year 2004 can be translated into fewer asthma attacks, reduced incidence of coughing and chest discomfort, reduced number of days of restricted activity, and the like on the basis of predicted population of that area in that year. For example, the clinical dose-response function we used to predict the reduced incidence of coughs for an entire population at a given reduction in ambient O_3 concentrations is

$$\Delta C = \left\{ \left[\frac{1}{1 + \exp(\gamma - \beta\chi_1)} \right] - \left[\frac{1}{1 + \exp(\gamma - \beta\chi_0)} \right] \right\} \theta(\text{pop}) \quad (1)$$

where ΔC is the change in number of cough episodes for a 2-hour period; χ_0 is the average concentration for a 2-hour period, baseline; χ_1 is the average concentration for a 2-hour period, postcontrol; $\gamma = 1.742$; $\beta = 14.100$ per parts per million; and θ is the percentage of time that the population is engaged in exercise.

Because clinical and epidemiologic studies provide quite different types of information for use in estimating health effects, we computed two separate estimates of these physical improvements on health. Clinical studies are relatively precise in quantifying relationships between O_3 exposure in the laboratory and either respiratory symptoms or changes in lung function. The most useful clinical studies estimate dose-response functions by observing the presence or absence of specific symptoms in a small group (from 20 to 135 individuals) of heavily but intermittently exercising young adults (generally men) exposed to O_3 for 2 hours at various carefully controlled concentrations. Extrapolating from these clinical studies to determine health benefits to the general population from reduced exposures to ambient O_3 is difficult. Among other things, it requires adjusting for the exposures and the time that people spend indoors and outdoors as well as the time that they spend either at rest or exercising at low rates. Information about such behavior, particularly at alternative O_3 levels, is sparse.

Epidemiologic studies have the advantage of not requiring these extrapolations and adjustments. Nevertheless, they can show only statistical associations that may not have causal connections between pollution concentrations or exposure and adverse health effects (18). The studies that we used associate ambient O_3 concentrations, measured at fixed monitoring stations located close to a person's home, to daily or 2-week records of illness, symptoms, or days of restricted activity that the person experienced.

We estimated the reduced incidence of the quantifiable adverse health effects in the year 2004 accompanying a 35% reduction in emissions of VOCs for an estimated 129 million people living in the 94 metropolitan areas (322 counties) predicted to be in nonattainment in 2004 (19, 20). For each metropolitan area, we made separate calculations on the basis of the predicted change in air quality there and then aggregated these estimates to obtain the national estimates.

From the epidemiologic studies, we found that the average asthmatic will experience about 0.2 fewer days per year on which he or she has an asthma attack and that the average nonasthmatic will experience about 0.1 fewer minor restricted activity days per year because of reduced VOC emissions and subsequently improved air quality (21). In addition, nonasthmatics will experience other minor health benefits as well in the form of reduced number of symptom days.

To convert these predicted changes in physical health into economic benefits, it is necessary to ascertain individuals' willingness to pay for a reduced incidence of illness and adverse symptoms. To do so, we drew on a number of studies designed to uncover these values, primarily through questioning of both healthy and infirm respondents with supplemental data on the out-of-pocket medical costs and lost income that may be associated with illness or symptomatic effects (22). These studies have found an average value of $25 for each asthma attack prevented, $20 for a reduction of one restricted activity day (on which an individual is neither bedridden nor forced to miss work but must alter his or her usual pattern in some way), and $5 for one fewer day of occasional coughing. When reduced incidence is combined with these values, the predicted aggregate dollar benefits across the United States from these improvements in individuals' acute health status amount to $250 million per year.

By using clinical rather than epidemiologic studies to estimate health benefits, we arrive at a somewhat larger value for acute health benefits. For example, we predict that the number of coughing spells of 2 hours' duration would be reduced by as much as 2.5 episodes per person per year. Also, fewer episodes of shortness of breath and pain on deep inspiration are predicted to occur. Both are important consequences of air pollution control. We estimate that the annual monetary benefits associated with these improvements in health would be on the order of $800 million annually (23).

COMPARISON.   To summarize, according to OTA, the costs associated with a 35% reduction in nationwide emissions of VOCs in nonattainment areas will be at least $8.8 billion annually by the year 2004 and could be as much as $12 billion. Yet the acute health improvements that we predict to result from these changes are valued at no more than $1 billion annually and could be as little as $250 million. The high estimate relies on the most generous of the four clinical studies that the EPA sanctioned in its staff paper on the health effects associated with O_3 and other photochemical oxidants (24). We also assumed that exercise rates would be high in the exposed population (which increases health benefits), and we included benefits even for those engaged in light or moderate exercise. Subject to the caveats discussed below, total health benefits are still relatively small.

In contrast to, say, the removal of lead from gasoline, for which estimated benefits are well in excess of costs (25), the benefit-cost comparison for national O_3 control is unfavorable. The reasons for this are, in part, the relatively small improvements in ambient O_3 levels that the controls effect (which in turn imply fairly small benefits) as well as the high costs of control.

THE LOS ANGELES PLAN

What about air pollution control efforts in Los Angeles, the nation's most notably polluted metropolis? In 1989, the supervisors of the South Coast Air Quality Management District (SCAQMD) approved an ambitious new plan designed to bring the four-county district into attainment with the NAAQS (26). The South Coast plan is designed to reduce ambient concentrations of particulates (such as sulfates), NO_x, SO_2, and CO in addition to O_3. To effect these reductions, the plan envisions three tiers of controls. Tier I consists mainly of the wider application of known pollution control technologies. The 120 measures identified in Tier I include such things as installing pollution control devices on equipment used in the manufacture of rubber products, substituting less polluting solvents in degreasing operations and in auto refinishing facilities, and adding new controls on electric power plants. Tier I also contains a number of less conventional pollution control measures, such as the banning of bias-ply tires (to improve mileage and to reduce particulate levels) and even restricting fuels used in backyard barbecues.

The 37 measures in Tier II of the South Coast's plan all require some advancement or extension of current pollution control techniques. They call for such measures as additional control of dust blown from roads and parking lots, incentives to reduce residential and industrial fuel consumption, and restrictions on automobile usage. The South Coast authorities envision implementing these measures over the next 10 to 15 years.

Tier III controls are more speculative still and are designed to bring about major technologic breakthroughs. The South Coast plan does not provide an explicit list of control measures for Tier III, but it does identify programs that together aim to eliminate almost all hydrocarbons from solvents, coatings, and motor vehicles and to convert all the region's vehicles to low-emitting vehicles. According to South Coast officials, enactment of all three sets of controls will, by the year 2010, bring the area into attainment with the NO_2 and CO standards and into virtual attainment for particulates and O_3.

does it generate $?

COSTS. The annual costs associated with 58 of the control measures in Tier I are estimated to be $3.4 billion per year and will obtain about one-third of the total emissions reductions under the whole plan (26). These control measures include those designed to reduce O_3 concentrations as well as particulates and CO. By itself, this estimate provides some insight into the sweep of the South Coast plan: In 1988, total spending to comply with all federal air pollution control regulations across the entire United States was approximately $30 billion (27).

The costs of the other elements of the plan to residents of the South Coast basin are difficult to assess because, among other things, they involve valuing time losses and inconvenience, such as those experienced in car-pools and those that require changes in driving, shopping, and even living habits. Although some of these measures involve relatively small out-of-pocket outlays, they will be costly in an economic sense if they increase inconvenience, waste time, or reduce well-being in other ways.

A recent study provides an indication of the potential costs of the entire Los Angeles air quality plan (28). The overall cost was estimated on the assumption that the per-ton costs of the control measures for which no cost data were provided were equal to the per-ton costs for the measures for which cost data were provided. The estimated annual cost of the entire plan is about $13 billion per year, or about $2700 per household in the Los Angeles basin.

Pollution control in the South Coast appears to be far more expensive on a per-ton basis than for the rest of the nation. For VOC control alone, the same study (28) estimated that the average cost is about $11,000 per ton compared to OTA's estimate of $1,800 to $2,500 per ton nationally (or about $9 billion to $12 billion per year). These differences occur mainly because the South Coast has already implemented far more stringent control sources than other parts of the country (29).

BENEFITS. Are the benefits valued at more than $10 billion to $13 billion annually? The South Coast authorities have made two sets of estimates of the benefits of the plan. The first, which was based on a study sponsored by the California Air Resources Board in 1985 (30), covered various health benefits (mortality and morbidity from particulates and acute morbidity from O_3) as well as materials, agricultural, and visibility effects from meeting the air quality standards in the basin in 2010. It estimated annual health benefits of $2.4 billion (a point estimate) with an upper bound of $6.4 billion annually and total benefits of $3.7 billion to $7.7 billion annually. A more recent estimate, sponsored by the South Coast authorities in 1989 (31), only addressed the above health effects, finding a best conservative estimate of $9.4 billion annually and health benefits as high as $20 billion or as low as $5 billion per year. Perhaps it is surprising, given the association of Los Angeles with O_3 problems, that $6 billion of the $9.4 billion in health benefits is for reduced mortality risks from meeting the particulate standards, whereas only $2.4 billion is for O_3-related benefits.

Because of the availability of newer studies on the health effects of air pollution, a number of serious methodologic problems with the second study (32), and the wide disparity in estimates issued by the South Coast, we have reexamined the benefits. On the basis of dose-response functions drawn from a recent epidemiologic study (33) and the South Coast's estimate that annual average ambient concentrations of sulfate (a proxy for acid aerosols) would be reduced by more than 7 $\mu g/m^3$, we estimate that premature mortality in the region might be reduced by 2000 cases per year with full implementation of the South Coast plan. Using results from several studies designed to ascertain the values that individuals place on reduced risk of premature mortality, we assign a value of $1000 to each reduction of 0.001 in annual mortality risk (34, 35). Combining the reduced mortality risk per individual, the expected population of the South Coast in the year 2010 (16 million), and the valuation of the risk reduction, we estimate possible mortality benefits of $2 billion annually associated with the pollution control plan.

To estimate benefits in the South Coast in the form of reduced acute morbidity, we use a procedure similar to that described in the discussion of nationwide pollution control benefits. That is, we translate the predicted reductions in airborne concentrations of O_3 and particulates into reduced illness and reduced frequency of respiratory symptoms by using epidemiologic and clinical studies. These are valued by means of willingness-to-pay estimates drawn from sources described above.

On the basis of this approach, we estimate that reduced O_3 concentrations will effect an annual reduction of 22 million person-days on which adverse respiratory or other symptoms will be experienced by South Coast residents. In dollar terms, these benefits amount to about $300 million annually. According to the South Coast officials, reduced ambient particulate concentrations will result in $700 million annually in reduced morbidity; reduced particulate loadings will provide $700 million annually in reduced materials damage, and another $130 million in materials damage will be saved as a result of reductions in ambient SO_2. We take these at face value.

In all, annual benefits to human health are predicted to be $3 billion ($2 billion in premature mortality, $0.3 billion in O_3-related morbidity, and $0.7 billion for particulate-related morbidity). If one includes the South Coast's estimates of materials damage, total annual benefits rise to about $4 billion. This is far short of the $13 billion per year that the plan may cost.

CAVEATS AND UNCERTAINTIES

To this point, we have presented benefits and costs as point estimates, but there are clearly great uncertainties in making such estimates. It is essential to understand them and to bear them in mind in interpreting the findings above.

With respect to our national comparison, OTA's estimate of control costs has a number of limitations. For instance, OTA did not estimate the cost of the mandatory introduction of alternative motor vehicle fuels (methanol, ethanol, or reformulated gasoline) such as is called for in the new Clean Air Act amendments. This will add approximately $3 billion to annual costs. Also, OTA did not anticipate the second round of vehicle emissions reductions that will almost surely be required under the amendments; this will add another $5 billion annually (36). Finally, no attempt was made to estimate nonpecuniary costs. For instance, OTA estimates that an enhanced motor vehicle inspection and maintenance program will cost about $50 per vehicle annually, including fees, administrative costs, and repair costs. The opportunity cost of people's time is ignored, however, even though the time spent can be significant. If this time were properly priced, it could add up to $7 per vehicle (37).

There is also great uncertainty about the costs of the South Coast plan. Marginal costs generally begin to rise sharply at higher levels of control, and VOCs have been controlled longer and more stringently in Southern California than in any other part of the country. For this reason, the controls envisioned in the South Coast plan could prove to be more expensive than anticipated. Also, we have little experience in the United States with stringent transportation control measures. If they are implemented and prove to be quite inconvenient to those affected, costs could be higher than those projected here. It is impossible to provide anything approaching statistical confidence intervals for either the national or the South Coast plan.

There are several respects in which costs could be much lower than those forecast here for both national and Los Angeles area air pollution control. For example, the cost of vehicle emissions controls are based on modest extensions of proven control technology (the catalytic converter). If, however, the electrically heated catalyst can be perfected and produced relatively inexpensively, control costs may be overstated here. Similarly, breakthroughs in reformulated gasoline or other alternative motor vehicle fuels could bring costs down considerably. Likewise, if the pace of technologic innovation accelerates sharply for VOC control from stationary sources, the same conclusion would apply. Finally (and particularly in Los Angeles), if driving restrictions eventually are imposed, and if commuters easily adapt to them, O_3 control costs may be lower than projected here.

Perhaps it is not surprising that uncertainties are greater concerning the benefit estimates presented here. These uncertainties arise from several sources, primarily the prediction of physical effects and the attribution of dollar values to them. For instance, if we had used the analysis of Whittemore and Korn (38) instead of that of Holguin et al. (17) to predict changes in asthma attacks, estimated benefits to asthmatics would be less than half those included above.

The largest such uncertainty concerns the link between particulate matter at current ambient concentrations and premature mortality. The statistical associations that epidemiologists and others have found between city mortality rates and annual particulate levels do not offer convincing evidence of the existence and magnitude of such effects; for instance, these effects become insignificant with minor changes in sample composition and model specification, and even the best of these studies uses a poor proxy (sulfates) for the particles now thought to be the causal agents (acid aerosols) (39). Because the total number of deaths from lung disease in the South Coast is 4000 annually, attributing a reduction of 2000 premature deaths to the South Coast's plan seems likely to be optimistic.

In monetizing the reduced frequency of respiratory symptoms or disease, a range of values could have been used. In the literature, the range cited for an asthma attack is $10 to $40; for a restricted-activity day, the corresponding range is $10 to $30; for a symptom day, it is $3 to $10.

The choice among epidemiologic and clinical studies, and among values to assign to physical effects, can have an important effect on estimated benefits. If we had used only upper bound estimates to predict each type of acute health effect from O_3 and, correspondingly, to attribute dollar values to reduced incidence of each, acute health benefits nationwide of a 35% VOC reduction would be $2 billion annually, and acute health benefits in the South Coast (of meeting only the O_3 standard) would be $2.4 billion per year. If we had used lower bound estimates, on the other hand, benefits would be 3% of our upper bound estimate.

There is another important caveat to be attached to the benefit estimates presented above, one that can only impart a downward bias to them. Specifically, we excluded certain types of benefits for which it was impossible to predict physical effects or to make reasonable dollar attributions. For instance, some animal toxicologic studies suggest that prolonged exposure to O_3 can permanently reduce the elasticity of the lung and, hence, initiate chronic respiratory illness (40). Although there is no convincing epidemiologic evidence for this potential effect in humans to date, such a finding would affect any benefit-cost analysis of efforts to control ground-level O_3 either nationally or locally (41). Similarly, we excluded in our estimates of national as well as South Coast benefits any improvements in forests or agricultural output in rural regions that might result from VOC control in urban areas because of the difficulty of translating emission reductions in urban areas into reduced ambient concentrations in agricultural regions. Also omitted are possible reductions in damage to rubber and other products exposed to O_3. Nevertheless, including such agricultural benefits would be unlikely to add more than $1 billion to the national total predicted here (15); South Coast benefits would increase minimally. Finally, the totals omit a dollar attribution for the improved visibility that should result from reduced ambient sulfate concentrations.

It is important to find ways to predict the physical likelihood of the exclusions identified here and to ascertain individuals' willingness to pay for any such improvements. These omitted categories would have to have large benefits associated with them, however, to tip the apparently unfavorable balance between benefits and costs for either the national or the regional air pollution control plans that we have examined.

CONCLUSIONS AND POLICY IMPLICATIONS

It is unpleasant to have to weigh in such a calculating manner the pros and cons of further air pollution control efforts. We would all prefer limitless resources so that every pollution control measure physically possible could be pursued. Because resources are scarce, however, the real cost of air pollution control is represented by the government programs or private expenditures that we forego by putting our resources into reducing VOC emissions. In the health area alone, $10 billion invested in smoking cessation programs, radon control, better prenatal and neonatal health care, or similar measures might contribute much more to public health and well-being (42).

Although we have discussed both national and regional air pollution control plans in all-or-nothing terms, neither plan is indivisible. Because the benefits and costs of air pollution control are sure to vary considerably among metropolitan areas, it may make economic sense to control a great deal in some places but little in others. Further controls will almost inevitably be justified in the Los Angeles area, where despite concerted efforts over the last 30 years air pollution is quite clearly unacceptable and adverse health effects are the most significant. On the basis of cost estimates made by the South Coast authorities in the Los Angeles area, particularly attractive VOC control possibilities include reformulating coatings used in the manufacture of wood furniture, modifying aircraft engines, and substituting less volatile cleaning solvents (26). By the same token, one must be especially careful in evaluating the benefits of mandatory van-pooling and other transportation control measures that have possibly large nonpecuniary costs. Even if such efforts temporarily relieve freeway congestion, new drivers may appear in the commuting brigade and wipe out apparent pollution reductions. The important point to emphasize is that all control measures must be viewed with an eye toward the good that they are likely to do and the costs that they are likely to impose.

Next, although smog is the pollution problem with which Los Angeles is most often associated, a substantial share of the benefits of further air pollution control there appears to arise from reduced particulate concentrations, according to the SCAQMD (30, 31). Controlling VOCs will have no direct effect on these particulates and will be quite expensive. It may make sense for authorities there to reorient their control plan toward particulate control to maximize health benefits per dollar of pollution control (32).

Finally, implicit in our discussion is discomfort with the premises on which our national air quality standards are now based. If, as seems likely, there are no pollution concentrations at which safety can be assured, the real question in ambient standard setting is the amount of risk that we are willing to accept. This decision must be informed by economics. Although such economic considerations should never be allowed to dominate air pollution control decisions, it is inappropriate and unwise to exclude them (43).

REFERENCES AND NOTES

1. These are the standards that apply to six common pollutants: sulfur dioxide, nitrogen dioxide, carbon monoxide, particulate matter (less than 10 μ in diameter), lead, and ground-level ozone.
2. *National Air Quality and Emissions Trends Report, 1989* (Document EPA-450/4-91-003, EPA, Washington, DC, February 1991). In 1988, because of less favorable weather conditions, 130 million people lived in areas exceeding the O_3 standard.
3. M. Russell, *Science* 241, 1275 (1988).
4. The national strategy being designed by Congress and the local plan being formulated by Los Angeles authorities focus almost exclusively on VOCs. For that reason, we concentrate on the costs and benefits of control measures for this pollutant. Some experts believe, however, that the best way to control tropospheric O_3 in some areas is by controlling NO_x, which represents the

other major precursor [W. L. Chameides, R. W. Lindsay, J. Richardson, C. S. Kiang, *Science* 241, 1473 (1988)].

5. E. M. Gramlich, *Benefit-Cost Analysis of Government Programs* (Prentice-Hall, Englewood Cliffs, NJ, 1981).
6. A. V. Kneese, *Measuring the Benefits of Clean Air and Water* (Resources for the Future, Washington, DC, 1984); A. M. Freeman III, *The Benefits of Environmental Improvement: Theory and Practice* (Resources for the Future, Washington, DC, 1979).
7. M. Hazilla and R. J. Kopp, *J. Polit. Econ.* 98, 853 (1990).
8. A. V. Kneese and W. D. Schulze, in *Handbook of Natural Resource and Energy Economics,* A. V. Kneese and J. L. Sweeney, Eds. (Elsevier, New York, 1985), pp. 191–220.
9. *National Air Quality and Emissions Trends Report, 1988* (Document EPA-450/4-90-002, EPA, Washington, DC, March 1990).
10. *Catching Our Breath: Next Steps for Reducing Urban Ozone* (OTA, Washington, DC, 1989).
11. The controls that OTA considered include reasonably available technology on all existing stationary sources not up to current practice; more state-of-the-art controls on new sources not currently being regulated; emission controls on all facilities that transport, store, or dispose of hazardous wastes; controls on applicators of surface coatings; added refueling controls on autos and gas pumps; strengthened vehicle inspection and maintenance programs; more stringent tailpipe standards for vehicles; restrictions on gasoline volatility; and required methanol use in fleet vehicles in selected cities.
12. Because some of these controls would also affect attainment areas, OTA dealt with them separately. By 2004, reductions of VOCs in attainment areas were estimated to be 3.1 million tons from a baseline of 10.9 million tons, or a reduction of 28%.
13. See J. H. Seinfeld, *J. Air Pollut. Control Assoc.* 38, 616 (1988), for a critical review of these and other, more sophisticated models.
14. A recent report by investigators at Resources for the Future basically supports the OTA conclusions. See A. J. Krupnick, M. Walls, M. Toman, *The Cost-Effectiveness and Energy Security Benefits of Methanol Vehicles* (final report to EPA, Resources for the Future, Washington, DC, September 1990).
15. A. J. Krupnick and R. Kopp, "The Health and Agricultural Benefits and Reductions in Ambient Ozone in the U.S.," report to OTA for *Catching Our Breath: Next Steps for Reducing Urban Ozone* (OTA, Washington, DC, 1989).
16. The O_3 season varies by state and generally includes the summer months and some portion of the spring and fall. States such as California feature a 12-month O_3 season, however.
17. The epidemiologic studies were chosen on the basis of criteria developed primarily by the EPA [see A. J. Krupnick, *An Analysis of Selected Health Benefits from Reductions in Photochemical Oxidants in the Northeastern United States,* report prepared for EPA, Ambient Standards Branch, Office of Air Quality Planning and Standards, Contract 68-02-4323 (Resources for the Future, Washington, DC, 1987)]. These studies include the following: A. H. Holguin *et al.,* in *Air Pollution Control Association Transactions on Ozone/Oxidants Standards* (Air Pollution Control Association, Houston, 1984), p. 262; A. J. Krupnick, W. Harrington, B. Ostro, *J. Environ. Econ. Manage.* 18,

1 (1990); P. R. Portney and J. Mullahy, *J. Urban Econ.* 20, 21 (1986); J. Schwartz, V. Hasselblad, H. Pitcher, *J. Air Pollut. Control Assoc.* 38, 158 (1989). The clinical study used was that of W. F. McDonnell *et al., J. Appl. Physiol.* 54, 1345 (1983), one of four key clinical studies identified by the EPA for standard setting. The others are E. L. Avol, W. S. Linn, T. G. Venet, D. A. Shamoo, J. D. Hackney, *J. Air Pollut. Control Assoc.* 34, 804 (1984); T. J. Kulle, L. R. Sauder, J. R. Hebel, M. D. Chatham, *Am. Rev. Respir. Dis.* 132, 36 (1985); W. S. Linn *et al., Toxicol. Ind. Health* 2, 99 (1986). McDonnell *et al.* provide larger estimates of health benefits than the other clinical studies.

18. A. R. Feinstein, *Science* 242, 1257 (1988).

19. There are few reliable studies linking O_3 to symptoms or restricted-activity days in children. Nevertheless, we assume that they respond like adults and include benefits to them in our calculations.

20. The 1985 urban population of 110.8 million was assumed to grow at 0.8% per year (Bureau of the Census projections). Only counties out of compliance in 1985 are included.

21. These are days that involve no work loss or time in bed but do involve some symptomatic distress.

22. E. T. Loehman *et al., J. Environ. Econ. Manage.* 6, 222 (1979); M. Dickie *et al.,* "Reconciling Averting Behavior and Contingent Valuation Benefit Estimates of Reducing Symptoms of Ozone Exposure" (draft), report to EPA, Washington, DC, 1987; G. S. Tolley *et al., Valuation of Reductions in Human Health Symptoms and Risks* (University of Chicago, final report to the EPA, Office of Policy Analysis, Washington, DC, 1986); R. D. Rowe and L. G. Chestnut, *Oxidants and Asthmatics in Los Angeles: A Benefits Analysis* (Energy and Resource Consultants, Inc., report to the EPA, Office of Policy Analysis, EPA-230-07-85-010, Washington, DC, 1985).

23. These estimates of health effects and benefits also depend heavily on assumptions about exercise patterns in the population and the more uncertain effect of O_3 on individuals exposed while moderately exercising.

24. *Review of the National Ambient Air Quality Standards for Ozone: Assessment of Scientific and Technical Information* [draft staff paper, EPA, Office of Air Quality Planning and Standards (OAQPS), Research Triangle Park, NC, November 1987]. See also *EPA Report of the Clean Air Scientific Advisory Committee Review of the NAAQS for O_3: Closure on the OAQPS Staff Paper (1988) Criteria Document Supplement (1988)* (EPA, Washington, DC, May 1989).

25. J. Schwartz *et al., Costs and Benefits of Reducing Lead in Gasoline* (Publication EPA-230-05-85-006, EPA, Washington, DC, 1985).

26. South Coast Air Quality Management District and Southern California Association of Governments, *Draft 1988 Air Quality Management Plan* (South Coast Air Quality Management District and Southern California Association of Governments, Los Angeles, 1989).

27. P. R. Portney, in *Public Policies for Environmental Protection,* P. R. Portney, Ed. (Resources for the Future, Washington, DC, 1990), p. 68.

28. D. Harrison, Jr., *Economic Impacts of the Draft Air Quality Management Plan Proposed by the South Coast Air Quality Management District* (National Economic Research Associates, Inc., Cambridge, 1988).

29. How can OTA peg the costs of a national O_3 control plan at $9 to $12 billion annually when the cost of air pollution control in the South Coast alone may

be on the same order? There are two explanations for this apparent inconsistency. First, the OTA study considered only the control of VOCs; the South Coast plan is aimed at VOCs primarily but also controls other air pollutants as well, as discussed above. Second, the OTA cost estimate is not predicated on attainment of the O_3 standard in the South Coast area (or most other areas, for that matter). Rather, it assumes but a 23% reduction in ambient O_3 levels in the Los Angeles area (from a design value of 0.36 ppm to a value of 0.28 ppm in the year 2010), stopping far short of ensuring attainment with the NAAQS. The South Coast estimate, on the other hand, is predicated on a set of controls designed to meet (or nearly meet) the current standards for O_3 and particulate matter.

30. R. D. Rowe *et al., The Benefits of Air Pollution Control in California* (report prepared for California Air Resources Board, Contract A2-118-32, Energy and Resources Consultants, Boulder, CO, 1986).

31. J. V. Hall *et al., Economic Assessment of the Health Benefits from Improvements in Air Quality in the South Coast Air Basin* (final report prepared for South Coast Air Quality Management District, California State University Fullerton Foundation, Fullerton, CA, June 1989).

32. A. Nichols and D. Harrison, Jr., *Benefits of the 1989 Air Quality Management Plan for the South Coast Air Basin: A Reassessment* (National Economic Research Associates, Inc., Cambridge, 1990).

33. H. Ozkaynak and G. D. Thurston, *Risk Anal.* 7, 449 (1987).

34. A. Fisher, L. Chestnut, D. Violette, *J. Policy Anal. Manage.* 8, 88 (1989).

35. The studies published in the literature ascertain such values from wage premia paid to workers in risky jobs and from direct survey techniques. These studies generally find that a reduction in annual mortality risk of 0.001 is valued at between $800 and $8000. Wage studies usually infer such values from cohorts of workers averaging 40 years of age, however. The premature mortality expected to result from ambient air pollution falls predominantly on the elderly. Thus fewer life years are likely to be saved per case of premature mortality avoided by air pollution control than through programs that reduce occupational risks, for instance. For this reason, we choose a value for reduced mortality risk that falls toward the lower end of the observed range.

36. These additional measures will produce added benefits as well. See P. Portney, *Econ. Perspect.* 4, 173 (1990).

37. V. D. McConnell, *Environ. Manage.* 30, 1 (1990).

38. A. S. Whittemore and E. L. Korn, *Am. J. Public Health* 70, 687 (1980).

39. F. W. Lipfert, S. C. Morris, R. E. Wyzga, *Environ. Sci. Technol.* 23, 11 (1989); *Envion. Health Perspect.* 79, 3 (1989).

40. D. Bartlett, Jr., C. S. Faulker II, K. Cook, Appl. Physiol..37, 92 (1974); B. E. Barry, F. J. Miller, J. D. Crapo, *Lab. Invest.* 53, 682 (1985).

41. Several preliminary epidemiologic analyses have found statistically significant associations between ambient air quality and some chronic respiratory illnesses. See P. R. Portney and J. Mullahy, *Reg. Sci. Urban Econ.* 20, 407 (1990); D. Abbey *et al.,* paper presented at the annual meeting of the Air Pollution Control Association, Dallas, TX, 1988.

42. P. R. Portney, *Issues Sci. Technol.* 4, 74 (1988).

43. We thank P. Abelson, R. Frank, A. M. Freeman III, R. Friedman, B. Goldstein, D. Harrison, Jr., M. Lippman, G. McRae, and our colleagues at Resources for the Future. We alone are responsible for the conclusions, however.

V
THE GLOBAL ENVIRONMENT

When environmental problems first attracted serious attention in the late 1960s, concern concentrated on pollution of rivers and harbors, lakes and streams, and of local or regional air mantles, most of whose damages were contained within national, or even regional, boundaries. During the 1980s, attention shifted increasingly to environmental degradation whose consequences transcend such boundaries and are capable in some instances of inflicting damages indiscriminately on life anywhere on the planet, now and into the indefinite future.

The crux of the problem of transnational environmental damages lies in the fact that while individual countries have insufficient incentives on their own to limit the damages they inflict, no international authority exists with the power to impose and enforce limits. Protection can be achieved only through negotiation or cooperation among countries. When only two or a few countries are involved—as in the case of SO_2 depositions in Canada emanating from utilities in the middlewestern United States—bargaining, along the lines that Ronald Coase describes, can, in principle, provide a practical solution. As the number of countries that are either producing or receiving emissions grows, agreeing upon and enforcing a policy can become vastly more complicated.

The ultimate in complexity is reached with problems of the global common such as the the warming of the earth's atmosphere due to continuing emissions of carbon dioxide (CO_2) and to a lesser extent CFCs and methane. This greenhouse problem, as Thomas C. Schelling points out here, "is truly one of the global common. A ton of carbon emitted anywhere on earth has the same effect as a ton emitted anywhere else." Another phenomenon of similarly global proportions is the destruction of the ozone layer due to emissions of CFCs and halons, which is capable of causing increases in the incidence of cancer and other health problems, as well as damages to the food chain of water-borne creatures, all over the earth.

What makes these two problems even more truly global is that not only are the damages distributed throughout the globe but so are the sources of emissions. Other environmental damages with global consequences, such as the destruction of a rain forest or of a particular plant or animal species, may emanate from selected countries. But, in either case, a resource that a very large number of people around the world depend on—such as the earth's atmosphere, its oceans, or its biological diversity—is degraded, without there existing any central authority to control its use or misuse.

In the first selection here, Scott Barrett considers the obstacles to achieving international cooperation in managing the global common and ways to overcome them. The opportunity for free riding, he finds, provides a strong incentive not to cooperate whenever enforcement is lacking. The theoretical arguments for supposing that cooperation will fail are, therefore, compelling. Nevertheless, cooperation does take place and is often codified into international agreements. The most recent and perhaps most notable instance is the 1987 Montreal Protocol on Substances that Deplete the Ozone Layer, which calls for the phasing out of emissions of CFCs and halons by the end of the century. As of 1992, over eighty countries had signed on, including the leading producers of the substances.

Using simple models and games, Barrett explores the conditions under which international cooperation might develop, and what the significance of different forms of cooperation might be. He concludes that cooperation is sometimes hardest to obtain when it is most needed.

In "Some Economics of Global Warming," Thomas C. Schelling is also concerned with finding a basis for international agreement, specifically for the control of the CO_2 emissions that give rise to the greenhouse effect. In pursuit of this goal, he analyzes the interests of different groups of countries in avoiding global warming and the ability of each to contribute to the solution. He begins by assessing the nature of the damages to be expected from global warming and their likely distribution around the globe.

Both the anticipated degree of global warming and its timing are matters of considerable uncertainty, although there is little doubt that warming will occur. Still greater uncertainty surrounds its climatic impacts in different parts of the world. Nevertheless, Schelling believes it is fairly obvious that the economic damages from global warming will fall mainly on agricultural production, and that this sector represents a major share of GDP only in the developing countries. Health damages, too, will be most severe among the vulnerable populations of poorer nations. It is, however, doubtful, he believes, that the developing countries can afford to incur the penalties in terms of economic growth that would be required to curtail their CO_2 emissions. Given a choice, they would probably prefer to have the money invested in their own development than in the reduction of global emissions.

On the other hand, Schelling concludes, the developed countries appear to have little economic self-interest in cutting emissions. Even so, he finds important reasons for them to invest in mitigating global warming, among them to ensure against the possibility that the CO_2 build-up could have far worse consequences than predicted. A reasonable target would be to postpone the doubling of atmospheric CO_2 for several decades. The cost will be high but not, he believes, prohibitive.

The cost will be lower, the more efficiently the emissions abatement is accomplished. However, countries best able to sustain the burden will not necessarily be those that can abate most efficiently. Various proposals for using CO_2 taxes or emissions permits are undergoing serious study as possible mechanisms for achieving an efficient allocation of abatement effort. Schelling rejects such schemes on the grounds that they would involve intolerably large transfers of revenues or of permit values. He proposes, instead, a regime for negotiation of targets among the wealthier countries modeled after the Marshall Plan, which allocated U.S. dollars among Western European countries after World War II.

Yoshiki Ogawa shows, in "Economic Activity and the Greenhouse Effect," that it would not be easy for the industrial nations to significantly curb global emissions of CO_2 without cooperation from the developing world. He decomposes trends in CO_2 emissions generated by commercial energy consumption for various groups of countries between 1973 and 1986 to indicate the relative contributions of changes in population, per-capita GDP, energy intensity of production, and CO_2 intensity of fuel. During that period, the contribution of developing countries to worldwide CO_2 emission rose from 17 to 28 percent, while that of industrial countries fell from 60 to 47 percent. The percentage for the former Soviet Union and Eastern Europe remained about the same.

Declining energy intensity of production and fuel switching kept total CO_2 emissions almost constant in the industrialized world, whereas population and economic growth were the main culprits in the developing world's 5.6 percent annual increase. The nonindustrialized economies have a lot of catching up to do, and their populations continue to grow alarmingly. Without a reversal of current population trends, merely keeping total global emissions of CO_2 from rising could be a tall order for the industrial economies on their own.

In "Adverse Environmental Consequences of the Green Revolution," the Pimentels paint a stark picture of the prospects for the developing countries to feed populations that continue to grow at 2 or 3 percent per year. The Green Revolution, which at the outset held the promise of continually enhanced and sustainable agricultural yields, is shown to have adverse environmental side effects that place that promise in serious jeopardy. These externalities, the authors argue, threaten both the physical basis of production and the broader physical setting in which the populations live. They also have implications for population growth and the potential for famine worldwide.

We end this volume with a discussion of the destruction of the tropical rain forests. No single environmental problem provides broader scope for applying the tools of analysis that have been developed in this collection. The forests combine aspects of both public and private goods; the consequences of their destruction are felt both locally and globally, and the root causes of their destruction are the failure of governments to properly value the long-run benefits and costs of their exploitation and to maintain policies that encourage users to internalize costs.

Kenton R. Miller, Walter V. Reid, and Charles V. Barber, in "Deforestation and Species Loss," examine the causes of tropical deforestation and its consequences for the indigenous populations and for the rest of the world. By far the leading cause of tropical deforestation is slash and burn agriculture practiced by growing populations of landless migrants who lack the traditional concern for a sustainable cycle of cultivation. Conversion to pastureland in the Amazon, unsustainable commercial logging in Asia and Africa, and overharvesting of firewood all add to the problem. Behind these trends, the authors say, lie overwhelming population pressures and government management policies that encourage exploitation for immediate gain while eliminating land tenure rules and local community controls that provided incentives to follow more sustainable land use policies in the past.

The authors paint a dire picture of the consequences of deforestation, both locally and globally. The forests, they say, are among the most valuable resources of the countries in which they lie, and the long-run economic costs to these countries of their depletion clearly offset any economic gain from their destruction. While forest dwellers are the first to bear the cost of degradation, the destruction of their habitat is already well on the way to creating a class of environmental refugees who increase pressures on remaining environmental preserves.

As a source of forest products and agricultural land, the tropical forest is a private good that can, in principle, be owned and exploited for maximum gain by indigenous populations. But its existence also provides public goods whose benefits are shared by people all over the world.

The primary public goods are the tropical forests' absolutely critical role in the preservation of species and their contribution to containing the release of greehouse gases. The problem is that beneficiaries of the public goods beyond the borders of the nations in which the forests lie lack the authority to prevent their destruction, whereas the owners of the forests consider their private returns from exploitation to exceed any possible gain from preservation. The authors are confident that this assessment is shortsighted. Nevertheless, given current population pressures and a high rate of time preference in the developing world, it will not be easy to persuade the owners of tropical rain forests to limit deforestation without compensation from other nations who stand to gain.

26
International Cooperation for Environmental Protection

SCOTT BARRETT[1]

Scott Barrett is Assistant Professor of Economics at the London Business School and Research Director of the Centre for Social and Economic Rresearch on the Global Environment in London.

Suppose that land is communally owned. Every person has the right to hunt, till, or mine the land. This form of ownership fails to concentrate the cost associated with any person's exercise of his communal right on that person. If a person seeks to maximize the value of his communal rights, he will tend to overhunt and overwork the land because some of the costs of his doing so are borne by others. The stock of game and the richness of the soil will be diminished too quickly. It is conceivable that those who own these rights, i.e. every member of the community, can agree to curtail the rate at which they work the lands if negotiating and policing costs are zero. . . . [However,] negotiating costs will be large because it is difficult for many persons to reach a mutually satisfactory agreement, especially when each hold-out has the right to work the land as fast as he pleases. [Furthermore,] even if an agreement among all can be reached, we must yet take account of the costs of policing the agreement, and these may be large, also.

Demsetz (1967, pp. 354–5)

Originally published as "The Problem of Global Environmental Protection," by Scott Barrett. Copyright © Oxford University Press and the Oxford Review of Economic Policy Limited 1990. Reprinted from the *Oxford Review of Economic Policy* Vol. 6 No. 1 by permission of Oxford University Press.

[1] I have benefited greatly from comments made by David Pearce on an earlier draft of this paper.

I. INTRODUCTION

Demsetz's influential paper on the development of private property rights makes depressing reading for anyone concerned about global common property resources such as the oceans and atmosphere. Demsetz's view—and it is one that is shared by many others—is that users of a communally owned resource will fail to come to an agreement on managing the resource even though it is in the interest of all users to co-operate and reduce their rates of use of the resource. The reason is that if this improved situation is attained, every user will earn even higher returns by free-riding on the virtuous behaviour of the remaining co-operators. As a consequence, united action on the part of users can be expected to be unstable; co-operative agreements, even if they are reached, will not persist. The only way out of the common property dilemma, as Demsetz makes clear, is intervention by '*the* state, *the* courts, or the leaders of *the* community' (emphasis added). In Demsetz's example, the intervention manifests itself in the development of private property rights to the resource, but the intervention could just as easily involve regulation.

The reason this view is disquieting is that for global common property resources there is no World Government empowered to intervene for the good of all. To be sure, there do exist international institutions—most notably the United Nations Environment Programme—which have been given the mandate to co-ordinate international environmental protection efforts. But none of these institutions can dictate what is to be done; that requires agreement by the parties concerned. The problem is perhaps best exemplified by the International Whaling Commission (IWC), which was established to conserve whale stocks, but whose best efforts in this regard have been repeatedly foiled. IWC membership is open to any country, and this leaves open the possibility that the whales could be protected for the global good. But any member can object to a majority decision, and hence render that decision meaningless. For example, a 1954 proposal to prohibit the taking of blue whales in the North Pacific was rejected by the only members who hunted blue whales in this ocean—Canada, Japan, the US, and the USSR—and hence did nothing to protect this species. In 1981 the IWC sought to ban the use of the non-explosive harpoon for killing minke whales. The ban was objected to by Brazil, Iceland, Japan, Norway, and the USSR. Since these were the only countries that hunted minke whales, the ban had no effect.[2]

Because national sovereignty must be respected, the problem of conserving global common property resources is no different from that described by Demsetz. The only way out of the global common property dilemma is agreement. Yet, just as in the situation Demsetz describes, there are strong incentives for governments not to co-operate, or to defect

[2]See Lyster (1985).

from an agreement should one be reached. This is the crux of the problem of managing global common property, and what distinguishes this problem from the long studied one of common property management under the jurisdictional control of a central authority.

Attempts to correct global, unidirectional externalities will encounter similar difficulties. Consider the problem where certain activities by one country harm all others. A good example is deforestation of Amazonia by Brazil. The rain forests play a crucial role in the protection of biological diversity and in the functioning of the carbon cycle. When standing, the rain forests serve as habitat to about a half of all wildlfe species and absorb carbon dioxide, one of the so-called greenhouse gases. When the forests are burned, masses of species can become extinct and substantial quantities of greenhouse gases are emitted. If the rights to generate these externalities are vested in the one country, as indeed they are in the case of Brazilian deforestation, then the others will have to pay this nation to cease its destructive activities. If the externality affected only one other country, then bargaining might be possible; the externalities might be internalized without outside intervention.[3] But in the case of global externalities, all countries except the generator suffer. All sufferers might be willing to bribe the generator to cease its harmful activities. But a contribution by any one country would confer benefits on all others and not just the one making the compensating payment. The others could therefore do better by free-riding. But then so too could the one that contemplated making the payment. Co-operation would again be foiled.[4] Mechanisms exist that can lead countries to reveal their preferences for global public goods truthfully (see, e.g., Groves and Ledyard, 1977), and hence for correcting global externalities. But in the absence of a World Government these mechanisms cannot be employed without the consent of the sovereign nations themselves. Every country would be better off if it agreed to participate in the revelation exercise. But each would do even better if others participated and it did not. All will therefore choose not to participate. The crux of the problem of correcting global externalities, like that of managing global common property, is that global optimality demands global co-operation, and yet the incentives facing individual countries work in the opposite direction.

[3]See Coase (1960). Bargaining has in fact taken place at the bilateral level. A famous example is the Trail Smelter case. The Canadian smelter emitted pollutants that crossed the US border. The case was arbitrated by an international tribunal comprised of an American, a Canadian, and a Belgian. The tribunal found that Canada was liable for damages, and also established emission regulations for the plant. The judicial decisions on this case make fascinating reading. See Trail Smelter Arbitral Tribunal (1939, 1941).
[4]Demsetz (1967, p. 357) argues that in the large numbers case, 'it may be too costly to internalize effects through the market-place'. Elsewhere, Demsetz (1964) argues that it might in fact be optimal for the externality not to be internalized since the costs of internalization should include the costs of transacting the agreement. But even then the free-rider problem would prevail; intervention, were it possible, might still be desired.

The theoretical arguments for supposing that co-operation will not develop are compelling. But they can hardly be complete. Co-operation *does* take place and is often codified in international agreements. Some of these are woefully ineffective—a famous example being the International Convention for the Regulation of Whaling (1946) which established the IWC. Others do appear to have achieved a great deal. Of these last, the Montreal Protocol on Substances that Deplete the Ozone Layer (1987) seems the most impressive, because it demands that its many signatories undertake substantial reductions in their emissions of ozone-depleting chlorofluorocarbons (CFCs) and halons. Though agreements dealing with unidirectional externalities are rare and almost invariably toothless, there is one—the World Heritage Convention—that at least holds some promise. This agreement places responsibility for safeguarding natural environments like the Serengeti and the Galapagos Islands on a community of nations, and could be invoked to protect the remaining tropical rain forests. There is clearly a need to explore why international co-operation might develop, and what the significance of particular forms of co-operation might be.

To make any progress we will need a basis from which to assess whether co-operation can in fact be expected to achieve much. Contrary to Hardin's (1968) famous allegory of the commons, the absence of co-operation need not lead to tragedy. Section II discusses some of the parameters that are important in determining the potential gains to co-operation. Having drawn the boundaries, we then consider how we might move closer to the full co-operative solution, the global optimum. Non-co-operation may sometimes wear the disguise of co-operation, and section III shows that an outcome better than the purely nationalistic one can emerge even where binding agreements are absent. Effective management of global environmental resources does however seem to rely on the more formal institution of international law. Section IV discusses the rudiments of a model that explains why countries would co-operate when the free-rider problem must surely bite, and what international agreements mean for global social welfare and the welfare of citizens of individual countries. Just as failure to co-operate may not lead to tragedy, so co-operation may not buy us very much. Indeed, combining the analysis of the potential gains to co-operation with this model of formal agreements, it can be shown that co-operation is sometimes hardest to obtain when it is most needed.

II. THE POTENTIAL GAINS TO CO-OPERATION

Where a global externality is unidirectional, the country causing the externality will, without a negotiated settlement, ignore the damages its activities impose on other countries. This is the full *non-co-operative* outcome.

The full *co-operative* outcome is found by internalizing the externality. Here the country inflicting the externality chooses its actions so as to maximize the net benefits of all countries, including itself. Global net benefits will of course be higher in this case. The difference between the global net benefits for the co-operative and non-co-operative outcomes defines the potential gains to co-operation.

Where the externalities of concern are reciprocal in nature, every country has some incentive to take unilateral action even in the absence of a binding agreement. Furthermore, the strength of this incentive will depend on the actions taken by all other countries. An example of a reciprocal externality is the emission of a global pollutant. If one country reduces its emissions, it will benefit from the improved environmental quality, provided other countries do not increase their emissions so as to fully offset the one country's action. The other countries will benefit partly by being able to increase their emissions somewhat and partly by enjoying a cleaner environment (again, provided their increase in emissions does not entirely offset the one country's extra abatement). The extent of the benefit enjoyed by the conserving country will clearly depend on the actions taken by the other countries, and of course all countries are subject to a similar calculus. It is this interdependence which makes calculation of the potential gains to co-operative management of global common property more difficult. It is better, then, that we work with a specific model.

To fix ideas, reconsider the problem of global pollution. Suppose that the relevant number of countries is N. One might think that N would include all the world's countries, but that need not be so, a point we return to later. Let us however suppose for simplicity that N does include all countries and that each is identical. Each, therefore, emits the same quantity of a pollutant *ex ante*— that is, before the game is played—and each faces the same abatement cost and benefit functions. The problem is then perfectly symmetric. To simplify the analysis further, assume that the *marginal* abatement cost and benefit functions are linear. Clearly, the marginal abatement cost schedule for each country must depend on its own abatement level, while each country's marginal abatement benefit function must depend on *world-wide* abatement.

In the absence of any co-operation, each country will maximize its own net benefits of abatement and in so doing will choose a level of abatement at which its own marginal abatement cost equals its own marginal abatement benefit.[5] This is the non-co-operative (Nash equilibrium) solution to

---

[5] I am assuming here that every country believes that its choice of an abatement level will not alter the choices of the other countries; that is, I am assuming zero conjectural variations. One can impose positive or negative conjectures, but these assumptions would be *ad hoc*. Alternatively, we could determine a consistent conjectures equilibrium—that is, one in whose neighbourhood every country's conjectures are confirmed by the responses of the other countries. Comes and Sandler (1983) find that consistent conjectures can lead to even greater overuse of the resource compared with the Nash equilibrium.

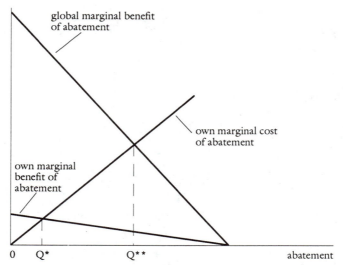

FIGURE 1. Graphical Illustration of the Potential Gains to Co-operation.

this game, and it is shown as abatement level Q^* in Figure 1. Were countries to co-operate fully, they would seek to maximize the global net benefits of abatement. Since we have assumed that all countries are identical, the global net benefits of abatement can be defined as the sum of every country's net benefits of abatement. In maximizing the global net benefits of abatement, each country will choose a level of abatement at which its own marginal costs of abatement equal the global marginal benefits of abatement, or the sum of the marginal abatement benefits enjoyed by all countries.[6] The full co-operative solution to this game is shown as abatement level Q^{**} in Figure 1.

One sees immediately that the full co-operative solution demands greater abatement but, equally, gives to every country a greater net benefit. For a given size of N, the difference between Q^* and Q^{**} can be shown to depend on the slopes of the marginal abatement benefit and cost curves. Denote the (absolute value of the) slope of each country's marginal abatement benefit curve by the letter b, and the slope of each country's marginal abatement cost curve by the letter c. Then it can be shown that the discrepancy will tend to be small whenever c/b is either 'large' or 'small' (see Barrett, 1991a). The approximate implication of this result is that fairly innocuous pollutants (that is, pollutants for which b is small) that

---

[6]This is of course nothing but a restatement of Samuelson's (1954) rule for the optimal provision of public goods. For an alternative presentation of these principles, see Dasgupta's (1982) model of a global fishery.

are very costly to control (that is, that have a large c) do not cause too great a problem. Nor do extremely hazardous pollutants (that is, pollutants associated with a large b) that are cheap to control (that is, pollutants that have a small c). In the former case, even collective action will not call for large abatement levels. In the latter, countries will want to abate substantial quantities of emissions unilaterally. The real problem is with pollutants whose marginal abatement benefit and cost curves are both either steep or flat—that is, hazardous pollutants that are costly to control, and mildly offensive pollutants that can be controlled at little cost. Of these, the former type naturally causes the greatest concern, for the cost of failing to co-operate in this case is very, very high.

Unidirectional and reciprocal externalities are plainly different in their effects. A country has no incentive to abate its emissions if the externality is unidirectional (provided side payments are ruled out), even if the pollutant is highly toxic. Not so if the externality is reciprocal in nature. For then the emitting country will have strong private incentives to control its emissions.

III. NON-CO-OPERATIVE ENVIRONMENTAL PROTECTION

The politics of global environmental protection are not as sterile as the above models would imply. Once we permit alternative strategies to be chosen, widen the choice sets themselves, and allow motivations other than self-interest (narrowly defined) to guide decision-making, global environmental protection can be enhanced.

Supergames

In the above games, strategies are chosen once, and the games are never repeated. But in the common property game countries are unavoidably locked in a continuing relationship, and this leaves open the possibility that they may retaliate and hence that co-operative strategies may be countenanced. Suppose then that all countries choose one of two strategies: they either choose the full co-operative abatement level Q^{**} initially, and continue to choose Q^{**} in every future period provided all other countries chose Q^{**} in every previous period; or they choose not to co-operate in any period.[7] Then the co-operative trigger strategy will

[7]If countries choose not to co-operate in the initial period, they need not choose Q^* in this period. The reason for this is that the optimal non-co-operative abatement level is contingent on the abatement levels chosen by the other countries.

constitute an equilibrium to this supergame provided the rate of discount is sufficiently low (for then the gains to choosing the non-co-operative strategy will be low, too; see Friedman, 1986). Although this supergame equilibrium entails co-operation, the co-operation is tacit and is enforced by means of a non-co-operative mechanism, retaliation; there is no explicit agreement, no open negotiation.[8]

One variant of this game involves countries adopting a convention that says that a subset of countries should co-operate in the face of free-riding by the others (see Sugden, 1986). It may, for example, be believed that the industrialized countries should co-operate to reduce emissions of CFCs or greenhouse gases (because it was their emissions that caused the environmental reservoirs to be filled in the first place), and that the poor countries should be allowed to free ride. In this game, each of the members of the subset could adopt the trigger strategy with respect to the subset, and each of the others could simply choose their optimal non-co-operative responses. This latter possibility may be fragile, because different countries may have different views about which should co-operate and which should not. This ambiguity may help explain why unilateral action to reduce CFC emissions was limited (US Environmental Protection Agency, 1988, p. 30576):

> In 1978 the United States restricted the use of CFCs in aerosols. While several nations adopted similar restrictions (e.g. Sweden, Canada, Norway) and others partially cut back their use (European nations, Japan), there was no widespread movement to follow the United States' lead. Concerns existed then that other nations had failed to act because the United States and a few other nations were making the reductions thought necessary to protect the ozone layer. Similar concerns exist today that unilateral action could result in 'free riding' by some other nations.

It is not obvious how sufferers of a unidirectional externality could punish the offending nation. But so long as the countries are engaged in some form of exchange, potent weapons may be at the sufferers' disposal. The Packwood-Magnuson Amendment to the US Fishery Conservation and Management Act (1976) requires the US government to retaliate whenever foreign nationals compromise the effectiveness of the IWC. An offending nation automatically loses half its allocation of fish products taken from US waters, and if the country refuses to improve its behaviour within a year, its right to fish in US waters is revoked.

---

[8]Axelrod's (1984) tournaments of the repeated prisoners' dilemma game suggest that behaviour in the disguise of co-operation may well emerge. Whether his findings carry over to the common property game with many players, however, is as yet unknown.

Matching

In the one-shot common property game, each country chooses an abatement level and nothing else. There is no reason why a country's choices need be limited in this way. We could, for example, allow countries to choose a 'base' abatement level—that is, a level of abatement which is not explicitly contingent on the abatement levels chosen by other countries—and a 'matching rate'—in our example, a fraction of the sum of all countries' base abatement levels. In effect, countries would then voluntarily subsidize each other's abatement levels. In the original game, one unit of abatement by country i buys country i one unit less of global emission (assuming that the emissions of all other countries are held fixed). In the matching game, one unit of abatement by country i may yield a much greater reduction in global emission. Matching might improve matters— Guttman (1978) shows that under certain conditions matching can sustain the full co-operative solution. But in the management of global common property, matching is rarely invoked. Recently, Norway announced that it would allocate one-tenth of one per cent of its GDP each year to a fund on climate change if other industrial countries matched its contribution (on a GDP percentage basis). The offer has yet to be taken up. Environmental groups in the United States have argued that the US should unilaterally surpass the reductions specified in the Montreal Protocol and at the same time impose restrictions on imports of products containing or made with CFCs from countries that fail to agree to make the same reductions. Their plea was rejected by the authorities (US Environmental Protection Agency, 1988, p. 30574).

Morality

In the models discussed thus far, the welfare of every country is assumed to depend solely on its *own* net benefits. An alternative way of looking at the problem is to assume that countries act according to some moral principle which requires that they take stock of the effect their actions have on the welfare of other countries. For example, suppose every country but one reduces its emissions of some global pollutant by at least x tonnes. Let us further suppose that the recalcitrant nation would like all others to reduce their emissions by y tonnes each. Then the leaders of the recalcitrant nation might feel compelled to obey the rule: if $y > x$ then we are morally obligated to reduce our emissions by at least x tonnes. If countries obey this rule, then the free-rider problem can be mitigated (see Sugden, 1984). That moral principles may guide non-co-operative abatement is suggested by the following remarks made by the House of Commons

Environment Committee (1984, p. lxxi) in its report recommending that the UK join the Thirty Per Cent Club:[9]

> As our inquiry has progressed the stance of the United Kingdom has become increasingly isolated by its refusal to legislate to reduce SO_2 and NO_x emissions. Since our work began three West European countries have joined those already in the 30 per cent club, and several Eastern European countries have committed themselves to reduce transfrontier emissions by 30 per cent. SO_2 emissions in the United Kingdom have indeed fallen by 37 per cent since 1970, but the levels of high-stack emissions which affect remote areas have not fallen. In 1970, when the 37 per cent fall began, we were the largest emitter in Western Europe. In 1984, we are still the largest emitter. NO_x emissions have not fallen. In Western Europe only West Germany deposits more SO_2 in other countries than does the United Kingdom, and further significant reductions cannot be achieved by either without controls.

The Committee's concern in this passage lies less with the net benefits to the UK of reducing emissions than with how UK abatement has lagged behind the rest of Europe. Indeed, the Committee's evaluation did not even consider whether the other European countries were abating more simply because it was in their own self interest to do so. The argument seems to be: 'The other European nations are reducing their emissions, so we should, too.' Compared with the supergame problem, co-operation in this case is not instrumentally important—the Environment Committee did not seek to reduce UK emissions so that others would reduce theirs even further—but intrinsically important. Concerns about fairness have been shown to militate against the free-rider problem in experimental tests (Marwell and Ames, 1981). However, in the next section we shall see that in a bargaining situation, obeying moral principles may serve only to undermine the cause of environmental protection.

IV. CO-OPERATIVE ENVIRONMENTAL PROTECTION

In the absence of co-operation, outcomes better than the full non-co-operative one can sometimes develop, at least in principle. But such instances seem to be rare. Even two close neighbours with strong trading ties can fail to arrive at a preferred solution, as the disagreement between the

[9]The Thirty Per Cent Club consists of the countries that have signed the Protocol to the 1979 Convention on Long-Range Transboundary Air Pollution on the Reduction of Sulphur Emissions or their Transboundary Fluxes by at least 30 Per Cent (1985). The UK has not joined this 'club' but it has committed itself to substantial reductions in sulphur dioxide and nitrogen oxides emissions by agreeing to comply with the European Community Large Combustion Plant Directive.

United States and Canada over the exploitation of the North Pacific fur seal illustrates. Following an initial conflict between the two nations over the pelagic seal hunt, a Tribunal Arbitration was convened at the request of the two parties (with Great Britain acting for Canada). In late 1893 the Tribunal decided that the United States did not have territorial jurisdiction over the Bering Sea, and hence could not keep Canadian sealing vessels out of these waters. This effectively sanctioned open-access harvesting of the species, and co-operation proved impossible to secure (Paterson and Wilen, 1977, p. 94):

> Following the decision of the Tribunal, the diplomatic efforts of both Great Britain and the United States had been directed to convincing the other to reduce its sealing in order to allow the herd to recover from earlier depradations. No agreement could be reached and in 1897 the United States unilaterally forbade its citizens to engage in pelagic sealing in the North Pacific. At the same time quota adjustments were made as the herd diminished in size. So strong was the reaction to the declining herd size and the continued Canadian pelagic hunt that a bill reached the [United States] Senate which called for the complete destruction of the herd. It did not pass. . . .

Better management of the population had to await the signing of the North Pacific Fur Seal Treaty by Great Britain, the United States, Japan, and Russia in 1911—a remarkable agreement that remains in force today.[10] Effective management of global environmental resources seems to demand that countries co-operate openly and put their signatures on international agreements, treaties, and conventions. Explanations for why co-operation of this kind might emerge are offered below.

International Environmental Agreements

Consider the following modification to the common property game described earlier. Suppose a subset of the N identical countries 'collude' by signing an international environmental agreement and that the remaining countries continue to act non-co-operatively. Suppose further that the signatories to the agreement choose their collective abatement level while taking as given the abatement decision functions of the non-signatories, while the latter countries continue to behave atomistically and choose their abatement levels on the assumption that the abatement levels of all other countries are fixed. That is, the signatories act as 'abatement leaders', and the non-signatories as 'abatement followers'. Quite clearly, we would like the number of signatories, the terms of the agreement, and the abatement levels of all non-signatories to be determined jointly. We also

[10]For legal background on this treaty and its successors, see Lyster (1985).

require that the agreement itself is stable. A stable agreement is one where non-signatories do not wish to sign the agreement and signatories do not wish to renege on their commitment. Then it can be shown that for identical countries with linear marginal abatement benefit and cost functions a stable international environmental agreement always exists (Barrett, 1991a).

The solution to this problem exhibits many of the features of actual agreements. The net benefits realized by both signatories and non-signatories are higher than in the earlier problem where negotiation was ruled out. What is more, the signatories would like the non-co-operators to sign the agreement. However, non-signatories do better by free-riding.

It is important to emphasize that the agreement is *self-enforcing*. Any signatory that renounces its commitment can reduce its abatement level and hence its costs. However, in pulling out of the agreement, the number of co-operators is reduced and the agreement itself is weakened; the remaining co-operators reduce their abatement levels, too. A signatory will want to pull out of an agreement only if the saving in abatement costs exceeds the resulting loss in benefits. Similarly, a country that joins an agreement will have to abate more and hence incur higher costs. But the very act of joining will strengthen the agreement; the other co-operators will also increase their abatement levels. Joining appears attractive if the resulting increase in benefits realized by the new signatory exceeds the increase in costs that this country must incur in committing itself to the terms of the agreement.

Real treaties are not rewritten with every defection or accession, but mechanisms are at work that have a similar effect. It is common for treaties to come into force only after being ratified by a minimum number of signatories. The Montreal Protocol did not come into force until it had been ratified by at least eleven countries representing at least two-thirds of global consumption of the controlled substances. It is also common for treaties to be reviewed and altered when necessary and often at regular intervals. The Montreal Protocol was renegotiated in June 1990, and was significantly strengthened. That this agreement is self-enforcing is suggested by a comment made by the US Environmental Protection Agency (EPA) (1988, p. 30573):

> EPA judged that the obvious need for broad international adherence to the Protocol counseled against the United States' deviating from the Protocol, because any significant deviation could lessen other countries' motivation to participate.

Self-enforcement is essential in any model of international environmental agreements because nation states cannot be forced to perform their legal obligations. A country can be taken to the International Court of Justice for failing to comply with the terms of a treaty, but only with the defendant's permission. Even then, the disputing countries cannot be forced to comply with the Court's decision.[11]

What are the gains to having international environmental agreements? The answer depends partly on the number of relevant and potential signatories. When N is large, international environmental agreements can achieve very little no matter the number of signatories. The reason, quite simply, is that when N is large, defection or accession by any country has only a negligible effect on the abatement of the other co-operators.

Determination of N is not always a trivial matter. Some treaties do not restrict participation, but in these cases many of the signatories may have no effective say in environmental protection. Over 100 countries have signed the 1963 Partial Nuclear Test Ban Treaty, but only a few signatories possess nuclear weapons technology. The 1967 United Nations Treaty on Principles Governing the Activities of States in the Exploration and Use of Outer Space including the Moon and Other Celestial Bodies has been ratified by scores of countries but not by the two with space technology capabilities—the US and USSR. Other treaties explicitly restrict participation. The Agreement on the Conservation of Polar Bears can only be signed by five circumpolar countries (Canada, Denmark [including Greenland], Norway, the US, and the USSR). To become a signatory to the Antarctic Treaty of 1959 a country must maintain a scientific research station in the Antarctic and be unanimously accepted by existing parties to the agreement. In these cases non-signatories may quite clearly be affected by how the signatories manage the resource. Signatories to the Antarctic treaty voted recently to allow mineral exploration, despite appeals by non-signatories to designate Antarctica a nature reserve.

In the above model, N was assumed to represent both the number of countries that emit a (uniformly mixed) pollutant into the environment and the number harmed as a consequence. However, for some problems the number of emitters may be less than the number of sufferers (for global pollutants, all countries). When the number of emitters is small, an international environmental agreement signed by a subset of emitters may well have a significant effect on the welfare of these countries. However, the effect on global welfare may still be small because the emitters have no incentive to take into account the welfare losses suffered by non-emitting nations. The appropriate way to account for countries that do not emit the pollutant but nevertheless suffer the consequences of others' emissions is to admit side payments—payments which induce emitting nations to undertake greater abatement but which leave all parties no worse off compared to the situation where side payments are forbidden. We return to the side payments issue later.

The gains to international co-operation can also be shown to depend on c, the slope of each country's marginal abatement cost curve; and b, the

---

[11]See Lyster (1985) for a discussion of other compliance mechanisms.

(absolute value of the) slope of each country's marginal abatement benefit curve. For a given size of N, the number of signatories to a treaty increases as c/b falls. This suggests that we should expect to observe a large number of signatories (in absolute terms) when N is 'large', the marginal abatement cost curve is flat, and the marginal abatement benefit curve steep. However, we already know that when c/b is 'small' the benefit of having an agreement is diminished. It is commonly asserted that treaties signed by a large number of countries accomplish little of substance: 'The greater the number of participants in the formulation of a treaty, the weaker or more ambiguous its provisions are likely to be since they have to reflect compromises making them acceptable to every State involved' (Lyster, 1985, p. 4). This analysis suggests that the reason treaties signed by a large number of countries appear to effect little additional abatement is not that the signatories are heterogeneous—although that may be a contributing factor. Nor is the reason solely that in these cases N is also large. A major insight of the model is that a large subset of N will sign an agreement only when the non-co-operative and full co-operative outcomes are already close.

This latter observation may not seem consistent with all the evidence. The Montreal Protocol, for example, demands of its signatories significant reductions in the production and consumption of the hard CFCs and halons, and about 46 countries have already signed the agreement—a fairly large number by any standard. As Table 1 shows, the effect of the agreement on ozone depletion is estimated to be very significant. Percentage ozone depletion is estimated to be reduced from 50 to 2 per cent in 2075 as a result of the agreement. But of course each country has some incentive to take unilateral action in reducing emissions; in doing so all other countries will benefit, but so too will the country taking the action. Furthermore, non-signatories to the agreement may well face an incentive to abate less than they would otherwise for the simple reason that greater abatement on the part of signatories improves the environment for non-signatories as well. Hence it is by no means clear that the agreement necessarily means that the environment and global welfare will be significantly better off, contrary to what the figures in Table 1 imply. What the model does suggest is that so many countries would not have committed themselves to the agreement in the first place unless they already intended

TABLE 1. Estimates of the Reduction in Percentage Ozone Depletion Effected by the Montreal Protocol

CASE	2000	2025	2050	2075
No controls	1.0	4.6	15.7	50.0
Montreal Protocol	0.8	1.5	1.9	1.9

Source: US EPA (1988), Table 3, p. 30575.

to take substantial unilateral action. In other words, although the agreement itself may effect only little additional abatement, the very fact that so many countries have signed the agreement suggests that the potential gains to co-operation were in this instance not very great.

What does the model predict about the prospects of an agreement being reached on global warming? N will again be large, and this will militate against significant united action. However, in this case c/b will be large, too; the marginal costs of abating carbon dioxide emissions will rise very steeply as fossil fuels must be substituted for and energy is conserved. This suggests that the number of signatories to an agreement would be small, and that little additional abatement could be effected by co-operation. The tragedy is that in the case of global warming fuller co-operation could potentially result in huge gains in global welfare. [12]

Leadership

It is sometimes asserted that countries should, on their own, do more than the non-co-operative solution demands of them. US environmental groups, for example, have argued that the US should have taken greater unilateral action before the Montreal Protocol was drafted, that it should now comply with the terms of the Protocol in advance of the deadlines, and that it should exceed the agreed emission reductions and phase out production and consumption of these chemicals entirely. The House of Commons Energy Committee (1989, p. xvii), in its recent investigation on the greenhouse effect, recommended ' . . . that the UK should . . . consider setting an example to the world by seriously tackling its own emission problems in advance of international action, especially where it is economically prudent to do so'.

We have already seen that such 'unselfish' unilateral actions need not be matched by other countries. The United States, Canada, Sweden, and Norway banned the use of CFCs in non-essential aerosols in the late 1970s, and yet other countries did not reciprocate. [13] Unilateral restrictions on pelagic sealing in the North Pacific by the US were not duplicated by Canada. We have also seen that countries may wish to give in to their moral beliefs and embrace a less insular view of their responsibilities. An important question is whether 'unselfish' unilateral action can be expected to have a positive influence on international negotiations. If one country (or group of countries) abates more than the Nash non-co-operative solu-

[12]The analysis given here is by necessity over-simplified. For a more detailed analysis, see Barrett (1991a).
[13]Reciprocity was certainly not full. The European Community, for example, passed two decisions limiting production capacity of the so-called hard CFCs (CFC-11 and -12) and reducing their use in aerosols by 30 per cent.

tion demands, and all others choose the abatement levels that are optimal for them in a non-co-operative setting, will the environment be any better protected when international treaties are later negotiated?

In a two-country analysis, Hoel (1989) shows that the answer depends on whether the unilateral action is taken before agreement is reached and is not contingent on that agreement or whether the action is a commitment to abate more than the negotiated agreement requires. Hoel shows that in the former case, 'unselfish' unilateral action may compromise negotiations and lead, ultimately, to *greater* emissions than would have occurred had both countries behaved 'selfishly'. In the latter case, however, the country's announced commitment to overfulfil its negotiated abatement level can be expected to reduce total emissions.

There is an obvious incentive compatability problem with this tactic, for the 'unselfish' country could do better by reneging on its commitment (the agreement is therefore *not* self-enforcing). Nevertheless, the analysis shows that the desire by environmentalists and others to reduce total emissions may not be well served by their calls for 'unselfish' unilateral action, a point that the EPA stressed in defending its ozone depletion policy (1988, p. 30574): 'Unilateral action by the United States would not significantly add to efforts to protect the ozone layer and could even be counter productive by undermining other nations' incentive to participate in the Protocol.'

It is important to note that the US and the European Community announced their intentions to phase out production and consumption of the ozone-depleting chemicals by the end of the century *after* the Montreal Protocol came into force but before renegotiation talks had started. It would be wrong, however, to ascribe these developments simply to 'unselfish' behaviour. After all, the world's largest manufacturer of CFCs, US-based Du Pont, announced its intention to phase out production of CFCs by the end of the century *before* the phase-out decisions were taken by the US and EC. Three days after the EC decided to phase out CFCs, the chairman of the leading European producer of CFCs, ICI, declared that production of CFCs should cease 'as soon after 1998 as is practicable'. Much more is at work here. [14]

Efficient Co-operation

Signatories to an international environmental agreement are assumed to maximize the net benefits accruing to the *group*. This means, among other things, that the marginal abatement costs of every signatory must be equal; the abatement undertaken by the group must be achieved at minimum total cost.

[14]See Barrett (1991*b*) for an analysis of these developments.

How realistic is this assumption? In the case of the Montreal Protocol, the assumption is not very wide of the mark. The Protocol imposes on every industrial country signatory an obligation to reduce its production and consumption of CFCs by an equal percentage. This requirement on its own is inefficient because at the margin the costs of complying with the Protocol will surely vary. For example, the UK can apparently meet its obligations by simply prohibiting the use of CFCs in aerosols—an action that is nearly costless. The US banned the use of CFCs in aerosols many years ago, and hence can meet the terms of the Protocol only by instituting more costly measures. However, the Protocol allows limited international *trading* in emission reductions. For any signatory, CFC production through mid-1998 can be 10 per cent, and from mid-1998 onwards 15 per cent, higher than it would have been without trading provided the increase in production by this signatory is offset by a decrease in production by another signatory. Furthermore, trades of consumption (but, strangely, not production) quotas are permitted by the Protocol within the European Community. These provisions will help increase the efficiency of attaining the total emission reduction implicit in the agreement, although they almost certainly do not go far enough.

Side Payments

The equilibrium in the model of international environmental agreements is determined by a concept of stability that prohibits side payments. An important question is whether side payments might effect a Pareto improvement. To investigate this issue, reconsider the concept of stability employed in the model. In equilibrium, non-signatories do better than signatories, but no country can do better by changing its status. Signatories want non-co-operators to sign the agreement, because their net benefits would then increase. But non-signatories do worse by signing. Hence, without compensating payments, non-signatories will not want to sign the agreement. It is in this sense that the agreement is stable.

However, the very fact that signatories do better if non-co-operators sign the agreement suggests that trade might be possible. In particular, it might be possible for signatories to make side payments to a subset of non-co-operators to encourage them to sign the agreement. All might be made better off. It is in fact very easy to show that this can happen, that an international environmental agreement that specifies abatement levels *and* side payments can manage the global common property resource better than one that prohibits side payments.

An important feature of the World Heritage Convention is that it does admit side payments. The Convention established a World Heritage Fund that is used to help protect natural environments of 'outstanding universal value'. Each party to the Convention (there are over 90 signatories) is required to provide the Fund every two years with at least one per cent of its contribution to the regular budget of UNESCO.[15] In practice this means that the Fund is almost entirely financed by the industrial countries. Clearly, both the industrial and poor countries benefit from the Convention—otherwise they would not have signed it—but the poor countries may not have signed the Convention were it not for the Fund. Though the Fund is small, the mechanism could prove instrumental in protecting many of the world's remaining natural environments, including the tropical rain forests.

The World Heritage Convention is unique among international environmental agreements for incorporating side payments. But the need for side payments is not unique to the conservation of natural environments and wildlife. It has, for example, become increasingly clear that the success of the Montreal Protocol will ultimately hinge on the accession of non-signatories to the Protocol. Concern has specially been voiced about the need to get the poor countries—and, in particular, China and India—to sign the Protocol. These countries have declared their need for financial assistance, and the issue of setting up a global fund for protecting the ozone layer is likely to dominate discussion at the next meeting of the Protocol's signatories. It would seem to be something that cannot be avoided.

BIBLIOGRAPHY

Axelrod, R. (1984), *The Evolution of Cooperation*, New York, Basic Books.
Barrett, S. (1991*a*), 'The Paradox of International Environmental Agreements', mimeo, London Business School.
——— (1991*b*), 'Environmental Regulation for Competitive Advantage', *Business Strategy Review*, 2(1), 1–15.
Coase, R. H. (1960), 'The Problem of Social Cost', *Journal of Law and Economics*, 3, 1–44.
Cornes, R. and Sandler, T. (1983), 'On Commons and Tragedies', *American Economic Review*, 73, 787–92.
Dasgupta, P. (1982), *The Control of Resources*, Cambridge, MA., Harvard University.
Demsetz, H. (1964), 'The Exchange and Enforcement of Property Rights', *Journal of Law and Economics*, 7, 11–26.
——— (1967), 'Toward a Theory of Property Rights', *American Economic Review*, 57, 347–59.
Friedman, J. W. (1986), *Game Theory with Applications to Economics*, Oxford, Oxford University Press.

[15]The United States and the UK continued to contribute to the Fund even after withdrawing their funding from UNESCO.

Groves, T. and Ledyard, J. (1977), 'Optimal Allocation of Public Goods: A Solution to the "Free Rider" Problem', *Econometrica*, 45, 783–809.

Guttman, J. M. (1978), 'Understanding Collective Action: Matching Behavior', *American Economic Review Papers and Proceedings*, 68, 251–5.

Hardin, G. (1968), 'The Tragedy of the Commons', *Science*, 162, 1243–8.

Hoel, M. (1989), 'Global Environmental Problems: The Effects of Unilateral Actions Taken by One Country', Working Paper No. 11, Department of Economics, University of Oslo.

House of Commons Energy Committee (1989), *Energy Policy Implications of the Greenhouse Effect*, vol. 1, London, HMSO.

House of Commons Environment Committee (1984), *Acid Rain*, vol. 1, London, HMSO.

Lyster, S. (1985), *International Wildlife Law*, Cambridge, Grotius.

Marwell, G. E. and Ames, R. E. (1981), 'Economists Free Ride, Does Anyone Else?', *Journal of Public Economics*, 15, 295–310.

Paterson, D. G. and Wilen, J. (1977), 'Depletion and Diplomacy: The North Pacific Seal Hunt, 1886–1910', *Research in Economic History*, 2, 81–139.

Samuelson, P. (1954), 'The Pure Theory of Public Expenditure', *Review of Economics and Statistics*, 36, 387–9.

Sugden, R. (1984), 'Reciprocity: The Supply of Public Goods through Voluntary Contributions', *Economic Journal*, 94, 772–87.

—— (1986), *The Economics of Rights, Co-operation and Welfare*, Oxford, Basil Blackwell.

Trail Smelter Arbitral Tribunal (1939), 'Decision', *American Journal of International Law*, 33, 182–212.

—— (1941), 'Decision', *American Journal of International Law*, 35, 684–736.

US Environmental Protection Agency (1988), 'Protection of Stratospheric Ozone; Final Rule', *Federal Register*, 53, 30566–602.

27

Some Economics of Global Warming*

THOMAS C. SCHELLING

Thomas C. Schelling is Professor of Economics at the
University of Maryland.

Global warming from carbon dioxide was an esoteric topic fifteen years
ago, unknown to most of us. But in a few years, helped along by some hot
summers, it has climbed to the top of the international agenda. Cabinets,
Parliaments, and heads of government have issued pronouncements on
reducing carbon emissions, and in June of 1992 more than 100 governments will be represented by ministers or heads of government at a great
United Nations Conference on Environment and Development to be held
in Rio. Together with nongovernmental organizations representing labor,
business, students, environmentalists, scientists, and groups concerned
with health and child development and family planning, these representatives are expected to need 25,000 hotel rooms. A "framework agreement"
is widely anticipated, together with some institutional arrangements that
will keep global environmental issues permanently on every government's
agenda. And at the center of these issues will be the phenomenon that has
come to be known as the "greenhouse effect."

The greenhouse effect itself is simple enough to understand and is not
in any real dispute. What is in dispute is its magnitude over the coming
century, its translation into changes in climates around the globe, and the

"Some Economics of Global Warming," by Thomas C. Schelling,
from *American Economic Review* (May 1992). Reprinted by
permission.

*This paper was presented as the presidential address at the 104th annual meeting of the
American Economics Association in New Orleans, January 1992.

impacts of those climate changes on human welfare and the natural environment. And these are beyond the professional understanding of any single person. The sciences involved are too numerous and diverse. Demography, economics, biology, and the technology sciences are needed to project emissions; atmospheric chemistry, oceanography, biology, and meterology are needed to translate emissions into climates; biology, agronomy, health sciences, economics, sociology, and glaciology are needed to identify and assess impacts on human societies and natural ecosystems. And those are not all.

There are expert judgments on large pieces of the subject but no single person clothed in this panoply of disciplines has shown up or is likely to. So I venture to offer my judgment.

I

First on the principle. The metaphor of the greenhouse is not quite appropriate, but the basic idea is not in dispute. The earth is bathed in sunlight, some reflected and some absorbed. If the absorption is not matched by radiation back into space, the earth gets warmer until the intensity of that thermal radiation matches the absorbed incoming sunlight. Some gases in the atmosphere that are transparent to sunlight absorb radiation in the infrared spectrum, blocking that outward radiation and warming the atmosphere. When the atmosphere has warmed enough to intensify the thermal radiation so that it matches the absorbed incoming sunlight, equilibrium is achieved at the higher temperature. These so-called "greenhouse" gases can be identified in the laboratory. Carbon dioxide is one of them; methane is another, as is nitrous oxide, as are the chlorofluorocarbons (CFC's).

The principle has been in practice for decades. On a clear day in January, the earth and its adjacent air in Orange County, California, warm nicely, but the warmth radiates rapidly away during the clear nights, and frost may threaten the trees. Smudgepots, burning cheap oil on a windless night, produce substances, mainly carbon dioxide, that absorb the radiation and protect the trees with a blanket of warm air. Greenhouses, in contrast, mainly trap the air warmed by the earth's surface and keep it from rising to be replaced by cooler air. The phenomenon should have been called the "smudgepot effect," but it is too late to do anything about it.

A first step in pursuing this phenomenon is to assess how much warming might go with an enhanced concentration of these gases. That cannot be done in the laboratory; there are too many feedbacks. A warmer atmosphere will contain more water vapor; water vapor itself is a greenhouse gas. Changes in temperature and humidity will change cloud cover;

clouds can reflect or absorb incoming or outgoing light according to their composition and altitude. The average temperature is only one dimension; temperatures at different altitudes and different latitudes matter. But a starting point has been the expected change in average surface atmospheric temperature to accompany a specified increase in the concentration of greenhouse gases; and arbitrarily, but reasonably, the base case is taken as a doubling of the concentration.

A moment on why a doubling is the benchmark. To compare estimates of warming, people must use the same hypothesized concentration of greenhouse gases in the atmosphere. (Alternatively, they could use the same hypothesized temperature increase and estimate the corresponding concentration.) Doubling, like a half-life in reverse, is a natural unit if it is within the range of practical interest, and it is. A doubling is expected sometime in the next century, so it is temporally relevant; and a doubling is estimated to make a substantial but not cataclysmic difference. If fixation on a doubling seems to imply an upper limit on any expected increase, the implication is unfortunate: enough fossil fuel exists to support several doublings.

In 1979 a committee of the National Academy of Sciences (NAS, 1979, p. 2) estimated the change in average temperature to accompany a doubling of CO_2 in the atmosphere: 3 degrees Celsius, with a range of 1.5 degrees either side. (In the last fifteen years other greenhouse gases have received attention; these other gases can be converted to their carbon dioxide equivalents and the original estimate applied to the mixture.) That Academy appointed another committee a few years later to reexamine that estimate, and the new committee saw no reason to change it (NAS, 1982, p. 51). An Intergovernmental Panel on Climate Change (IPCC) consisting of scientists from many nations revisited the estimate in 1990 and concluded, from the several climate models they had examined, that "the models' results do not justify altering the previously accepted range of 1.5 to 4.5 degrees C" (IPCC, 1990, p. xxv). Thus the estimate appears robust over time, but the spread of uncertainty remains large: the upper limit is three times the lower. (No quantitative interpretation of these upper and lower "limits" has been made public. Both National Academy reports referred to them as "probable error.")

II

But the uncertainties are even greater in translating a temperature change into climates. The media support a popular view that things will just get hotter; a news magazine cover showed a sweating global face. But the laboratories that do the meteorology do not simply predict warming; they do not even predict that the most noticeable effects will necessarily be temperature changes. Among the great driving forces of weather and climate is the temperature differential between equatorial and polar regions; convection currents coupled with the rotation of the earth are engines of atmospheric circulation and, ultimately, ocean circulation. The models predict greater temperature change in the polar regions than near the equator. This change in gradient can drive changes in circulation. The results may be warmer in some places and colder in others, wetter in some places and drier in others, cloudier in some places and sunnier in others, stormier in some places and less stormy in others—generally a complex of changes that would bear no easy relation to an average change in global temperature.

The change in average temperature is useful as an *index* of climate change. It is thought, and the models demonstrate, that the greater the change in average temperature the greater the departure of current climates from what they are now. Thus, while it is wrong to think that what is going to happen can be readily characterized as "warming," it is not erroneous to take that average warming as a rough measure of the extent or severity of change to be expected. Unfortunately the widespread reference to "global warming" promotes the notion that things will simply get hotter. (Interestingly, virtually all public discussion is on hotter summers, not warmer winters; 100 years ago popular discussion of a warming trend would likely have concentrated on the milder winters to be expected.)

If 3 degrees is taken as an index of climate change to come within the next century or so, how big is that compared with what has happened within the last century, or the last 10,000 years? From what I have just said, this cannot be answered by whether anyone would notice the difference if every night and every morning, every winter and every summer, temperatures were exactly 3 degrees higher than they otherwise would have been. The question is: how would a 3-degree change in a global average compare with what has been experienced in the past?

The answer is that for 10,000 years, since the disappearance of the last ice age, average temperature appears never to have varied over anything like 3 degrees. A band of 1 degree would cover the current estimates of what average temperatures have been since the dawn of recorded history. We will be moving into a climatic regime that has never been experienced in the current interglacial period.

"Mankind will undergo greater climate change in the next 100 years than has been experienced in the last 10,000." Properly qualified, the statement is true; what it neglects is that peoples have been migrating over great distances for at least several thousand years. Goths and Vandals, Huns, West Europeans who populated North and South America, southerners who went north during the depression and northeasterners who moved southwest after World War II, all experienced changes in climate greater than any being forecast by the models. Almost everybody who attends this lecture in New Orleans will have undergone a greater change in the past few days than is expected to occur in any fixed locality during the coming century.

The changes that the models produce are *gradual* both in time and in space. The models do not produce discontinuities. Climates will "migrate" slowly. The climate of Kansas may become like Oklahoma's, Nebraska's like that of Kansas, South Dakota's like Nebraska's, but none of these is expected to become like the climates of Oregon, Louisiana, or Massachusetts.

A caution: the models probably cannot project discontinuities—just gradual change—because nothing goes into the models that will produce *catastrophes.* There may be phenomena that could produce drastic change but they are not known with enough confidence to introduce into the models. So the reassuring gradualness may be an artifact of the methodology. I will return to this point later.

This greenhouse problem, if problem it proves to be, is truly one of the "global common." A ton of carbon emitted anywhere on earth has the same effect as a ton emitted anywhere else. And carbon dioxide has a long residence time in the atmosphere, a century or more. There may be ways to remove it, but it doesn't disappear. The greenhouse influence on any national territory depends solely on the global concentration, not in any way on what part of the total is due to a nation's own emissions.

As I shall detail later, the costs of reducing carbon emissions will be large compared with any other emissions that have caused concern. The costs of phasing out CFC's will be in the billions per year for some years, and complete elimination is expected to be feasible. The cost of reducing sulfuric acid may be in the tens of billions. Proposals to hold emissions of carbon dioxide constant (with a linear increase of concentration in perpetuity), or to reduce emissions by 50% below what they would otherwise be, beginning perhaps in 2010, are expected to cost in the hundreds of billions in perpetuity.

There are a few numbers worth carrying in mind. There are 700 billion tons of carbon in the atmosphere. (Quotations are sometimes in tons of carbon dioxide, rather than carbon; the figure is then three and two-thirds times as large, about 2,600 billion.) Annual emissions are 6 billion tons. Close to half disappears somewhere, a little over half remains in the atmosphere; so the concentration is increasing by one-half percent per year. It has increased 25% in the last 100 years. (Concentration is reported more often than tonnage; it is currently about 350 parts per million.) And there are upwards of 10 trillion tons of carbon fuels out there to be burned; if it were all burned and half stayed in the atmosphere, the concentration could double at least three times.

If the carbon in the atmosphere has already increased by a quarter, has the average temperature gone up as predicted? And were the recent hot American summers that stirred popular interest harbingers of greenhouse summers to come?

To the first question the answer is that average global temperature—

summer and winter, both hemispheres, night and day—has apparently risen by half a degree in the last 100 years, but whether "as predicted" depends on what qualifications one reads into the predictions. The pattern differed between the northern and southern hemispheres. The global average rose during the first forty years of this century, was level for the next forty, and rose during the past decade. This pattern demonstrates that whether or not we are witnessing the greenhouse effect, there are other decades-long influences that can obscure any smooth greenhouse trend. (The carbon concentration is not at issue; it is well measured and shows steady rise on a decadal scale.) There are known phenomena that could account for the irregular temperature increase of the past century, and whether we are witnessing the "signal" probably depends on whether you want high confidence to reject a null hypothesis or are about to bet money on whether, another twenty-five years from now, looking back, all doubt will have been removed. I don't know what bets are being placed by "greenhouse scientists," but they are cautious in public on the question.

To the second question—do the hot American summers of the past few years announce the arrival of a greenhouse, confirming predictions?—the answer is in two parts. Maybe it's the greenhouse. But it's not what the greenhouse models predict. The global average in the four hot years of the past seven was only .2 degree above the level of the preceding forty years; and sudden hot American summers are not what the models predict.

III

In anticipating the impact on human welfare or natural systems two kinds of uncertainly are unlikely to be dispelled soon. One is simply the question of what the changes will be in each region or locality. Current models are severely limited in their agreement with each other, in their handling of such topographical variables as mountain ranges, and in the fineness of the grids they superimpose on the globe. And there is no great confidence that the models will be greatly improved within the next decade or two. A chaos-like process may defeat efforts to improve local predictions; and uncertainties in gross phenomena, such as the behavior of ocean currents under changed climatic conditions, may not be much better understood soon.

But even if we had confident estimates of climate change for different regions of the world, there would still be uncertainties about the kind of world it is going to be 50, 75, or 100 years from now.

Imagine it were 1900 and the climate changes associated with a 3-degree average temperature increase were projected to 1992. On what kind of world would we superimpose either a vaguely described potential change in climate or even a specific description of changes in the weather in all the seasons of the year, even for our own country. There would have been no way to assess the impact of changing climates on air travel, electronic communication, the construction of skyscrapers, or the value of California real estate. Most of us worked outdoors; life expectancy was 47 years (it is now 75); barely a fifth of us lived in cities of 50,000. Anticipating the automobile we might have been concerned whether wetter and drier seasons would bring more or less mud, not anticipating that the nation would become thoroughly paved. The assessment of effects on health would be without antibiotics or inoculation. And in contrast to most contemporary concern with the popular image of hotter summers to come, I think we'd have been more concerned about warmer winters, later frost in autumn and earlier thaw in the spring.

If the world, both in North America and in the other continents, is going to change as much in the next ninety years as it has changed in the ninety just past, we are going to be hard put to imagine the effects of climate changes.

Another thought experiment: suppose the kind of climate change expected between now and, say, 2080 had already taken place, since 1900. Ask somebody 50, 60, or 80 years old what is different compared with when he or she was a child. Would the climate change be noticed? Even ask a 70-year-old farm couple living on the same farm where they were born: would the change in climate be among the most dramatic changes in either their farming or their lifestyle? I expect changing from horses to tractors and from kerosene to electricity, the arrival of the telephone and the automobile and the paving of roads, the development of pesticides and artificial fertilizer, the discovery of soy beans and the development of hybrid corn, and even improvements in outdoor clothing, veterinary medicine, and agricultural practices generally would swamp the climate change. And if instead of living and working conditions we inquire about changes in wildlife and natural ecosystems, changes in regional climates would have been competing, in their impact on nature, with population growth and economic development.

A conclusion we might reach is that a climate change would have appeared to make a vastly greater difference to the way people lived and earned their living in 1900 than to the way people live and earn their living today. Today very little of our gross domestic product is produced outdoors, susceptible to climate. Agriculture and forestry are less than 3%, and little else is much affected. Some activities—tourism and holidays, professional sports, and schoolteaching—are seasonal, but many of the seasonalities are conventions that reflect the influence of climate in earlier times. (Children were needed in the fields in summer and could start school when the harvest was in; hockey and basketball used to be winter sports because one depended on ice and the other could fit in a building.)[1]

Manufacturing rarely depends on climate, and where temperature and

---

[1] An imaginative discussion is in Ausubel (1991).

humidity used to make a difference air conditioning has intervened. When Toyota chooses among Ohio, Alabama, and Southern California for locating an automobile assembly, geographical considerations are important but not because of climate. Minerals are extracted where they happen to occur, and oil fields and coal mines inhabit all kinds of climates and are little affected. The U.S. Postal Service vow that neither snow nor rain nor heat nor gloom of night will "stay these couriers from the swift completion of their appointed rounds" sounds quaint in the era of EMAIL and FAX.

Finance is little affected by climate, or health care, or education, or broadcasting. Transportation can be affected, but improvements in all-weather landing and takeoff in the last thirty years are greater than any differences that climate makes. If the average effect is a warming, iced waterways and snow removal may decline in importance. Construction is affected, mainly by cold, and if the average effect is in the direction of warming, construction may benefit slightly.

It is really agriculture that is affected. But even if agricultural productivity declined by a third over the next half century, the per capita GNP we might have achieved by 2050 we would achieve only in 2051. Considering that in most of the developed countries—the United States, Japan, France, Britain, Netherlands, Israel—the agricultural problem has been protecting farmers; that agricultural productivity in most parts of the world continues to improve; and that many crops and cultivated plants will benefit directly from enhanced photosynthesis due to increased carbon dioxide; one cannot be certain that the net impact on agricultural productivity will be negative or, if negative, will be noticed in the developed world.

I conclude that in the United States, and probably Japan, western Europe, and other developed countries, the impact on economic output will be negligible and unlikely to be noticed. And there is no reason to believe that in these countries there could be a noticeable impact on health. Any influence of climate on health in this country would be more in the regional distribution of the population than in changes in local and regional climates. [2]

Comfort is worth considering. Fortunately the climate models predict a greater warming in winter than in summer. Most people in the United States and Japan and western Europe go south for vacation, both summer and winter; and when people move upon retiring in the United States, they typically move toward warmer climates. In future years elderly people may suffer more heat stroke in summer in St. Louis, but we can hope fewer broken bones from ice in Boston. (Inhaling air richer in carbon dioxide has no effect on health.)

---

[2] A comprehensive discussion of both impacts and costs of abatement is Nordhaus (1991a). A carefully argued opposing view is Cline (forthcoming).

IV

This complacent assessment cannot be extended to the much larger population of the underdeveloped world. The livelihoods earned in agriculture and other climate-sensitive outdoor activities, 3% in the United States, is 30% and more in most of the developing world. Reliable forecasts of likely climate changes in the different areas so dependent on agriculture are simply not available, so no assessment, region by region, of the effect on productivity can be provided. There is no strong presumption that the climates prevailing in different regions 50 or 100 years from now will be less conducive to food production. But there is also no assurance that climate changes will not be harmful, and even if on balance the impact is neutral, there may be large areas with large populations that suffer severely. Those people are vulnerable in a way that Americans and western Europeans and Japanese are not.

Nor can the impact on health be dismissed or readily subsumed among generally improving health conditions, as for the developed world. Numerous parasitic and other vector-borne diseases affecting hundreds of millions of people are sensitive to climate. Again, there is no strong presumption that malaria mosquitos, to take an example, will on balance benefit from climate changes, but the risk is there.

It is with the less developed countries that we have to be most careful about superimposing the climates of the future on the economies and societies of today. As it was in our own country during this century, the trend in developing countries is to be less dependent on agriculture and less vulnerable to climate in transportation and other activities and health. If per capita income growth in the next forty years compares with the forty years just past, vulnerability to climate change should diminish and the resources available for adaptation should be greater. I say this not to minimize concern about climate change but to anticipate the question whether developing countries should make sacrifices in their development to minimize the emission of gases that may change climate to their disadvantage. Their best defense against climate change may be their own continued development.

This is a point worth emphasizing. Some environmentalists argue that developing countries should sacrifice some of their hopes for economic development in the interest of slowing the climate change that may prove disastrous. But the advice contains a contradiction. Any disaster to developing countries from climate change will be a disaster to their economic development. What is desired is to optimize development by investing in greenhouse gas abatement only when that appears, subject to all the uncertainties, to contribute more to their development in the future than the alternative direct investment in development. It is not economic growth versus environment; it is growth with the environment taken into account.

A related point: population growth is important for climate change, in two respects. One is that carbon emissions in developing countries are positively driven by population; population growth does not merely dilute carbon emissions per capita, but for a number of reasons more people means more carbon. If China succeeds in holding population growth to near zero for the next couple of generations, it may do as much for the earth's atmosphere as would a heroic Chinese anticarbon program coupled with 2% annual population growth.

The other population effect is simply that the most likely adverse impact of climate change on human productivity and welfare would be on food production. And in the poorest parts of the world the adequacy of food depends on the number of mouths and stomachs. In 100 years, adverse changes in climate for food production would be far more tragic if the countries we now associate with the developing world had populations totalling 12 billion than if they totalled 9 billion. For the developing world, the increasing concentration of people is probably more serious than the increasing concentration of carbon dioxide.

At this point I appear to have reached the conclusion that the *developed* world has no self-interest in expensively curtailing carbon consumption and that the *developing* world cannot afford to incur economic penalties to slow the greenhouse effect. There is a mismatch between those who may be vulnerable to climate change and those who can afford to do anything about it.

V

Why should the rich developed countries care enough about climate change to do anything about it? The answer must depend partly on how expensive it is going to be to do anything about it. Abatement programs have been examined in a number of econometric models that suggest we might want to treat as pertinent the sacrifice of perhaps 2% of world GNP in perpetuity.

A strong argument for trying seriously to slow climate change is that the developing countries are vulnerable and we care. Developed countries are currently providing $50 billion per year of assistance to the developing world; we would be talking about expending or foregoing perhaps four to eight times that much to slow emissions and slow climate change. Whether people in the developed democracies could be mobilized to contribute so much to benefit, half a century from now, the people in the countries we now call developing, I do not know, but I believe that if the developed countries were prepared to invest, say, $200 billion a year in greenhouse gas abatement, explicitly for the benefit of developing countries fifty years or more from now, the developing countries would clamor to receive the resources immediately in support of their continued development. There would undoubtedly be abatement opportunities so cheap that they could compete with direct aid to developing countries, but it would be hard to make the case that the countries we now perceive as vulnerable would be better off fifty or seventy-five years from now if $10 or $20 trillion had been invested in carbon abatement rather than in their economic development.

A second argument for an expensive program of carbon abatement is that while our production of material goods and services may not suffer from climate change, our natural environment may be severely damaged. Natural ecosystems will be destroyed; plant and animal species will become extinct. Places of natural beauty will be degraded. Valuable chemistries of plant and animal life will be lost before we learn their genetic secrets. And the earth itself deserves our respect. For many people something close to religious values are at stake.

This issue is doubly difficult to assess. It is difficult to know how to value what is at risk, and it is difficult to know just what is at risk. Even if climate changes at each point in time could be predicted accurately, the impacts on natural ecosystems could not yet be determined. And the benefits of slowing climate change by some particular amount would be even more uncertain. We know that carbon fuels are not going to be discontinued; the issue is the marginal gains, from carbon abatement and a slowing of climate change, in the survival of species and ecosystems and the preservation of enjoyable environments. This is an issue that simply has not been joined.

The third argument for spending heavily to slow climate change is that the conclusions I reported earlier may be quite wrong. I said that the climate models predict that climates will change slowly and not much; the models do not produce discontinuities, surprises, *catastrophes*. What is known about weather and climate constitutes an equilibrium system.

The possibility has to be considered that if global temperature increases not by the median estimate of 3 degrees for a doubling of carbon in the atmosphere but by 4 or 5 degrees, and continues to rise beyond the doubling because carbon fuels are still in use worldwide, some atmospheric or oceanic circulatory systems may flip to alternative equilibria, producing regional changes that are both sudden and extreme.

Have any such possibilities been thought of? One that was thought of but that yielded upon further investigation was the possibility that the west Antarctic ice sheet might glaciate into the ocean and raise the sea level by 20 feet. As recently as fifteen years ago the best scientific judgment was that that could happen within seventy-five years as a result of global warming. This prospect naturally attracted attention, and further investigation with the help of newly available satellite sensing of glacial movement led to reassuring estimates that if that catastrophic rise in sea level were to happen, it would take at least a few hundred years and be gradual, not sudden. But there isn't any scientific principle according to which all alarming possibilities prove upon further investigation to be benign.

A currently discussed likely source of discontinuous change is in the way oceans behave. Amsterdam is north of Newfoundland, courtesy of the Gulf Stream. There is some indication that in earlier interglacial periods ocean currents may have pursued different courses. If a current like the Gulf Stream, or the Japanese current for the United States, flipped into an alternative pattern, the climatic consequences might be both sudden and severe. (Paradoxically, global warming might freeze western Europe.)

Insurance against catastrophes is thus an argument for doing something expensive about greenhouse emissions. But to pay a couple percent of GNP as insurance premium one would hope to know more about the risk to be averted. I believe research to improve climate predictions should be concentrated on the extreme possibilities, not on modest improvements to median projections.

I said that current estimates suggest that it might cost a couple percent of GNP to postpone the doubling of carbon in the atmosphere by several decades. Is 2% a big number or a small one?

That depends on your perspective and on what the comparison is. In recent years $100 billion per year in budgets or taxes has been a politically unmanageable magnitude in the United States. On the other hand, subtracting 2% from GNP in perpetuity lowers the GNP curve by not much more than the thickness of a line drawn with a number two pencil, or, to formulate it as I did earlier, it postpones the GNP of 2050 until 2051. I say this not to belittle the loss of $10 trillion from the American GNP over the next sixty years, only to point out that the insurance premium, if we choose to pay it, will not send us to the poorhouse. The proper question is whether, if we were prepared to spend 2% of our GNP in the interest of protecting against damage due to climate change, we might find better use for the money.

One use I have mentioned—directly investing to improve the economies of the poorer countries. Another would be direct investment in preserving species or ecosystems or wilderness areas. There is concern that many ecosystems could not migrate as rapidly as climate may change in the coming century; there has been little investigation of what might be done to facilitate the migration of ecosystems if the alternative is to invest $5 or $10 trillion in the reduction of carbon emissions.

VI

What can be done to reduce or offset carbon emissions? Reducing energy use and the carbon content of energy have received, I believe properly, most of the attention, especially the attention of economists. There are other possibilities to mention.

Trees store carbon. In growing, they take it out of the atmosphere.

When they rot or burn, it goes back into the atmosphere. A new forest will absorb carbon until it reaches maturity, i.e., maximum carbon density, in 75 or 100 years. If it then merely replenishes itself, with new growth replacing the oxidized dead trees, it holds its carbon but does not absorb more. If trees are harvested, the part that becomes house frames or furniture may last 100 years or more; removing mature trees and storing them anaerobically is possible but expensive. The most recent report of the National Academy of Sciences considered that reforestation in the United States might sequester 2% or 3% of current global carbon dioxide emissions.[3] The prospects for that kind of reforestation in the rest of the world are not nearly so promising, and we should conclude that reforestation can contribute but not greatly.

Stopping or slowing *deforestation* is important for reasons other than carbon emissions, but is quantitatively more important than reforestation. Reforestation is unlikely to take up as much as 100 billion tons of carbon; deforestation, in areas where deforestation is likely, could contribute several hundred billion tons of carbon, partly because forest subsoils contain carbon typically greater than the amount in the trees themselves, and is subject to oxidation when the trees are removed.

Carbon can be "scrubbed" from stack gases, probably not with any known technology that would make such removal economically competitive with reducing emissions. (Part of the expense is disposing of sludge; where gaseous carbon might be pumped into the ocean or into underground cavities, economical disposal may prove feasible.) Parallel to reforestation is the idea of enhancing oceanic photosynthesis, by "fertilizing" the oceans, possibly with iron, if enough of the carbon residues from the enhanced growth will sink rather than remain near the surface. Experiments would probably be reversible and modest in scale; their political acceptability may be tested in the near future.

Finally—although nothing is final in a subject as new as the one we are talking about—there are numerous possibilities for putting substances or objects in orbit or in the stratosphere to reflect something like 1% of incoming sunlight to offset a large part of the radiation imbalance caused by greenhouse gases. Some of these are as apparently innocuous as stimulating cloud formation and some as dramatic as huge mylar balloons in low earth orbit. Until very recently these possibilities were nearly unmentionable but they have recently been dignified by inclusion, along with caveats about "large unknowns concerning possible environmental side effects," in the 1991 report of the National Academy of Sciences. I shall

[3]Their estimate is 10% of U.S. emissions at "low to moderate cost" on economically or environmentally marginal crop and pasture lands and nonfederal forest lands in the United States (NAS, 1991, p. 57). A review of the issues in both afforestation and deforestation by Andrew Plantinga is in Darmstadter (1991); an optimistic estimate of the afforestation option is Moulton and Richards (1990).

not pursue them here except for two observations. First, if in decades to come the greenhouse impact begins to confirm the more alarmist expectations, and if the economic sacrifices required to reduce emissions prove unmanageable for economic or political reasons, some of these "geoengineering" options will invite attention. And second, if they do, and especially if they prove to be within the budgetary capabilities of individual nations, international greenhouse diplomacy will be transformed.

VII

What remains nearly certain is that the main responses to the greenhouse threat will be adapting to climate as climate changes, and reducing carbon emissions. (CFC's are potent greenhouse gases and if unchecked might have rivaled carbon dioxide in decades to come. But international actions are making good progress and are among the cheapest ways of reducing greenhouse emissions.)

Like estimates of warming, estimates of the costs of reducing emissions require some common but arbitrary objective to be comparable. A doubling of carbon became the conventional benchmark for warming estimates; no such benchmark for reduced carbon emissions has been adopted for estimating costs. (In principle the estimates could adopt that doubling: the issue could be formulated as the cost of retarding the doubling time by a decade, two decades, or half a century.) Most estimates take as their target a reduction of emissions either to a specified fraction of what they would be in the absence of controls, or to some fixed ratio to the projected emissions of 1990, 2000, or 2010. The estimates examine minimum-cost trajectories, implicitly or explicitly assuming something like a uniform tax on the carbon content of fuel as the policy instrument. They typically make some assumption about a "fallback" energy technology, at least for electricity, available at some price in some decade of the next century. They have to project estimates of non-price-induced improvements in the use or avoidance of energy by industries and households. And if they deal with global emissions, they have to make some assumption about the distribution of abatement efforts among nations, especially among the developing countries, which, including China, account for about a quarter of emissions now and would be expected to account for half by the middle of the next century.

Any estimate of the cost of abatement needs therefore to specify at least half a dozen target assumptions. Furthermore, the estimates are produced by people and institutions that do not simultaneously estimate the costs associated with climate change, either damages or costs of adapting; the estimates do not optimize the combined costs of abatement and climate change. A "not unreasonable" target for reduction might be delaying a doubling by, say, four decades. One decade might be too trivial, a century too ambitious, and four decades an objective in which most audiences would be interested. But nobody who makes such an estimate wishes to be interpreted as proposing that when all the uncertainties about climate changes and their impacts have been resolved, if they ever are resolved, the optimum reduction in emissions will be found to retard doubling by forty years, or any other specified period of time.

All I can do to summarize a multitude of estimates is to specify an order of magnitude that many economists and the Congressional Budget Office would not consider outrageous. That is the figure I mentioned earlier, possibly 2% of GNP for the developed countries and a like, but even much more uncertain, percent of GNP for the developing world. The uncertainty for the developing world is partly due to the estimates' being mainly derived from the American economy.[4]

Two characteristics of these estimates need to be emphasized. One is that they tend to assume optimal technological adjustment, as in response to a carbon tax. To the extent that carbon emissions are controlled by direct regulatory measures there may be the usual expected inefficiencies, and I leave the reader to make his or her own adjustment.

The second is that since the early years of the energy crisis, in the 1970s, there have been enthusiastic portrayals of currently available technologies, ranging from light bulbs to electric motors, double-glazed windows and improved internal combustion engines, that for some reason have not been successfully marketed. The interest continues, and the recent National Academy study gave sympathetic attention, but no analysis, to a number of proposals for residential, commercial, industrial, and transportation energy management, and for improved electricity production and fuel supply, and concluded that, including reductions in CFCs, "the United States could reduce its greenhouse gas emissions by between 10 and 40 percent of the 1990 level at very low cost. Some reductions may even be at a net savings if the proper policies are implemented."

All of these ideas are completely orthogonal to the econometric estimates. The Academy Panel that produced the report was unable to offer an explanation for why these low-costs or negative-cost technologies have not caught on. Its quantitative assessment, including an allowance for elimination of CFC's, ranged from as little as 10 percent to as much as 40 percent of current U.S. emissions; CFC's aside, their range of possibility is from zero to about 30 percent. Whatever the correct figure, this is probably a once-for-all backlog of accumulated technologies, and once exploited may be permanent but not progressive. But the strong suggestion is that there is a lot to be accomplished in the next two or three decades.

---

[4]Several critiques and surveys of different abatement-cost estimates are available. See William D. Nordhaus (1991b), Joel Darmstadter (1991), and Congressional Budget Office (1990).

VIII

With these qualifications, let us look at that 2% of GNP as a permanent reduction over the coming century. I consider it altogether improbable that the developing world, at least for the next several decades, will incur any significant sacrifice in the interest of reduced carbon (nor would I advise them to). Anything done to reduce emissions in China, India, or Nigeria will be at the expense of the richer countries.

Financing energy conservation, energy efficiency, and switching from high-carbon to lower-carbon or non-carbon fuels in Asia and Africa would not only be a major economic enterprise but a complex effort in international diplomacy and politics. If successful it would increase the costs to the developed world by at least another percent or two on top of the 2% I mentioned. It is furthermore not easy to hide the transfer of resources on the order of a couple of hundred billion dollars, dollars "budgeted" somehow or other, compared with hiding some of the costs due to regulation, such as automobile fuel efficiency standards in the United States. The kind of thing we are talking about is inducing the Chinese, through our somehow offsetting their cost, to forego a massive electrification based on coal and the cheapest coal-combustion technology. Without engaging in blackmail the Chinese can assert that it is not in their interest to do that at their own expense, even if they are the keystone of a "social contract," and no other nation will do anything unless the Chinese fully participate.

I shall sketch what I can imagine as a major attack on the greenhouse problem. And I should be explicit about what I cannot imagine. For reasons that I would be delighted to elaborate but for which I cannot take space here, a universal uniform carbon tax is not a solution that I can imagine. My reason is simple. A carbon tax sufficient to make a big dent in the greenhouse problem would have to be roughly equivalent at least to a dollar per gallon on motor fuel, and for the United States alone such a tax on coal, petroleum, and natural gas would currently yield close to half a trillion dollars per year in revenue. No greenhouse taxing agency is going to collect $1 trillion per year in revenue; and no treaty requiring the United States to levy internal carbon taxation at that level, keeping the proceeds, would be ratified by the Senate. Reduce the tax by an order of magnitude and it becomes imaginable, but then it becomes trivial as greenhouse policy.[5]

Tradeable permits have been proposed as an alternative to the tax. There are two main possibilities, estimating "reasonable" emissions country by country and establishing commensurate quotas, or distributing tradeable rights in accordance with some "equitable" criterion, such as equal emissions per capita (a possibility that has actually been discussed).

[5] A careful treatment of the universal carbon tax is Poterba (1991).

Depending on how restrictive the aggregate of such tradeable emission rights might be, the latter is tantamount to distributing trillions of dollars in discounted value and making, for a country like Nigeria, the outcome of its population census the country's major economic policy. But if instead quotas are negotiated to correspond to every country's currently "reasonable" emissions level, they will surely be renegotiated every five or ten years, and selling an emissions right will be perceived as evidence that a quota was initially too generous. It is unlikely that governments will engage in trades that acknowledge excessive initial quotas.

I do not foresee negotiated national quotas subject to serious enforcement, especially enforcement through financial penalties. Any international regime for carbon abatement I think can seriously include only the developed countries, and I exclude from this category the countries that we used to call the Eastern bloc. I can easily imagine institutional arrangements that are universalist, some kind of "framework agreement" to which every country subscribes, with specific commitments to be negotiated later. But I expect serious commitments to be undertaken only by the countries that can afford to, and I am undecided whether an institutional pretense of a universalist system has advantages, or instead the developed world should proceed independently and unencumbered by the need for a universalist facade.

The model that I find most helpful in conceptualizing a greenhouse regime among the richer countries is the negotiations among the countries of western Europe for distributing Marshall Plan dollars among themselves and the negotiations, beginning in 1951, on "burden sharing" in NATO. There was never a formula for distributing Marshall Plan dollars; there was never an explicit criterion, such as equalizing living standards, equalizing growth rates, maximizing aggregate output or growth, or establishing a floor under levels of living. Baseline dollar-balance-of-payments deficits were a point of departure, but the negotiations took into account investment needs, traditional consumption levels, war-induced capital needs, opportunities for import substitution and export promotion, and opportunities to substitute intra-European trade for trade with hard-currency countries.

The United States insisted that the recipients argue out and agree on shares. In the end they did not quite make it, the United States having to make the final allocation. But all the submission of data and open argument led, if not to consensus, to a reasonable appreciation of each nation's needs. The negotiations were professional; they were assisted by a proficient secretariat. The resources involved for most recipient countries were immensely important. Good relations were observed throughout; and proficiency in debate, acceptance of criteria, and negotiating etiquette steadily improved.

That is the only model I find plausible, and I believe distribution of Marshall Plan and defense-support funds to Europe is the only model of multilateral negotiation involving resources commensurate with the cost of greenhouse abatement. (In the first year, Marshall Plan funds were about 1.5% of United States GNP and—adjusting for overvalued currencies—probably 5% of OEEC GNP.)

What that model suggests is that the main participating countries in a greenhouse abatement regime would submit for each other's scrutiny and cross examination plans for reducing carbon emissions. The plans would be accompanied by estimates of emissions, or emissions reduction from some projected level, but any commitments undertaken would be to the policies, not the emissions. And not all of the plans would necessarily be commitments.

The United States, for instance, could present a plan for the introduction of a new generation of nuclear power reactors beginning sometime in the next century, but it is difficult to see how the federal government can commit itself to what reactors public utilities will be purchasing twenty years from now. The United States can have a plan to mandate fuel-efficiency standards for automobiles, but it takes ten years for the standards to work their way into the automobile fleet, and there is no accounting procedure that will estimate the effect on motor fuel consumption of any level of average fuel efficiency a decade from now.

The current popular expectation is that participation in any greenhouse regime will take the form of commitments to specified percentage reductions of emissions below those of some specified year, like 1990 or 2000. I cannot help believing that adoption of such a commitment is an indication of insincerity. A serious proposal would specify policies and programs, like taxes and regulations and subsidies, and research and development, accompanied by very uncertain estimates of their likely effect on emissions. In an international public forum governments could be held somewhat accountable for the policies they had or had not put into effect, but probably not for the emissions levels achieved.

Such a modest beginning will require finding a way to sublimate the current international enthusiasm for a new universalist greenhouse regime into institutional arrangements that are helpful but noncommittal, and require an understanding among the developed countries that it is initially up to them to find a way to mobilize their populations in support of national greenhouse policies.

IX

A major commitment to financing abatement in the developing world is surely too far away to need specific plans now. A developing-world carbon abatement effort would, in principle, be altogether different from foreign aid as we have known it since World War II. In principle it would all be directed, from whatever sources and through whatever channels, to protecting that same global common. There would be, for the first time, a single criterion—economizing carbon. In the abstract, aid recipients in the war on greenhouse gases would not compete; they would not make India–Pakistan comparisons, or Arab–Israel, or Poland–Czechoslovakia. All would in principle benefit equally from maximum carbon conservation wherever it could be achieved. Trees may grow more rapidly, in carbon content, in Madras or Szechuan or Borneo or Alaska or South Carolina, but if someone were willing to finance the growth of a tree to absorb carbon dioxide, the citizens of those states should not have the slightest care where the tree were to be planted; they all benefit solely from the carbon fixed in the tree and benefit more the faster the tree grows, no matter where it grows.

It wouldn't work that way, of course. Somebody gets the shade, or leases land for the tree; and if it's not a tree but a nuclear power plant to supplant coal, there are local impacts that make huge differences, and negotiations over sharing the cost differential between the coal and the nuclear plants. But it is worth noticing that if there were a "pure" carbon abatement or carbon-absorbing technology that accomplished nothing else, there should be no dispute about locating it wherever it would be most effective. That is new in foreign aid and foreign investment.

If the developed countries ever manage to act together toward the developing countries, their bargaining position is probably enhanced by the fact that cleaner fuels and more efficient fuel technologies bring a number of benefits other than reduced carbon, and recipients of greenhouse aid will be actively interested parties, not merely neutral agents attending to the global atmosphere. At the same time large nations like India and China will be aware of the extortionate power that resides in ambitious coal-development projects.

On a greatly reduced scale, there may be something constructive to do more immediately. There is huge difference between transferring "technology" and transferring capital goods that embody technology or, going further, financing entire investments (local construction, etc.) in which the technology is embedded. The difference in cost is at least an order of magnitude. While the developed countries are feeling their way into some common attack on their own carbon emissions, a tangible expression of their interest and an effective first step would be to establish a permanent means of funding technical aid and technology transfer for developing countries, as well as research, development, and demonstration in carbon-saving technologies suitable to those countries. Eventually the rural Chinese household may cook more efficiently with nuclear-powered electricity, but for another generation or two what is important is less carbon-wasteful ways of cooking and heating.

Maybe there is a role here for the carbon tax. Western Europe, North America, and Japan will be burning 3 or 4 billion tons of carbon per year

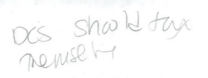

[handwritten: DCs should tax themselves]

for the next decade. Taxing themselves, that is, contributing in proportion to the carbon they consume, at one, two, or three dollars per ton, they could contribute to a fund that might begin at $3 billion per year and grow to $10 billion. The carbon tax is a little arbitrary here, and a U.S. administration may be wary about a precedent that carries over when the tax rises an order of magnitude, but compared with alternative criteria for sharing costs it might not even be a bad precedent.

[handwritten: He emphasizes nuclear power]

REFERENCES

Ausubel, Jesse H., "Does Climate Still Matter?" *Nature,* vol. 350, 25 April 1991, 649–52.

Cline, William R., *Global Warming: The Economic Stakes.* Washington, D.C.: Institute for International Economics, 1992.

Congressional Budget Office, *Carbon Charges as a Response to Global Warming: The Effects of Taxing Fossil Fuels,* August 1990.

Darmstadter, Joel, "The Economic Cost of CO_2 Mitigation: A Review of Estimates for Selected World Regions." Discussion Paper ENR91-06, Resources for the Future, Washington, D.C., 1991.

Energy Modelling Forum, "Global Climate Change: Energy Sector Impacts of Greenhouse Gas Control Strategies," Energy Modelling Forum Report No. 12, Stanford University, forthcoming.

Intergovernmental Panel on Climate Change, *Climate Change: The IPCC Scientific Assessment.* J. T. Houghton, G. L. Jenkins, and J. J. Ephraums, eds., Cambridge: Cambridge University Press, 1990.

Moulton, Robert J., and Kenneth R. Richards, "Costs of Sequestering Carbon Through Tree Planting and Forest Management in the United States." U.S. Department of Agriculture, Forest Service, General Technical Report WO-58, December 1990.

National Academy of Sciences, *Carbon Dioxide and Climate: A Scientific Assessment.* Washington, D.C.: National Academy Press, 1979.

———, *Carbon Dioxide and Climate: A Second Assessment.* Washington, D.C.: National Academy Press, 1982

———, *Policy Implications of Greenhouse Warming.* Washington, D.C.: National Academy Press, 1991.

Nordhaus, William D. (1991a), "Economic Approaches to Greenhouse Warming," in *Global Warming: Economic Policy Responses,* Rudiger Dornbusch and James M. Poterba, eds. Cambridge: The MIT Press, 1991, pp. 33–66.

——— (1991b), "The Cost of Slowing Climate Change: A Survey," *The Energy Journal,* vol. 12, 1991, 37–66.

Poterba, James M., "Tax Policy to Combat Global Warming: On Designing a Carbon Tax," in *Global Warming: Economic Policy Responses,* Dornbusch and M. Poterba, eds., Cambridge: The MIT Press, 1991, pp. 71–97.

28

Economic Activity and the Greenhouse Effect

YOSHIKI OGAWA

Yoshiki Ogawa is a Senior Economist in the Energy and Environment Group of the Research Division of the Institute for Energy Economics in Japan.

INTRODUCTION

Global warming is now becoming recognized as one of the most important issues in international politics, although many uncertainties remain concerning both its scientific causes and its impacts. One of these uncertainties is our grasp of the role that various socio-economic factors play in global warming under a variety of conditions. Specifically, it is important to understand the extent to which population, economic activity, energy utilization, and fuel composition are responsible for generating the critically important carbon dioxide.

CO_2 is estimated to have been responsible for 66 percent of the incremental greenhouse effect between 1880 and 1980, and for 49 percent of the incremental greenhouse effect during the 1980s.[1] (The declining share in the most recent period is due to the large increase in CFCs, which are now coming under control.) Over the past century, fossil fuel burning was responsible for an estimated 57 percent of CO_2 emissions, forest destruc-

"Economic Activity and the Greenhouse Effect," by Yoshiki Ogawa, from *The Energy Journal,* v. 12 (1992), pp. 23–35. Reprinted by permission.

[1]US EPA, 1989.

tion for 20 percent, farming/livestock activities for 9 percent, and natural sources for 14 percent.[2]

Thus, fossil fuel burning is the major cause of global warming, and proposed measures must focus on the utilization of fossil fuels. However, such measures may have serious implications for both energy use and economic growth. Thus, it is important to analyze the historical record of CO_2 emissions from fossil fuel burning in an effort to evaluate the consequences of various policy options for dealing with global warming.

The paper is organized as follows. The next section outlines world CO_2 emissions and economic growth over the 14-year period 1973–1987. Then two sections discuss energy conservation and fuel switching. This is followed by a factor analysis of emissions for selected industrial and developing countries. After a brief examination of non-commercial energy consumption in LDCs, the final section summarizes our conclusions.

ECONOMIC GROWTH AND CO_2 EMISSIONS

CO_2 emissions for countries and regions have been estimated from data published by the International Energy Agency[3] and British Petroleum Company.[4] Emission factors used are as follows:

Coal	1.880 T-C/TOE
Oil	0.864 T-C/TOE
Natural Gas	0.616 T-C/TOE

where T-C/TOE is tons of carbon equivalent per ton of oil equivalent, which takes account of the net calorific value of each fuel. The IEA study treats, in detail, eight industrial countries (United States, Britain, West Germany, France, Italy, Sweden, Japan and the Soviet Union) and seven developing countries (Republic of Korea, Malaysia, Indonesia, Thailand, People's Republic of China, India and Bangladesh). It also combines all countries for which data are available into three groups: industrial market countries (IMCs), USSR/Eastern Europe (UEE), and less developed countries, including the People's Republic of China (LDCs).

In 1973, CO_2 emissions resulting from energy use amounted to 4.66 billion T-C worldwide (Table 1 and Figure 1). Of this, the IMCs were responsible for 60 percent, UEE for just under one-quarter, and the LDCs

[2]Ibid.
[3]IEA, 1976; IEA, 1987; IEA, "Energy Balances of OECD Countries 1986/1987," 1989; IEA, "World Energy Statistics and Balances 1971–1987," 1989.
[4]British Petroleum, 1989.

TABLE 1. Worldwide CO$_2$ Emissions from Energy Use
(Billions of Tons of Carbon)

	1973	(%)	1987	(%)	ANNUAL CHANGE 87/73
Industrial countries	2.79	60	2.82	47	0.1%
USSR/Eastern Europe	1.08	23	1.51	25	2.4%
Developing countries	0.79	17	1.70	28	5.6%
World total	4.66	100	6.03	100	1.9%

for one-sixth. By 1987, the total had increased by 29 percent to 6.03 billion T-C. Emissions in the IMCs were virtually unchanged, but emissions in the UEE block had risen by 40 percent and the LDCs' emissions had more than doubled. As a result, the LDC share of the total had risen to well over one-fourth while that of the IMCs had fallen to less than one-half.

The IMCs, because they were the first to industrialize, have in the past been the more important contributors to world CO$_2$ emissions. However, the LDCs, and potentially the UEE group (depending on events there), are likely to have a greater impact on emissions in the future. In any case, it is clear that any effective action to control the problem must be global in scope, with all groups of countries participating in the effort.

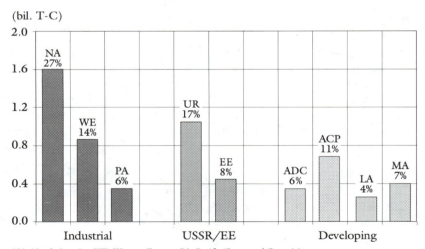

NA: North America, WE: Western Europe, PA: Pacific (Japan and Oceania),
UR: USSR, EE: Eastern Europe, ADC: Asia Developing Countries,
ASP: Asia Centrally planned Economy, LA: Latin, MA: Middle East and Africa.

FIGURE 1. CO$_2$ Emission in selected IMCs, LDCs, and USSR/EE in 1987.

In the developing countries that we have examined, GNP per capita was as low as ¹⁄₂₀ to ¹⁄₅₀ of a typical high-income IMC (using Sweden as a base of comparison, see Figure 2). Economic differentials in the late 1980s, although slightly smaller than in 1971, remained marked. Also, during that period, population growth in the LDCs was high (2–3 percent per annum) compared to the much lower rates in the IMCs (0.5 percent per annum or lower, in most cases).

For the LDCs, economic growth is indispensable for narrowing income differentials with advanced industrial countries, and increased energy use is unavoidable, though its rate of growth may be subject to modification. Thus, it is clear that the solution of the global warming problem is intricately tied to north-south economic problems.

ENERGY INTENSITY OF CONSUMPTION

From the viewpoint of energy consumption, two factors are vital to analyze: (i) energy intensity, and (ii) the possibility of fuel switching. Between 1960 and 1987 energy consumption in IMCs has shifted dramatically downward, with the turning point occurring around 1973, the year of the first oil crisis. These changes reflected both energy conservation (chiefly greater energy efficiency) and structural changes in the countries' economies from higher to lower energy-intensive industries.

Examination of Energy changes (GDP ratios for selected countries are shown in Figures 3 and 4) by economic sector reveals several interesting trends. In the United States the ratio has fallen in all major sectors— residential/commercial, industrial, and transportation. In Japan the in-

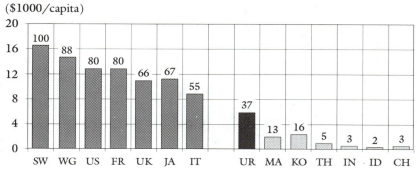

FIGURE 2.   Per capita GDP, in dollars and as ratio to Sweden, selected countries, 1987.

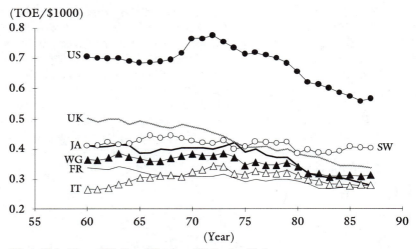

(TOE/$1000)

US: the United States, UK: United Kingdom, SW: Sweden, JA: Japan,
WG: West Germany, FR: France, IT: Italy

FIGURE 3. Energy/GDP Ratio in Industrial Countries.

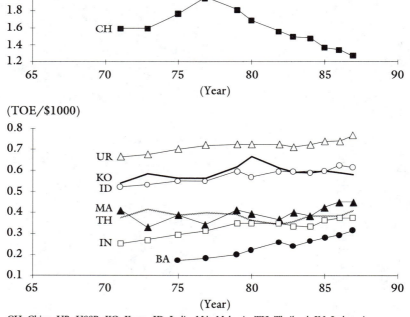

(TOE/$1000)

(TOE/$1000)

CH: China, UR: USSR, KO: Korea, ID: India, MA: Malaysia, TH: Thailand, IN: Indonesia,
BA: Bangladesh

FIGURE 4. Energy/GDP Ratio in the USSR & Developing
Countries.

dustrial sector registered particularly marked declines. In the United Kingdom the industrial ratio fell prior even to 1973, reflecting the long-term decline of that country's energy-intensive industries. In Sweden energy intensity decreased in the industrial and residential/commercial sectors, but this was offset by the substitution of electricity for direct fuel burning in the residential/commercial sector. This demonstrates the difficulty of reducing the overall E/GDP ratio when rapid electrification of major sectors is occurring. In the Soviet Union and India energy intensity grew in the industrial and power generating sectors, but did not grow significantly in transportation or residential/commercial sectors, which affect people's lifestyles most directly.

FUEL SWITCHING AND CO_2 EMISSIONS

Changes in CO_2 emissions per unit of energy consumption (CO_2/E) are attributable to changes in the composition of the energy mix, i.e., fuel switching. In the IMCs, changes in fuel patterns have varied greatly among countries (see Figure 5). In the United Kingdom and West Germany CO_2/E started falling in the 1960s as a result of shifts from coal to oil and natural gas. In France and Sweden the ratio fell sharply due to the

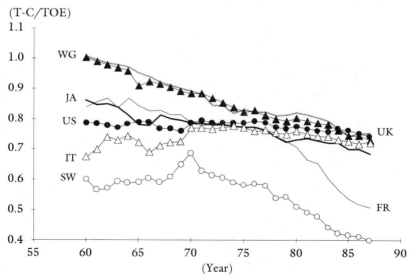

UK: United Kingdom, WG: West Germany, JA: Japan, FR: France, US: the United States, IT: Italy, SW: Sweden

FIGURE 5. CO_2 Emissions Per Unit of Energy Consumption in Industrial Countries.

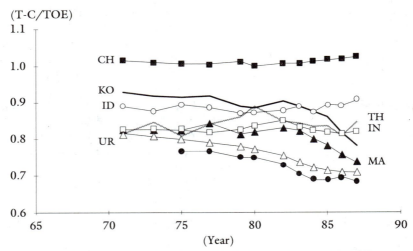

CH: China, KO: Korea, ID: India, MA: Malaysia, TH: Thailand, IN: Indonesia, UR: USSR,
BA: Bangladesh

FIGURE 6. CO_2 Emissions Per Unit of Energy Consumption in
USSR/Developing Countries.

expansion of hydro power and the growing role of nuclear power. Japan,
too, experienced fuel switching with an expanding nuclear program and
introduction of LNG, but the fall in CO_2/E was not as sharp as in France
or Sweden. The United States and Italy showed little change because
offsetting trends were at work (e.g., expansion of nuclear but decline in
natural gas in the US).

As shown in Figure 6, the emissions ratio declined in the USSR, largely
due to expansion of natural gas consumption. In the LDCs, differences
among individual countries were again marked. In the Republic of Korea
and Malaysia the ratio decreased, chiefly because of the growth of nuclear
power in Korea and expanded use of natural gas in Malaysia. In the
People's Republic of China, India, Indonesia and Thailand, the ratio has
been virtually constant—in the former two countries because expansion of
clean sources was counterbalanced by growing use of coal.

Examination of the energy mix in power generation also reveals various
trends, as shown in Table 2. All countries examined, except India, showed
a rapid expansion in the share of nuclear, and reductions in the share of
hydro. Moreover, in the United States the use of oil and natural gas in
electricity generation fell, while that of coal grew. In Japan the role of oil
fell dramatically while natural gas rose. France reduced the use of all fossil
fuels and nuclear became the dominant electricity source. In the USSR the
role of coal fell and natural gas became the leading power generation
source. In India coal became even more dominant.

TABLE 2. Energy Mix in Power Input and Final Consumption
(percent of country total)

		POWER INPUT					FINAL CONSUMPTION			
		COAL	OIL	GAS	NUCLEAR	HYDRO	COAL	OIL	GAS	ELECTRIC
USA	1973	46	18	19	4	13	9	49	30	12
	1987	58	5	11	17	9	10	50	24	16
Japan	1973	12	67	2	2	17	16	68	1	15
	1987	15	26	18	29	12	16	59	3	22
France	1973	25	37	5	8	25	12	70	8	10
	1987	8	2	1	70	19	8	56	16	20
Sweden	1973	1	14	0	3	82	13	69	0	18
	1987	4	2	0	45	49	19	43	1	37
USSR	1973	43	19	25	1	12	26	37	17	20
	1987	26	18	38	8	10	16	31	30	23
India	1973	55	9	1	3	32	49	39	1	11
	1987	73	4	2	2	19	38	41	5	16

Table 2 also shows the mix between electricity and direct use of fossil fuels in final consumption. Here the changes were less dramatic. The most pervasive was an increased role for electricity in all countries, most marked in Sweden and France. The share of oil consumption fell in Japan, France, Sweden and the USSR, but not in the US and India. The role of coal decreased in France, the USSR and India, but rose in Sweden.

FACTOR ANALYSIS OF CO_2 EMISSIONS

CO_2 emissions may be expressed in the following equation:

$$\ln C = \ln U + \ln S + \ln G + \ln P \tag{1}$$

where U indicates the fuel mix, S denotes energy intensity, G represents economic growth and P stands for population changes.

Differentiating equation (1) and rearranging yields,

$$dC = (C/U)\,dU + (C/S)\,dS + (C/G)\,dG + (C/P)\,dP \tag{2}$$
$$\text{(a)} \qquad \text{(b)} \qquad \text{(c)} \qquad \text{(d)} \qquad \text{(e)}$$

where

  (a) = change in CO_2 emissions
  (b) = contribution of fuel switching
  (c) = contribution of energy conservation
  (d) = contribution of economic growth
  (e) = contribution of population growth

This equation is in a form which is useful for analyzing the historical contribution each factor has made to CO_2 emissions as shown in Figure 7. Population and real GDP data are taken from publications of the IMF,[5] the World Bank[6] and the US CIA.[7] Real GNP in domestic currency units was converted to US dollars, using fixed exchange rates for 1980.

Examining the results in Figure 7, it is clear that both economic growth and population expansion contributed to increasing CO_2 emissions. Not unexpectedly, population growth played a greater role in the LDCs than in the IMCs—for example, in India. The United States was an exception among industrial countries since its population expansion was partially sustained by immigration.

[5]IMF, 1988.
[6]World Bank, 1989.
[7]CIA, 1978 to 1988 editions.

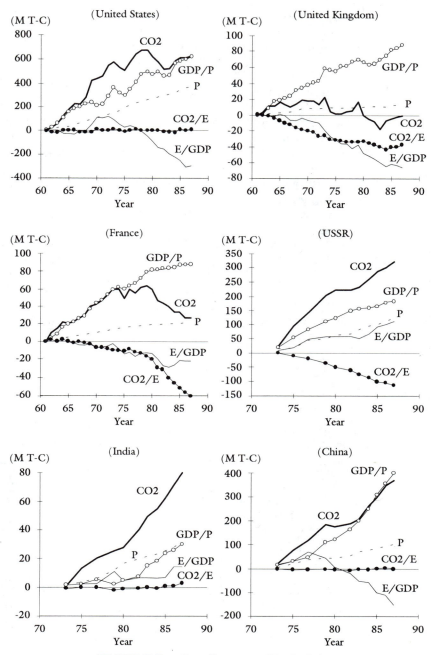

FIGURE 7. Contributors to CO_2 Emissions.

In contrast to population and economic growth, changes in energy intensity and fuel composition tended to push emissions down in the IMCs. In the United States and Japan the greatest contributor was reduced energy intensity, while in France and Sweden fuel switching was the major factor. Since the 1950s, reduced energy intensity and fuel switching have also made contributions in the United Kingdom.

On the other hand, in the Soviet Union and some LDCs, energy intensity and fuel composition changes did not always contribute to reducing CO_2 emissions. In the USSR and India, for example, energy intensity rose significantly. A major exception was the People's Republic of China, where the E/GDP ratio fell sharply. Fuel switching contributed to lowering CO_2 emissions in the Soviet Union, the Republic of Korea and Malaysia, but not in other LDCs.

Energy intensity and fuel composition are the only two factors which can be directly modified by policies aimed at reducing CO_2 emissions. In the IMCs both factors have already been operating, keeping CO_2 emissions flat or down to small increases. These two factors will of course continue to be important in the industrial countries. In the LDCs population control will be a key factor in any effective policy to counteract global warming. Finally, changes in industrial structure can contribute to reducing CO_2 emissions in a particular country. Globally, of course, they are not an answer if they merely shift the location of energy intensive industries elsewhere.

NON-COMMERCIAL ENERGY CONSUMPTION IN LDCS

The discussion thus far has dealt with commercial energy only, but LDCs also consume large quantities of non-commercial fuels like wood, bagasse and other agricultural wastes. When these are taken into account, total

TABLE 3. Commercial and Non-Commercial Energy Consumption in Developing Countries in 1987

				MIL. TOE
	COMMERCIAL	NON-COMMERCIAL	TOTAL	N-C SHARE
Bangladesh	4.5	6.2	10.7	58%
India	147.7	57.4	205.1	28%
Indonesia	34.9	31.5	66.4	47%
Thailand	17.0	11.4	28.4	40%
Malaysia	14.1	1.9	16.0	12%
Korea	60.2	1.5	61.7	2%

Source: United Nations, "Energy Balances and Electricity Profiles 1986," 1988.

energy consumption turns out to be 10 percent to 140 percent larger (see Table 3). Considering that firewood burning is often responsible for forest destruction, it is clear that the LDCs hold an even more critical position in the global warming problem than previously indicated.

CONCLUDING REMARKS

Among the factors examined, the burning of fossil fuels bears the greatest responsibility for global CO_2 emissions. In the IMCs this has been a major element for well over a century, while in the USSR and the LDCs its dominant role is of more recent origin. Given the growth in emissions in the LDCs, which will probably continue in the future, global action to regulate emissions cannot be effective without their full participation. North-south problems thus need to be addressed simultaneously, or before, the problem of global warming. Energy intensity and changes in fuel composition have helped to curb CO_2 emissions in IMCs since the early 1970s, and they will continue to be important in the future. In LDCs, population growth is an additional key factor that requires analysis for any effective CO_2 emissions policy.

The high temperatures and serious drought conditions in the late 1980s led many to view global warming as an issue that arose suddenly. This is incorrect: attention was first called to global environmental problems before the first energy crisis by the Club of Rome.[8] Soon thereafter however energy supply concerns were dominant and environmental concerns were pushed to the background. Since the collapse of oil prices in 1986, however, energy demand has begun to increase again and action to curb CO_2 emissions is becoming ever more urgent, even if it will not be easy to get countries with very different economic structures and energy patterns to agree on a common course of action.

REFERENCES

British Petroleum (1989). *Statistical Review of World Energy.*
IEA (1976). *Energy Balances of OECD Countries 1960/1974.*
IEA (1987). *Energy Balances of OECD Countries 1970/1985.*
IEA (1989). *Energy Balances of OECD Countries 1986/87.*
IEA (1989). *World Energy Statistics and Balances 1971–1987.*
IMF (1988). *International Financial Statistics—Yearbook 1988.*

[8]See D. H. Meadows et al., *The Limits to Growth* (1972).

Meadows, D. H., D. L. Meadows, J. Randers and W. W. Behrens III (1972). *The Limits to Growth.* University Books.

United Nations (1988). *Energy Balances and Electricity Profiles.*

US CIA (1978–1988). *Handbook of Economic Statistics.*

US EPA (1989). *Policy Options for Stabilizing Global Climate.* Chapter 2.

World Bank (1989). *World Tables 1988–1989 Edition.*

29

Adverse Environmental Consequences of the Green Revolution

DAVID PIMENTEL

MARCIA PIMENTEL
.

David Pimentel is Professor of Insect Ecology and
Agricultural Science at the College of Agriculture and Life
Sciences, and Marcia Pimentel is a Senior Lecturer in the
Division of Nutritional Sciences, both at Cornell University.

The new agricultural technologies, energy inputs, and crop varieties that
together are known as the Green Revolution are widely accepted as a
resounding success in raising agricultural yields in much of the Third
World. The Green Revolution seemingly banished what had been projected as imminent food shortages in many countries as populations continued to grow at annual rates of 2 to 3 percent. Optimists could envisage
increases in yields continuing indefinitely, as agricultural production was
put on a sound technological footing supported by a modern research and
development infrastructure. Early fears of labor displacement were shown
to be misplaced: the crop regime required as much or more labor as before.

Implicit in such rosy scenarios, however, is the assumption that the

"Adverse Environmental Consequences of the Green Revolution," by
David Pimentel and Marcia Pimentel, from *Resources, Environment,
and Population, Present Knowledge, Future Options* edited by Kingsley
Davis and Mikhail S. Bernstam. Copyright © 1991 by The Population
Council, Inc. Reprinted by permission of Oxford University Press,
Inc.

Green Revolution does not have adverse side effects on the environment—
on the immediate physical bases of production and on the broader physical setting in which the rural population lives. That assumption, we argue,
is false. This comment briefly examines the environmental impacts of the
Green Revolution and suggests that they call into question the long-run
sustainability, let alone the long-run further growth, of high agricultural
yields. In some instances they are potentially damaging to public health.
We conclude that this technological route cannot be seen as an alternative
to urgent action to stem population growth.

Focused mainly on enhanced yield, the Consultative Group on International Agricultural Research that developed the crop plant improvement
for the Green Revolution paid little or no attention to how this new
technology would affect the quality of the environment (Baum, 1987).
Pesticide use in Green Revolution rice production, for example, was reported to increase sevenfold over levels used in traditional rice production
(Subramanian et al., 1973). Despite this increased use of insecticides there
is no proof that losses due to insects in rice have been reduced; rather, at
best such losses have remained around 27 percent (Oka, 1987). One reason
for this is that more insect-susceptible varieties of rice have been planted.
Further, the use of 2,4-D herbicide has increased the level of attack of
insects on rice. One investigation documented that the rice stem-borer
grew 45 percent larger in size on 2,4-D-treated rice and as a result, the pest
consumed 45 percent more rice (Pimentel, 1971).

Perhaps the most serious problem associated with pesticide use is
human poisonings. The World Health Organization (1981) estimates that
annually there are 500,000 human pesticide poisonings and about 10,000
fatalities worldwide (see also Loevinsohn, 1987). In addition, heavy application of pesticides causes numerous other environmental problems. For
example, beneficial natural enemies of pests are killed and in some cases
this may lead to new pest outbreaks. In Indonesia in the early 1980s, when
the Ministry of Agriculture encouraged farmers to use more insecticide on
rice, the brown planthopper pest increased in numbers because of the
destruction of its natural enemies; the result was a significant decline in
Indonesian rice production during the mid-1980s (Oka, 1987). The situation became so desperate that in March 1985 President Suharto banned 56
of 57 insecticides used on rice in Indonesia. This drastic action helped rice
yields to return to the high levels farmers were achieving before the heavy
use of insecticides.

Pesticides have also reduced fish and shrimp populations in the paddy
fields that provide an important food source for poor people (Oka, 1987).
The pesticides lower the quantity of fish and shrimp that can be harvested
and contaminate the harvest, posing a serious threat to public health
(ICAITI, 1977).

Another major problem associated with the treatment of aquatic environments, either intentionally as with rice or unintentionally (by drift) as

with dryland crops, is the contamination of aquatic ecosystems with insecticides. The spread of insecticides into aquatic environments has contributed to the increased resistance observed in mosquito populations, resulting in the spread of malaria in Asia, Africa, and Latin America (ICAITI, 1977). Research on pesticide use on cotton in Central America revealed that malaria increased threefold during a two-year-period, principally due to this practice (ibid.).

In addition to the heavy reliance on pesticides, the Green Revolution made necessary increased fertilizer requirements, in some cases by 20- to 30-fold (Wen and Pimentel, 1984; Wittwer et al., 1987). Fertilizer washed and leached into lakes and streams enriches the aquatic system and may result in heavy algal growths that kill fish, shrimp, and other beneficial organisms (Pimentel, 1989). This eutrophication is a serious problem in industrialized countries where fertilizer use is very heavy, and it can be expected to grow in developing nations. Higher levels of fertilizer use, especially nitrogen, also has increased the susceptibility of rice to more intense insect and disease attack. It is common knowledge that when the nutrient level of crops is increased, insect and disease populations also increase (Pimentel, 1977).

As with all agriculture, mismanagement of soils and water results in diminishing yields, lack of future sustainability, and increased costs of production. Thus in the vast areas where the Green Revolution predominates, soil erosion and water runoff are ever-present problems. In India, for example, the average rate of erosion is 30 tonnes/hectare/year (Pimentel et al., 1987). There the amount of fertilizer nutrients lost with the eroded soil is about equal to the total amount of commercial fertilizer applied each year (Khoshoo and Tejwani, 1989). The loss of water associated with erosion (Pimentel et al., 1987) reduces crop productivity even more than does the loss of nutrients. The reduced productivity requires added fertilizer, irrigation, and pesticides to offset soil and water degradation. This starts a cycle of greater agricultural chemical use, increased reliance on energy (especially fossil-based), and further increases the production costs the farmer must bear. Poorer farmers are finding it difficult to afford the fertilizers and pesticides needed to sustain yields of their Green Revolution crops.

We conclude that considering the Green Revolution the answer to providing adequate food supplies for the ever-expanding human population is shortsighted. Unless the basic resources of agriculture, such as arable land, pure and adequate water supplies, and use of chemicals, are better managed, agricultural technology of the sort represented by the Green Revolution will be unable to meet the challenge.

REFERENCES

Baum, W. C. (with the collaboration of Michael L. Lejeune). 1987. *Partners Against Hunger.* Published for the Consultative Group on International Agricultural Research. Washington, D.C.: The World Bank.

ICAITI. 1977. *An Environmental and Economic Study of the Consequences of Pesticide Use in Central American Cotton Production.* Guatemala City: Instituto Centro Americano de Investigación y Tecnología Industrial.

Khoshoo, T. N., and K. G. Tejwani. 1989. "Soil erosion and conservation in India (status and policies)," *World Soil Erosion and Conservation.* Gland, Switzerland: International Union for the Conservation of Nature.

Loevinsohn, M. E. 1987. "Insecticide use and increased mortality in rural Central Luzon, Phillippines," *The Lancet* 8546 (13 June): 1350–1362.

Oka, I. N. 1987. Personal communication, Bogor Research Institute for Food Crops, Indonesia.

Pimentel, D. 1971. *Ecological Effects of Pesticides on Non-Target Species.* Washington, D.C.: US Government Printing Office.

———. 1977. "Ecological basis of insect pest, pathogen and weed problems," in *The Origins of Pest, Parasite, Disease and Weed Problems,* ed. J. M. Cherrett and G. R. Sagar. Oxford: Blackwell Scientific Publishers, pp. 3–31.

———, et al. 1987. "World agriculture and soil erosion," *BioScience* 37:277–283.

Pimentel, M. 1989. "Food as a resource," in *Food and Natural Resources,* ed. D. Pimentel and C. W. Hall, New York: Academic Press, pp. 409–437.

Subramanian, S. R., K. Ramamoorthy, and S. Varadarajan. 1973. "Economics of I. R. 8 paddy—a case study," *Madras Agricultural Journal* 60:192–195.

Troeh, F. R., J. A. Hobbs, and R. L. Donahue. 1980. *Soil and Water Conservation for Productivity and Environmental Protection.* Englewood Cliffs, N.J.: Prentice Hall.

Wen, D., and D. Pimentel. 1984. "Energy inputs in agricultural systems of China," *Agriculture, Ecosystems and Environment* 11:29–35.

World Health Organization. 1981. "Pesticide deaths: What's the toll?" *Ecoforum* 6: 10.

Wittwer, S., et al. 1987. *Feeding a Billion.* East Lansing: Michigan State University Press.

30

Deforestation and Species Loss*

KENTON R. MILLER

WALTER V. REID

CHARLES V. BARBER

Kenton R. Miller is Director of the Program in Forests and
Biodiversity of the World Resources Institute: Walter V.
Reid and Charles V. Baker are Associates in the program.

In the millennia since the world's forest mantle was first breached to make
way for systematic agriculture and human settlement, it has changed
markedly. While considerable tracts of temperate forests in Europe, North
Africa, North America, and the Orient long ago fell to the ax, until
recently a vast reservoir of forested lands remained in most of the tropics.
But today the erosion of forest ecosystems due to deforestation is rampant
in tropical Asia, Africa, and America.

This rapid decline represents an unprecedented raid on the planet's
biological wealth. Earth's tropical forests comprise a living treasure-
house—a trove of uncounted habitats and species as diverse and individual as snowflakes, linked in the complex webs of interaction that define
local ecosystems. Researchers estimate that as many as 25 percent of all

"Deforestation and Species Loss: Responding to the Crisis," by
Kenton R. Miller, Walter V. Reid, and Charles V. Barber, in the
American Assembly, *Preserving the Global Environment,* ed., Jessica
Tuchman Matthews, pp. 78–111. New York, The American Assembly.

*We have omitted the final section of this article, entitled "Addressing the Loss: Policies for
Change"—*Eds.*

species inhabiting earth in the mid-1980s will have disappeared by 2015 if current deforestation trends hold. For this reason, tropical deforestation and species loss are indissolubly linked.

These forest communities and the diversity of life they harbor represent an irreplaceable asset to the biosphere and humankind. They are great engines of biological productivity, considerable sources of climatic stability, and home to at least 500 million people. They also contribute directly to human survival as sources of food, fiber, medicines, industrial products, and the genes needed to breed the improved crop varieties upon which the world's food security rests. Scientists believe that the wealth of the planet's forests has barely been tapped. Indeed, only a tiny fraction of the potentially useful species that reside in tropical rainforests has even been identified.

As the twenty-first century nears, we are "eating our seed corn," squandering in a heedless evolutionary moment the forest's genetic capital, evolved over billions of years. The price for doing so is biological impoverishment in the years ahead and a consequent ecological decline that will threaten the health, commerce, and quality of life enjoyed by developed and developing nations alike.

Tropical forest loss will also have serious impacts on global, regional, and local climate. The loss of forest cover reduces transpiration of water vapor into the atmosphere, changes the albedo (reflectivity) of the earth's surface, removes an important "sink" for ozone, and, through burning, contributes to the greenhouse effect through the release of carbon dioxide and other greenhouse gases into the atmosphere. At the same time, the essential services provided by intact forest ecosystems—such as watershed protection and regulation, the storage of carbon in plant tissues, or the absorption and breakdown of pollutants—are being degraded or destroyed.

In spite of these dangerous trends, it is not too late to address global forest decline. But the steps needed to prevent the loss of forests and the diversity they harbor will not be easy ones. They must address the combination of population pressures, migration of the rural poor, misguided government policies, and false economic premises that are the root causes of modern deforestation. Complemented by a well-managed system of critical forest preserves, strategies that combat these factors could allow realistic and sustainable development without plundering the forest resource and its genetic wealth.

DIMENSIONS OF THE PROBLEM

The Scope of Forest Loss

Since the dawn of agriculture about 10,000 years ago, when humans first began forest clearing in earnest, the world's forests have declined in area by about a fifth, from 50 to 40 million square kilometers (5 to 4 billion hectares). Until recently, temperate forests had suffered much of this loss, losing about a third of their area, and tropical evergreen rainforests— largely inaccessible—had suffered least, declining only about 4 to 6 percent. Today, forests and woods still cover two-fifths of the earth's land surface and account for about 60 percent of the net productivity of all terrestrial ecosystems (including agriculture) through biomass production—plant growth.

But the remaining forests are under increasing stress worldwide, of which the worst is outright deforestation, which is proceeding fastest in the tropics, where just over half the remaining forests are located. Together with the soil erosion, fuelwood loss, decline in agricultural productivity, and species extinction it causes, tropical deforestation can rightfully be considered the developing world's single worst environmental problem.

The extent of tropical deforestation and the area of remaining forests have not been accurately measured. The best available evidence indicates that these forests have lost from a quarter to a half of their former area. Most of this decline has taken place within the last forty years. Between 1950 and 1983, for example, forest and woodland areas dropped 38 percent in Central America and 24 percent in Africa. The rate of decline seems to be increasing.

The only worldwide estimates of the deforestation rate—calculated by the United Nations Food and Agriculture Organization (FAO) in 1980— showed an annual loss of some 11.3 million hectares worldwide. These numbers reflect only those areas completely cleared for other uses; they do not include the additional 10 million hectares grossly disrupted annually through logging or other activities that destroy the forest integrity and its habitat potential, even if what remains is nominally still called forest.

More recent studies using satellite data indicate that the widely cited FAO estimate is seriously understated. Brazil provides the most dramatic example. It has the largest remaining tracts of tropical forest and by far the greatest annual deforestation rate. Studies conducted by the National Space Research Institute of Brazil conclude that 8 million hectares of virgin rainforest—more than five times the official calculation—were burned and cleared in the Amazon region in 1987 alone.

These data are controversial, since they rely on satellite photos of the smoke that rises when the forest is cleared through burning. In addition, 1987 may have been an anomalously high year for deforestation because it was the last year tax credits were offered to Brazilian landholders who cleared their Amazon holdings. Follow-up studies showed a decline in the rate of clearing in 1988 and 1989, perhaps brought on by wetter weather and/or changes in government policy. Nonetheless, at least 7 percent of Amazon forests had been cleared by 1988.

Other recent studies also reveal sharply higher annual rates of deforestation in India (1.5 million hectares), Indonesia (900,000 hectares), Mayanmar (Burma) (667,000 hectares), the Philippines (143,000 hectares), Costa Rica (124,000 hectares), and Cameroon (100,000 hectares). Less recent statistics show high yearly losses in such nations as Colombia, Mexico, Ecuador, Peru, Malaysia, Thailand, Nigeria, Ivory Coast, and some thirty other countries in tropical Asia, Africa, and Latin America. Extrapolating from these data, two 1990 estimates of worldwide tropical forest loss put the total in the range of 20–25 million hectares annually—approximately double the FAO figure.

Since the data on deforestation rates and the amount of remaining forest area are sketchy at best, it is hard to make reliable predictions of how fast tropical forests will disappear. Nonetheless, it is evident that if present trends continue, significant—probably catastrophic—declines in tropical forest extent and quality will take place within thirty years, especially in accessible lowland forests.

In fact, the World Commission on Environment and Development concluded in 1987 that, given the current deforestation rate and the expected growth in world population and economic activity, little virgin rainforest would survive beyond the turn of the century outside of forest preserves, except in portions of the Zaire Basin, northeast Brazilian Amazonia, western Amazonia, the Guianan tract of forest in northern South America, and New Guinea. Even these tracts are not expected to last more than a few decades longer because their commodity value is soaring and pressures to cut or settle them are rising.

Such predictions do not ignore reforestation and natural regrowth. However, much tropical forestland, once converted to other uses, does not easily revert to healthy forest, even if such regrowth is allowed. Full forest regeneration requires almost one hundred years, even when pastures are only lightly used. If such pastures are more heavily used, or repeatedly burned over to stimulate grass growth, recovery time is much longer. As a practical matter, forest soils usually are seriously degraded after conversion. Moreover, studies have shown that the changes in regional climate following large-scale deforestation could make reestablishing the forest on any time scale even more difficult.

The Scope of Species Loss

Although closed tropical forests cover only 7 percent of the earth's land surface, they contain at least half—and possibly up to 90 percent—of the world's species. It is impossible to estimate the exact proportion of species that reside there, since so little is known about tropical organisms and ecosystems. Perhaps only 10 percent of all tropical species have been described to date. Current estimates place about half of all vascular plant and vertebrate species in tropical forests, and among invertebrates, the percentage may be considerably higher. Species diversity in temperate forests differs strikingly from that in tropical forests. Typically, forty to one hundred species of trees occur on a single hectare of tropical rainforest, compared to only ten to thirty on a hectare of forest in the eastern United States. Some 700 species of trees can be found in ten hectares in Borneo, the same as in all of North America. One tree in Peru may contain as many species of ants as occur in all of the British Isles.

With such exceptional species richness, tropical ecosystems (and a few others, such as coral reefs) lie at the heart of the world's biological diversity—or biodiversity, the blanket term for the variety of the world's organisms, including their genetic make-up and the communities and interrelationships they develop. Forest losses on the order of those occurring in the tropics—which radically modify or eliminate whole ecosystems—thus directly threaten the biodiversity that undergirds the well-being of the living earth.

Species loss through extinction is certainly not a new phenomenon in nature. The 10 million or so species that populate the earth today are the survivors of the several billion that evolution has produced since life began. Over the history of the planet, there have been several mass extinctions—relatively short periods when significant fractions of the world's species died out. The last such mass extinction was some 65 million years ago. Since then, global biodiversity has rebounded and is now close to its all-time high.

Humanity's impact on species extinction rates goes back thousands of years, but over the last century—especially over the last several decades—the human factor has increased dramatically. For instance, among all birds and mammals, we would expect an extinction only once every 100 to 1,000 years in the absence of humans. However, the actual extinction rate for birds and mammals between 1850 and 1950 was one per year—as much as 1,000 times greater than the background rate.

Predicting future extinction rates in response to the massive habitat disruption that accompanies deforestation is difficult at best. However, a useful rule of thumb is that if a habitat is reduced by 90 percent in area, roughly half of its species will be lost.

Between 1990 and 2020, tropical deforestation may extinguish 5 to 15 percent of the world's species. This percentage translates into an annual loss of 15,000 to 50,000 species—about 50 to 150 per day. According to another less conservative estimate, 5 to 10 percent of world species may be lost *per decade* over the next quarter century. In theory, if deforestation were to continue until all forests in the Amazon Basin were eliminated except those now legally protected from harvest, 66 percent of plant species and 69 percent of bird species in these forests would disappear.

Judging the effects of such massive losses on the functioning of the remaining species is difficult because of the complex dynamics among plant and animal species in even the simplest of ecosystems. In a given ecosystem, the extinction of certain key species may cause a cascade effect upon the populations of many other species. For example, during the dry season in Manu National Park in Peru, only 12 out of 2,000 plant species support as much as 80 percent of the park's mammals and a major fraction of the park's birds. Clearly, the loss of one or more of these key species would significantly affect a host of dependent species.

Habitat disruption through deforestation is not the only factor endangering biodiversity today. Overharvesting, ubiquitous chemical pollution, competition from introduced species, and climatic change also take a toll. Nor is outright extinction the only concern. Many species not in imminent danger of demise are nonetheless suffering from reduced populations and declining genetic variability that make them more vulnerable to disease, inbreeding, hunting, and other environmental stresses that can threaten their survival.

The Mechanisms of Tropical Deforestation

Tropical deforestation has four direct causes: slash-and-burn cultivation by a growing army of landless migrant poor; conversion of forests to cattle pastureland; wasteful and unsustainable commercial logging; and overharvesting of subsistence fuelwood and fodder. Often these factors work in concert to strip and degrade forest lands so that regrowth occurs slowly, if at all, even while the agricultural potential of the cleared land drops rapidly.

By far the most important source of forest loss is conversion for subsistence agriculture. Traditional "slash-and-burn" agriculture has been practiced for millennia by indigenous cultures in an environmentally sound fashion. Most tropical forest soils are not suited to continuous cultivation because of low fertility and other physical limitations, so forest dwellers have developed various systems of shifting cultivation.

In these traditional systems, most of the woody vegetation is cleared from a patch of land—usually by burning—and crops are planted for several years. Over time, declining soil organic matter, reduced nutrient levels, and competition from re-sprouting vegetation and weeds cause the farmer to move elsewhere, abandoning the plot to lie fallow. If the fallow time is long enough and the cultivated plot not overly degraded, the forest restores itself and replenishes soil productivity.

Unfortunately, this sustainable cycle has broken down in many forest areas because of population growth and a rapid influx of landless peasants. These displaced or "shifted" cultivators now far outnumber indigenous forest farmers in most areas, and they bring with them dietary preferences and farming practices often ill-suited to the forest. Lacking an understanding of the traditional sustainable cycle and driven by short-term needs, these forest settlers almost inevitably pursue continuous cropping until soil fertility is lost and the land fully degraded, before moving on to the next plot.

The Brazilian state of Rondonia in southern Amazonia exemplifies the impact of displaced cultivators. In 1975, when the Brazilian government began encouraging the urban poor to settle in Rondonia, the area's population stood at 110,000 and little forest was cleared. By 1986 the population had reached more than 1 million and the cleared area had expanded to 2.8 million hectares, with smallholder clearings and conversion to cattle ranches accounting for most of the loss.

Displaced rural farmers and urban poor are by no means confined to Brazil. Slash-and-burn agriculture by displaced peasants is also the primary agent of forest loss in the Philippines, Indonesia, Thailand, India, Madagascar, Tanzania, Kenya, Nigeria, Ivory Coast, Colombia, Peru, Ecuador, and Bolivia. The link between land degradation, burgeoning population growth, and rural poverty is incontrovertible, and can only worsen unless current trends are reversed.

The conversion of forest to pastureland for cattle ranching is a second major contributor to tropical deforestation, but one that is confined almost entirely to Latin America. There, more than 20 million hectares of rainforest have been converted to cattle pastures over the past twenty years. At least half of this conversion has taken place in the Brazilian Amazon, about one-fourth in Mexico, and the rest in Colombia (which contains about one-fourth of the Amazon Basin), Peru, Venezuela, and Central America.

The cattle surge in the Brazilian Amazon began in the late 1960s, fueled by a growing network of paved roads, government tax incentives and subsidized credit, and land speculation. By 1980 almost 9 million head of cattle grazed Amazonian pastures, accounting for 72 percent of all forest conversion in the region to that point. Although Brazil's government no longer subsidizes it, the conversion of forest to pasture continues as a means of securing and retaining land title through "improvement" of the forest land.

In part, this emphasis on cattle raising stemmed from increased demand for cheap beef in North America—the infamous "hamburger connection." Ironically, while beef production in Central America more than tripled between 1955 and 1980, beef consumption among Central Americans actually fell during this period.

To form a pasture, forest cover is cut and burned, and African forage grasses are often planted. These usually flourish for a few years until soil nutrients wane and shrubs take over. Reburning the pasture revitalizes it for a time, but the response to successive fires diminishes until the soil is completely exhausted. In this manner, much of the converted pastureland has become degraded over the years.

Commercial logging operations are a third potent factor in forest loss. Logging is essentially a "mining" operation throughout the tropics: harvested timber is not replaced either by natural regeneration or forest plantations. A 1988 survey found that the amount of land under sustained yield management was negligible—only about 4.4 million hectares out of a total of 828 million hectares of exploitable forest.

Most tropical logging practices severely disrupt the forest's ecological integrity. Typically, commercially valuable species make up a small percentage of the stand. Only 10 to 20 percent of the trees are cut, but another 30 to 50 percent are destroyed or fatally injured, and the soil is disturbed enough to impede regeneration, even in the long run.

Some 4.4 million hectares of closed forest were logged annually between 1981 and 1985, mostly in Southeast Asia and West Africa. By 1985 about half of the productive closed forests in tropical Asia had been logged, and about one-fourth of tropical Africa's timber was gone. Countries such as the Philippines, once a leading exporter, have nearly exhausted their lowland forests. Indonesia, Malaysia, and the Philippines have all totally or partially banned log exports to reduce overharvesting and develop domestic processing facilities in order to capture more of the value of their timber.

The Amazonian region and Central Africa have been exploited significantly less, largely because of difficult access. However, logging is bound to accelerate in these regions as better roads and harvesting techniques become available and more accessible regions are cut over. Indeed, Latin America, where only 10 percent of exploitable forests have been logged, is expected to become the world's major source of tropical hardwoods within the next decade.

The impact of logging is greatly magnified by the subsequent conversion of forestland to agriculture. Logging roads open forest areas to subsequent encroachment. According to one 1982 report, nearly 70 percent of annual clearance of closed forests is in logged-over areas.

Overcutting of fuelwood, construction wood, and fodder is the fourth major source of deforestation. In open forests near population centers, these activities can destroy tree cover, especially if combined with overgrazing and repeated burning. About 80 percent of all wood harvested in developing countries is used for fuelwood and charcoal, and wood and other biomass fuels figure centrally in many countries' energy economies.

Problems of woodfuel scarcity are most severe in densely populated portions of Africa and Asia, and severe land degradation is the common result.

THE ROOT CAUSES OF DEFORESTATION AND SPECIES LOSS

Population Pressures and Flawed Government Policies

A complex array of social, economic, and political forces is behind the present crisis of forest and species loss. These powerful forces are often deeply rooted in past development patterns and frequently reflect the imbalances that may exist within developing countries; rapid population growth that strikes hardest at those with fewest resources; concentration of land in the hands of a few, leaving millions landless; slow growth of job opportunities in both city and countryside that induces migration; and colonial patterns of resource exploitation that emphasize maximum short-term gain.

The rapid population increases typical of developing countries place ever greater demands on their resource bases and increase pressure to liquidate forest capital to finance immediate development needs. Since 1960 Brazil's population has climbed from 73 million to 147 million today; India's population from 442 million to 813 million, and Indonesia's from 96 million to 178 million, to cite just a few examples. Such population growth is already a prominent factor not only in the increase in landless forest settlers, but also in many government development policies that result in deforestation.

Flawed government management policies constitute a second major force in forest demise. Such policies include asserting government control over forest areas once governed—and protected—by local communities; establishing land tenure rules that encourage forest settlement, and clearing without providing the security that might encourage sustainable land use; building roads or making other improvements in forest areas aimed at extending government control into unsecured regions; and granting economic incentives that encourage forest exploitation.

Until relatively recently, most forestlands in developing countries were the communal property of local tribes, and farming and pastoral communities. These groups developed a diverse array of hunting and gathering, farming, and grazing practices based on traditional custom that effectively regulated forest access and use. However, in the past forty years, more than 80 percent of the world's tropical forests have been brought under government ownership, overruling traditional rights of forest control. Deprived of legal authority, tribal heads and community leaders no longer have a strong incentive to conserve the forest resource by, say, limiting shifting agriculture or timber operations. As customary conservation values erode, forests and wildlife have thus been put at greater risk.

For their part, governments have found themselves unable to defend the rights they have asserted: the forest areas are too vast, most national forest agencies are underfunded and understaffed, and encroachment pressures are too strong. Yet having once asserted control, governments are reluctant to relinquish ownership and management rights to other parties. One exception is timber concessionaires, who are viewed as contractors able to generate much needed state revenue. As a result, no party with a vested interest in the long-term forest resource remains to regulate forest use.

The issue of land ownership—or tenure—is especially important in determining whether forestland use will be sustainable over the longer term. Under the rules of tenure in many countries, title to public domain lands is granted to those who "improve" it through conversion to another use. This arrangement has triggered deforestation in such regions as the Amazon Basin, the Philippines, and Malaysia, among others.

A corollary to this rule is that those who do not "improve" the land, such as many indigenous peoples who traditionally exercised communal ownership, are generally not granted direct title, though such ownership might be more conducive to sound forest management. Just as important, landless immigrants whose tenure is insecure are less likely than landowners to adopt sustainable practices, preferring instead to farm intensively for immediate gains in case they are forced to move on.

Governments often use the opening and development of forest areas to further their internal security and national defense goals. Such concerns have loomed large in the forest development policies of Brazil, Burma, Indonesia, Peru, Thailand, and Venezuela. Political insurgencies are often based in inaccessible forest areas. And since forest-dwelling peoples number among those least integrated into the social and political mainstream, governments seek to bring them into the fold by exerting political controls and extending government services to them.

External defense priorities also frequently motivate governments to open and secure the forest periphery. National borders in many tropical countries are ill-defined and frequently disputed because they lie in inaccessible forest areas. Building roads, railways, and other infrastructure elements near border areas tends to delineate and secure these areas.

In Brazil, Bolivia, Indonesia, and other countries, security concerns have also helped stimulate agricultural resettlement schemes in forest areas. While these projects are obviously intended to serve economic development goals as well, security concerns often shape their implementation.

Finally, many governments are directly or indirectly subsidizing forest exploitation through investment incentives, tax and revenue systems, and other policies that invite mismanagement. Generally, tropical timber harvesters work under regulatory regimes that provide incentives for the rapid exploitation of a few commercially valuable species and discourage investments in sustainable forestry.

Economic Factors

At the root of many flawed government policies are faulty economic analyses and unsupported economic assumptions about the use of forest resources. In essence, neither the true value of the forest—both its biological resources and the environmental services it renders—nor the true cost of exploiting it is accurately calculated. This pervasive misvaluation makes it hard to abandon abusive forest practices since their real costs remain hidden and the benefits of sustainable practices undervalued.

Typically, the value assigned to forests is limited to the "stumpage value" of their timber or the agricultural potential of the cleared lands. Even if their usefulness as watershed, tourist attractions, pollution removers, or repositories of biodiversity is recognized, it is difficult to assign a monetary value to these environmental services.

Biological diversity, for example, is a "public good" whose benefits are largely intangible but nonetheless essential to the enjoyment of various other products, including most of the world's major food crops and many pharmaceutical products. The costs of depleting biodiversity are equally diffuse and difficult to calculate, but very real. On the other hand, the immediate benefits of logging and other forms of forest exploitation are easily measurable and naturally weigh more heavily in land use decisions.

Often ignored in the forest planning of developing nations is the capacity of the intact forest to supply a perpetual stream of valuable non-wood products that can be harvested without cutting down the trees. Frequently, these wild foods and materials contribute greatly to rural livelihoods. Worldwide, there are more than 500 million forest dwellers (10 percent of global population), many of whom depend on nuts, berries, game, fish, honey, and other forest products to help them survive. The value of these products is largely ignored because they rarely reach the marketplace and mostly benefit weak minorities in the hinterlands.

However, resins, essential oils, medicinals, rattan, flowers, and other products do flow into commercial channels, producing considerable income for those who collect and trade them. Although they are usually considered "minor" products, their aggregate value can be quite large. For instance, by the early 1980s Indonesia exported $125 million worth of such products annually. These products constitute a small fraction of the *potential* sustainable yield of tropical forests. Indeed, in some cases, the value of non-timber forest products may greatly exceed the value of a one-time timber harvest. One study in Peru found that the net revenue obtained from harvesting fruit and latex from the forest was thirteen times greater than the revenue that could be obtained by using the forest for timber production.

Even as many forest resources are undervalued, all too often the anticipated benefits of forest exploitation—especially the expected agricultural yields of converted forestland—are overvalued. The soils underlying 95 percent of the remaining tropical forests are infertile and easily degraded through erosion, laterization, or other processes once the vegetative cover is stripped. Unlike temperate forests, where organic matter can build up in the soil, copious rains and high temperatures quickly leach nutrients from tropical soils. In the intact tropical forest, most of the nutrients are held in the plant tissues themselves and efficiently recycled before they can be lost. Most agricultural systems cannot duplicate the rainforest's complex recycling ability, so the few nutrients in the soil are lost to leaching and erosion within a few years, and agricultural productivity drops rapidly.

The economic benefits of timber harvests are also widely overstated because the costs of logging—to the environment and to forest dwellers— are rarely included in economic evaluations. In addition, most of the profits from timber harvests have gone to private logging companies— many of them foreign—and the politicians and military officers who are often their silent partners. Government treasuries realize a surprisingly small net gain from depleting their forest base, often less than the administrative and infrastructure costs of the timber harvest (a situation paralleled in many U.S. national forests, where "below-cost" timber sales are common and quite controversial). Even the employment benefits of logging enterprises have usually failed to live up to expectations, and local communities are often left with little to show for the destruction of the nearby forest.

Finally, since the costs of unsustainable forest practices such as watershed degradation or species loss are not explicitly recognized, those who benefit from such practices are seldom the ones to pay these costs. Instead, they are passed on to innocent parties or to the society as a whole—often to future generations. For example, the logging company does not bear the cost of the siltation it causes; that is borne by downstream farmers or communities whose use of the waterways is compromised. Likewise, the cattle rancher does not have to concern himself with the species he wipes out to clear his pasture; that cost is borne by the nation or the world at large. If the costs of these inadvertent losses fell to the parties that caused them, a much different forest economy would emerge as the value of sustainable practices took on tangible dimensions.

International Factors

Among international economic factors that play a major role in forest and species declines, the crushing debt load carried by many developing countries is the most important. Half of the Third World external debt and over two-thirds of global deforestation occur in the same fourteen developing countries. These nations owe some $800 billion to public and private banks, and much of their working capital goes to service this debt. This burden operates as a powerful inducement to liquidate forest capital since it is one of the only resources readily convertible to much needed cash. The need for foreign exchange also stimulates the planting of cash crops for export, often grown on converted forestland.

Most countries cannot hope to halt and reverse deforestation and other environmental problems as long as a significant portion of their financial resources is siphoned off to repay foreign debt. Forestry and other low-priority sectors are often hardest hit by cutbacks in staff and expenditures imposed by economic austerity programs, and these programs, combined with economic stagnation, also intensify pressures on forests through their impacts on the poor. The economic adjustment process almost invariably hits the poor hardest as unemployment increases and incomes decline. This forces people back into subsistence agriculture and the conditions of poverty that lead to migration and forest degradation.

A second important international factor is the high demand in industrialized nations for tropical timber and other commodities grown at the expense of forests. Roughly half of all tropical hardwood timber produced is exported to the developed world, with a net worth of some $7 billion annually. Japan, in particular, has been a major tropical timber importer, accounting for nearly one-third of the world trade. Exports of pasture-fed beef to North America and plantation-grown cassava (commonly used as a cheap and nutritious animal feed) to Europe are also important influences in tropical forest conversion.

THE IMPLICATIONS OF FOREST AND SPECIES DECLINE

National Implications

Forests and their biological wealth are among the most valuable productive assets that developing countries possess. They contribute to many sectors of the national economy and provide vital sources of food, fuel, fodder, medicines, building materials, and many other necessities for hundreds of millions of people in developing nations. They help make farming and pastoral systems stable and provide a major source of income to those who gather and process forest products.

Extensive deforestation and species loss thus represent a serious threat to the economies of many nations. In essence, the net income derived from exploiting forest resources disguises the actual loss of real wealth. Even as gross national product (GNP) increases, total assets of the economy are declining. The estimated economic costs of unsustainable forest depletion range from 4 to 6 percent of the GNP in major tropical hardwood-exporting countries—clearly, enough to offset any economic growth that exploitation might allow.

Growth built on such resource depletion is almost certain to be unsustainable. While international trade in hardwoods is expected to double by the year 2000 as the timber boom reaches its peak, an even more precipitous decline is anticipated thereafter. By 2020 exports will likely be only 70 percent of their 1980 level and declining.

Once depleted, biological wealth (especially biodiversity) is all but impossible to recapture. Its loss will leave developing countries asset poor precisely when their resource needs are rapidly accelerating. This progressive impoverishment has implications in every sector of government—from national security to energy policy and environmental protection.

The potential for societal upheaval is also profound. Indigenous peoples will suffer greatly as traditional sources of wild food, fodder, and building materials dwindle and the mainstays of social and spiritual life unravel. Forest immigrants will also be squeezed as soil fertility on converted ground is lost and forest areas available for new settlement shrink.

Forest dwellers will be the first to bear the brunt of the degradation of environmental services. Local erosion, flooding, siltation, susceptibility to fire, and even desertification in some places will render forest areas less hospitable and less able to support increasing populations. But these impacts will not be locally confined. The creation of a class of environmental refugees is already well advanced in some Asian and African nations, and will increase pressures on remaining undisturbed forests and reserves, international borders notwithstanding.

The prospects for energy security in developing countries will also dim as forest areas disappear, because wood is such a large component of these nations' budgets. In Africa, for example, 76 percent of the energy consumed for all purposes is supplied by wood. Scarcity of woodfuel is already a problem of major dimensions and a potent force in forest decline. According to the FAO, 1.5 billion of the 2 billion people who rely on wood for fuel are cutting wood faster than it is growing back, a figure that is projected to double in less than a decade.

The implications for national security are also sobering. Deforestation directly erodes a nation's productive capacity as surely as the occupation of a foreign power does. Impoverished governments will be less able to provide even basic services to the growing ranks of landless rural and urban poor, with increasing discontent the inevitable result.

Many countries use forests as "safety valves" to cope with demographic and economic pressures. Promoting migration to forest areas relieves overcrowding and landlessness in prime agricultural regions, and is considerably easier than creating jobs, ameliorating rural poverty, or pushing for land reform. As deforestation proceeds, forest areas will be less able to serve this function, thus perhaps exacerbating internal frictions and inciting international ones as well.

All these facets of deforestation and species loss will probably take place on a regional basis too. Although the severity of their effects will vary from country to country, they will likely exact a cumulative, regional toll far beyond that felt by individual nations. Environmental degradation may proceed synergistically as ecosystem services erode over ever wider areas and spillover effects begin to accumulate.

On a hopeful note, the prospect of a regional threat also brings the possibility of regional response, which may be far more effective than unilateral efforts in meeting the common challenge of deforestation. The development of common approaches and policies that reach beyond country borders could help stretch limited conservation budgets and could lend credibility and momentum to national programs.

Global Implications

Tropical forests provide goods and services whose loss will have global repercussions. As one example, these forests contain genetic material that plant breeders use to confer resistance to disease and pests on some of the world's most important food crops, such as rice and cassava, as well as on such other important commodity crops as coffee, cocoa, bananas, and pineapples. In addition, many natural insecticides (such as pyrethrins and rotenoids) derive from tropical plants, and insect predators and parasites found in tropical forests control at least 250 kinds of agricultural pests. The wholesale loss of species due to deforestation thus puts global food security at risk, as well as the international commerce based on the trade of agricultural commodities.

Tropical plants also represent a veritable pharmacopeia. Indians dwelling in the Amazon Basin make use of some 1,300 medicinal plants, including antibiotics, narcotics, abortifacients, contraceptives, antidiarrheal agents, fungicides, anesthetics, muscle relaxants, and many others—most of which have not yet been investigated by researchers. As it is, one-fourth of all prescription drugs sold in the United States originate in wild plants and animals, and 70 percent of the more than 3,000 plant species known to produce anticancer agents are found in tropical plants. Obviously, current uses of tropical species represent only a small fraction of their potential uses.

The loss will not be confined to useful products. Such environmental services as primary production of food, fuel, and fiber; decomposition;

nutrient and water cycling; soil generation; erosion control; pest control; and climate regulation have a global reach—some directly, like climate regulation, and others indirectly through their contribution to the smooth functioning of the biosphere.

In this regard, perhaps the most serious global threat that deforestation poses is its contribution to the greenhouse effect and global warming. Deforestation is second only to fossil fuel combustion as a human source of atmospheric carbon dioxide, currently the most important heat-trapping, or greenhouse, gas. (It is also a substantial contributor to methane and tropospheric ozone concentrations: two other important greenhouse gases.) According to the most recent estimates of deforestation, forest clearing accounts for the release of some 2.8 million metric tons of carbon dioxide per year—about a third of the total annual carbon dioxide emissions caused by humans—and almost all of this load originates in the tropics.

In 1987 eleven countries were responsible for 82 percent of this net carbon release: Brazil, Indonesia, Colombia, Ivory Coast, Thailand, Laos, Nigeria, Viet Nam, Philippines, Burma, and India. During that year, when land clearing by fire peaked in the Amazon Basin, more than 1.2 million metric tons of carbon are believed to have been released. By comparison, the United States, by far the world's largest greenhouse emitter, released 1.2 million metric tons of carbon in 1987 through fossil fuel combustion and cement plant emissions.

Although uncertainties abound related to the timing, intensity, and distribution of the effects of global warming, its disruptive potential is generally accepted. Climate change could alter the constitution and functioning of ecosystems worldwide, with consequences for agriculture, marine and forest productivity, and the survival of native flora and fauna. In turn, these changes pose fundamental threats to the economic and environmental security of all nations, north and south.

In light of these many impacts, a good deal of international attention has been focused on deforestation and species loss in the last decade. Vigorously stated international concern evoked a cool response at first from many governments in the tropics, which expressed irritation that northern politicians and nongovernmental organizations should tell them to preserve forests instead of developing as temperate countries have. But international concern has generally echoed and strengthened national concerns with forest and development policies that were being voiced both within and outside of governments, and many developing nations now embrace the concept of minimizing damage to forests—even if strong programs to address the problem are not yet in place—so their economies can grow more sustainably.

The debate over the fate of tropical forests has become a topic of fruitful discussion and the subject of possible global cooperation in the future, including the funding of critical conservation programs through international development aid, and the establishment of demonstration projects based on sustainable agricultural and forestry methods. Likewise, the development of strategies to combat the greenhouse effect has already become an international priority, with some developing nations as active participants.

*   *   *